Plant Proteomic Research

Special Issue Editors

Setsuko Komatsu

Zahed Hossain

Special Issue Editors
Setsuko Komatsu
National Institute of Crop Science
Japan

Zahed Hossain
University of Kalyani
India

Editorial Office
MDPI AG
St. Alban-Anlage 66
Basel, Switzerland

This edition is a reprint of the Special Issue published online in the open access journal *International Journal of Molecular Sciences* (ISSN 1422-0067) from 2015–2017 (available at: http://www.mdpi.com/journal/ijms/special_issues/plant-proteomic).

For citation purposes, cite each article independently as indicated on the article page online and as indicated below:

Author 1; Author 2; Author 3 etc. Article title. *Journal Name*. **Year**. Article number/page range.

ISBN 978-3-03842-428-4 (Pbk)
ISBN 978-3-03842-429-1 (PDF)

Table of Contents

About the Guest Editors vii

Preface to "Plant Proteomic Research"
Reprinted from: *Int. J. Mol. Sci.* **2017**, 18(1), 88; doi:10.3390/ijms18010088
http://www.mdpi.com/1422-0067/18/1/88 ix

Khawaja Ghulam Rasool, Muhammad Altaf Khan, Abdulrahman Saad Aldawood,
Muhammad Tufail, Muhammad Mukhtar and Makio Takeda
Identification of Proteins Modulated in the Date Palm Stem Infested with Red Palm Weevil
(*Rhynchophorus ferrugineus* Oliv.) Using Two Dimensional Differential Gel Electrophoresis and
Mass Spectrometry
Reprinted from: *Int. J. Mol. Sci.* **2015**, 16(8), 19326–19346; doi:10.3390/ijms160819326
http://www.mdpi.com/1422-0067/16/8/19326 1

Qin Li, Juan Li, Shuoqian Liu, Jianan Huang, Haiyan Lin, Kunbo Wang, Xiaomei Cheng and
Zhonghua Liu
A Comparative Proteomic Analysis of the Buds and the Young Expanding Leaves of the Tea Plant
(*Camellia sinensis* L.)
Reprinted from: *Int. J. Mol. Sci.* **2015**, 16(6), 14007–14038; doi:10.3390/ijms160614007
http://www.mdpi.com/1422-0067/16/6/14007 18

Daqiu Zhao, Saijie Gong, Zhaojun Hao, Jiasong Meng and Jun Tao
Quantitative Proteomics Analysis of Herbaceous Peony in Response to Paclobutrazol Inhibition of
Lateral Branching
Reprinted from: *Int. J. Mol. Sci.* **2015**, 16(10), 24332–24352; doi:10.3390/ijms161024332
http://www.mdpi.com/1422-0067/16/10/24332 48

Mohammad-Zaman Nouri, Ali Moumeni and Setsuko Komatsu
Abiotic Stresses: Insight into Gene Regulation and Protein Expression in Photosynthetic Pathways
of Plants
Reprinted from: *Int. J. Mol. Sci.* **2015**, 16(9), 20392–20416; doi:10.3390/ijms160920392
http://www.mdpi.com/1422-0067/16/9/20392 66

Zahed Hossain, Ghazala Mustafa and Setsuko Komatsu
Plant Responses to Nanoparticle Stress
Reprinted from: *Int. J. Mol. Sci.* **2015**, 16(11), 26644–26653; doi: 10.3390/ijms161125980
http://www.mdpi.com/1422-0067/16/11/25980 86

Olga Pechanova and Tibor Pechan
Maize-Pathogen Interactions: An Ongoing Combat from a Proteomics Perspective
Reprinted from: *Int. J. Mol. Sci.* **2015**, 16(12), 28429–28448; doi: 10.3390/ijms161226106
http://www.mdpi.com/1422-0067/16/12/26106 96

Chen Liu, Shixue Li, Jing Yue, Wenhan Xiao, Qian Zhao, Dengyun Zhu and Jingjuan Yu
Microtubule-Associated Protein SBgLR Facilitates Storage Protein Deposition and Its Expression
Leads to Lysine Content Increase in Transgenic Maize Endosperm
Reprinted from: *Int. J. Mol. Sci.* **2015**, 16(12), 29772–29786; doi: 10.3390/ijms161226199
http://www.mdpi.com/1422-0067/16/12/26199 115

Yun-Peng Wang, Zheng-Yi Wei, Xiao-Fang Zhong, Chun-Jing Lin, Yu-Hong Cai, Jian Ma,
Yu-Ying Zhang, Yan-Zhi Liu and Shao-Chen Xing
Stable Expression of Basic Fibroblast Growth Factor in Chloroplasts of Tobacco
Reprinted from: *Int. J. Mol. Sci.* **2016**, *17*(1), 19; doi: 10.3390/ijms17010019
http://www.mdpi.com/1422-0067/17/1/19 130

Juanjuan Yu, Jinzheng Zhang, Qi Zhao, Yuelu Liu, Sixue Chen, Hongliang Guo, Lei Shi and
Shaojun Dai
Proteomic Analysis Reveals the Leaf Color Regulation Mechanism in Chimera *Hosta*
"Gold Standard" Leaves
Reprinted from: *Int. J. Mol. Sci.* **2016**, *17*(3), 346; doi: 10.3390/ijms17030346
http://www.mdpi.com/1422-0067/17/3/346 139

May Alqurashi, Chris Gehring and Claudius Marondedze
Changes in the *Arabidopsis thaliana* Proteome Implicate cAMP in Biotic and Abiotic Stress
Responses and Changes in Energy Metabolism
Reprinted from: *Int. J. Mol. Sci.* **2016**, *17*(6), 852; doi: 10.3390/ijms17060852
http://www.mdpi.com/1422-0067/17/6/852 165

Huan Li, Lin-Tong Yang, Yi-Ping Qi, Peng Guo, Yi-Bin Lu and Li-Song Chen
Aluminum Toxicity-Induced Alterations of Leaf Proteome in Two Citrus Species Differing in
Aluminum Tolerance
Reprinted from: *Int. J. Mol. Sci.* **2016**, *17*(7), 1180; doi: 10.3390/ijms17071180
http://www.mdpi.com/1422-0067/17/7/1180 176

Xia An, Jingyu Zhang, Lunjin Dai, Gang Deng, Yiwen Liao, Lijun Liu, Bo Wang and
Dingxiang Peng
Isobaric Tags for Relative and Absolute Quantitation (iTRAQ)-Based Comparative Proteome
Analysis of the Response of Ramie under Drought Stress
Reprinted from: *Int. J. Mol. Sci.* 2016, *17*(10), 1607; doi: 10.3390/ijms17101607
http://www.mdpi.com/1422-0067/17/10/1607 195

Zhujia Ye, Sasikiran Sangireddy, Ikenna Okekeogbu, Suping Zhou, Chih-Li Yu, Dafeng Hui,
Kevin J. Howe, Tara Fish and Theodore W. Thannhauser
Drought-Induced Leaf Proteome Changes in Switchgrass Seedlings
Reprinted from: *Int. J. Mol. Sci.* **2016**, *17*(8), 1251; doi: 10.3390/ijms17081251
http://www.mdpi.com/1422-0067/17/8/1251 215

Xiaoli Wang, Xiaofeng Cai, Chenxi Xu, Quanhua Wang and Shaojun Dai
Drought-Responsive Mechanisms in Plant Leaves Revealed by Proteomics
Reprinted from: *Int. J. Mol. Sci.* **2016**, *17*(10), 1706; doi: 10.3390/ijms17101706
http://www.mdpi.com/1422-0067/17/10/1706 233

Qingzhu Hua, Qianjun Zhou, Susheng Gan, Jingyu Wu, Canbin Chen, Jiaqiang Li,
Yaoxiong Ye, Jietang Zhao, Guibing Hu and Yonghua Qin
Proteomic Analysis of *Hylocereus polyrhizus* Reveals Metabolic Pathway Changes
Reprinted from: *Int. J. Mol. Sci.* **2016**, *17*(10), 1606; doi: 10.3390/ijms17101606
http://www.mdpi.com/1422-0067/17/10/1606 262

Xiao-Ling Zhang, Jin Zhang, Ying-Hua Guo, Pei Sun, Hui-Xia Jia, Wei Fan, Meng-Zhu Lu and
Jian-Jun Hu
Comparative Proteomic Analysis of Mature Pollen in Triploid and Diploid *Populus deltoides*
Reprinted from: *Int. J. Mol. Sci.* **2016**, *17*(9), 1475; doi: 10.3390/ijms17091475
http://www.mdpi.com/1422-0067/17/9/1475 275

Xinliang Liu, Wanwen Yu, Guibin Wang, Fuliang Cao, Jinfeng Cai and Huanli Wang
Comparative Proteomic and Physiological Analysis Reveals the Variation Mechanisms of Leaf Coloration and Carbon Fixation in a Xantha Mutant of *Ginkgo biloba* L.
Reprinted from: *Int. J. Mol. Sci.* **2016**, *17*(11), 1794; doi: 10.3390/ijms17111794
http://www.mdpi.com/1422-0067/17/11/1794 293

About the Guest Editors

Setsuko Komatsu is Chief of Field Omics Research Unit at the National Institute of Crop Science and a Professor at the University of Tsukuba, Japan. She obtained her Ph.D. from Meiji Pharmaceutical University, and her Ph.D. thesis work focused on the role of protein kinase dependance on phosphorylation during mammalian fertilization. She was employed at the Meiji Pharmaceutical University and then the Keio University, School of Medicine. Since 1990, she has been working on plant proteomics using a protein sequencer and mass spectrometry at the National Institute of Agrobiological Sciences. From 2006, she is overseeing research in her current official position at the National Institute of Crop Science. Her main research interests are within the field of crop proteomics, biochemistry, and molecular biology with a special focus on signal transduction in cells. Furthermore, she is trying to promote the development of agriculture proteomics as the president of the Asia Oceania Agricultural Proteomics Organization.

Zahed Hossain currently holds the position of *Associate Professor* at the Department of Botany, University of Kalyani, West Bengal, India. Prior to joining, he served West Bengal State University as *Assistant Professor* in Botany from 2009 to 2015. Dr. Hossain has been working in the field of plant stress physiology for the past 15 years. He pursued his Ph.D. research work on *Plant Salinity Stress* from the National Botanical Research Institute, Lucknow as CSIR-JRF and SRF. Dr. Hossain is the recipient of several prestigious National, International Awards and Fellowships, such as JSPS Invitation Fellowship, DST-BOYSCAST Fellowship, Visiting Research Fellowship from Generalitat Valenciana, Spain for pursuing research at the Universitat Jaume I, Castello de la Plana, Castello, Merit Scholarships from University of Kalyani for securing top position in B.Sc. and M.Sc. levels. His research group investigates the plant response towards different environmental challenges at the physiological, biochemical, transcript and proteomic levels. His current research work has focused on exploring the miRNA-mediated response to heavy metal stress in maize.

International Journal of
Molecular Sciences

Editorial

Preface—Plant Proteomic Research

Setsuko Komatsu [1,*] and Zahed Hossain [2,*]

[1] National Institute of Crop Science, National Agriculture and Food Research Organization, Kannondai 2-1-18, Tsukuba 305-8518, Japan
[2] Department of Botany, University of Kalyani, West Bengal 741235, India
*   Correspondence: skomatsu@affrc.go.jp (S.K.); zahed_kly@yahoo.com (Z.H.); Tel.: +81-29-838-8693 (S.K.); +91-33-2582-8750 (ext. 318) (Z.H.)

Academic Editor: Marcello Iriti
Received: 17 November 2016; Accepted: 30 December 2016; Published: 4 January 2017

Plants, being sessile in nature, are constantly exposed to environmental challenges resulting in substantial yield loss. To cope with harsh environments, plants have developed a wide range of adaptation strategies involving morpho-anatomical, physiological, and biochemical traits. In recent years, there has been phenomenal progress in the understanding of plant responses to environmental cues at the protein level. This progress has been fueled by the advancement in mass spectrometry techniques, complemented with genome-sequence data and modern bioinformatics analysis with improved sample preparation and fractionation strategies. As proteins ultimately regulate cellular functions, it is perhaps of greater importance to understand the changes that occur at the protein-abundance level, rather than the modulation of mRNA expression. This special issue on "Plant Proteomic Research" brings together a selection of insightful papers that address some of these issues related to applications of proteomic techniques in elucidating master regulator proteins and the pathways associated with plant development and stress responses. This issue includes four reviews and 13 original articles primarily on environmental proteomic studies.

The first review by Hossain et al. [1] summarizes the recent contributions of plant proteomic research to comprehend the complex mechanism of plant response to nanoparticles stress. Pechanova and Pechan [2] present an overview of maize-pathogen interactions at the proteome level, emphasizing the application of various mass spectrometry-based high-throughput proteomic techniques in identifying possible candidate proteins involved in maize pathogen resistance. Wang et al. [3] summarize the recent proteomic studies related to drought sensing and signaling mechanisms for better understanding the molecular basis of plant drought tolerance. Nouri et al. [4] present a comprehensive picture about the fine tuning of photosynthetic pathways at the protein level linked to plant adaptation to abiotic stresses.

Among the 13 original articles, six articles highlight iTRAQ-based proteomic approaches. Ye et al. [5] present a deep and extensive research work on drought-induced leaf proteome modulation in Switchgrass (*Panicum virgatum*) using the iTRAQ labeling method followed by nano-scale liquid chromatography mass spectrometry analysis. Li et al. [6] provide an overview of aluminum stress-mediated alterations of leaf proteome in two contrasting citrus species differing in aluminum tolerance. An et al. [7] present comparative proteomic analysis of ramie plants under PEG-mediated drought stress. Hua et al. [8] emphasize the application of proteomic analysis in unraveling the molecular mechanism of betalain biosynthesis in *Hylocereus polyrhizus* fruits at the posttranscriptional level. Li et al. [9] present comparative proteomic analyses of buds and young expanding leaves of the tea plant (*Camellia sinensis* L.), highlighting the molecular mechanism involved in secondary metabolite production. Zhao et al. [10] perform quantitative proteomics analysis of herbaceous peony (*Paeonia lactiflora* Pall.) in response to Paclobutrazol, a triazole compound inhibiting growth of lateral branching. By using gel-free proteomics, Alqurashi et al. [11] analyze the changes in the *Arabidopsis thaliana* proteome composition implicating the role of cAMP in biotic and abiotic stress responses by inducing complex changes in cellular energy homeostasis. Rasool et al. [12] present the comprehensive expression pattern of peptides in the date palm stem infested with Red Palm Weevil (*Rhynchophorus ferrugineus* Oliv.) using two-dimensional gel electrophoresis (2-DE) and MALDI-TOF mass spectrometry. Liu et al. [13] provide new insights into the molecular regulation of leaf color variation and carbon fixation in a xantha mutant of *Ginkgo biloba* L. by exploiting 2-DE coupled with MALDI-TOF/TOF mass spectrometry. Yu et al. [14] present comparative proteomic analysis of chimera *Hosta* "Gold Standard" leaves from various regions at different development stages and under excess nitrogen fertilization using 2-DE coupled MALDI-TOF/TOF MS. Findings provide new insights towards understanding the mechanisms of leaf color regulation in variegated leaves. Zhang et al. [15] unravel the protein regulation mechanism of pollen infertility and allergenicity in triploid and diploid poplar (*Populus deltoids*) plants using the 2-DE technique followed by MALDI-TOF-TOF mass spectrometry analysis. Wang et al. [16] demonstrate the stable expression of basic fibroblast growth factor in chloroplasts of tobacco providing an additional option for the production of chloroplast-produced therapeutic proteins. Integration of the foreign expression cassette into the plastid genome of transformants is confirmed by PCR and Southern hybridization and expression is quantified by ELISA. Liu et al. [17] provide an overview of the seed-specific expression of microtubule-associated protein SBgLR in transgenic maize (*Zea mays*), resulting in increased seed protein and lysine contents. The zein, non-zein, and total protein extracts of the seeds of transgenic plants are analyzed by sodium dodecyl sulfate-polyacrylamide gel electrophoresis (SDS-PAGE).

This special issue on "Plant Proteomic Research" is an attempt to provide researchers with a glimpse of advanced mass spectrometry techniques with a special emphasis on candidate proteins and pathways associated with plant development and stress responses. We believe that this special issue reflects the current perspective and state of the art of plant proteomics, which would not only enrich us in understanding the plant's response to environmental clues but would further help us in designing better crops with the desired phenotypes. The articles in this issue will be of general interest to proteomic researchers, plant biologists, and environmental scientists.

We would like to express our gratitude to all authors for their high quality contributions and numerous peer reviewers for their critical evaluation and valuable suggestions. Moreover, we render our heartiest thanks to the Managing Editor Yong Ren and Section Managing Editor Yue Chen for giving us the opportunity to serve "Plant Proteomic Research" as Guest Editors and Editorial Office, a special mention goes to Sophie Suo for her untiring efforts in coordinating with authors and keeping us updated about the manuscript submission and review process, which helped us in completing the surmount task on time. Finally, we extend our sincere thanks to those professionals whose expertise in proofreading and formatting greatly improved the quality of this special issue.

Conflicts of Interest: The authors declare no conflict of interest.

References

1. Hossain, Z.; Mustafa, G.; Komatsu, S. Plant Responses to Nanoparticle Stress. *Int. J. Mol. Sci.* **2015**, *16*, 26644–26653. [CrossRef] [PubMed]
2. Pechanova, O.; Pechan, T. Maize-Pathogen Interactions: An Ongoing Combat from a Proteomics Perspective. *Int. J. Mol. Sci.* **2015**, *16*, 28429–28448. [CrossRef] [PubMed]
3. Wang, X.; Cai, X.; Xu, C.; Wang, Q.; Dai, S. Drought-Responsive Mechanisms in Plant Leaves Revealed by Proteomics. *Int. J. Mol. Sci.* **2016**, *17*, 1706. [CrossRef] [PubMed]
4. Nouri, M.-Z.; Moumeni, A.; Komatsu, S. Abiotic Stresses: Insight into Gene Regulation and Protein Expression in Photosynthetic Pathways of Plants. *Int. J. Mol. Sci.* **2015**, *16*, 20392–20416. [CrossRef] [PubMed]
5. Ye, Z.; Sangireddy, S.; Okekeogbu, I.; Zhou, S.; Yu, C.-L.; Hui, D.; Howe, K.J.; Fish, T.; Thannhauser, T.W. Drought-Induced Leaf Proteome Changes in Switchgrass Seedlings. *Int. J. Mol. Sci.* **2016**, *17*, 1251. [CrossRef] [PubMed]

6. Li, H.; Yang, L.-T.; Qi, Y.-P.; Guo, P.; Lu, Y.-B.; Chen, L.-S. Aluminum Toxicity-Induced Alterations of Leaf Proteome in Two Citrus Species Differing in Aluminum Tolerance. *Int. J. Mol. Sci.* **2016**, *17*, 1180. [CrossRef] [PubMed]

7. An, X.; Zhang, J.; Dai, L.; Deng, G.; Liao, Y.; Liu, L.; Wang, B.; Peng, D. Isobaric Tags for Relative and Absolute Quantitation (iTRAQ)-Based Comparative Proteome Analysis of the Response of Ramie under Drought Stress. *Int. J. Mol. Sci.* **2016**, *17*, 1607. [CrossRef] [PubMed]

8. Hua, Q.; Zhou, Q.; Gan, S.; Wu, J.; Chen, C.; Li, J.; Ye, Y.; Zhao, J.; Hu, G.; Qin, Y. Proteomic Analysis of *Hylocereus polyrhizus* Reveals Metabolic Pathway Changes. *Int. J. Mol. Sci.* **2016**, *17*, 1606. [CrossRef] [PubMed]

9. Li, Q.; Li, J.; Liu, S.; Huang, J.; Lin, H.; Wang, K.; Cheng, X.; Liu, Z. A Comparative Proteomic Analysis of the Buds and the Young Expanding Leaves of the Tea Plant (*Camellia sinensis* L.). *Int. J. Mol. Sci.* **2015**, *16*, 14007–14038. [CrossRef] [PubMed]

10. Zhao, D.; Gong, S.; Hao, Z.; Meng, J.; Tao, J. Quantitative Proteomics Analysis of Herbaceous Peony in Response to Paclobutrazol Inhibition of Lateral Branching. *Int. J. Mol. Sci.* **2015**, *16*, 24332–24352. [CrossRef] [PubMed]

11. Alqurashi, M.; Gehring, C.; Marondedze, C. Changes in the *Arabidopsis thaliana* Proteome Implicate cAMP in Biotic and Abiotic Stress Responses and Changes in Energy Metabolism. *Int. J. Mol. Sci.* **2016**, *17*, 852. [CrossRef] [PubMed]

12. Rasool, K.G.; Khan, M.A.; Aldawood, A.S.; Tufail, M.; Mukhtar, M.; Takeda, M. Identification of Proteins Modulated in the Date Palm Stem Infested with Red Palm Weevil (*Rhynchophorus. ferrugineus* Oliv.) Using Two Dimensional Differential Gel Electrophoresis and Mass Spectrometry. *Int. J. Mol. Sci.* **2015**, *16*, 19326–19346. [CrossRef] [PubMed]

13. Liu, X.; Yu, W.; Wang, G.; Cao, F.; Cai, J.; Wang, H. Comparative Proteomic and Physiological Analysis Reveals the Variation Mechanisms of Leaf Coloration and Carbon Fixation in a Xantha Mutant of *Ginkgo biloba* L. *Int. J. Mol. Sci.* **2016**, *17*, 1794. [CrossRef] [PubMed]

14. Yu, J.; Zhang, J.; Zhao, Q.; Liu, Y.; Chen, S.; Guo, H.; Shi, L.; Dai, S. Proteomic Analysis Reveals the Leaf Color Regulation Mechanism in Chimera Hosta "Gold Standard" Leaves. *Int. J. Mol. Sci.* **2016**, *17*, 346. [CrossRef] [PubMed]

15. Zhang, X.-L.; Zhang, J.; Guo, Y.-H.; Sun, P.; Jia, H.-X.; Fan, W.; Lu, M.-Z.; Hu, J.-J. Comparative Proteomic Analysis of Mature Pollen in Triploid and Diploid *Populus deltoides*. *Int. J. Mol. Sci.* **2016**, *17*, 1475. [CrossRef] [PubMed]

16. Wang, Y.-P.; Wei, Z.-Y.; Zhong, X.-F.; Lin, C.-J.; Cai, Y.-H.; Ma, J.; Zhang, Y.-Y.; Liu, Y.-Z.; Xing, S.-C. Stable Expression of Basic Fibroblast Growth Factor in Chloroplasts of Tobacco. *Int. J. Mol. Sci.* **2016**, *17*, 19. [CrossRef] [PubMed]

17. Liu, C.; Li, S.; Yue, J.; Xiao, W.; Zhao, Q.; Zhu, D.; Yu, J. Microtubule-Associated Protein SBgLR Facilitates Storage Protein Deposition and Its Expression Leads to Lysine Content Increase in Transgenic Maize Endosperm. *Int. J. Mol. Sci.* **2015**, *16*, 29772–29786. [CrossRef] [PubMed]

International Journal of
Molecular Sciences

Article

Identification of Proteins Modulated in the Date Palm Stem Infested with Red Palm Weevil (*Rhynchophorus ferrugineus* Oliv.) Using Two Dimensional Differential Gel Electrophoresis and Mass Spectrometry

Khawaja Ghulam Rasool [1,2,*], Muhammad Altaf Khan [3], Abdulrahman Saad Aldawood [1], Muhammad Tufail [1,2], Muhammad Mukhtar [4,*] and Makio Takeda [2]

[1] Department of Plant Protection, College of Food and Agriculture Sciences, King Saud University, Riyadh 11451, Saudi Arabia; aldawood@ksu.edu.sa (A.S.A.); mtufail@ksu.edu.sa (M.T.)
[2] Graduate School of Agricultural Science, Kobe University, Kobe 657-8501, Japan; mtakeda@kobe-u.ac.jp
[3] Department of Plant Production, College of Food and Agriculture Sciences, King Saud University, Riyadh 11451, Saudi Arabia; altafksu@gmail.com
[4] Department of Biotechnology, American University of Ras Al Khaimah, Ras Al Khaimah 10021, United Arab Emirates
* Correspondence: krasool@ksu.edu.sa (K.G.R.); muhammad.mukhtar@aurak.ac.ae (M.M.); Tel.: +966-567-461-455 (K.G.R.); +971-7221-0900 (ext. 1174) (M.M.); Fax: +966-114-678-423 (K.G.R.); +971-7221-0300 (M.M.)

Academic Editor: Setsuko Komatsu

Received: 9 April 2015; Accepted: 3 August 2015; Published: 17 August 2015

Abstract: A state of the art proteomic methodology using Matrix Assisted Laser Desorption/Ionization-Time of Flight (MALDI TOF) has been employed to characterize peptides modulated in the date palm stem subsequent to infestation with red palm weevil (RPW). Our analyses revealed 32 differentially expressed peptides associated with RPW infestation in date palm stem. To identify RPW infestation associated peptides (I), artificially wounded plants (W) were used as additional control beside uninfested plants, a conventional control (C). A constant unique pattern of differential expression in infested (I), wounded (W) stem samples compared to control (C) was observed. The upregulated proteins showed relative fold intensity in order of I > W and downregulated spots trend as W > I, a quite interesting pattern. This study also reveals that artificially wounding of date palm stem affects almost the same proteins as infestation; however, relative intensity is quite lower than in infested samples both in up and downregulated spots. All 32 differentially expressed spots were subjected to MALDI-TOF analysis for their identification and we were able to match 21 proteins in the already existing databases. Relatively significant modulated expression pattern of a number of peptides in infested plants predicts the possibility of developing a quick and reliable molecular methodology for detecting plants infested with date palm.

Keywords: date palm; stem; red palm weevil; infestation; differential expression; proteins; two dimensional differential in-gel electrophoresis (2D-DIGE); Matrix Assisted Laser Desorption/Ionization-Time of Flight (MALDI TOF)

---

1. Introduction

The red palm weevil (RPW) (*Rhynchophorus ferrugineus* Oliv., Coleoptera: Curculionidae) has become the most destructive pest of date palm trees in several regions of the world including Saudi Arabia. This palm-damaging pest was first reported in Southeast Asia on coconut palm [1]. Since its discovery in the Gulf Region in the year 1980s, the insect has been spreading rapidly and reported from almost every palm growing country in the World [2]. Bulk movement of date palm offshoots for planting is blamed on the invasion source of RPW in the Middle East [3]. The RPW has been reported to infest 26 palm species belonging to 16 different genera worldwide [4]. Although it is difficult to evaluate the overall actual global damages caused by RPW, in Saudi Arabia along with just ~5% infestation; management and eradication of RPW in date plantation cause more than 8.69 million USD of economic loss [5].

It is also worth mentioning that RPW larval stage is the most destructive, and responsible for damaging the palm. The larvae feed within the date palm trunk until they are fully developed [6,7]. This insect completes several generations within the same palm without any obvious symptoms in the plant until the tree finally collapses [8,9]. This cryptic feeding behavior of the RPW makes it difficult to detect infestations at early stages, and severe decaying of the internal tissues leads to the death of the tree [3,10].

In Saudi Arabia, the Ministry of Agriculture has launched a national campaign for controlling RPW to avoid losses inflicted on the production of dates. The campaign includes removal of infested plants, pesticide application through injection and spraying in severely infested and newly infested areas, and the use of pheromone traps for monitoring and decreasing RPW populations [11]. It has been observed that infested plants can be recovered if infestation is detected early. Currently available detection techniques, including visual inspections, acoustic sensors [12,13], sniffer dogs [14], and pheromone traps [15], are in practice to identify infestations at early stages. However, the development of an effective and efficient high throughput screening procedure is still needed for the early detection of RPW.

We employed proteomic methodologies to identify responses associated with RPW infestation. It has been previously reported that plants have evolved various innate and acquired defense mechanisms against visible/invisible injuries afflicted by insect pests [16]. Innate or direct defense mechanisms in plants include specialized characteristics, such as thorns, trichomes, and primary and secondary metabolites [16]. Some herbivores feeding induce proteinase inhibitors in plants that prevent digestive enzymes required for insect's proper digestion thus limiting invasion [17]. Acquired defenses involve release of volatile organic compounds that attract arthropod predators and parasitoids to control herbivore populations [18]. Herbivores oral secretions specially discharged into plant tissues during feeding induce a number of specific responses [19,20]. The herbivores regurgitate and other oral secretions trigger plant defense related proteins (parallel to acquired immune system of mammals) or activate the plant defense system releasing volatile compounds to attract predators [21,22].

Studies also confirm differential molecular defensive responses upon infestation with a wide variety of insects/pests. As the plant genomics/proteomics related to plant defense responses is comparatively a new field of study, more data is emerging to better understand plant responses against biotic and abiotic stresses [23]. The insect plant interactions itself have great impact on plant defense responses [16,24].

Proteomics strategies have been extensively used for identifying infections/diseases among humans; however their uses for plants have been relatively less. A few proteomic studies involving plants encouraged us to embark on utilizing these methodologies for saving beneficial date palm plants from RPW infestation. For example, proteome analysis of brittle leaf diseased date palm leaves when compared with that of their normal counterparts revealed quantitative differences in several proteins. Of the differentially expressed proteins, Manganese (Mn-binding) PSBO and PSBP proteins were downregulated; whereas, several other proteins were upregulated in diseased samples [25]. Likewise, proteome analysis of brittle leaf disease affected date palm leaves indicated changes in the proteome at early disease stage where the decrease in Mn deficiency associated with MSP-33 kDa subunit protein was considered as brittle leaf disease biomarker [26]. Moreover, Gómez-Vidal *et al.* [27] have evaluated the plant defense/stress, photosynthesis and energy metabolism related proteins, using 2-dimensional electrophoresis (2DE) proteomic techniques, that were differentially expressed in the date palm (*Phoenix dactylifera*) leaves in response to the attack by entomopathogenic fungi (*Beauveria bassiana, Lecanicillium dimorphum* and *L. cf. psalliotae*) as compared to control samples. In another study, 2DE of date palm sap 52 identified spots among 100 were related to *Saccharomyces cerevisiae*, supposed to be natural microflora of date palm sap while others were related to vegetable proteins playing a role in the vascular system of the plant [28]. Several changes in date palm leaf proteome were observed when red palm weevil infested and healthy date palm leaves proteome were compared [29,30]. Furthermore, we have a highly integrated proteomic initiative to identify molecular markers associated with RPW infestation and other diseases to this plant. As far as the proteome defenses between leaf and stem are concerned, we observed different protein moieties modulated in stem and leaves. However, an extensive analysis regarding this aspect is ongoing. Modulations in the date palm fruit proteome have also been reported during development and ripening stages [31]. In another study two dimensional differential gel electrophoresis and mass spectrometry also revealed changes in the proteome of salinity and drought-stressed palm seedlings when compared with non-stressed plants [32].

Very few studies have addressed the date palm proteomics association with physiological or induced changes in date palm. Our study is unique in a way that it provides proteomic changes associated with RPW infestation and its comparison with artificially induced injury to plant.

As such, the objective of the present study was to characterize the proteome changes occurring in date palm stem infested with RPW using 2D-DIGE and mass spectrometry so as to explore the biomarker for the early detection of RPW for its effective management.

2. Results and Discussion

2.1. Date Palm Proteome Analysis by 2D-DIGE

Plants respond to injuries/infestations and other abiotic stresses by activating a broad range of acquired defense system, including activation of pathogenesis-related (PR) genes both at local and systemic sites [16], crosslinking of cell wall proteins, generation of reactive oxygen species (ROS), and local programmed cell death. This study reports our ongoing efforts to understand proteomic modulation associated with RPW infestation. Previously, we optimized protein isolation from date palm leaves and its utilization in proteomic evaluations. This study extrapolates our work on the stem part of the date palm plants. As expected, different peptides were modulated in stem as the metabolic activities of leaves are quite different. We identified eleven peptides modulated in the leaves of date palm belonging to three main categories *i.e.*, stress/defense, photosynthesis, and ion transport [29]. For evaluating differential proteomic responses subsequent to infestation stem samples from infested, control, and artificially wounded plants were subjected to protein isolation followed by differential expression analyses. Proteins were extracted from stem samples using phenol- sodium dodecyl sulfate (SDS) extraction method [33]. For reducing internal variations three replicates for each plant were used. The extracted proteins were then quantified using 2D quant kit after solubilizing in 2D-rehyration buffer. Ten µg aliquots of each sample was solubilized in SDS loading buffer and separated on 12.5% sodium dodecyl sulfate polyacrylamide gel electrophoresis (SDS-PAGE) before staining. Protein profile after staining with Coomassie showed good reproducibility among replicates, consistent solubilization and reproducible extraction methods (Figure 1).

Moreover, 2D DIGE (General Electric (GE) Healthcare, Buckinghamshire, United Kingdom) was run to compare differences among control, infested, and wounded samples. These samples were labeled with either Cy3 or Cy5 dyes while the internal standard was consistently labeled with Cy2. The experimental design for 2D-DIGE experiments is shown in (Table 1).

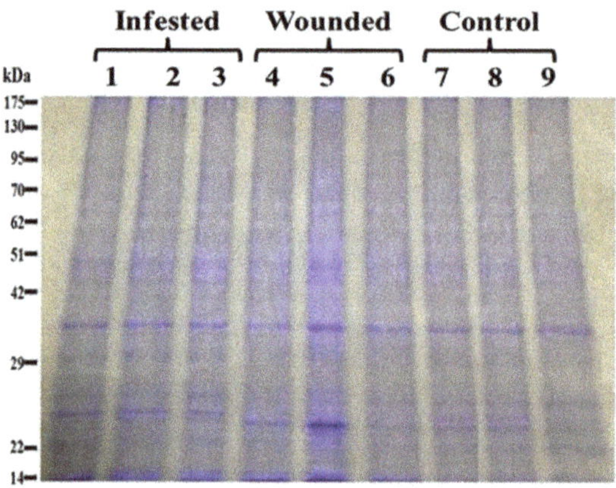

Figure 1. Comparative protein expression profiling of the control, infested and wounded date palm samples using SDS-PAGE. Lanes 1–3 represent total cell proteins from 3-infested replicates; while lanes 4–6 represent proteins from wounded date palm samples; and lanes 7–9 represent proteins from control date palm samples.

Table 1. Experimental design for 2D-DIGE. Three replicates from each control, infested, and wounded protein samples were labeled and combined for 2D-DIGE.

Gel Number	Protein Samples Labeling		
	Cy2	Cy3	Cy5
1	Pooled sample	Infested R1	Control R3
2	Pooled sample	Infested R2	Wounded R1
3	Pooled sample	Infested R3	Wounded R2
4	Pooled sample	Control R1	Wounded R3
5	Pooled sample	–	Control R2

R = Replications of the treatments.

After labeling with Cy dyes, two samples were mixed with different combinations along with internal standard and electrophoresed on the same gel except one gel contained a single sample with internal standard. The representative gels of 2D-DIGE after scanning with a fluorescence gel scanner, Typhoon imager (Trio) (GE Healthcare), are shown in Figure 2.

Progenesis Samespots software version 3.3 (Nonlinear Dynamics Ltd., Newcastle, United Kingdom) was used to statistically analyze the protein expression among control, infested, and wounded samples.

A total of 522 well-resolved protein spots were observed on each gel, and out of them, 32 spots showed statistically significant differences ($p \leq 0.05$, and intensity fold change ≥ 1.5) among expressions of proteins in either of this combination. Among 32 differential expressed spots, 11 were upregulated in infested and eight in wounded. However, there was downregulation of 13 spots in both infested and wounded when compared with control (Figure 3).

Interestingly, spots from the wounded samples showed the same trend of up or downregulation compared to control, but fold change intensity is lower than the infested samples. Differential protein patterns were also observed in the infested and wounded samples that are of our interest. The same spot (protein) was upregulated in the infested and downregulated in the wounded compared to control and *vice versa*. A majority of the upregulated spots (spot numbers 414, 243,554, 300, 447, 994, 1281, 1073 and 438) followed a unique trend of upregulation like (I > W > C) except two spots (1308 and 1287) where the pattern was observed like (I > C > W). Similar to upregulation, almost the same trend was observed among majority of downregulated spots in reverse order C > W > I (spots numbers 1027, 987,976, 781, 798, 1023, 951, 1020, 683, 1086, 1163 and 884). However, some deviations were noticed in some spots (221, 235, 234, 506, 822 and 318) where the expression pattern was observed as W > C > I, except in two spots (137 and 313) this pattern was like W > I > C. Our data clearly demonstrates that RPW infestation relatively enhances the modulation of proteins upregulated or downregulated with few minor variations. The relative expression patterns are quite intriguing and previous reports also suggest such trends in plants [34,35].

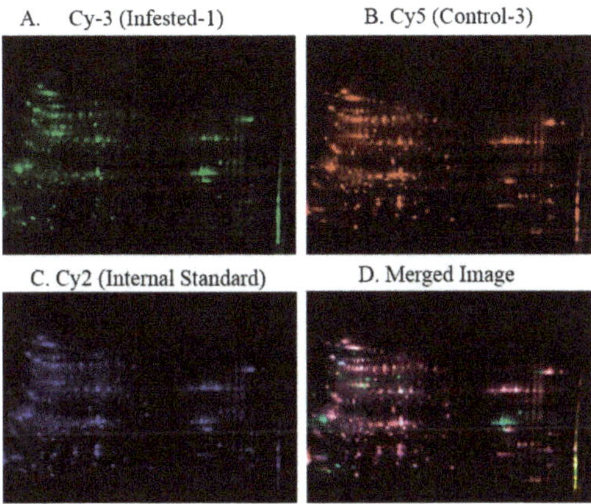

Figure 2. 2D-DIGE images of date palm proteins. The protein sample of Infested-1, Control-3 and internal standard (pooled of all the samples) are individually labeled with Cy dyes, mixed together and separated by 2D-DIGE followed by image scanning. (**A**) Image of date palm infested sample and labeled with Cy3 dye; (**B**) Image of date palm control sample labeled with Cy5 dye; (**C**) Image of date palm sample pooled from all and labeled with Cy2 dye; and (**D**) Overlay gel of infested, and control samples along with internal standard.

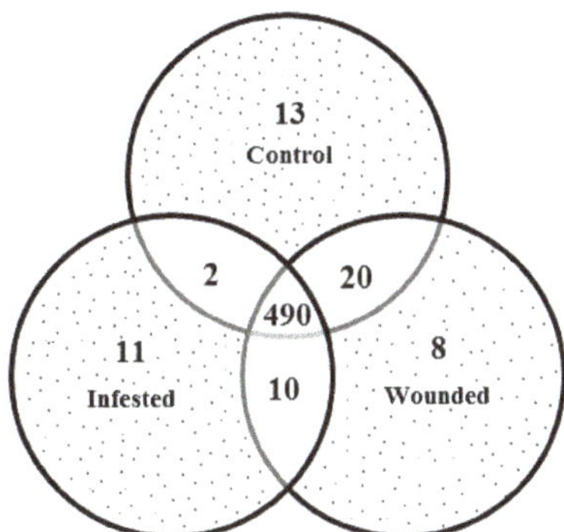

Figure 3. Venn diagram for the relative distribution of differentially expressed proteins spots in control, mechanically wounded and RPW infested date palm samples. The non-overlapping segments of diagram represent the number of proteins that were significantly upregulated (>1.5-fold) in the corresponding group when compared with the other two groups. The overlapping region between any two groups represents the number of proteins spots significantly upregulated (>1.5-fold) compared to the third one. While the central overlapping region represents the protein spots where no any statistically significant change was observed.

2.2. Protein Identification by Mass Spectrometry

Proteomic methodologies for differential expression are quite tedious by nature; however, it provides highly reliable information once a modulated peptide has been identified. Final identification subsequent to proteomic methodology involves identification procedures involving mass spectrometric analyses. A state of the art proteomic methodology MALDI-TOF used for the identification of proteins by peptide mass fingerprinting was employed to characterize peptides modulated in date palm stems subsequent to infestation with RPW. The above-described 32 differentially expressed spots were excised from preparative gels, digested enzymatically with trypsin, and subjected to mass spectrometry MS analysis. Data were examined using BioTools 3.2 (Bruker Daltonics, Bremen, Germany) in combination with the Mascot search algorithm (version 2.0. 04) against green plants database. MS results showed that identified 21 spots matched with previously reported proteins found in databases (Table 2). We were unable to characterize 11 protein spots as there was no protein matching proteins in the database. It is quite possible that date palm plant has some unique proteins not found in other plants. Once the date palm genome is deciphered, these unique proteins will be of much interest. In protein identification a small number of different spots showed the same identified proteins, indicating the presence of isoforms of a specific gene or these emerged because of post-translational modifications (PTMs) differences in molecular weight and pI. In our case V-type proton ATPase catalytic subunit A (spot numbers 221 and 243), and probable glycerol-3-phosphate acyltransferase 8 (spot numbers 683 and 781) belong to these categories.

The identities of differentially expressed proteins were confirmed by their comparison with the previously characterized proteins in Swiss-Prot database. Among these 33% were characterized based on their homology with *Arabidopsis thaliana* proteins, 14.2% *Oryza sativa* (*subspecies Japonica*) (rice), and 9.5% *Sorghum bicolor* (Sorghum), respectively. Only 4.7% proteins were confirmed from *Zea mays*, *Solanum lycopesicum*, *Eucalyptus gunnii* (Cider gum), *Hevea brasiliensis* (Para rubber tree), *Citrus unshiu* (Satsuma mandarin), *Carica papaya* (papaya), *Medicago truncatula* (Barrel medic), *Spinacia oleracea* (Spinach), and *Hordeum vulgare* (Barley) proteome.

Moreover, differentially expressed proteins had been characterized as ion transport (6), lipid biosynthesis (2), protein folding (1), plant defense (1), ethylene signaling pathway (1), carbohydrate metabolism (1), proteolysis (1), *S*-adenosylmethionine biosynthetic process (1), nitrate assimilation (1), porphyrin biosynthesis (1), transcription (1), lignin biosynthesis (1), cytoskeletal protein (1), calcium ion binding (1) and unknown (1).

It is worth mentioning that among the 21 proteins identified through homology comparison, six belong to the ion transport family of proteins that are stress related, while several others could potentially be involved in response to RPW infestation. These were V-type proton ATPase catalytic subunit A (*Zea mays* (Maize) (spot numbers. 221, 243). The spot 221 were suppressed 1.5-fold in the infested samples relative to control. However, spot 243 was much higher in infested samples (5.3-fold) compared to control. V-type proton ATPase is a proton pump present inside plant vacuoles which plays an important role in plant salinity tolerance [36,37]. V-type proton ATPase plays a crucial role in maintenance of ion homeostasis inside plant cells by acidifying compartments of the vacuole [38,39]. It maintains electrochemical H^+-gradient to drive the transport of Na^+ into the vacuole lumen, compartmentalizing this toxic ion from the cytoplasm and maintaining low cytoplasmic Na^+ concentrations. Upregulation of this enzyme was also reported in response to stress in other plants as well [40–43]. The upregulation of enzyme V-type proton ATPase might provide insights into understanding the infestation of date palm and could be a useful marker for early diagnosis of RPW. Another spot (spot number 414) identified as a differential protein and upregulated in infested samples corresponded to ATP synthase subunit β, mitochondrial.

Table 2. Differentially expressed proteins between control, wounded and infested date palm stem by MALDI-TOF peptide mass fingerprinting after 2D-DIGE.

Spot Number	FC (I/C)	FC (W/C)	Accession (Uniprot)	Protein Description	Function	pI	M_W	Coverage	Score	Organism
1308	1.60↑	3.03↓	P27161	Calmodulin	M: Calcium ion binding	4.18	16,950	69	229	*Solanum lycopersicum* (Tomato)
781	2.42↓	1.67↓	Q5XF03	Probable glycerol-3-phosphate acyltransferase 8	Lipid biosynthesis	9.14	56,345	25	60	*Arabidopsis thaliana* (Mouse-ear cress)
221	1.50↓	1.17↑	P49087	V-type proton ATPase catalytic subunit A	Ion transport	5.89	62,198	25	62	*Zea mays* (Maize)
798	2.08↓	1.48↓	O04854	Caffeoyl-CoA *O*-methyltransferase	Lignin biosynthesis	5.02	28,010	12	72	*Eucalyptus gunnii* (Cider gum)
414	2.77↑	1.28↑	P29685	ATP synthase subunit β, mitochondrial	ATP hydrolysis coupled proton transport	5.98	60,335	35	75	*Hevea brasiliensis* (Para rubber tree)
994	1.68↑	1.56↑	Q94FT8	Non-symbiotic hemoglobin 3	Iron and oxygen transport	9.83	18,614	65	65	*Oryza sativa* subsp. japonica (Rice)
1281	1.76↑	1.58↑	Q94FT8	Non-symbiotic hemoglobin 3	Iron and oxygen transport	9.83	18,614	83	62	*Oryza sativa* subsp. japonica (Rice)
137	1.66↑	1.99↑	P22953	Heat shock 70 kDa protein 1	Stress response, plant defense,	5.03	71,712	38	88	*Arabidopsis thaliana* (Mouse-ear cress))
1073	2.07↑	1.50↑	Q9SVQ0	Ethylene-responsive transcription factor ERF062	Ethylene signaling pathway	9.46	44,283	37	68	*Arabidopsis thaliana* (Mouse-ear cress)
243	5.31↑	1.77↑	Q9SM09	V-type proton ATPase catalytic subunit A	Hydrogen ion transport	5.29	68,923	34	70	*Citrus unshiu* (Satsuma mandarin)
554	1.99↑	1.39↑	Q1PFG1	F-box protein At1g66490	Uncharacterized	8.75	43,577	26	65	*Arabidopsis thaliana* (Mouse-ear cress)
300	2.98↑	2.13↑	Q39251	Actin-depolymerizing factor 2	Cytoskeletal protein	5.24	15,963	36	65	*Arabidopsis thaliana* (Mouse-ear cress)
447	2.36↑	1.17↑	B1A986	NAD(P)H-quinone oxidoreductase subunit 4L, chloroplastic	Oxidoreductaseactivity, ATP synthesis coupled electron transport	9.65	11,327	54	67	*Carica papaya* (papaya)

Table 2. *Cont.*

Spot Number	FC (I/C)	FC (W/C)	Accession (Uniprot)	Protein Description	Function	pI	M_W	Coverage	Score	Organism
1020	2.81↓	2.35↓	Q9SKQ0	Peptidyl-prolyl *cis-trans* isomerase CYP19-2	B: protein folding	8.33	1868	48	63	*Arabidopsis thaliana* (Mouse-ear cress)
683	2.58↓	1.69↓	Q5XF03	Probable glycerol-3-phosphate acyltransferase 8	Lipid biosynthesis	9.14	56,345	29	60	*Arabidopsis thaliana* (Mouse-ear cress)
506	1.36↓	1.65↑	Q8W0A1	β-galactosidase 2	Carbohydrate metabolic process	5.59	92,630	19	62	*Oryza sativa* subsp. japonica (Rice)
822	1.92↓	1.06↑	C5X3M7	Putative uncharacterized protein Sb02g009233	Nucleic acid binding	5.60	14,533	88	73	*Sorghum bicolor* (Sorghum)
318	1.48↓	1.62↑	A5C9D0	Putative uncharacterized protein	Proteolysis	9.41	119,434	20	71	*Sorghum bicolor* (Sorghum)
438	2.24↑	1.02↑	A4PU48	S-adenosylmethionine synthase	Adenosylmethionine biosynthesis	5.59	43,708	29	63	*Medicago truncatula* (Barrel medic)
884	1.61↓	1.45↓	P23312	S Nitrate reductase [NADH]	Nitrite assimilation	6.25	104,703	68	62	*Spinacia oleracea* (Spinach)
313	2.44↑	2.47↑	Q42836	Delta-aminolevulinic acid dehydratase, chloroplastic)	Porphyrin biosynthesis	6.05	46,639	46	80	*Hordeum vulgare* (Barley)

Arrows indicate the proteins up (↑) and down (↓) regulations, FC = Fold change, I = RPW Infested samples, W = Mechanically Wounded samples; pI = Isoelectric point, M_W = Molecular Weight.

8

The expression of calcium binding protein, calmodulin (spot no. 1308) increased (1.6-fold) in response to the infestation of RPW. Calmodulin is a calcium-binding messenger protein and regulates downstream functions in response to Ca^{2+}. It has several targets including ion channel and a large number of enzymes (i.e., kinases, phosphatases, cytoskeletal proteins, synaptic proteins, cell cycle proteins) [44–47] and also is involved in heat-shock signal transduction [48]. RPW infestation associated upregulation of this protein probably is associated with enhanced signaling activity associated with infestation. The signaling cascade involves a number of proteins, and its interactions with several above-described proteins have been previously reported.

Protein belonging to S-adenosylmethionine synthase (spot number 438) showed the increased expression (2.2-fold) in infested sample relative to control. This enzyme catalyzes the synthesis of S-adenosylmethionine from methionine and ATP. Also, expression was upregulated by 2.4-fold in spot number 313, homologous to Delta-aminolevulinic acid dehydratase, chloroplastic) as compared to control. This enzyme is involved in the formation of porphobilinogen [49]. The expression of the spot number 137, homologous to heat shock 70 KDa proteins (HSP 70), was increased (1.66-fold) in the infested samples as compared to control. However, its expression was increased more in the wounded (~2-fold) relative to control. HSP 70 protein is a chaperone that assists the folding process of newly synthesized proteins and minimizes aggregation [50]. High expression of this protein under stress conditions has also been reported previously [50,51]. Several other HSPs have also been reported in response to different stress conditions like because of pea and *Erysiphe pisi* interaction [52], pea and *Mycosphaerella pinodes* interaction [53], or triticale under low N fertilization level [54]. The over-expression of these stress responsive proteins should not be surprising, because it is a natural defense mechanism that responds, as mentioned above, to any outer biotic/abiotic stress. The over expression of HSP 70 in present study could be a useful marker against RPW infestation.

The 2-spots (781 and 683) matching glycerol-3-phosphate acyltransferase, an essential enzyme for glycerolipid biosynthesis, were found to be down regulated in infested samples. Another down regulated spot (506) matched with protein β-galactosidase 2 was identified. This enzyme is responsible for non-reducing β-D-galactose residues in β-D-galactosides.

Furthermore, spot number 1073 was identified as ethylene response factor (ERF1) and its expression was upregulated (2.07-fold) in infested date palm sample relative to control. ERF1 is a member of the novel family of plant-specific transcriptional factors in *Arabidopsis thaliana* [55], and it is activated by either ethylene (ET) or jasmonate (JA) and is activated synergistically by both hormones [56]. ERF1 regulates defense response genes to the necrotrophic fungi *B. cinerea* and *P. cucumerina* by integrating ET and JA defense responses in *Arabidopsis* [57,58]. ERFs have been reported to affect a number of developmental processes, and are also differentially adapted to biotic or abiotic stresses such as pathogen attack, wounding, extreme temperature, and drought [59,60]. Our results suggest that ERF1 could be a key element in defense against RPW attack.

Another spot 884 matching with nitrate reductase was suppressed (1.6-fold) in date palm infested sample compared to control. This enzyme plays a key role in the synthesis of nitric oxide (NO) [61], an important signaling molecule mediating physiological and developmental processes. Also, NO plays an important role in plant responses to biotic and abiotic stresses [62–64]. Furthermore, NO was also reported to modulates ethylene, salicylic acid, and jasmonic acid-signaling pathways and abscisic acid (ABA)-induced stomatal closure [65,66].

Another important protein identified as nonsymbiotic hemoglobins 3 (spot numbers 994 and 1281) was upregulated in infested and wounded samples that are induced in plants during hypoxic stress. Nonsymbiotic hemoglobin AHb1 plays an important role in NO detoxification in *Arabidopsis* by scavenging NO and reducing its emission under hypoxic stress. These proteins were reported to protect plants during hypoxia or other similar stresses [67,68] and scavenge NO produced in stress conditions [69]. Nonsymbiotic hemoglobin AHb1 plays an important role in NO detoxification in *Arabidopsis thaliana* by scavenging NO and reduces NO emission under hypoxic stress [70].

A significant increase (2.98-fold) of actin depolymerizing factors (ADFs) (spot number 300) was found in infested samples compared to control. ADF is a small actin-binding protein and involved in plant growth, development stress response, and pathogen defense [71–73]. The role of actin had also been reported in response to plant hormones and biotic or abiotic stresses [71,74]. The ADFs have been found related to plant resistance to various pathogens in *Arabidopsis* and barley [75,76]. The energy produced by the depolymerization and polymerization of actin is used for the directional movement of cells, which is necessary for wound healing, immune response, embryonic development and development of tissues [77–79]. In this process a number of actin binding proteins, such as profillin, actin depolymerizing factor (ADF)/cofilin, myosin, fibrin and villin, are also involved.

Spot number 447 corresponding to NAD(P)H-quinone oxidoreductase subunit 4L, chloroplastic showed a significant increase (2.34-fold) in infested date palm samples as compared to control. NAD(P)H-quinone oxidoreductase is an important enzyme that is involved in the detoxification of quinones and their derivatives [80,81]. Taken together, this enzyme increased to reduce oxidative stress and to detoxify toxic molecules produced by the stress.

The spot no. 1020 corresponding to peptidyl-prolyl *cis-trans* isomerase (PPIase) CYP19-2 was downregulated (2.8-fold) in the infested samples. This enzyme catalyses the *cis-trans* isomerisation process of proline residues during protein folding [82]. Important groups of proteins, such as cyclophilin proteins, have this PPIase domain. Cyclophilins have been shown to be involved in a wide range of cellular processes like stress tolerance [83], cell division [84], transcriptional regulation [85], protein trafficking, [86] cell signaling [87], pre-mRNA splicing [88], and molecular chaperoning [89]. Taken together, a majority of identified proteins were directly or indirectly related to defense of date palm and could be exploited for the early diagnosis of RPW infestation. The information obtained here must contribute to clarification of plant defensive mechanism and RPW management.

Our current study also reveals that 21 differentially expressed peptides identified through our proteomic analyses have a major group of ion transport proteins. Furthermore, proteins belonging to various groups like lipid biosynthesis, protein folding, plant defense, carbohydrate metabolism, lignin biosynthesis, cytoskeletal proteins and a few others in plant stem, manifest integrated signaling mechanisms involved in injury/infestation processes. Once the date palm genome is revealed completely, study of proteomics networks involved upon infestation or injury will be of great interest. If we compare our current findings with the previous proteomic optimization from plant leaves [29]), it is clear that stem and leaves have different metabolic activities. Particularly, the stem part of the date palm plant serves several purposes such as storage, transport to aerial parts and maintaining overall plant integrity. Our report on differential proteomics of leaves showed 11 modulated peptides falling into only three categories, *i.e.* stress/defense, photosynthetic and ion transport. These findings make more sense as the aerial part of this plant is involved with photosynthetic processes whereas stem mainly contributes to plant overall structure and defense. The 11 peptides having no homologous counterparts in the plant proteome will be part of our future research interests. Diversity of modulated proteins either upon artificial injury or red palm weevil infestation shows integrated signaling in various metabolic aspects once the plant is exposed to unnatural circumstances either through injury or infestations.

3. Experimental Section

3.1. Date Palm Material and Infestation with RPW Larvae

Six-year old tissue cultured date palm plants of Khudry cultivar were obtained from Al Rajhi Tissue Culture Laboratory, Riyadh, Saudi Arabia. These plants were divided into 3 groups, with three replicates each, and were then used for mechanical wounding and infestation with RPW as described previously [90]. Group one was artificially infested (I) with RPW larvae, group two artificially wounded (W), and the third group was kept without any treatment as control (C). For artificial infestation each plant was introduced with 5 second instar RPW larvae by making tiny holes in the stem using drill machine with 6-mm size bit to accommodate larva. After treatment the stem part of the plants were wrapped up with fine steel mesh. The stem samples were taken after 3-days of infestation and stored at $-80\,^{\circ}$C until use.

3.2. Protein Extraction and SDS-PAGE

Total proteins from control, infested, and wounded date palm stem samples (three replicates from each sample) were extracted using phenol/SDS extraction method as described by Gomez-Vidal *et al*, [33] with minor modifications [29]. Briefly, stem tissues (one gram) were ground into fine powder in liquid nitrogen in pestle/mortar and suspended in 5 mL phenol and 5 mL dense SDS buffer (30% *w/v* sucrose, 2% *w/v* SDS, 0.1 M Tris-HCl, pH 8.0, 5% *v/v* 2-mercaptoethanol). After mixing and vortexing, mixture was centrifuged for 5 min at 10,000 rpm at 4 $^{\circ}$C. The upper phenolic phase was collected and precipitated with five volumes of cold 0.1 M methanolic ammonium acetate. After incubating at -20 $^{\circ}$C for 30 min, precipitated proteins were recovered by centrifugation at 1000 rpm for 5 min at 4 $^{\circ}$C and then washed two times with cold methanol solution containing 0.1 M ammonium acetate and then two times with cold 80% *v/v* acetone. The protein pellet was recovered each time by centrifugation at 8000 rpm for 5 min. The final protein pellet was air-dried at room temperature and suspended in 100 mM Tris buffer pH 8.0 and the then added equal volume of 2X SDS-reducing buffer (100 mM Tris–Cl (pH 6.8), 4% SDS, 0.2% bromophenol blue, 20% glycerol) containing 200 mM mercaptoethanol. SDS-PAGE analysis was carried out as described by Laemmli [91]. The gel was stained with Coomassie brilliant blue G-250 with constant and gentle agitation overnight. After staining the gel was placed in destining solution until the background becomes clear.

3.3. Two-Dimensional Difference Gel Electrophoresis (2D-DIGE)

The dried protein sample for 2D-DIGE was solubilized in rehydration buffer containing the chaotropic agent urea, alongside surfactants 3-[(3-cholamidopropyl)dimethylammonio]-1-propanesulfonate (CHAPS) and thiourea (7 M urea, 2 M thiourea, 2% CHAPS *w/v*, 2% Dithiothreitol (DTT), 0.5% Immobilized pH Gradient (IPG) buffer pH 3–11, 0.002% bromophenol blue) by shaking at 150 rpm for 1 h at 25 $^{\circ}$C. The insoluble residue was removed by centrifugation at 12,000 rpm for 10 min. Protein concentration was measured using 2-D Quant kit (GE Healthcare, Buckinghamshire, United Kingdom) in accordance with to the manufacturer's protocol and using bovine serum albumin (BSA) as a reference standard. The samples were further cleaned for 2D using the 2D Clean-Up Kit (GE Healthcare, Buckinghamshire, United Kingdom) and solubilized in buffer (7 M urea, 2 M thiourea, 2% CHAPS, 30 mM Tris-Cl, pH 8.5) without DTT and IPG buffer, and then they were quantified again using 2-D Quant kit (GE healthcare, Buckinghamshire, United Kingdom). Protein was aliquoted to required amount (300 µg) and frozen.

After adjusting the pH to 8.5, each protein sample was labeled with CyDye Fluor minimal dyes (GE healthcare, Buckinghamshire, United Kingdom) according to manufacturer's recommendation. Briefly, 50 µg each protein sample was incubated with 400 pmol CyDye Fluor minimal dyes on ice for 30 min in the dark. The control, wounded, and infested samples were labeled alternatively with Cy3 or Cy5 (Table 1). Internal standard containing equal amount of proteins from each sample was labeled with Cy2. The reaction was stopped by adding 1 µL of 10mM lysine solution and incubated 10 min on ice and proteins samples were combined according to experimental design as shown in Table 1.

For 2D-DIGE, five IPG Immobiline DryStrips 24 cm pH 3-11 (GE Healthcare, Buckinghamshire, United Kingdom) were rehydrated for 16 h with rehydration buffer (7 M urea, 2 M thiourea, 2% CHAPS *w/v*, 0.2% DTT, 0.5% IPG buffer pH 3–11, 0.002% bromophenol blue) containing the protein samples for each gel. Isoelectric focusing was performed with Ettan IPGphor3 IEF unit (GE Healthcare, Buckinghamshire, United Kingdom) at 50 µA per strip at 20 $^{\circ}$C according to following programme: (1) 500V for 1 h, (2) 1000V for 1 h, (3) 8000V for 3 h, (4) 8000 V for 45,000Vh. Strips were immediately equilibrated after Isoelectric focusing for 15 min in equilibration buffer 1 (50 mM Tris-HCl, pH 8.8; 6 M urea; 20% (*v/v*) glycerol and 2% (*w/v*) SDS) containing 2% DTT at room temperature under gentle agitation, and then by equilibration buffer 2 (50 mM TrisHCl, pH 8.8; 6 M urea; 20% (*v/v*) glycerol and 2% (*w/v*) SDS) containing 2.5% iodoacetamide. Second-dimension sodium dodecyl sulfate polyacrylamide gel electrophoresis (SDS-PAGE) was performed on 5%–20% polyacrylamide gradient gels using the Ettan DALT six vertical unit (GE Healthcare, Little Chalfont, UK) at 15 °C for 1W per gel for 1 h and then 2W per gel until the bromophenol blue dye reached the end of gel. Polyacrylamide gradient gels were prepared on low fluorescence glass using 2D Optimizer (Nextgen Sciences).

3.4. Image Acquisition and Analysis

The 2D gels were scanned using fluorescence gel scanner, Typhoon imager (Trio) (GE Healthcare, Uppsala, Sweden), using appropriate wavelengths and filters for Cy2, Cy3 and Cy5 dyes according to manufacturer's recommended protocol. The 2D gels images were analyzed using Progenesis Same Spot software version 3.3 (Nonlinear Dynamics Ltd, Newcastle upon Tyne, UK). The differential expression was ascertained using normalized protein spots in the Cy5 and Cy3 channels compared to the internal standard (Cy2). The spots of infested and wounded date palm samples were compared to control samples. One-way ANOVA was used to calculate the fold difference values and *p*-values. A threshold level was set at 1.5-fold up or downregulation, at $p \leq 0.05$ level.

3.5. Protein Identification by Mass Spectrometry

Preparative 2D gels were run using 700 µg total protein sample obtained by pooling all the samples present in the experimental design. The gels were fixed in ethanol (35% *v/v*), with phosphoric acid (2% *v/v*) overnight and then washed three times with water for 30 min each time. Then the gels were incubated for 1 h in methanol (34% *v/v*) containing ammonium sulfate (17% *w/v*) and phosphoric acid (3% *v/v*) for I h and after that 0.5% g/L Coomassie G-250 were added. The gels were stained for 5 days followed by rinsing in Milli Q water and stored until spots were picked and identified by Mass spectrometry (UltraFlexTrem, Bruker Daltonics, Bremen, Germany). Corresponding differential spots were matched to a colloidal Coomassie-stained image of the preparative gel, which was first mapped to the reference image.

The differential protein spots were manually excised form Coomassie-stained preparative gels and washed with solution containing 50 mM Ammonium bicarbonate. Then, the gel pieces were destained with 50 mM ammonium bicarbonate and 50% acetonitrile followed by 100% acetonitrile. Once destained, gel pieces were dried using vacuum centrifugation at 40 °C for 5 min. Dried gel pieces were rehydrated and digested with 10 µL trypsin at a concentration of 2ng/µL (Promega, Madison, WI, USA) in 25 mM NH_4HCO_3 pH 8.0 at 4 °C for 60 min, and digestion was continued for an additional 16–24 h at 37 °C. The digestion was stopped and digest mixtures were transferred to 0.5 mL tube, and the peptides were extracted by adding 50% acetonitrile having 0.1% Trifluoroacetic acid followed by drying to 10 µL using vacuum centrifugation. The 0.5 µL peptides was mixed with the matrix (10 mg α-Cyano-4-hydroxycinnamic acid in 1 mL of 30% acetonitrile containing 0.1% TFA) and applied on MALDI-target and dried before MS analysis, and then it was subjected to MALDI-TOF-MS (UltraFlexTrem, Bruker Daltonics, Bremen, Germany) in a positive mode. Peptide mass fingerprints were processed using flex analysis software (version 2.4, Bruker Daltonics, Bremen, Germany). MS data were interpreted by BioTools 3.2 (Bruker Daltonics, Bremen, Germany) in combination with the Mascot search algorithm (version 2.0. 04) against Swiss-Prot database for green plants. The validity/accuracy of identified proteins was only accepted when the mascot score was ≥ 60.

4. Conclusions

Date palm infestation with RPW is a key threat to the survival of date palm trees in the Kingdom of Saudi Arabia and other Gulf countries. We have utilized proteomic methodologies to identify molecular changes associated with RPW infestation of this plant. Though our interest is to identify RPW-specific molecular responses, it was however interesting to note that similar molecular moieties are upregulated in artificial wounding as well as RPW infestation. Nevertheless, relative modulation (downregulation/upregulation) remained quite differential. As both mechanical injury and RPW infestation impinges upon the same molecular moieties of date palm stem, we would have to establish certain baselines of proteomic changes for characterizing RPW-specific molecular changes. This study mainly provides a proof of concept that the stem, the hard part of the date palm tree, is amenable to molecular analytical procedures for understanding infestation with RPW. Furthermore, this work opens new avenues for understanding the proteome of this important tree. Techniques established and data generated will be crucial for date palm scientists in understanding diseases/infestation associated with the physiological process in this plant, in addition to developing new cultivars.

Acknowledgments: This Project was funded by the National Plan for Science, Technology and Innovation (MAARIFAH), King Abdulaziz City for Science and Technology, Kingdom of Saudi Arabia, Award Number (09-BIO900-02). We gratefully acknowledge support from Assim Abdulaziz Alfadda, Hicham Benabdelkamel and Afshan Masood Obesity Research Center, College of Medicine, King Saud University for mass spectrometry facilities. The authors gratefully acknowledge Osayimwense Osa and Larry Griffin from department of Languages and Literature at the American University of Ras Al-Khaimah for their support in proof reading and editing of this manuscript. Also we thank economic entomology research unit (EERU) members for their support in rearing RPW and other technical assistance.

Author Contributions: Khawaja Ghulam Rasool, Abdulrahman Saad Aldawood; and Muhammad Mukhtar conceived the project; Khawaja Ghulam Rasool, Muhammad Altaf Khan, Abdulrahman Saad Aldawood, Muhammad Tufail, Muhammad Mukhtar and Makio Takeda designed the experiments; Khawaja Ghulam Rasool and Muhammad Altaf Khan performed the experiments; Khawaja Ghulam Rasool drafted the manuscript; Muhammad Altaf Khan, Abdulrahman Saad Aldawood, Muhammad Tufail, Muhammad Mukhtar and Makio Takeda reviewed the manuscript. All authors read and approved the final manuscript.

Conflicts of Interest: The authors declare no conflict of interest.

References

1. Ferry, M.; Gómez, S. The red palm weevil in the Mediterranean area. *Palms Lawrence* **2002**, *46*, 172–178.
2. Aldawood, A.S.; Rasool, K.G. Rearing optimization of red palm weevil: *Rhynchophorus ferrugineus* (Coleoptera: Curculionidae) Olivier on Date Palm: *Phoenix dactylifera**. *Fla. Entomol.* **2011**, *94*, 756–760. [CrossRef]
3. Abraham, V.A.; Al Shuaibi, M.A.; Faleiro, J.R.; Abuzuhairah, R.A.; Vidyasagar, P.S.P.V. An integrated management approach for red palm weevil, *Rhynchophorus ferrugineus* Oliv., a key pest of date palm in the Middle East. *Sultan Qaboos Univ. J. Sci. Res. Agric. Sci.* **1998**, *3*, 77–84.
4. Dembilio, O.; Jacas, J.A. Bio-ecology and integrated management of the red palm weevil, *Rhynchophorus ferrugineus* (Coleoptera: Curculionidae), in the region of Valencia (Spain). *Hell. Plant Prot. J.* **2012**, *5*, 1–12.
5. El-Sabea, A.M.R.; Faleiro, J.R.; Abo-El-Saad, M.M. The threat of red palm weevil *Rhynchophorus ferrugineus* to date plantations of the Gulf region of the Middle East: An economic perspective. *Outlooks Pest Manag.* **2009**, *20*, 131–134. [CrossRef]
6. Faleiro, J.R. A review of the issues and management of red palm weevil, *Rhynchophorus ferrugineus* (Coleoptera: Rhynchophoridae) in coconut and date palm during the last one hundred years. *Int. J. Trop. Insect Sci.* **2006**, *26*, 135–154.
7. Kaakeh, W. Longevity, fecundity, and fertility of the red palm weevil, *Rynchophorus ferrugineus* Olivier (Coleoptera: Curculionidae) on natural and artificial diets. *Emir. J. Agric. Sci.* **2005**, *17*, 23–33. [CrossRef]
8. Rajamanickam, K.; Kennedy, J.S.; Christopher, A. Certain components of integrated management for red palm weevil, *Rhynchophorus ferrugineus* F. (Curculionidae: Coleoptera) on coconut (in South India). *Meded. Fac. Landbouwkd. Toegep. Biol. Wet. Univ. Gent.* **1995**, *60*, 803–805.
9. Avand Faghih, A. The biology of red palm weevil, *Rhynchophorus ferrugineus* Oliv. (Coleoptera: Curculionidae) in Saravan region (Sistan and Balouchistan Province, Iran). *Appl. Entomol. Phytopathol.* **1996**, *63*, 16–18.
10. Gadelhak, G.G.; Enan, M.R. Genetic diversity among populations of Red Palm Weevil, *Rhynchophorus ferrugineus* Olivier (Coleoptera: Curculionidae), determined by random amplified polymorphic DNA polymerase chain reaction (RAPD-PCR). *Int. J. Agric. Biol.* **2005**, *7*, 395–399.

11. Mukhtar, M.; Rasool, K.G.; Parrella, M.P.; Sheikh, Q.I.; Pain, A.; Lopez-Llorca, L.V.; Aldryhim, Y.N.; Mankin, R.W.; Aldawood, A.S. New initiatives for management of red palm weevil threats to historical Arabian date palms. *Fla. Entomol.* **2011**, *94*, 733–736. [CrossRef]
12. Mankin, R.W.; Hagstrum, D.W.; Smith, M.T.; Roda, A.L.; Kairo, M.T.K. Perspective and promise: A century of insect acoustic detection and monitoring. *Am. Entomol.* **2011**, *57*, 30–44. [CrossRef]
13. Potamitis, I.; Ganchev, T.; Kontodimas, D. On automatic bioacoustic detection of pests: The cases of *Rhynchophorus ferrugineus* and *Sitophilus oryzae*. *J. Econ. Entomol.* **2009**, *102*, 1681–1690. [CrossRef] [PubMed]
14. Nakash, J.; Osem, Y.; Kehat, M. A suggestion to use dogs for detecting *Rhynchophorus ferrugineus* infestation in date palms in Israel. *Phytoparasitica* **2000**, *28*, 153–155. [CrossRef]
15. Faleiro, J.R.; Kumar, J.A. A rapid decision sampling plan for implementing area-wide management of red palm weevil, *Rhynchophorus ferrugineus*, in coconut plantations of India. *J. Insect Sci.* **2008**, *8*, 9. [CrossRef] [PubMed]
16. Kessler, A.; Baldwin, I.T. Plant responses to insect herbivory: The emerging molecular analysis. *Annu. Rev. Plant Biol.* **2002**, *53*, 299–328. [CrossRef] [PubMed]
17. Tamayo, M.C.; Rufat, M.; Bravo, J.M.; San Segundo, B. Accumulation of a maize proteinase inhibitor in response to wounding and insect feeding, and characterization of its activity toward digestive proteinases of *Spodoptera littoralis* larvae. *Planta* **2000**, *211*, 62–71. [CrossRef] [PubMed]
18. Dicke, M.; van Loon, J.J.A. Multitrophic effects of herbivore-induced plant volatiles in an evolutionary context. *Entomol. Exp. Appl.* **2000**, *97*, 237–249. [CrossRef]
19. Felton, G.W.; TumLinson, J.H. Plant–insect dialogs: Complex interactions at the plant–insect interface. *Curr. Opin. Plant Biol.* **2008**, *11*, 457–463. [CrossRef] [PubMed]
20. Halitschke, R.; Schittko, U.; Pohnert, G.; Boland, W.; Baldwin, I.T. Molecular interactions between the specialist herbivore *Manduca sexta* (Lepidoptera, Sphingidae) and its natural host *Nicotiana attenuata*. III. Fatty acid-amino acid conjugates in herbivore oral secretions are necessary and sufficient for herbivore-specific plant responses. *Plant Physiol.* **2001**, *125*, 711–717. [PubMed]
21. Korth, K.L.; Dixon, R.A. Evidence for chewing insect-specific molecular events distinct from a general wound response in leaves. *Plant Physiol.* **1997**, *115*, 1299–1305. [PubMed]
22. Turlings, T.C.; TumLinson, J.H.; Lewis, W. Exploitation of herbivore-induced plant odors by host-seeking parasitic wasps. *Science* **1990**, *250*, 1251–1253. [CrossRef] [PubMed]
23. Bricchi, I.; Leitner, M.; Foti, M.; Mithofer, A.; Boland, W.; Maffei, M.E. Robotic mechanical wounding (MecWorm) *versus* herbivore-induced responses: Early signaling and volatile emission in Lima bean (*Phaseolus lunatus* L.). *Planta* **2010**, *232*, 719–729. [CrossRef] [PubMed]
24. Mewis, I.; Tokuhisa, J.G.; Schultz, J.C.; Appel, H.M.; Ulrichs, C.; Gershenzon, J. Gene expression and glucosinolate accumulation in *Arabidopsis thaliana* in response to generalist and specialist herbivores of different feeding guilds and the role of defense signaling pathways. *Phytochemistry* **2006**, *67*, 2450–2462. [CrossRef] [PubMed]
25. Marqués, J.; DuranVila, N.; Daròs, J.A. The Mn-binding proteins of the photosystem II oxygen-evolving complex are decreased in date palms affected by brittle leaf disease. *Plant Physiol. Biochem.* **2011**, *49*, 388–394. [CrossRef] [PubMed]
26. Sghaier-Hammami, B.; Redondo-López, I.; Maldonado-Alconada, A.M.; Echevarría-Zomeño, S.; Jorrín-Novo, J.V. A proteomic approach analysing the *Arabidopsis thaliana* response to virulent and avirulent Pseudomonas syringae strains. *Acta Physiol. Plant* **2012**, *34*, 905–922. [CrossRef]
27. Gómez-Vidal, S.; Salinas, J.; Tena, M.; Lopez-Llorca, L.V. Proteomic analysis of date palm (*Phoenix dactylifera* L.) responses to endophytic colonization by entomopathogenic fungi. *Electrophoresis* **2009**, *30*, 2996–3005. [CrossRef] [PubMed]
28. Ben Thabet, I.; Francis, F.; de Pauw, E.; Besbes, S.; Attia, H.; Deroanne, C.; Blecker, C. Characterisation of proteins from date palm sap (*Phoenix dactylifera* L.) by a proteomic approach. *Food Chem.* **2010**, *123*, 765–770. [CrossRef]
29. Rasool, K.G.; Khan, M.A.; Aldawood, A.S.; Tufail, M.; Mukhtar, M.; Takeda, M. Optimization of protein isolation from date palm plants and its utilization in differential proteomics associated with red palm weevil infestation. *Pak. J. Agric. Sci.* **2014**, *51*, 907–917.

30. Mukhtar, M.; Rasool, K.G.; Aldawood, A.S. Integrating genomic and proteomic efforts in date palm research. In proceedings of the 3rd International Conference on Global Warming: Food Security, EPDA, Al Hamra Convention Center, Ras Al Khiamh, United Arab Emirates, 5–7 May 2015.

31. Marondedze, C.; Gehring, C.; Thomas, L. Dynamic changes in the date palm fruit proteome during development and ripening. *Hortic. Res.* **2014**, *1*, 1–15. [CrossRef]

32. El Rabey, H.A.; Al-Malki, A.L.; Abulnaja, K.O.; Rohde, W. Proteome Analysis for understanding abiotic stress (salinity and drought) tolerance in date palm (*Phoenix dactylifera* L.). *Int. J. Genom.* **2015**, *2015*, 1–11. [CrossRef] [PubMed]

33. Gómez-Vidal, S.; Tena, M.; Lopez-Llorca, L.V.; Salinas, J. Protein extraction from *Phoenix dactylifera* L. leaves, a recalcitrant material, for two-dimensional electrophoresis. *Electrophoresis* **2008**, *29*, 448–456. [CrossRef] [PubMed]

34. Turlings, T.C.J.; Bernasconi, M.; Bertossa, R.; Bigler, F.; Caloz, G.; Dorn, S. The induction of volatile emission in maize by three herbivores species with different feeding habits: Possible consequences for their natural enemies. *Biol. Control* **1998**, *1*, 122–129. [CrossRef]

35. Duceppe, M.O.; Cloutier, C.; Michaud, D. Wounding, insect chewing and phloem sap feeding differentially alter the leaf proteome of potato, *Solanum tuberosum* L. *Proteome Sci.* **2012**, *10*, 73. [CrossRef] [PubMed]

36. Kluge, C.; Lahr, J.; Hanitzsch, M.; Bolte, S.; Golldack, D. New insight into the structure and regulation of the plant vacuolar H^+-ATPase. *J. Bioenerg. Biomembr.* **2003**, *35*, 377–388. [CrossRef] [PubMed]

37. Barkla, B.J.; Vera-Estrella, R.; Hernández-Coronado, M.; Pantoja, O. Quantitative proteomics of the tonoplast reveals a role for glycolytic enzymes in salt tolerance. *Plant Cell* **2009**, *21*, 4044–4058. [CrossRef] [PubMed]

38. Schnitzer, D.; Seidel, T.; Sander, T.; Golldack, D.; Dietz, K.J. The cellular energization state affects peripheral stalk stability of plant vacuolar H^+-ATPase and impairs vacuolar acidification. *Plant Cell Physiol.* **2011**, *52*, 946–956. [CrossRef] [PubMed]

39. Silva, P.; Gerós, H. Regulation by salt of vacuolar H^+-ATPase and H^+-pyrophosphatase activities and Na^+/H^+ exchange. *Plant Signal. Behav.* **2009**, *4*, 718–726. [CrossRef] [PubMed]

40. Silva, P.; Façanha, A.R.; Tavares, R.M.; Gerós, H. Role of tonoplast proton pumps and Na^+/H^+ antiport system in salt tolerance of *Populus euphratica* Oliv. *J. Plant Growth Regul.* **2010**, *29*, 23–34. [CrossRef]

41. Qiu, N.; Chen, M.; Guo, J.; Bao, H.; Ma, X. Coordinate up-regulation of V-H^+-ATPase and vacuolar Na^+/H^+ antiporter as a response to NaCl treatment in a C3 halophyte *Sueda salsa*. *Plant Sci.* **2007**, *172*, 1218–1225. [CrossRef]

42. Kabala, K.; Klobus, G. Modification of vacuolar proton pumps in cucumber roots under salt stress. *J. Plant Physiol.* **2008**, *165*, 1830–1837. [CrossRef] [PubMed]

43. Otoch, M.L.O.; Sobreira, A.C.M.; Aragão, M.E.F.; Orellano, E.G.; Lima, M.G.S. Salt modulation of vacuolar H^+-ATPase and H^+-Pyrophosphatase activities in *Vigna unguiculata*. *J. Plant Physiol.* **2001**, *158*, 545–551. [CrossRef]

44. Yap, K.L.; Kim, J.; Truong, K.; Sherman, M.; Yuan, T.; Ikura, M. Calmodulin target database. *J. Struct. Funct. Genom.* **2000**, *1*, 8–14. [CrossRef]

45. Hoeflich, K.P.; Ikura, M. Calmodulin in action: Diversity in target recognition and activation mechanisms. *Cell* **2002**, *108*, 739–742. [CrossRef]

46. Kahl, C.R.; Means, A.R. Regulation of cell cycle progression by calcium/calmodulin-dependent pathways. *Endocr. Rev.* **2003**, *24*, 719–736. [CrossRef] [PubMed]

47. Calabria, L.K.; Garcia Hernandez, L.; Teixeira, R.R.; de Souza, M.V.; Espindola, F.S. Identification of calmodulin binding proteins in brain of worker bees. *Comp. Biochem. Physiol. Part B* **2008**, *15*, 41–45. [CrossRef] [PubMed]

48. Borisjuk, N.; Sitailo, L.; Adler, K.; Malysheva, L.; Tewes, A.; Borisjuk, L.; Manteuffel, R. Calreticulin expression in plant cells: Developmental regulation, tissue specificity and intracellular distribution. *Planta* **1998**, *206*, 504–514. [CrossRef] [PubMed]

49. Von Wettstein, D.; Gough, S.; Kannangara, C.G. Chlorophyll biosynthesis. *Plant Cell* **1995**, *7*, 1039–1057. [CrossRef] [PubMed]

50. Fink, A.L. Chaperone-mediated protein folding. *Physiol. Rev.* **1999**, *79*, 425–449. [PubMed]

51. De Rocher, A.E.; Vierling, E. Developmental control of small heat shock protein expression during pea seed maturation. *Plant J.* **1994**, *5*, 93–102. [CrossRef]

52. Curto, M.; Camafeita, L.E.; Lopez, J.A.; Maldonado, A.M.; Rubiales, D.; Jorrı, J.V. A proteomic approach to study pea (*Pisum sativum*) responses to powdery mildew (*Erysiphe pisi*). *Proteomics* **2006**, *6*, S163–S174. [CrossRef] [PubMed]

53. Castillejo, M.A.; Kirchev, H.; Jorrin, J.V. Differences in the Triticale (X *Triticosecale* Wittmack) flag leaf 2-DE protein profile between varieties and nitrogen fertilization levels. *J. Agric. Food Chem.* **2010**, *58*, 5698–5707. [CrossRef] [PubMed]

54. Castillejo, M.A.; Curto, M.; Fondevilla, S.; Rubiales, D.; Jorrin, J.V. Two-dimensional electrophoresis based proteomic analysis of the pea (*Pisum sativum*) in response to *Mycosphaerella pinodes*. *J. Agric. Food Chem.* **2010**, *58*, 12822–12832. [CrossRef] [PubMed]

55. Nakano, T.; Suzuki, K.; Fujimura, T.; Shinshi, H. Genome-wide analysis of the ERF gene family in *Arabidopsis* and rice. *Plant Physiol.* **2006**, *40*, 411–432. [CrossRef] [PubMed]

56. Lorenzo, O.; Piqueras, R.; Sánchez-Serrano, J.J.; Solano, R. Ethylene response factor1 integrates signals from ethylene and jasmonate pathways in plant defense. *Plant Cell* **2003**, *15*, 165–178. [CrossRef] [PubMed]

57. Berrocal-Lobo, M.; Molina, A.; Solano, R. Constitutive expression of ethylene-response-factor1 in *Arabidopsis* confers resistance to several necrotrophic fungi. *Plant J.* **2002**, *29*, 23–32. [CrossRef] [PubMed]

58. Ecker, J.R. The ethylene signal transduction pathway in plants. *Science* **1995**, *268*, 667–675. [CrossRef] [PubMed]

59. Penninckx, I.A.; Eggermont, K.; Terras, F.R.; Thomma, B.P.; de Samblanx, G.W.; Buchala, A.; Métraux, J.P.; Manners, J.M.; Broekaert, W.F. Pathogen induced systemic activation of a plant defensin gene in *Arabidopsis* follows a salicylic acid-independent pathway. *Plant Cell* **1996**, *8*, 2309–2323. [CrossRef] [PubMed]

60. O'Donnell, P.J.; Calvert, C.; Atzorn, R.; Wasternack, C.; Leyser, H.M.O.; Bowles, D.J. Ethylene as a signal mediating the wound response of tomato plants. *Science* **1996**, *274*, 1914–1917. [CrossRef] [PubMed]

61. Rockel, P.; Strube, F.; Rockel, A.; Wildt, J.; Kaiser, W. Regulation of nitric oxide (NO) production by plant nitrate reductase *in vivo* and *in vitro*. *J. Exp. Bot.* **2002**, *53*, 103–110. [CrossRef] [PubMed]

62. Qiao, W.; Fan, L.M. Nitric oxide signaling in plant responses to abiotic stresses. *J. Integr. Plant Biol.* **2008**, *50*, 1238–1246. [CrossRef] [PubMed]

63. Corpas, F.J.; Leterrier, M.; Valderrama, R.; Airaki, M.; Chaki, M.; Palma, J.M.; Barroso, J.B. Nitric oxide imbalance provokes a nitrosative response in plants under abiotic stress. *Plant Sci.* **2011**, *181*, 604–611. [CrossRef] [PubMed]

64. Moreau, M.; Lindermayr, C.; Durner, J.; Klessig, D.F. NO synthesis and signaling in plants—Where do we stand? *Physiol. Plant* **2010**, *138*, 372–383. [CrossRef] [PubMed]

65. Lamattina, L.; Garcia-Mata, C.; Graziano, M.; Pagnussat, G. Nitric oxide: the versatility of an extensive signal molecule. *Annu. Rev. Plant Biol.* **2003**, *54*, 109–136. [CrossRef] [PubMed]

66. Wendehenne, D.; Durner, J.; Klessig, D.F. Nitric oxide: A new player in plant signaling and defence responses. *Curr. Opin. Plant Biol.* **2004**, *7*, 449–455. [CrossRef] [PubMed]

67. Hunt, P.W.; Klok, E.J.; Trevaskis, B.; Watts, R.A.; Ellis, M.H.; Peacock, W.J.; Dennis, E.S. Increased level of hemoglobin 1 enhances survival of hypoxic stress and promotes early growth in *Arabidopsis thaliana*. *Proc. Natl. Acad. Sci. USA* **2002**, *99*, 17197–17202. [CrossRef] [PubMed]

68. Dordas, C.; Hasinoff, B.B.; Igamberdiev, A.U.; Manac'h, N.; Rivoal, J.; Hill, R.D. Expression of a stress-induced hemoglobin affects NO levels produced by alfalfa root cultures under hypoxic stress. *Plant J.* **2003**, *35*, 763–770. [CrossRef] [PubMed]

69. Dordas, C.; Rivoal, J.; Hill, R.D. Plant haemoglobins, nitric oxide and hypoxic stress. *Ann. Bot.* **2003**, *91*, 173–178. [CrossRef] [PubMed]

70. Perazzolli, M.; Dominici, P.; Romero-Puertas, M.C.; Zago, E.; Zeier, J.; Sonoda, M.; Lamb, C.; Delledonne, M. Arabidopsis non-symbiotic hemoglobin AHb1 modulates nitric oxide bioactivity. *Plant Cell* **2004**, *16*, 2785–2794. [CrossRef] [PubMed]

71. Drobak, B.K.; Franklin-Tong, V.E.; Staiger, C.T. The role of the actin cytoskeleton in plant cell signaling. *New Phytol.* **2004**, *163*, 13–30. [CrossRef]

72. Staiger, C.J.; Blanchoin, L. Actin dynamics: Old friends with new stories. *Curr. Opin. Plant Biol.* **2006**, *9*, 554–562. [CrossRef] [PubMed]

73. Hussey, P.J.; Ketelaar, T.; Deeks, M.J. Control of the actin cytoskeleton in plant cell growth. *Annu. Rev. Plant Biol.* **2006**, *57*, 109–125. [CrossRef] [PubMed]

74. Solanke, A.U.; Sharma, A.K. Signal transduction during cold stress in plants. *Physiol. Mol. Biol. Plants* **2008**, *14*, 69–79. [CrossRef] [PubMed]

75. Miklis, M.; Consonni, C.; Bhat, R.A.; Lipka, V.; Schulze-Lefert, P.; Panstruga, R. Barley MLO modulates actin-dependent and actin-independent antifungal defense pathways at the cell periphery. *Plant Physiol.* **2007**, *144*, 1132–1143. [CrossRef] [PubMed]

76. Clement, M.; Tijs, K.; Natalia, R.; Mohamed, Y.B.; Andrei, S.; Gilbert, E.; Pierre, A.; Patrick, J.H.; de Janice, A.E. Actin-depolymerizing factor 2-mediated actin dynamics are essential for root-knot nematode infection of *Arabidopsis. Plant Cell* **2009**, *21*, 2963–2979. [CrossRef] [PubMed]

77. Tian, M.; Chaudhry, F.; Ruzicka, D.R.; Meagher, R.B.; Staiger, C.J.; Day, B. *Arabidopsis* actin-depolymerizing factor AtADF4 mediates defense signal transduction triggered by the *Pseudomonas syringae* effector AvrPphB. *Plant Physiol.* **2009**, *150*, 815–824. [CrossRef] [PubMed]

78. Pollard, T.D.; Borisy, G.G. Cellular motility driven by assembly and disassembly of actin filaments. *Cell* **2003**, *112*, 453–65. [CrossRef]

79. Radjendirane, V.; Joseph, P.; Lee, Y.H.; Kimura, S.; KleinSazanto, A.J.P.; Gonzalez, F.J.; Jaiswal, A.K. Disruption of the DT diaphorase (*NQO1*) gene in mice leads to increased menadione toxicity. *J. Biol. Chem.* **1998**, *273*, 7382–7389. [CrossRef] [PubMed]

80. Schuler, F.; Yano, T.; Bernardo, S.D.; Yagi, T.; Yankovskaya, V.; Singer, T.P.; Casida, J.E. NADH-quinone oxidoreductase: PSST subunit couples electron transfer from iron–sulfur cluster N2 to quinine. *Proc. Natl. Acad. Sci. USA* **1999**, *96*, 4149–4153. [CrossRef] [PubMed]

81. Gaikwad, A.; Long, J.D., II; Stringer, J.L.; Jaiswal, A.K. *In vivo* role of NAD(P)H:Quinone Oxidoreductase 1 (NQO1) in the regulation of intracellular redox state and accumulation of abdominal adipose tissue. *J. Dial. Chem.* **2001**, *276*, 22559–22564. [CrossRef] [PubMed]

82. Reimer, U.; Fischer, G. Local structural changes caused by peptidyl-prolyl *cis/trans* isomerization in the native state of proteins. *Biophys. Chem.* **2002**, *96*, 203–212. [CrossRef]

83. Andreeva, L.; Heads, R.; Green, C.J. Cyclophilins and their possible role in the stress response. *Int. J. Exp. Pathol.* **1999**, *80*, 305–315. [CrossRef] [PubMed]

84. Schreiber, S.L. Chemistry and biology of the immunophilins and their immunosuppressive ligands. *Science* **1991**, *251*, 283–287. [CrossRef] [PubMed]

85. Shaw, P.E. Peptidyl-prolyl isomerases: A new twist to transcription. *EMBO Rep.* **2002**, *3*, 521–526. [CrossRef] [PubMed]

86. Price, E.R.; Jin, M.; Lim, D.; Pati, S.; Walsh, C.T.; McKeon, F.D. Cyclophilin B trafficking through the secretory pathway is altered by binding of cyclosporin. *Proc. Natl. Acad. Sci. USA* **1994**, *91*, 3931–3935. [CrossRef] [PubMed]

87. Freeman, B.C.; Toft, D.O.; Morimoto, R.I. Molecular chaperone machines: chaperone activities of the cyclophilin Cyp-40 and the steroid aporeceptor-associated protein p23. *Science* **1996**, *274*, 1718–1720. [CrossRef] [PubMed]

88. Horowitz, D.S.; Lee, E.J.; Mabon, S.A.; Misteli, T. A cyclophilin functions in pre-mRNA splicing. *EMBO J.* **2002**, *21*, 470–480. [CrossRef] [PubMed]

89. Weisman, R.; Creanor, J.; Fantes, P. A multicopy suppressor of a cell cycle defect in S. pombe encodes a heat shock-inducible 40 kDa cyclophilin-like protein. *EMBO J.* **1996**, *15*, 447–456. [PubMed]

90. Lippert, D.; Chowrira, S.; Ralph, S.G.; Zhuang, J.; Aeschliman, D.; Ritland, C.; Ritland, K.; Bohlmann, J. Conifer defense against insects: Proteome analysis of Sitka spruce (*Picea sitchensis*) bark induced by mechanical wounding or feeding by white pine weevils (*Pissodes strobi*). *Proteomics* **2007**, *7*, 248–270. [CrossRef] [PubMed]

91. LaemmLi, U.K. Cleavage of structural proteins during the assembly of the head of bacteriophage T4. *Nature* **1970**, *227*, 680–685. [CrossRef] [PubMed]

 International Journal of
Molecular Sciences

Article

A Comparative Proteomic Analysis of the Buds and the Young Expanding Leaves of the Tea Plant (*Camellia sinensis* L.)

Qin Li [1,2,3,4,†], Juan Li [1,†], Shuoqian Liu [2], Jianan Huang [1,3,4], Haiyan Lin [2,4], Kunbo Wang [2], Xiaomei Cheng [1] and Zhonghua Liu [1,2,3,4,*]

[1] Key Laboratory of Tea Science of Ministry of Education, Hunan Agricultural University, Changsha 410128, Hunan, China; liqinvip@126.com (Q.L.); xixi_lj@126.com (J.L.); jian7513@gmail.com (J.H.); xiaomeichengvip@163.com (X.C.)

[2] National Research Center of Engineering Technology for Utilization of Functional Ingredients from Botanicals, Hunan Agricultural University, Changsha 410128, Hunan, China; shuoqianliuvip@126.com (S.L.); linhaiyanvip@126.com (H.L.); wkboo163@163.com (K.W.)

[3] Collaborative Innovation Centre of Utilisation of Functional Ingredients from Botanicals, Hunan Agricultural University, Changsha 410128, Hunan, China

[4] Hunan Provincial Key Laboratory for Germplasm Innovation and Utilization of Crop, Hunan Agricultural University, Changsha 410128, Hunan, China

* Correspondence: larkin-liu@163.com; Tel.: +86-731-8463-5306; Fax: +86-731-8463-5304

† These authors contributed equally to this work.

Academic Editor: Setsuko Komatsu
Received: 15 April 2015; Accepted: 19 May 2015; Published: 18 June 2015

Abstract: Tea (*Camellia sinensis* L.) is a perennial woody plant that is widely cultivated to produce a popular non-alcoholic beverage; this beverage has received much attention due to its pleasant flavor and bioactive ingredients, particularly several important secondary metabolites. Due to the significant changes in the metabolite contents of the buds and the young expanding leaves of tea plants, high-performance liquid chromatography (HPLC) analysis and isobaric tags for relative and absolute quantitation (iTRAQ) analysis were performed. A total of 233 differentially expressed proteins were identified. Among these, 116 proteins were up-regulated and 117 proteins were down-regulated in the young expanding leaves compared with the buds. A large array of diverse functions was revealed, including roles in energy and carbohydrate metabolism, secondary metabolite metabolism, nucleic acid and protein metabolism, and photosynthesis- and defense-related processes. These results suggest that polyphenol biosynthesis- and photosynthesis-related proteins regulate the secondary metabolite content of tea plants. The energy and antioxidant metabolism-related proteins may promote tea leaf development. However, reverse transcription quantitative real-time PCR (RT-qPCR) showed that the protein expression levels were not well correlated with the gene expression levels. These findings improve our understanding of the molecular mechanism of the changes in the metabolite content of the buds and the young expanding leaves of tea plants.

Keywords: *Camellia sinensis* L.; proteome; iTRAQ

---

1. Introduction

Tea (*Camellia sinensis* L.) is a perennial woody plant that is widely cultivated to produce a popular non-alcoholic beverage; this beverage has received much attention due to its pleasant flavor and bioactive ingredients, particularly several key secondary metabolites [1]. Tea leaves contain important secondary metabolites, including polyphenols (catechins, flavones, anthocyanidin and phenolic acid), alkaloids (theobromine, theophylline and caffeine), and theanine, which not only contribute to tea quality but also have important human health benefits [2].

The changes in the chemical composition of the buds and the young expanding leaves of tea have been extensively studied. A previous study showed that during seeding development, total catechins, epigallocatechin gallate (EGCG) and epicatechin gallate (ECG) decreased, whereas the epigallocatechin (EGC) content increased [3]. As the shoots matured, the total flavonol glycoside and myricetin contents increased, but the kaempferol content decreased [4]. Purine alkaloid metabolism also appears to be closely associated with leaf development and aging in tea seedlings. In addition, the expression levels of several genes related to metabolite synthesis in tea leaves were analyzed. A positive correlation was found between the catechin concentration and the expression of flavanone 3-hydroxylase (F3H) in tea leaves at different developmental stages [5]. A study has shown that most catechins accumulate to higher levels in the shoots than in the mature leaves; similarly, the genes involved in catechin synthesis, including phenylalanine ammonia-lyase 1 (PAL1), chalcone synthase (CHS), dihydroflavonol 4-reductase (DFR), leucoanthocyanidin reductase (LCR), and F3H are more highly expressed in the shoots than in the mature leaves [6]. Zhang *et al.* also found that the content of non-galloylated catechins—except gallocatechin (GC)—as well as the activity of DFR and anthocyanidin reductase (ANR), gradually increased from the buds to the mature leaves [7]. An analysis of purine alkaloids in different parts of the seedlings showed that the caffeine and theobromine content was greater in young leaves and decreased with increasing leaf maturity, and the levels of tea caffeine synthase (TCS) transcripts were also highest in young leaves and declined markedly during leaf development [8,9]. Different levels of metabolites in tea leaves are likely characterized by diverse gene and protein expression profiles at each developmental stage.

Despite studies on the metabolite synthesis-related genes in tea plants, the molecular mechanisms underlying the changes in metabolite content have not yet been examined in detail. In this study, isobaric tags for relative and absolute quantitation (iTRAQ) analysis were first used to separate the differentially expressed proteins. In addition, the content of a set of important metabolites was studied, and the expression of the genes associated with the differentially expressed proteins was also measured. The purpose of this study is to provide an improved understanding of the molecular mechanisms behind the change in the metabolite content between the apical buds and the young expanding leaves of tea plants.

2. Results

2.1. Analysis of Metabolite Profiles

To further investigate the important changes in metabolite content, the polyphenol, catechin, and flavonoid contents of the buds and the young expanding leaves of tea plants were analyzed (Figure 1). As shown in Figure 1A, the concentration of total catechin in young expanding leaves (132.507 ± 3.889 mg/g) was 0.839-fold lower ($p < 0.05$) than that in the buds (150.851 ± 3.640 mg/g). The total polyphenol content of the young expanding leaves (329.395 ± 6.984 mg/g) was 0.951-fold lower than that of the buds (346.219 ± 8.609 mg/g), but this difference was not significant ($p > 0.05$). However, the total flavonoid content of the young expanding leaves (44.754 ± 3.731 mg/g) was 1.734-fold higher than that of the buds (25.803 ± 2.619 mg/g) ($p < 0.01$).

The levels of non-galloylated catechins, including EGC, epicatechin (EC) and DL-catechin (DL-C), were significantly greater in the young expanding leaves (13.280 ± 0.338 mg/g) than in the buds (7.574 ± 0.053 mg/g) ($p < 0.01$). However, the contents of galloylated catechins, including EGCG, GCG and ECG, were significantly lower in the young expanding leaves (119.226 ± 0.997 mg/g) than in the buds (143.277 ± 0.823 mg/g) ($p < 0.05$) (Figure 1A). In both the buds and the young expanding leaves the most abundant individual catechin was EGCG, and the least abundant individual catechin was GCG (gallocatechin gallate). The relative concentrations of each individual catechins in both the buds and the young expanding leaves were EGCG > ECG > EGC > EC > DL-C > GCG. The concentrations of EGC and EC in the young expanding leaves (EGC: 7.626 ± 0.859 mg/g, EC: 4.244 ± 0.060 mg/g) were greater than those in the buds (EGC: 3.167 ± 0.034 mg/g, EC: 3.127 ± 0.044 mg/g) ($p < 0.01$ for EGC and $p < 0.05$ for EC), and the level of DL-C was slightly higher ($p > 0.05$) in the young expanding leaves (1.410 ± 0.095 mg/g) than in the buds (1.280 ± 0.081 mg/g). However, the concentrations of EGCG and ECG were lower in the young expanding leaves (EGCG: 80.292 ± 2.216 mg/g, ECG: 38.646 ± 0.769 mg/g) than in the buds (EGCG: 101.169 ± 2.343 mg/g, ECG: 41.705 ± 1.204 mg/g) ($p < 0.05$), and the GCG level was also slightly lower ($p > 0.05$) in the young expanding leaves (0.288 ± 0.008 mg/g) than in the buds (0.403 ± 0.051 mg/g) (Figure 1B). In the young expanding leaves, the levels of individual flavonols, including myricetin, quercetin and kaempferol, were all greater than those in the buds (1.181 ± 0.026 mg/g myricetin, 3.627 ± 0.051 mg/g quercetin, and 4.441 ± 0.063 mg/g kaempferol in the leaves compared with 0.635 ± 0.017 mg/g myricetin, 1.767 ± 0.021 mg/g quercetin, and 3.193 ± 0.038 mg/g kaempferol in the buds, $p < 0.01$ for myricetin and quercetin, and $p < 0.05$ for kaempferol) (Figure 1C). Three types of alkaloids, including theobromine, theophylline and caffeine, were also detected via HPLC analysis. The theobromine and caffeine levels were lower in the young expanding leaves than in the buds (theobromine: 23.165 ± 0.213 mg/g in leaves and 29.418 ± 0.299 mg/g in buds, $p < 0.01$; caffeine: 38.167 ± 0.704 mg/g in leaves and 40.484 ± 0.396 mg/g in buds, $p < 0.05$), and the theophylline levels were slightly higher in the young expanding leaves (0.247 ± 0.017 mg/g) compared with the buds (0.235 ± 0.013 mg/g) ($p > 0.05$) (Figure 1D). Due to the significant changes in the metabolite contents of the buds and the young expanding leaves of tea plants, iTRAQ analysis was performed to determine the molecular mechanisms behind this change.

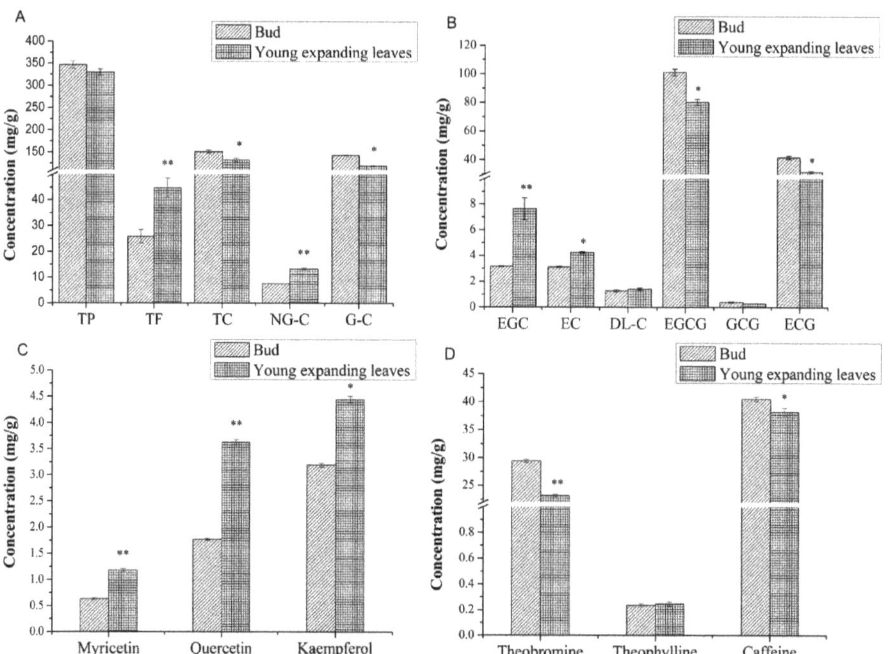

Figure 1. Changes in the levels of secondary metabolites in the buds and the young expanding leaves of tea. (**A**) Total polyphenols (TP), total flavonoids (TF), total catechins (TC), non-galloylated catechins (NG-C) and galloylated catechins (G-C); (**B**) Individual catechins; (**C**) Myicetin, quercetin and kaempferol; and (**D**) Individual alkaloids. Statistical significance: * $p < 0.05$ and ** $p < 0.01$.

2.2. Protein Identification

To explore the correlation between the proteomic and metabolite profiles of buds and young expanding leaves, samples were analyzed by iTRAQ proteomics coupled with LC-MS/MS. A total of 60,820 spectra were generated from the iTRAQ experiment and the data were analyzed using Mascot software. A total of 8015 spectra were matched to known spectra, 6974 spectra were matched to unique spectra, 4746 were matched to peptides, 4260 were matched to unique peptides and 2507 were matched to proteins (Figure 2A). The distribution of the number of peptides defining each protein is shown in Figure 2B; over 55% of the proteins were represented by at least two peptides.

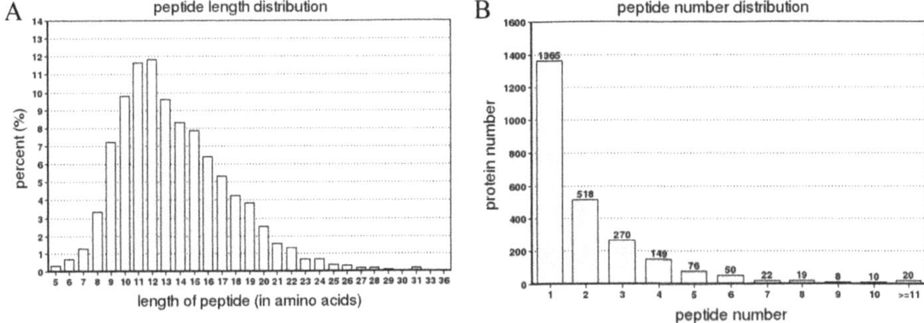

Figure 2. The spectra, peptides, and proteins, as well as the number of peptides in the iTRAQ proteomic analysis identified as matching proteins. The spectra, peptides and proteins were identified by searching against a database (**A**); and The number of peptides matched to proteins using MASCOT (**B**).

2.3. Functional Classification of the Differentially Expressed Proteins

The proteins whose levels changed more than 1.5-fold and had a *p*-values of less than 0.05 were considered differentially expressed. Based on these two criteria, 233 proteins were differentially expressed between the buds and the young expanding leaves, and these proteins were isolated and quantified using comparative proteomics via iTRAQ. Of the 233 differentially expressed proteins, 116 were more abundant and 117 were less abundant in the young expanding leaves compared with the buds. GO analysis revealed that the differentially expressed proteins participated in several biological processes ($p < 0.05$), as shown in Table S1. KEGG enrichment analysis suggested that the differentially expressed proteins are involved in several pathways ($p < 0.05$), including phenylalanine metabolism (Table S2).

The proteins were classified into seven functional categories based on their functional biological properties and pathways: metabolism (58, 25.11%), nucleic acid metabolism (33, 14.04%), protein metabolism (59, 25.11%), biological regulation and signal transduction (24, 10.21%), stress/defense/detoxification (19, 8.09%), transport (7, 2.55%), and unknown function (35, 14.89%) (Figure 3A). Of the up-regulated proteins, 25.00% (29 proteins) function in metabolism, 16.38% (19 proteins) function in nucleic acid metabolism, 16.38% (19 proteins) are involved in protein metabolism, 7.76% (nine proteins) have biological regulation and signal transduction function, 9.58% (11 proteins) function in stress/defense/detoxification, 4.31% (5 proteins) are involved in transport and 20.69% of them (24 proteins) were of unknown function (Figure 3B). Among the down-regulated proteins, 24.37% (29 proteins) function in metabolism, 11.76% (14 proteins) function in nucleic acid metabolism, 33.61% (40 proteins) have a role in protein metabolism, 12.61% (15 proteins) are involved in biological regulation and signal transduction, 6.72% (8 proteins) are involved in stress/defense/detoxification, 1.68% (two proteins) function in transport and 9.24% (11 proteins) were of unknown function (Figure 3C). More detailed information can be found in Table 1.

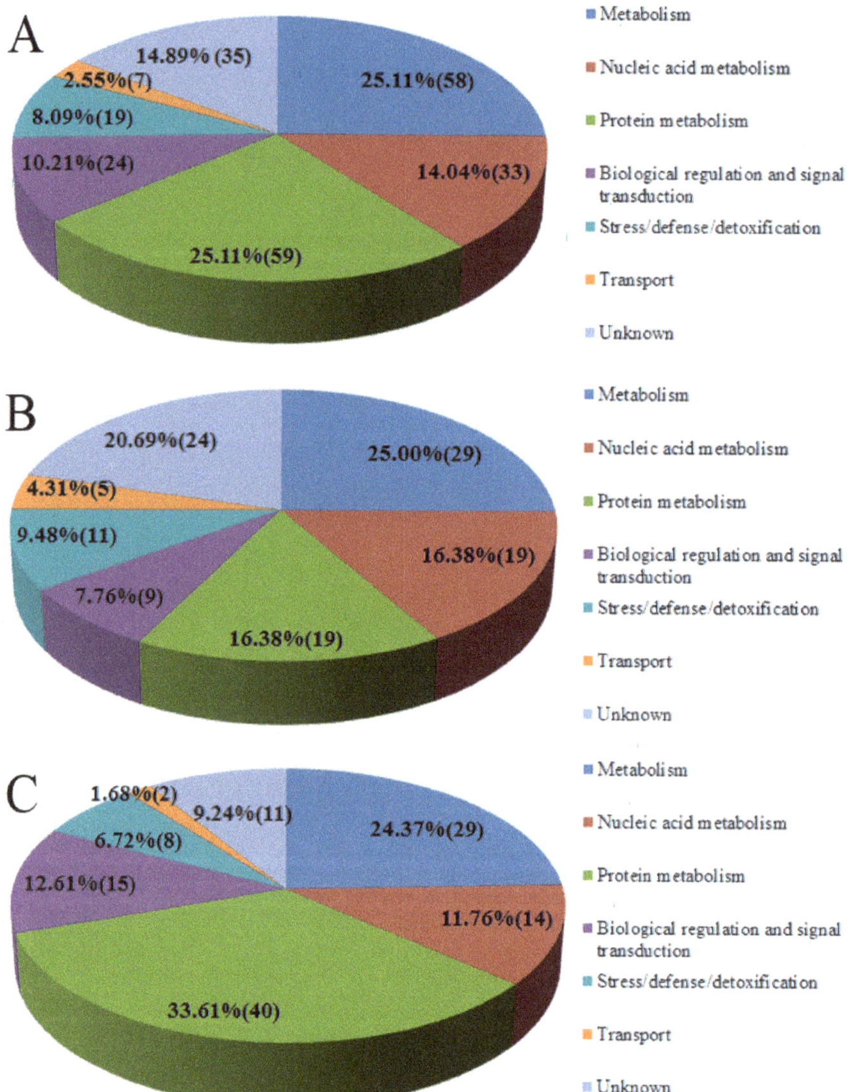

Figure 3. Functional classification of the differentially expressed proteins. Functional groups and the numbers of proteins of all 233 differentially expressed proteins that fall into each group (**A**); categorization of the 116 up-regulated proteins (**B**); and categorization of the 117 down-regulated proteins (**C**). The number in each functional category represents the number of proteins in that category.

Table 1. List of proteins that are differentially expressed between the buds and the young expanding leaves of tea plants.

Accession Number	Proteins Name and Species	Score	Mass (Da)	Coverage	Peptide Count	Fold Change (Leaves/Bud)	Function
gi│350536667│	Dihydrolipoamide dehydrogenase precursor [*Solanum lycopersicum*]	202	68,166	13.9	5	2.164	Metabolism
gi│15081610│	Xyloglucan endotransglycosylase XET2 [*Vitis vinifera*]	137	39,952	12.4	3	1.614	Metabolism
gi│76786311│	Flavonol synthase [*Camellia sinensis*]	288	45,768	27.9	6	1.788	Metabolism
gi│225458243│	PREDICTED: isoflavone reductase homolog P3 [*Vitis vinifera*]	315	37,529	31.6	7	2.649	Metabolism
gi│359491464│	PREDICTED: lysosomal α-mannosidase [*Vitis vinifera*]	111	38,176	12.7	3	1.825	Metabolism
gi│71535021│	α-glucosidase [*Medicago sativa*]	204	86,320	6	4	2.353	Metabolism
gi│255578100│	Dihydrolipoamide succinyltransferase component of 2-oxoglutarate dehydrogenase, putative [*Ricinus communis*]	57	38,166	6.9	2	1.878	Metabolism
gi│225426623│	PREDICTED: 2-keto-3-deoxy-L-rhamnonate aldolase-like [*Vitis vinifera*]	78	21,742	10.2	1	2.281	Metabolism
gi│193290728│	Putative pyruvate dehydrogenase E3 subunit [*Capsicum annuum*]	74	42,069	8.3	2	1.567	Metabolism
gi│255566959│	NADH-cytochrome B5 reductase, putative [*Ricinus communis*]	32	11,336	11.4	1	1.58	Metabolism
sp│Q9SVG4│	Reticuline oxidase-like protein [*Spinacia oleracea*]	108	30,376	6.9	1	1.729	Metabolism
sp│Q9M069│	Glucan endo-1,3-β-glucosidase 7 [*Arabidopsis thaliana*]	139	40,171	16.2	3	1.568	Metabolism
gi│147767550│	Hypothetical protein VITISV_013343 [*Vitis vinifera*]	159	20,700	22.4	2	1.542	Metabolism
gi│193290702│	Putative 3-isopropylmalate dehydrogenase small subunit [*Capsicum annuum*]	354	25,595	36.2	4	1.681	Metabolism
gi│225451235│	PREDICTED: cysteine synthase isoform 2 [*Vitis vinifera*]	321	47,206	14.9	4	2.216	Metabolism
gi│225454278│	PREDICTED: cysteine synthase, chloroplastic/chromoplastic isoform 1 [*Vitis vinifera*]	60	40,113	10.5	2	1.905	Metabolism
sp│P50246│	Adenosylhomocysteinase [*Medicago sativa*]	186	65,326	11.9	5	1.551	Metabolism
sp│P47999│	Cysteine synthase, chloroplastic/chromoplastic [*Arabidopsis thaliana*]	111	51,056	14.5	4	1.856	Metabolism
sp│P94170│	Carbonic anhydrase [*Nostoc* sp.]	191	12,938	27.6	2	2.527	Metabolism
sp│Q42876│	Leucine aminopeptidase 2, chloroplastic [*Solanum lycopersicum*]	124	74,254	3.9	2	1.852	Metabolism
gi│75755999│	TO87b-13 [*Taraxacum officinale*]	260	83,317	6	3	2.18	Metabolism

Table 1. *Cont.*

Accession Number	Proteins Name and Species	Score	Mass (Da)	Coverage	Peptide Count	Fold Change (Leaves/Bud)	Function
gi|330318698|	Light-inducible protein atls1 [*Camellia sinensis*]	115	18,370	8.8	1	2.68	Metabolism
gi|224098154|	predicted protein [*Populus trichocarpa*]	145	29,956	28.8	5	1.504	Metabolism
gi|225433426|	PREDICTED: 3-ketoacyl-CoA thiolase 2, peroxisomal isoform 2 [*Vitis vinifera*]	197	7506	51.9	2	1.568	Metabolism
gi|225462452|	PREDICTED: GDSL esterase/lipase At5g45670 [*Vitis vinifera*]	250	22,819	30.5	4	1.79	Metabolism
sp|Q93YW8|	GDSL esterase/lipase At4g18970 [*Arabidopsis thaliana*]	250	22,819	30.5	4	1.79	Metabolism
sp|Q9SYF0|	GDSL esterase/lipase 2 [*Arabidopsis thaliana*]	316	46,565	14.2	5	2.203	Metabolism
sp|Q9LZS7|	GDSL esterase/lipase At5g03610 [*Arabidopsis thaliana*]	80	45,316	11.8	3	3.399	Metabolism
gi|296088201|	Unnamed protein product [*Vitis vinifera*]	73	30,392	9.5	2	1.55	Metabolism
gi|18140567|	Ribulose-1,5-bisphosphate carboxylase/oxygenase large subunit [*Camellia japonica*]	518	52,867	15.8	6	0.357	Metabolism
gi|156106226|	Rubisco activase [*Camellia sinensis*]	646	43,205	31.4	7	0.564	Metabolism
gi|20257362|	Ribulose 1,5-bisphosphate carboxylase/oxygenase, partial (chloroplast) [*Schima superba*]	303	24,090	10.3	2	0.465	Metabolism
gi|255553993|	Phosphoenolpyruvate carboxylase, putative [*Ricinus communis*]	108	32,666	14.7	3	0.469	Metabolism
gi|169807676|	NADP-dependent glyceraldehyde-3-phosphate dehydrogenase [*Platanus x acerifolia*]	322	69,081	17	6	0.543	Metabolism
gi|356524319|	PREDICTED: probable glycerophosphoryl diester phosphodiesterase 1-like [*Glycine max*]	212	56,841	8.8	3	0.552	Metabolism
gi|2266947|	Phosphoenolpyruvate carboxylase 1 [*Gossypium hirsutum*]	173	94,629	7.7	5	0.477	Metabolism
gi|255581778|	chlorophyll A/B binding protein, putative [*Ricinus communis*]	96	41,261	3.8	1	0.664	Metabolism
sp|P81833|	Thylakoid lumenal 29 kDa protein, chloroplastic (Fragment) [*Spinacia oleracea*]	176	43,057	22.2	4	0.651	Metabolism
sp|Q8H1Q1|	Thylakoid lumenal protein At1g12250, chloroplastic [*Spinacia oleracea*]	254	35,331	21.5	4	0.655	Metabolism
sp|O04138|	Chitinase 4 [*Oryza sativa* subsp. *Japonica*]	223	30,457	19.4	3	0.466	Metabolism
sp|Q9FKK7|	Xylose isomerase [*Spinacia oleracea*]	172	63,511	13	4	0.575	Metabolism
gi|27804768|	Sedoheptulose-1,7-bisphosphatase precursor [*Oryza sativa Indica* Group]	268	54,158	8.7	3	0.58	Metabolism
gi|380508822|	Putative hydroxycinnamoyl-CoA:shikimate/quinate hydroxycinnamoyltransferase [*Camellia sinensis*]	48	7531	32.7	2	0.493	Metabolism

Table 1. *Cont.*

Accession Number	Proteins Name and Species	Score	Mass (Da)	Coverage	Peptide Count	Fold Change (Leaves/Bud)	Function
gi|255603771|	DNA binding protein, putative [*Ricinus communis*]	250	22,819	30.5	4	1.79	Nucleic acid metabolism
gi|255603771|	DNA binding protein, putative [*Ricinus communis*]	177	41,602	16.9	3	2.491	Nucleic acid metabolism
sp|Q9S7C9|	Putative DNA-binding protein ESCAROLA [*Spinacia oleracea*]	250	38,185	17	4	1.664	Nucleic acid metabolism
gi|79596510|	AT hook motif DNA-binding family protein [*Spinacia oleracea*]	137	29,457	21.1	4	2.576	Nucleic acid metabolism
sp|Q9S7C9|	Putative DNA-binding protein ESCAROLA [*Spinacia oleracea*]	250	38,185	17	4	1.664	Nucleic acid metabolism
gi|45533923|	Glycine-rich RNA-binding protein RGP-1c [Nicotiana sylvestris]	487	21,422	28.1	4	2.945	Nucleic acid metabolism
sp|Q9SVM8|	Glycine-rich RNA-binding protein 2, mitochondrial [*Spinacia oleracea*]	410	20,091	17.1	2	1.506	Nucleic acid metabolism
gi|225440996|	PREDICTED: histone deacetylase HDT1-like [*Vitis vinifera*]	99	37,638	12.2	3	3.522	Nucleic acid metabolism
sp|Q9XI36|	Methyl-CpG-binding domain-containing protein 10 [*Spinacia oleracea*]	285	42,453	37.6	7	1.364	Nucleic acid metabolism
gi|225457458|	PREDICTED: transcription factor BTF3 [*Vitis vinifera*]	170	25,374	35.6	4	1.829	Nucleic acid metabolism
gi|297723091|	Os04g0385700 [*Oryza sativa Japonica* Group]	56	34,525	4.3	1	2.34	Nucleic acid metabolism
gi|296081863|	Unnamed protein product [*Vitis vinifera*]	177	38,697	15.7	3	2.094	Nucleic acid metabolism
gi|297744195|	Unnamed protein product [*Vitis vinifera*]	155	29,045	22.4	3	2.146	Nucleic acid metabolism
gi|255642098|	Unknown [*Glycine max*]	119	51,914	10.3	4	2.23	Nucleic acid metabolism
sp|Q9LFN6|	DEAD-box ATP-dependent RNA helicase 56 [*Spinacia oleracea*]	220	54,506	18.6	5	0.649	Nucleic acid metabolism
sp|Q84UQ1|	DEAD-box ATP-dependent RNA helicase 42 [*Oryza sativa* subsp. *Japonica*]	98	120,162	2.2	2	0.325	Nucleic acid metabolism

Table 1. *Cont.*

Accession Number	Proteins Name and Species	Score	Mass (Da)	Coverage	Peptide Count	Fold Change (Leaves/Bud)	Function
sp|B6EUA9|	Pre-mRNA-processing protein 40A [*Spinacia oleracea*]	242	83,063	9.7	5	0.233	Nucleic acid metabolism
sp|O22315|	Pre-mRNA-splicing factor SF2 [*Spinacia oleracea*]	126	7308	35.7	2	0.432	Nucleic acid metabolism
gi|374095609|	Spliceosomal-like protein [*Camellia sinensis*]	23	4551	17.1	1	0.385	Nucleic acid metabolism
sp|Q9S709|	Splicing factor U2af small subunit A [*Spinacia oleracea*]	71	29,349	7.9	1	0.637	Nucleic acid metabolism
sp|P81766|	Nucleoside diphosphate kinase 3 [*Spinacia oleracea*]	61	31,625	7.5	2	0.508	Nucleic acid metabolism
gi|224117596|	Predicted protein [*Populus trichocarpa*]	368	54,505	13.4	5	0.398	Nucleic acid metabolism
gi|225462994|	PREDICTED: DNA replication licensing factor mcm5-A-like [*Vitis vinifera*]	259	94,881	13.6	8	0.582	Nucleic acid metabolism
sp|O04716|	DNA mismatch repair protein MSH6 [*Spinacia oleracea*]	159	50,441	4.6	1	0.276	Nucleic acid metabolism
gi|359386142|	RNA recognition motif protein 1 [*Citrus sinensis*]	155	14,712	42.7	3	0.492	Nucleic acid metabolism
gi|195626496|	Glycine-rich RNA-binding protein 2 [*Zea mays*]	318	21,175	33.1	4	0.435	Nucleic acid metabolism
sp|Q9FLH0|	PUTATIVE nuclear matrix constituent protein 1-like protein [*Spinacia oleracea*]	69	78,347	5.6	2	0.66	Nucleic acid metabolism
gi|385213056|	20S proteasome β2 subunit, partial [*Oryza brachyantha*]	163	40,674	14.1	4	2.297	Protein metabolism
gi|49175785|	26S proteasome β subunit [*Pisum sativum*]	187	35,781	16	4	1.632	Protein metabolism
gi|16225442|	26S proteasome regulatory subunit S12 isolog-like protein [*Castanea sativa*]	144	38,542	10.6	3	2.263	Protein metabolism
gi|225431100|	PREDICTED: 26S proteasome non-ATPase regulatory subunit 4 [*Vitis vinifera*]	73	8946	16.9	1	1.553	Protein metabolism
gi|24473796|	60s acidic ribosomal protein [*Prunus dulcis*]	208	14,983	15.8	2	2.924	Protein metabolism

Table 1. *Cont.*

Accession Number	Proteins Name and Species	Score	Mass (Da)	Coverage	Peptide Count	Fold Change (Leaves/Bud)	Function
gi|330318716|	60S acidic ribosomal protein p2 [*Camellia sinensis*]	156	15,062	16.2	2	4.223	Protein metabolism
sp|Q8LEQ0|	60S acidic ribosomal protein P1-3 [*Spinacia oleracea*]	185	19,758	10.1	1	1.928	Protein metabolism
sp|Q9SVZ6|	60S acidic ribosomal protein P3-1 [*Spinacia oleracea*]	666	15,391	13.7	1	1.866	Protein metabolism
gi|255574159|	Proteasome subunit β type 6,9, putative [*Ricinus communis*]	368	31,544	22.1	5	1.644	Protein metabolism
gi|255564428|	Elongation factor 1-β, putative [*Ricinus communis*]	62	33,531	5.7	1	1.899	Protein metabolism
gi|255539639|	Cucumisin precursor, putative [*Ricinus communis*]	86	56,932	2.8	1	1.535	Protein metabolism
gi|14594919|	Putative α5 proteasome subunit [*Nicotiana tabacum*]	170	30,889	13.3	3	2.021	Protein metabolism
gi|356549495|	PREDICTED: heat shock 70 kDa protein, mitochondrial-like [*Glycine max*]	62	12,864	10.9	1	1.605	Protein metabolism
gi|272716096|	Disulfide isomerase-like protein [*Gloeospermum blakeanum*]	87	43,279	12.1	2	1.781	Protein metabolism
gi|272716065|	Disulfide isomerase [*Gloeospermum blakeanum*]	250	22,819	30.5	4	1.79	Protein metabolism
sp|Q8VX13|	Protein disulfide isomerase-like 1-3 [*Spinacia oleracea*]	165	80,245	10.1	5	1.625	Protein metabolism
sp|O65351|	SUBTILISIN-like protease [*Spinacia oleracea*]	128	32,198	16.7	3	1.556	Protein metabolism
gi|359473000|	PREDICTED: aspartic proteinase nepenthesin-1-like [*Vitis vinifera*]	250	22,819	30.5	4	1.79	Protein metabolism
sp|P81898|	Peptide-N4-(N-acetyl-β-glucosaminyl)asparagine amidase A [*Prunus dulcis*]	49	28,569	5.3	1	1.823	Protein metabolism
gi|7141245|	26S proteasome regulatory ATPase subunit S10b [*Vitis riparia*]	164	54,246	13.8	4	0.539	Protein metabolism
gi|56481167|	40S ribosomal protein S3a [*Pseudotsuga menziesii* var. *menziesii*]	119	40,753	17.6	3	0.437	Protein metabolism

Table 1. *Cont.*

Accession Number	Proteins Name and Species	Score	Mass (Da)	Coverage	Peptide Count	Fold Change (Leaves/Bud)	Function
sp|Q9SCM3|	40S ribosomal protein S2-4 [*Spinacia oleracea*]	253	38,711	12.9	3	0.548	Protein metabolism
gi|241865275|	40S RPS3B [*Sonneratia alba*]	150	30,886	27.4	5	0.536	Protein metabolism
gi|255569736|	40S ribosomal protein S6, putative [*Ricinus communis*]	88	41,992	8.8	2	0.574	Protein metabolism
gi|330318726|	40S ribosomal protein s9 [*Camellia sinensis*]	126	28,301	14	3	0.513	Protein metabolism
gi|357444481|	40S ribosomal protein S18 [*Medicago truncatula*]	223	25,387	24.2	3	0.414	Protein metabolism
gi|255544840|	40S ribosomal protein S2, putative [*Ricinus communis*]	202	35,538	13.4	3	0.64	Protein metabolism
gi|255549228|	40S ribosomal protein S4, putative [*Ricinus communis*]	277	39,687	25.6	6	0.415	Protein metabolism
gi|241865275|	40S RPS3B [*Sonneratia alba*]	209	31,709	27.4	5	0.538	Protein metabolism
sp|Q9ZNS1|	40S ribosomal protein S7 [*Avicennia marina*]	58	32,180	12.8	3	0.61	Protein metabolism
sp|O80360|	50S ribosomal protein L3, chloroplastic (Fragment) [*Nicotiana tabacum*]	179	36,343	19.5	4	0.591	Protein metabolism
gi|255551787|	60S ribosomal protein L22, putative [*Ricinus communis*]	119	22,528	22.1	3	0.615	Protein metabolism
gi|148466442|	60S ribosomal protein L21 [*Paeonia suffruticosa*]	56	26,661	9.8	2	0.467	Protein metabolism
sp|P51413|	60S ribosomal protein L17-2 [*Spinacia oleracea*]	48	28,359	5.1	1	0.525	Protein metabolism
sp|Q6UNT2|	60S ribosomal protein L5 [*Cucumis sativus*]	90	44,735	6.7	2	0.661	Protein metabolism
sp|Q9SPB3|	60S ribosomal protein L10 [*Vitis riparia*]	189	33,718	12.9	3	0.532	Protein metabolism
sp|P30707|	60S ribosomal protein L9 [*Pisum sativum*]	184	33,248	26.4	4	0.641	Protein metabolism
gi|225427377|	PREDICTED: 60S ribosomal protein L37a-like [*Vitis vinifera*]	93	15,986	16.3	1	0.64	Protein metabolism

Table 1. *Cont.*

Accession Number	Proteins Name and Species	Score	Mass (Da)	Coverage	Peptide Count	Fold Change (Leaves/Bud)	Function
gi|330318574|	Ribosomal petrp-like protein [*Camellia sinensis*]	48	28,359	5.1	1	0.525	Protein metabolism
gi|3885519|	Similar to ribosomal protein L32 [*Medicago sativa*]	86	23,581	13.9	2	0.363	Protein metabolism
gi|209922600|	Elongation factor 1-α [*Prunus persica*]	351	80,605	24.2	11	0.624	Protein metabolism
gi|225452282|	PREDICTED: elongation factor Tu, chloroplastic-like isoform 1 [*Vitis vinifera*]	313	57,834	26.6	8	0.593	Protein metabolism
gi|356524672|	PREDICTED: eukaryotic translation initiation factor 3 subunit C-like [*Glycine max*]	58	6862	27.3	1	0.545	Protein metabolism
gi|71534902|	Histidyl-tRNA synthetase [*Medicago sativa*]	71	41,277	10.8	2	0.603	Protein metabolism
sp|P31542|	ATP-dependent Clp protease ATP-binding subunit clpA homolog CD4B, chloroplastic [*Solanum lycopersicum*]	497	55,091	24.1	8	0.562	Protein metabolism
gi|356516495|	PREDICTED: chaperone protein ClpC, chloroplastic-like [*Glycine max*]	497	55,091	24.1	8	0.562	Protein metabolism
gi|52075839|	Putative chloroplast protease [*Oryza sativa Japonica* Group]	340	85,534	15.4	8	0.523	Protein metabolism
sp|Q8VY06|	Presequence protease 2, chloroplastic/mitochondrial [*Spinacia oleracea*]	85	35,368	13.8	3	0.622	Protein metabolism
sp|Q75GT3|	Chaperone protein ClpB2, chloroplastic [*Oryza sativa* subsp. *Japonica*]	385	130,278	16.2	11	0.504	Protein metabolism
gi|225431090|	PREDICTED: proteasome subunit α type-7 [*Vitis vinifera*]	292	35,407	21.6	4	0.646	Protein metabolism
gi|225457058|	PREDICTED: T-complex protein 1 subunit gamma [*Vitis vinifera*]	354	76,271	13.4	6	0.552	Protein metabolism
gi|225459806|	PREDICTED: T-complex protein 1 subunit β [*Vitis vinifera*]	975	60,327	38.8	11	0.482	Protein metabolism

Table 1. *Cont.*

Accession Number	Proteins Name and Species	Score	Mass (Da)	Coverage	Peptide Count	Fold Change (Leaves/Bud)	Function
gi	255567297		chaperonin containing t-complex protein 1, α subunit, tcpa, putative [*Ricinus communis*]	84	28,459	18.8	3	0.622	Protein metabolism
sp	P32955		Cysteine proteinase 2 (Fragment) [*Carica candamarcensis*]	385	130,278	16.2	11	0.504	Protein metabolism
sp	P35016		Endoplasmin homolog [*Catharanthus roseus*]	416	123,589	18.5	13	0.657	Protein metabolism
sp	P38661		Probable protein disulfide-isomerase A6 [*Medicago sativa*]	213	54,232	24.9	8	0.665	Protein metabolism
sp	Q5Z974		ATP-dependent zinc metalloprotease FTSH 1, chloroplastic [*Oryza sativa* subsp. *Japonica*]	281	42,007	19.7	4	0.427	Protein metabolism
gi	147766666		Hypothetical protein VITISV_035841 [*Vitis vinifera*]	177	44,924	17.6	4	0.557	Protein metabolism
gi	224141163		Predicted protein [*Populus trichocarpa*]	60	36,379	10.7	2	0.56	Protein metabolism
gi	59797458		Superoxide dismutase [*Lilium hybrid cultivar*]	223	21,087	29.1	3	1.849	Stress/ defense/detoxification
sp	Q93VQ9		Thioredoxin O2, mitochondrial [*Spinacia oleracea*]	80	25,925	12	2	1.907	Stress/ defense/detoxification
gi	536838		NADPH thioredoxin reductase, partial [*Helianthus annuus*]	207	45,192	17.4	4	1.778	Stress/ defense/detoxification
sp	Q9LS40		protein aspartic protease in guard cell 1 [*Rabidopsis thaliana*]	193	48,663	17.6	5	1.635	Stress/ defense/detoxification
sp	Q96520		Peroxidase 12 [*Spinacia oleracea*]	132	41,132	15.2	3	1.899	Stress/ defense/detoxification
gi	3201547		Endochitinase [*Persea americana*]	79	18,633	4.3	1	1.971	Stress/ defense/detoxification
sp	Q06015		Endochitinase 3 (Fragment) [*Arachis hypogaea*]	167	39,946	13.3	3	1.691	Stress/ defense/detoxification
gi	215398978		Dehydrin [*Camellia sinensis*]	44	20,578	11	2	5.811	Stress/ defense/detoxification
gi	15637350		Glutaredoxin [*Tilia platyphyllos*]	150	18,171	10.9	1	1.74	Stress/ defense/detoxification
sp	P13240		Disease resistance response protein 206 [*Pisum sativum*]	167	39,946	13.3	3	1.691	Stress/ defense/detoxification

Table 1. *Cont.*

Accession Number	Proteins Name and Species	Score	Mass (Da)	Coverage	Peptide Count	Fold Change (Leaves/Bud)	Function
sp⏐O80934⏐	Uncharacterized protein At2g37660, chloroplastic [*rabidopsis thaliana*]	161	35,389	21.8	4	1.963	Stress/defense/detoxification
gi⏐75138338⏐	Peroxiredoxin Q, chloroplastic [*Gentiana triflora*]	95	27,886	11.1	3	0.595	Stress/defense/detoxification
sp⏐O23044⏐	Peroxidase 3 [*Spinacia oleracea*]	241	36,174	19.8	5	0.584	Stress/defense/detoxification
sp⏐A7NY33⏐	Peroxidase 4 [*Vitis vinifera*]	119	33,610	21.9	4	0.599	Stress/defense/detoxification
sp⏐P22242⏐	Desiccation-related protein PCC13-62 [*Craterostigma plantagineum*]	695	21,349	33.3	4	0.615	Stress/defense/detoxification
gi⏐270064305⏐	Abscisic stress ripening [Musa ABB Group]	243	26,152	15.3	2	0.265	Stress/defense/detoxification
sp⏐Q41328⏐	Pto-interacting protein 1 [*Solanum lycopersicum*]	52	42,570	13.3	3	0.63	Stress/defense/detoxification
sp⏐Q9FM19⏐	Hypersensitive-induced response protein 1 [*Spinacia oleracea*]	124	37,917	9.1	2	0.517	Stress/defense/detoxification
sp⏐P85524⏐	kirola [*Actinidia deliciosa*]	136	24,392	19.3	3	0.599	Stress/defense/detoxification
gi⏐15637165⏐	Voltage-dependent anion channel [β *vulgaris*]	340	39,615	13.9	4	2.321	Transport
gi⏐225439482⏐	PREDICTED: importin subunit β-1 [*Vitis vinifera*]	65	90,956	3.9	2	2.014	Transport
gi⏐526118004⏐	Probable E3 ubiquitin-protein ligase HERC1 [*Vitis vinifera*]	106	55,419	8	2	1.845	Transport
gi⏐147859669⏐	Hypothetical protein VITISV_026572 [*Vitis vinifera*]	105	32,300	9.1	2	1.988	Transport
gi⏐147842983⏐	Hypothetical protein VITISV_024360 [*Vitis vinifera*]	41	29,785	4	1	3.304	Transport
sp⏐Q41009⏐	Translocase of chloroplast 34 [*Pisum sativum*]	42	12,006	28.9	2	0.599	Transport
gi⏐87247471⏐	Putative glutathione *S*-transferase [*Populus x canadensis*]	295	31,506	16.9	2	0.577	Transport
gi⏐8896066⏐	FtsZ1 [*Tagetes erecta*]	71	30,227	11.2	2	2.163	Biological regulation and signal transduction

Table 1. *Cont.*

Accession Number	Proteins Name and Species	Score	Mass (Da)	Coverage	Peptide Count	Fold Change (Leaves/Bud)	Function
gi	71535005		Zinc finger Glo3-like protein [*Medicago sativa*]	151	53,757	9.9	3	1.887	Biological regulation and signal transduction
sp	P93508		Calreticulin [*Ricinus communis*]	250	22,819	30.5	4	1.79	Biological regulation and signal transduction
gi	255562771		STS14 protein precursor, putative [*Ricinus communis*]	165	20,709	22.1	3	4.166	Biological regulation and signal transduction
gi	40807639		Cystatin [*Actinidia eriantha*]	250	22,819	30.5	4	1.79	Biological regulation and signal transduction
gi	359497545		PREDICTED: leucine-rich repeat receptor-like serine/threonine-protein kinase BAM1-like [*Vitis vinifera*]	250	22,819	30.5	4	1.79	Biological regulation and signal transduction
sp	Q8H100		Probable ADP-ribosylation factor GTPase-activating protein AGD8 [*Spinacia oleracea*]	382	58,424	15.4	5	2.137	Biological regulation and signal transduction
gi	359495838		PREDICTED: uncharacterized protein LOC100264206 [*Vitis vinifera*]	58	30,115	10.4	2	3.438	Biological regulation and signal transduction
sp	O23193		CBS domain-containing protein CBSX1, chloroplastic [*Spinacia oleracea*]	138	27,995	20.6	3	1.672	Biological regulation and signal transduction
sp	P93654		Syntaxin-22 [*Spinacia oleracea*]	35	30,780	8.7	2	0.522	Biological regulation and signal transduction
gi	350534900		14-3-3 protein 3 [*Solanum lycopersicum*]	368	38,590	27.9	7	0.599	Biological regulation and signal transduction
gi	359492889		PREDICTED: 14-3-3 protein [*Vitis vinifera*]	244	40,230	20.7	6	0.525	Biological regulation and signal transduction

Table 1. *Cont.*

Accession Number	Proteins Name and Species	Score	Mass (Da)	Coverage	Peptide Count	Fold Change (Leaves/Bud)	Function
sp | Q9FM65 |	Fasciclin-like arabinogalactan protein 1 [*Spinacia oleracea*]	247	36,736	18.8	3	0.559	Biological regulation and signal transduction
gi | 95116526 |	Ethylene inducible protein hever [*Theobroma cacao*]	131	37,972	15.2	4	0.619	Biological regulation and signal transduction
sp | A1Y2B7 |	Protein suppressor of gene SILENCING 3 homolog [*Zea mays*]	86	51,127	7.8	2	0.602	Biological regulation and signal transduction
gi | 255587170 |	Minichromosome maintenance protein, putative [*Ricinus communis*]	63	16,167	7.5	1	0.547	Biological regulation and signal transduction
gi | 255571471 |	Systemin receptor SR160 precursor, putative [*Ricinus communis*]	200	108,992	6.6	5	0.483	Biological regulation and signal transduction
gi | 359494860 |	PREDICTED: protein MOR1-like, partial [*Vitis vinifera*]	102	17,742	8.8	1	0.603	Biological regulation and signal transduction
gi | 359495954 |	PREDICTED: syntaxin-51-like [*Vitis vinifera*]	318	19,583	10.5	1	0.243	Biological regulation and signal transduction
sp | Q94KK7 |	Syntaxin-52 [*Spinacia oleracea*]	40	31,732	6.3	1	0.594	Biological regulation and signal transduction
gi | 359493650 |	PREDICTED: early nodulin-like protein 2-like [*Vitis vinifera*]	214	26,770	23	4	0.538	Biological regulation and signal transduction
gi | 225456479 |	PREDICTED: signal recognition particle 68 kDa protein [*Vitis vinifera*]	52	30,059	4.6	1	0.623	Biological regulation and signal transduction
sp | O22126 |	Fasciclin-like arabinogalactan protein 8 [*Spinacia oleracea*]	256	45,342	16.4	5	0.51	Biological regulation and signal transduction

Table 1. *Cont.*

Accession Number	Proteins Name and Species	Score	Mass (Da)	Coverage	Peptide Count	Fold Change (Leaves/Bud)	Function
gi|388491766|	Unknown [*Lotus japonicus*]	171	15,992	18.3	1	0.538	Biological regulation and signal transduction
gi|255545216|	Conserved hypothetical protein [*Ricinus communis*]	85	37,085	13	3	2.209	Unknown biological processes
gi|255547524|	Conserved hypothetical protein [*Ricinus communis*]	50	41,524	3.3	1	2.09	Unknown biological processes
sp|Q6YYB0|	UNCHARACTERIZED protein Os08g0359500 [*Oryza sativa* subsp. *Japonica*]	56	15,660	12.7	1	1.541	Unknown biological processes
gi|225423539|	PREDICTED: uncharacterized protein LOC100262861 [*Vitis vinifera*]	152	24,072	16	2	2.042	Unknown biological processes
gi|330318602|	Hypothetical protein [*Camellia sinensis*]	40	19,541	6	1	1.809	Unknown biological processes
gi|224070853|	Predicted protein [*Populus trichocarpa*]	104	23,465	22.3	3	1.788	Unknown biological processes
gi|224144195|	Predicted protein [*Populus trichocarpa*]	93	28,934	6.2	1	1.668	Unknown biological processes
gi|147818671|	Hypothetical protein VITISV_014852 [*Vitis vinifera*]	43	14,360	10.5	1	1.715	Unknown biological processes
gi|359488537|	PREDICTED: uncharacterized protein LOC100853981 [*Vitis vinifera*]	71	21,254	8.3	1	1.921	Unknown biological processes
gi|147818796|	Hypothetical protein VITISV_021596 [*Vitis vinifera*]	625	66,920	20.2	9	1.795	Unknown biological processes
gi|225452887|	PREDICTED: uncharacterized protein At5g39570 [*Vitis vinifera*]	141	28,611	31.3	4	1.902	Unknown biological processes
gi|359491847|	PREDICTED: uncharacterized protein LOC100240982 [*Vitis vinifera*]	78	22,451	15.1	2	1.893	Unknown biological processes
gi|225463725|	PREDICTED: uncharacterized protein LOC100261025 [*Vitis vinifera*]	79	82,239	7.5	3	1.632	Unknown biological processes
gi|147818796|	Hypothetical protein VITISV_021596 [*Vitis vinifera*]	625	66,920	20.2	9	1.795	Unknown biological processes
gi|356551464|	PREDICTED: uncharacterized protein LOC100807412 [*Glycine max*]	53	57,807	3	1	2.315	Unknown biological processes

Table 1. *Cont.*

Accession Number	Proteins Name and Species	Score	Mass (Da)	Coverage	Peptide Count	Fold Change (Leaves/Bud)	Function
gi︱298205066︱	Unnamed protein product [*Vitis vinifera*]	293	59,849	8	3	1.506	Unknown biological processes
gi︱358249210︱	Uncharacterized protein LOC100818758 [*Glycine max*]	74	26,221	15.1	2	2.26	Unknown biological processes
gi︱224089721︱	Predicted protein [*Populus trichocarpa*]	64	16,865	21	2	1.824	Unknown biological processes
gi︱225443833︱	PREDICTED: uncharacterized protein LOC100253185 [*Vitis vinifera*]	155	13,410	33	3	1.815	Unknown biological processes
gi︱297743302︱	Unnamed protein product [*Vitis vinifera*]	269	40,524	24.2	5	1.913	Unknown biological processes
gi︱296081618︱	Unnamed protein product [*Vitis vinifera*]	280	16,151	51.6	4	2.009	Unknown biological processes
gi︱225459322︱	PREDICTED: uncharacterized protein LOC100260886 isoform 2 [*Vitis vinifera*]	80	43,415	11.6	3	1.733	Unknown biological processes
sp︱Q6ID70︱	Uncharacterized protein At3g03773 [*Spinacia oleracea*]	189	28,352	12.6	2	1.772	Unknown biological processes
gi︱225451915︱	PREDICTED: uncharacterized protein LOC100244706 [*Vitis vinifera*]	181	28,654	12.2	2	2.492	Unknown biological processes
gi︱297738842︱	Unnamed protein product [*Vitis vinifera*]	68	67,341	4.9	2	0.657	Unknown biological processes
gi︱359489218︱	PREDICTED: uncharacterized protein LOC100232913 [*Vitis vinifera*]	95	31,024	10.9	2	0.565	Unknown biological processes
gi︱17863981︱	Unknown [*Davidia involucrata*]	150	95,611	9.6	6	0.598	Unknown biological processes
gi︱224056457︱	Predicted protein [*Populus trichocarpa*]	43	32,607	7.2	1	0.25	Unknown biological processes
gi︱359488731︱	PREDICTED: uncharacterized protein LOC100264617 [*Vitis vinifera*]	111	34,149	5.2	1	0.446	Unknown biological processes
gi︱359476152︱	PREDICTED: uncharacterized protein LOC100260975 [*Vitis vinifera*]	142	40,178	10.4	3	0.374	Unknown biological processes

Table 1. *Cont.*

Accession Number	Proteins Name and Species	Score	Mass (Da)	Coverage	Peptide Count	Fold Change (Leaves/Bud)	Function
gi｜297742161｜	Unnamed protein product [*Vitis vinifera*]	141	34,984	19.3	3	0.436	Unknown biological processes
gi｜225449483｜	PREDICTED: uncharacterized protein LOC100244410 [*Vitis vinifera*]	73	23,689	7.1	1	0.65	Unknown biological processes
gi｜225463406｜	PREDICTED: uncharacterized protein LOC100250442 [*Vitis vinifera*]	85	20,570	7.9	1	0.631	Unknown biological processes
gi｜351722061｜	Uncharacterized protein LOC100305495 precursor [*Glycine max*]	236	23,212	10.4	1	0.52	Unknown biological processes
gi｜224141949｜	Predicted protein [*Populus trichocarpa*]	101	18,222	19	1	0.514	Unknown biological processes

2.4. RT-qPCR Analysis and Enzyme Activity Assay

To evaluate the iTRAQ results, RT-qPCR analysis and enzyme activity assays were performed. Five proteins were selected for RT-qPCR analysis; three were up-regulated (flavonol synthase, FLS; dehydrin, DHN; and 60S acidic ribosomal protein p2, RPLP2), and two were down-regulated (phenylalanine ammonia-lyase, PAL; photosystem I reaction center subunit XI, PRC subunit XI) in the young expanding leaves compared with the buds. As shown in Figure 4, the expression levels of FLS and DHN were significantly up-regulated in the young leaves compared with the buds (FLS: 2.01 ± 0.06-fold, $p < 0.01$. DHN: 3.33 ± 0.34-fold, $p < 0.01$). However, the expression levels of PAL and PRC subunit XI were significantly down-regulated in the young leaves compared with the buds (PAL: 0.51 ± 0.04-fold, $p < 0.05$. PRC subunit XI: 0.41 ± 0.02-fold, $p < 0.05$). The expression of RPLP2 was also down-regulated in the young expanding leaves compared with the buds (0.64 ± 0.05-fold), but no significant difference was observed ($p > 0.05$). The transcription levels of FLS, DHN, PAL and PRC subunit XI were closely correlated with the levels of their translation products in the buds and the young expanding leaves, whereas the RPLP2 transcript levels did not correspond with those of its translation products. As shown in Figure 5, PAL activity was significantly lower in young expanding leaves than in buds, which is consistent with its gene and protein expression levels in the buds and the young expanding leaves of tea plants.

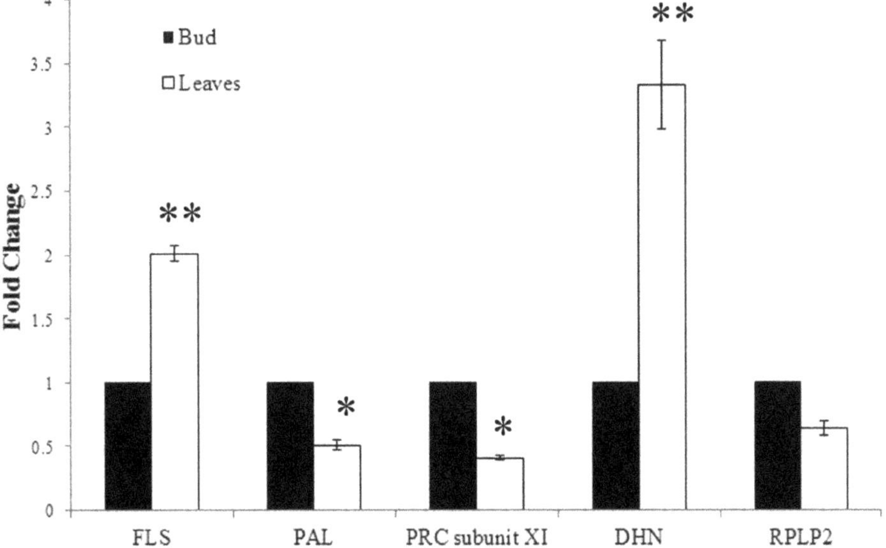

Figure 4. RT-qPCR analysis of the transcript levels of the differentially expressed proteins. FLS: flavonol synthase; PAL: phenylalanine ammonia-lyase; PRC subunit XI: photosystem I reaction center subunit XI; DHN: dehydrin; RPLP2: 60S acidic ribosomal protein p2. Statistical significance: * $p < 0.05$ and ** $p < 0.01$.

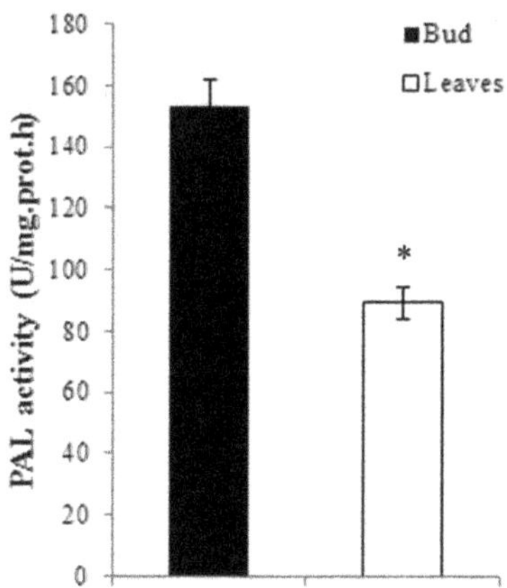

Figure 5. PAL activity in the buds and in young expanding leaves. Statistical significance: * $p < 0.05$.

3. Discussion

A previous study used subtractive cDNA library analysis to reveal the genes involved in the production of polyphenols and other secondary metabolites that are relatively abundant in young leaves [2]. However, because of post-transcriptional regulation, protein expression levels cannot always be predicted from quantitative mRNA data; the mRNA level does not always correlate with the protein level [10]. Therefore, proteomic analysis could improve our understanding of the molecular mechanisms underlying the change in the metabolite contents of the apical buds and the young expanding leaves of tea plants.

3.1. Changes in Secondary Metabolites

Tea leaves contain large amounts of flavonoids, including flavanones, flavones, flavonols, flavan-3-ols, and anthocyanidins. The predominant flavonoid in tea is catechin, which distinguishes tea from other plants and is an important determinant of tea quality and taste. A previous study showed that the concentrations of total catechins and polyphenols in tea leaves declined with leaf age, but changes in individual catechins varied [11]. Our HPLC analysis showed that EGCG and ECG were the most abundant catechins in both the buds and the young expanding leaves. These compounds exist in the green parts of tea seedlings but were not detected in the roots or cotyledons [12]. The catechins index [(EGCG + ECG)/EGC] was positively correlated with the sensory evaluation of brewed green tea [13]. Based on HPLC results, the green tea quality indexes of the buds and the young leaves were 45.11 and 15.59, respectively. These results were consistent with previous research [7,11]. Historically, tea has been valued for its purine alkaloids, including theobromine, theophylline and caffeine [14]. Theobromine is formed as part of the caffeine biosynthetic pathway and is produced in abundance if the methylation pathway of caffeine biosynthesis is absent [14]. An analysis of purine alkaloids in different tea seedling organs showed that more than 99% of the caffeine was in the leaves, with older leaves containing more per gram of fresh weight. Theobromine was found only in the younger leaves, and theophylline was either not present or present only in trace amounts [8]. Our study showed that the concentrations of theobromine and caffeine were lower in young expanding leaves, but no significant difference in theophylline levels was observed. Purine alkaloid metabolism also appears to be closely associated with leaf development and aging in tea seedlings [9,15]. The major biosynthetic route for caffeine is thought to be xanthosine→7-methyxanthosine→7-methylxanthine→theobromine →caffeine, and previous studies have indicated that caffeine biosynthesis was primarily controlled by the first N-methyl-transfer reaction, which is catalyzed by 7-methylxanthosine synthase [16,17]. Hence, the relatively lower caffeine and theobromine contents of young expanding leaves found in this study may be attributable either to a smaller supply of xanthosine for caffeine biosynthesis or to the lower activity of 7-methylxanthosine synthase in young expanding tea leaves.

3.2. Proteins Involved in Carbohydrate and Energy Metabolism

In plants, glycolysis and the tricarboxylic acid (TCA) cycle provide not only energy and cofactors but also important substrates for the synthesis of metabolites, as well as feedback signals [18]. Dynamic proteomic analysis revealed that the levels of glycolysis- and TCA cycle-related proteins increased during early-stage seed development in rice [19]. Our present results show that a subset of the differentially expressed proteins were involved in glycolysis and TCA, such as NADP-dependent glyceraldehyde-3-phosphate dehydrogenase (NADP-dependent GAPDH), dihydrolipoyl dehydrogenase (DLD), pyruvate dehydrogenase E3 subunit (PDE3), dihydrolipoamide succinyltransferase component of 2-oxoglutarate dehydrogenase (DLST) and phosphoenolpyruvate carboxylase (PEPC); these proteins were present at higher levels in the young, expanding leaves than in the buds. These results indicated that glycolysis and the TCA cycle increased in the young, expanding leaves and that more energy and substrates were required during the developmental stage at which young, expanding leaves are present.

3.3. Proteins Related to Secondary Metabolism

Polyphenols are the most important chemical compounds in tea plants, and have received increasing attention in recent years because of their benefits to human health [20–23]. The polyphenols in tea are predominantly members of three subclasses: flavanols, flavones and flavonols [24]. Four major catechins (flavanols), (−)-epicatechin (EC), (−)-epicatechin gallate (ECG), (−)-epigallocatechin (EGC), and (−)-epigallocatechin gallate (EGCG), constitute approximately one-third of the dry weight of green tea [25]. Quercetin, kaempferol, myricetin and their glycosides (flavonols), as well as apigenin glycosides (flavones), are also present, but at much lower concentrations [24]. Several proteins related to polyphenol biosynthesis were differentially expressed between the buds and the young, expanding leaves. Flavonol synthase (FLS), a dioxygenase that converts dihydroflavonols into flavonols, was initially found in parsley and was shown to require 2-oxoglutarare and Fe/ascorbate for full activity [26]. In FLS-silenced tobacco, there was a 25%–93% reduction in the flavonoid (quercetin) content and an increase in the catechin and epicatechin content [27,28]. Our previous study also indicated that FLS expression was a negative regulator of catechin biosynthesis, and especially of ECG and EGCG [29]. In our proteomic analysis, the expression of FLS was increased at the stage of young, expanding leaves, which indicated that at this stage, flavonol biosynthesis was enhanced and catechin biosynthesis was inhibited. These results also agree with our metabolic data, which show that compared with the buds, the flavonol content was greater and the total catechin content was lower in the young, expanding leaves. Isoflavone reductase homolog P3 belongs to the NmrA-type oxidoreductase family and the isoflavone reductase subfamily. Isoflavone reductase (IFR) specifically recognizes isoflavones and catalyzes a stereospecific, NADPH-dependent reduction to (3R)-isoflavanone [30]. In tea plants, IFR catalyzes the conversion of leucocyanidin and leucodelphinidin to (+)-catechin and (+)-gallocatechin, respectively. In our proteomic analysis, the expression of IFR homolog P3, which is involved in the accumulation of high levels of catechins, was more highly expressed in the buds compared with the young, expanding leaves. Phenylalanine ammonia-lyase (PAL) is an enzyme that catalyzes the conversion of L-phenylalanine to ammonia and *trans*-cinnamic acid [31]. PAL resides at a metabolically important position, linking secondary metabolism to primary metabolism. PAL is part of the first committed step in the phenylpropanoid pathway and is a key enzyme in the allocation of significant amounts of carbon from phenylalanine into the biosynthesis of several important secondary metabolites, such as lignins, flavonoids, and coumarins [32,33]. The overall flux into phenylpropanoid metabolism has been suggested to be regulated by PAL, which acts as a rate-limiting enzyme [34]. Park *et al.* found that PAL gene expression and catechin content were also reduced in mature leaves compared with young leaves [2]. A positive correlation between catechin content and the gene expression of PAL was observed under drought stress, after wounding and after abscisic acid treatment [35]. In the present study, the expression of both the PAL gene and protein were inhibited, and the catechin content was also reduced in young, expanding leaves. These results indicated that the carbon flux from phenylalanine into the biosynthesis of secondary metabolites was inhibited in the young, expanding leaves compared with the buds. Hydroxycinnamoyl-CoA: shikimate/quinate hydroxycinnamoyltransferase (HCT), which converts p-coumarate from CoA to shikimate/quinate esters, has been described as reversible enzyme [36]. It is involved in a step in lignin synthesis, and its down-regulation affects lignin content and composition [37,38]. In our proteomic analysis, the expression level of HCT was lower in the young, expanding leaves than in the buds. Arabidopsis plants in which HCT is silenced or lignin is repressed direct the metabolic flux into flavonoids through chalcone synthase [39], which may explain why the non-galloylated catechin content increased in the young, expanding tea leaves.

3.4. Photosynthetic Proteins

Photosynthesis is a key biological process in plant growth and development. In the present study, the abundance of several proteins involved in photosynthesis differed between the buds and the young expanding leaves. These proteins include ribulose-1,5-bisphosphate carboxylase/oxygenase (Rubisco) and its large subunit (RubiscoL), sedoheptulose-1,7-bisphosphatase (SBPase) precursor, photosystem I reaction center subunit XI (PS I-E), thylakoid lumenal 29 kDa protein (TL29), peroxiredoxin Q (PRXQ) and chlorophyll A/B binding protein (CitCAB1,2). Several studies have shown that during leaf development, photosynthetic activity gradually increases, and photosynthetic enzymes slowly accumulate [40–43]. Correlations between the photosynthetic rate and the catechin content of the leaves of tea plants showed that there was a positive correlation between the photosynthetic rate and the EC and GCG contents but a negative correlation between the photosynthetic rate, the total catechin content and the galloylated catechin content [44]. A study focusing on the relationship between the synthesis and accumulation of phenolics and flavonoids and the photosynthetic rate in ginger showed that when photosynthesis decreased, the synthesis of flavonoids such as quercetin, catechin, epicatechin and naringenin increased, and the soluble carbohydrates and plant biomass decreased [45]. The results of our proteomic analysis also showed that the expression of photosynthetic proteins was down-regulated in the buds compared with the young, expanding leaves. We infer that in the buds, the rate of photosynthesis is lower, so the carbon flow shifts from photosynthesis to the shikimic acid pathway, thereby producing more phenolics and flavonoids.

3.5. Defense-Related Proteins

The cellular antioxidant system consists of different enzymes. In our proteomic analysis, antioxidant proteins, such as superoxide dismutase (SOD), thioredoxin O2 (TO2), NADPH thioredoxin reductase (NADPH-TR), and glutaredoxin (GRX), were more abundant in young, expanding leaves than in buds. The activity of antioxidant enzymes, such as SOD also increased at early stages of leaf expansion and was sustained throughout leaf expansion [46,47]. Therefore, the proteins involved in the antioxidant system may be related to leaf expansion. Another study also indicated that a certain concentration of reactive oxygen species (ROS) is necessary for leaf elongation, but it could not be determined if H_2O_2 or other ROS are the active agents [48]. We suggest that the accumulation of antioxidant proteins could dissipate excess excitation energy and protect leaves against photodamage, which can be caused by a certain levels of ROS in expanding tea leaves.

4. Experimental Section

4.1. Plant Materials

Tea plants were grown in the experimental tea garden of Hunan Agricultural University in Changsha, China. The apical buds and the first unfolding leaves were plucked from the same plants at different stages of development, briefly washed with sterile water, immediately frozen in liquid nitrogen and stored at $-80\ ^\circ$C prior to analysis (Figure 6).

Figure 6. The buds and young expanding leaves of tea plants.

4.2. Metabolic Analysis of Tea Samples

Total polyphenols, catechins and alkaloids were extracted from the samples and analyzed as previously described with a slight modification [29]; a total of 0.20 g of freeze-dried, ground leaves was accurately weighed and extracted twice with 5 mL of a 75:25 (v/v) ethanol:water solution at 80 $^\circ$C for 15 min. The extract was filtered through filter paper and then diluted to 50 mL. The total polyphenol and flavonoid content in the sample was determined using the ferrous tartrate method [49] and the aluminum trichloride method [50]. The catechin and alkaloid contents were determined with high-performance liquid chromatography (HPLC) according to Wang *et al.* [51] with slight modifications. A Shimadzu HPLC system (Shimadzu, Tokyo, Japan) with 10AD dual pumps was used with a reversed-phase column (Welchorm C18 200 $\times$ 4.6, 5 µm), a mobile phase of distilled water (A) and a mobile phase (B) of 40% *N,N*-dimethylformamide, 2% methanol and 1.5% acetic acid. The gradient was as follows: 0.01–13.00 min, linear gradient from 14% to 23% B; 13.00–25.00 min, linear gradient from 23% to 36% B; 25.00–28.00 min, 36% B; 28.00–30.00 min, linear gradient from 36%–14% B; 30.00–34.00 min, 14% B. The samples were eluted at 35 $^\circ$C and at a flow rate of 1.00 mL/min. The chromatograms were recorded at 278 nm. The peaks were identified by comparing the retention times of the sample to those of authentic standards. The extraction for flavone hydrolysis was carried out as follows: plant material (0.5 g dry weight) was mixed with 20 mL methanol and 2.0 mL HCl (6 M). After refluxing at 95 $^\circ$C for 1.5 h, the hydrolyzed solution was filtered through filter paper, then diluted to 25 mL with methanol. Flavonols were detected with the following HPLC method [52]: the mobile phase consisted of 30% acetonitrile in 0.025 M KH_2PO_4 buffer solution (v/v); the pH of the mobile phase was adjusted to 2.5 using H_3PO_4. The samples were eluted at 35 $^\circ$C at a flow rate of 1.00 mL/min and were monitored at 370 nm. The peaks were identified by comparing the retention times of the sample to those of authentic standards. All experiments included three separate biological replicates.

4.3. Protein Extraction

Leaf samples were weighed and ground in liquid nitrogen, then suspended in lysis buffer [7 M urea, 2 M thiourea, 4% 3-[(3-Cholamidopropyl)dimethylammonio]propanesulfonate (CHAPS), 40 mM *Tris*-HCl, pH 8.5, 1 mM Phenylmethanesulfonyl fluoride (PMSF), 2 mM Ethylene Diamine Tetraacetic

Acid (EDTA), 10 mM DL-Dithiothreitol (DTT)] and kept in an ice bath for 2 h. After this 2 h lysis, the samples were sonicated in an ice bath for 15 min and were clarified by centrifugation at 25,000× g. The supernatant was collected, and the protein concentration was determined with a 2D quantification kit (GE Healthcare, Chalfont St. Giles, Buckinghamshire, UK).

4.4. iTRAQ Analysis

iTRAQ analysis was performed at the Beijing Genomics Institute (BGI, Shenzhen, China). Protein samples were reduced with 10 mM DTT, alkylated with 55 mM iodoacetamide, digested using sequencing-grade trypsin (Promega, Madison, WI, USA), and labeled using an iTRAQ Reagent Multiplex Kit (AB SCIEX, Foster City, CA, USA) according to the manufacturer's protocol. The bud and leaf samples were labeled with 114 and 117 Da, respectively. After labeling, all samples were pooled and purified using a strong cation exchange chromatography (SCX) column (Phenomenex, Torrance, CA, USA) with an LC-20AB HPLC system (Shimadzu, Tokyo, Japan). The labeled peptides were separated with mobile phase B (2% water, 98% acetonitrile and 0.1% formic acid) at a flow rate of 300 nL/min, 0%–5% over 1 min, 5%–35% over 40 min and 35%–80% over 5 min on a nanoACQuity system (Waters, Milford, MA, USA). The LC fractions were analyzed using a Triple TOF 5600 mass spectrometer (AB SCIEX, Foster City, CA, USA) fitted with a Nanospray III source (AB SCIEX, Concord, MA, USA) and a pulled quartz tip (New Objectives, Woburn, MA, USA). The data were acquired using an ion spray voltage of 2.5 kV and an interface heater temperature of 150 °C. Curtain gas and nebulizer gas were delivered at 30 pounds per square inch (PSI) and 15 PSI, respectively. For information-dependent acquisition (IDA), survey scans were acquired in 250 ms, and once the detection of ions with a 2+ to 5+ charge state crossed a threshold of 150 counts per second, as many as 35 product ion scans were collected. The total cycle time was fixed at 2.5 s. A rolling collision energy setting was applied to all precursor ions for collision-induced dissociation (CID). Two independent biological experiments with three technical replicates each were performed.

4.5. Data Analysis

MS/MS data acquisition was performed with Analyst QS 2.0 software (AB SCIEX, Foster City, CA, USA). For protein identification, MS/MS data were searched against the "plant" subset of the National Center for Biotechnology Information Non-redundant protein sequences (NCBInr) database using Mascot version 2.3.02 (Matrix Science, London, UK). The search parameters were as follows: a peptide mass tolerance of 10 ppm was allowed for intact peptides and $\pm$ 0.05 Da for fragmented ions; a maximum of one missed cleavage was allowed in the trypsin digests; cysteine carbamidomethylation was considered a fixed modification; glutamine pyrophosphorylation variable oxidation of methionine and iTRAQ labeling of tyrosine were set as variable modifications; carbamidomethylation of cysteine and iTRAQ labeling of lysines and the N-terminal amino group of peptides were set as fixed modifications. Only peptides with significance scores greater than "identity score" were considered identified, and a protein was considered identified if at least one such unique peptide match was apparent for the protein. For protein quantitation, the peptide to be quantified was automatically selected using the Pro Group algorithm to calculate the reporter peak area, the error factor (EF), and the p-value. Proteins with a fold change of >1.5 and a two-tailed p-value of less than 0.05 were considered to have significantly different expression.

4.6. Bioinformatic Analysis of Proteins

Differentially expressed proteins were mapped to Gene Ontology Terms (GO) using a local Bell Labs Layered Space-Time (BLAST) against a reference database downloaded from the website (GO-Annotation@EBI). The Clusters of Orthologous Groups of Proteins system (COG) can be used to functionally annotate genes from new genomes and for research on genome evolution [53]. The Kyoto Encyclopedia of Genes and Genomes (KEGG) is an updated system that computerizes current knowledge on biochemical pathways and other types of molecular interactions and can be used as a reference for the systematic interpretation of sequencing data [54]. To augment the biological and functional properties of differentially expressed proteins, the proteins were further analyzed using the COG (http://www.ncbi.nlm.nih.gov/COG/) and KEGG databases (http://www.genome.jp/kegg/pathway.html).

4.7. Real-time Quantitative PCR Analysis

Total RNA for RT-qPCR analysis was extracted from leaves at the two developmental stages using an RNeasy Plant Mini Kit (Qiagen, Hilden, Germany) and an RNase-Free DNase Set (Qiagen, Hilden, Germany). cDNA was synthesized from the total RNA (1 µg) using oligo(dT)18 primers and Moloney murine leukemia virus reverse transcriptase (Promega, Madison, WI, USA) according to the manufacturer's instructions. The primers used for RT-qPCR (Table 2) were designed using Beacon Designer 7.0 software (Premier Biosoft, Palo Alto, CA, USA) and were based on the cDNA sequences. The reactions were carried out with a Rotor-Gene Q 6200 real-time PCR system (Qiagen, Hilden, Germany) using three-step cycling conditions of 95 °C for 10 min followed by 45 cycles of 95 °C for 10 s, 56 °C for 15 s and 72 °C for 20 s. The reaction mixture (20 µL) contained 1 µL of cDNA solution, 10 µL of Platinum SYBR® Green qPCR SuperMix-UDG (Invitrogen, Carlsbad, CA, USA) and primers at a concentration of 6 µM each. For each RT-qPCR sample, there were three biological replicates with three technical replicates. The GAPDH gene was used as an internal standard for the normalization of gene expression, and the tea buds were used as a reference sample whose value was set to 1. The relative gene expression was evaluated using the comparative cycle threshold method [55].

Table 2. Primers used in RT-qPCR analysis.

Gene Name	Accession Number	Primer Sequence (5′–3′)
Flavonol synthase	DQ198089	Forward: ggagaacagcaaggctatcg Reverse: tctcctcctgtgggagctta
Phenylalanine ammonia-lyase	D26596	Forward: tccgatcatcgacaaaatca Reverse: agctcagagaattgggcaaa
Photosystem I reaction center subunit XI	HM003371	Forward: tcaaagaaggagagccatcg Reverse: gcaagaaataggcccaaatg
Dehydrin	FJ436978	Forward: gaggagaggaccaacagcag Reverse: acgacaccgacacacacatt
60S acidic ribosomal protein p2	HM003314	Forward: gggtgctattgcagtgacct Reverse: attgggggagaaagaaggaa

4.8. PAL Extraction and Enzyme Assays

Tea samples (1 g) were ground into a fine powder with a mortar and pestle in liquid N_2. The powder was extracted with 5 mL of extraction buffer 50 mM Tris-HCl pH 8.9, 10 µM leupeptin, 5 mM EDTA, 15 mM β-mercaptoethanol, 5 mM Vc, 1 mM PMSF, 0.15% Polyvinylpyrrolidone (PVP)], and then was stirred on ice for 10 min. Subsequently, the mixture was centrifuged at 30,000× *g* for 30 min at 20 °C. The supernatant was stirred on Dowex (1 × 2) in the chloride (Cl) form for 30 min to remove residual phenolics. The cleared supernatant was used in a PAL enzyme assay. The protein concentrations in the enzyme extract were measured with a 2D quantification kit (GE Healthcare, Chalfont St Giles, Buckinghamshire, UK). PAL activity was assayed using the method of Solecka and Kacperska [29].

4.9. Statistical Analysis

Statistical analyses were performed using the Statistical Package for the Social Sciences software (SPSS; Chicago, IL, USA). ANOVA and Student's *t*-tests were used to determine significant differences between different groups. A *p*-value <0.05 was considered significant.

5. Conclusions

The quantitative protein expression data presented in this study provide a global overview of a set of proteins that are expressed in the buds and the young, expanding leaves of tea. A total of 233 proteins were identified as being differentially expressed between the buds and the young leaves. A large array of diverse functions, including energy metabolism and the metabolism of carbohydrates, secondary metabolites, nucleic acids and proteins, as well as photosynthesis and defense-related processes, were revealed. Based on these results, we infer that the proteins involved in polyphenol biosynthesis and photosynthesis may also mediate the secondary metabolite content in tea plants. The proteins related to energy and antioxidant metabolism may promote tea leaf development. However, the RT-qPCR results showed that the protein expression levels did not closely correlate with their gene expression levels. Overall, these findings improve our understanding of the molecular mechanisms underlying the change in the metabolite content from the buds to the young, expanding leaves of tea plants.

Supplementary Materials: Supplementary materials can be found at http://www.mdpi.com/1422-0067/16/06/14007/s1.

Acknowledgments: The work was financially supported by the National Natural Science Foundation of China (31200522 and 31470692), National "Five-twelfth" Plan for Science & Technology Support (2011BAD01B01), Program for Changjiang Scholar and Innovative Research Team in University (IRT0963), the Ministry of Science and Technology of China (NCET-11-0969), the Natural Science Foundation of Hunan Province (13JJ4067), Scientific Research fund of Hunan Agricultural University (13YJ13) and the "1515 Talent Project" of Hunan Agricultural University.

Author Contributions: Qin Li carried out the iTRAQ, HPLC and RT-qPCR experiments. Juan Li participated in the HPLC analysis. Shuoqian Liu participated in the design of this paper. Jianan Huang participated in the optimization of HPLC and RT-qPCR protocol. Haiyan Lin participated in the optimization of HPLC protocol. Kunbo Wang participated in the HPLC experiments. Xiaomei Cheng participated in the iTRAQ experiments. Zhonghua Liu conceived of the study, and participated in its design and coordination. All authors read and approved the final manuscript.

Conflicts of Interest: The authors declare no conflict of interest.

References

1. Khan, N.; Mukhtar, H. Tea and health: Studies in humans. *Curr. Pharm. Des.* **2013**, *19*, 6141–6147. [CrossRef] [PubMed]

2. Park, J.S.; Kim, J.B.; Hahn, B.S.; Kim, K.H.; Ha, S.H.; Kim, Y.H. EST analysis of genes involved in secondary metabolism in *Camellia sinensis* (tea), using suppression subtractive hybridization. *Plant Sci.* **2004**, *166*, 953–961. [CrossRef]

3. Shan, Y.; Li, W.; Wang, Y.; Liu, Y.; Wang, H.; Wang, X.; Lu, Z.; Tian, Y.; Gao, L.; Xia, T. Catechin synthesis and accumulation in tea seedlings at different development stages. *J. Anhui Agr. Univ.* **2011**, *38*, 600–605. (In Chinese)

4. Li, J.; Zhao, M.; Zhang, G.; Ding, X.; Hu, Y.; Shen, X.; Shao, W. HPLC analysis of myricetin, quercetin and kaempferol in fresh shoots of "Zijuan" tea, a new cultivar of *Camellia sinensis* var. *Assamica*. *J. Yunnan Agri. Univ.* **2012**, *27*, 235–240. (In Chinese)

5. Singh, K.; Rani, A.; Kumar, S.; Sood, P.; Mahajan, M.; Yadav, S.K.; Singh, B.; Ahuja, P.S. An early gene of the flavonoid pathway, flavanone 3-hydroxylase, exhibits a positive relationship with the concentration of catechins in tea (*Camellia sinensis*). *Tree Physiol.* **2008**, *28*, 1349–1356. [CrossRef] [PubMed]

6. Eungwanichayapant, P.D.; Popluechai, S. Accumulation of catechins in tea in relation to accumulation of mRNA from genes involved in catechin biosynthesis. *Plant Physiol. Biochem.* **2009**, *47*, 94–97. [CrossRef] [PubMed]

7. Zhang, X.; Gao, L.; Xia, T.; Liu, Y.; Gao, K. Study on the changes of non-galloylated catechins and relative enzymes in tea shoots. *J. Tea Sci.* **2009**, *29*, 365–371.

8. Suzuki, T.; Ashihara, H.; Waller, G.R. Purine and purine alkaloid metabolism in *Camellia* and *Coffea* plants. *Phytochemistry* **1992**, *31*, 2575–2584. [CrossRef]

9. Kato, M.; Kitao, N.; Ishida, M.; Morimoto, H.; Irino, F.; Mizuno, K. Expression for caffeine biosynthesis and related enzymes in *Camellia sinensis*. *Z. Naturforsch C* **2010**, *65*, 245–256. [CrossRef] [PubMed]

10. Yan, S.P.; Zhang, Q.Y.; Tang, Z.C.; Su, W.A.; Sun, W.N. Comparative proteomic analysis provides new insights into chilling stress responses in rice. *Mol. Cell. Proteomics* **2006**, *5*, 484–496. [CrossRef] [PubMed]

11. Mamati, G.E.; Liang, Y.; Lu, J. Expression of basic genes involved in tea polyphenol synthesis in relation to accumulation of catechins and total tea polyphenols. *J. Sci. Food Agric.* **2006**, *86*, 459–464. [CrossRef]

12. Ashihara, H.; Deng, W.W.; Mullen, W.; Crozier, A. Distribution and biosynthesis of flavan-3-ols in *Camellia sinensis* seedlings and expression of genes encoding biosynthetic enzymes. *Phytochemistry* **2010**, *71*, 559–566. [CrossRef] [PubMed]

13. Liang, Y.R.; Liu, Z.S.; Xu, Y.R.; Hu, Y.L. A study on chemical composition of two special green teas (*Camellia sinensis*). *J. Sci. Food Agric.* **1990**, *53*, 541–548. [CrossRef]

14. Harbowy, M.E.; Balentine, D.A.; Davies, A.P.; Cai, Y. Tea chemistry. *Crit. Rev. Plant Sci.* **1997**, *16*, 415–480. [CrossRef]

15. Ashihara, H.; Kubota, H. Patterns of adenine metabolism and caffeine biosynthesis in different parts of tea seedlings. *Physiol. Plant.* **1986**, *68*, 275–281. [CrossRef]

16. Deng, W.-W.; Li, Y.; Ogita, S.; Ashihara, H. Fine control of caffeine biosynthesis in tissue cultures of *Camellia sinensis*. *Phytochem. Lett.* **2008**, *1*, 195–198. [CrossRef]

17. Ashihara, H.; Sano, H.; Crozier, A. Caffeine and related purine alkaloids: Biosynthesis, catabolism, function and genetic engineering. *Phytochemistry* **2008**, *69*, 841–856. [CrossRef] [PubMed]

18. Gutierrez, L.; van Wuytswinkel, O.; Castelain, M.; Bellini, C. Combined networks regulating seed maturation. *Trends Plant Sci.* **2007**, *12*, 294–300. [CrossRef] [PubMed]

19. Ge, P.; Ma, C.; Wang, S.; Gao, L.; Li, X.; Guo, G.; Ma, W.; Yan, Y. Comparative proteomic analysis of grain development in two spring wheat varieties under drought stress. *Anal. Bioanal. Chem.* **2012**, *402*, 1297–1313. [CrossRef] [PubMed]

20. McKay, D.L.; Blumberg, J.B. The role of tea in human health: An update. *J. Am. Coll. Nutr.* **2002**, *21*, 1–13. [CrossRef] [PubMed]

21. Cabrera, C.; Artacho, R.; Gimenez, R. Beneficial effects of green tea–a review. *J. Am. Coll. Nutr.* **2006**, *25*, 79–99. [CrossRef] [PubMed]

22. Kao, Y.-H.; Chang, H.-H.; Lee, M.-J.; Chen, C.-L. Tea, obesity, and diabetes. *Mol. Nutr. Food Res.* **2006**, *50*, 188–210. [CrossRef] [PubMed]

23. Boon, N. Health potential for functional green teas? *Int. J. Vitam. Nutr. Res.* **2008**, *78*, 275–281. [CrossRef] [PubMed]

24. Beecher, G.R.; Warden, B.A.; Merken, H. Analysis of tea polyphenols. *Proc. Soc. Exp. Biol. Med.* **1999**, *220*, 267–270. [CrossRef] [PubMed]

25. Balentine, D.A.; Wiseman, S.A.; Bouwens, L.C.M. The chemistry of tea flavonoids. *Crit. Rev. Food Sci. Nutr.* **1997**, *37*, 693–704. [CrossRef] [PubMed]

26. Lin, G.-Z.; Lian, Y.-J.; Ryu, J.-H.; Sung, M.-K.; Park, J.-S.; Park, H.-J.; Park, B.K.; Shin, J.-S.; Lee, M.-S.; Cheon, C.-I. Expression and purification of His-tagged flavonol synthase of *Camellia sinensis* from *Escherichia coli*. *Protein Expr. Purif.* **2007**, *55*, 287–292. [CrossRef] [PubMed]

27. Mahajan, M.; Ahuja, P.S.; Yadav, S.K. Post-transcriptional silencing of flavonol synthase mRNA in tobacco leads to fruits with arrested seed set. *PLoS ONE* **2011**, *6*, e28315. [CrossRef] [PubMed]

28. Mahajan, M.; Joshi, R.; Gulati, A.; Yadav, S.K. Increase in flavan-3-ols by silencing flavonol synthase mRNA affects the transcript expression and activity levels of antioxidant enzymes in tobacco. *Plant Biol.* **2012**, *14*, 725–733. [CrossRef] [PubMed]

29. Xiong, L.; Li, J.; Li, Y.; Yuan, L.; Liu, S.; Huang, J.A.; Liu, Z. Dynamic changes in catechin levels and catechin biosynthesis-related gene expression in albino tea plants (*Camellia sinensis* L.). *Plant Physiol. Biochem.* **2013**, *71*, 132–143. [CrossRef] [PubMed]

30. Wang, X.; He, X.; Lin, J.; Shao, H.; Chang, Z.; Dixon, R.A. Crystal structure of isoflavone reductase from alfalfa (*Medicago sativa* L.). *J. Mol. Biol.* **2006**, *358*, 1341–1352. [CrossRef] [PubMed]

31. Camm, E.L.; Towers, G.H.N. Phenylalanine ammonia lyase. *Phytochemistry* **1973**, *12*, 961–973. [CrossRef]

32. Fritz, R.R.; Hodgins, D.S.; Abell, C.W. Phenylalanine ammonia-lyase. Induction and purification from yeast and clearance in mammals. *J. Biol. Chem.* **1976**, *251*, 4646–4650. [PubMed]

33. Tanaka, Y.; Matsuoka, M.; Yamanoto, N.; Ohashi, Y.; Kano-Murakami, Y.; Ozeki, Y. Structure and characterization of a cDNA clone for phenylalanine ammonia-lyase from cut-injured roots of sweet potato. *Plant Physiol.* **1989**, *90*, 1403–1407. [CrossRef] [PubMed]

34. Bate, N.J.; Orr, J.; Ni, W.; Meromi, A.; Nadler-Hassar, T.; Doerner, P.W.; Dixon, R.A.; Lamb, C.J.; Elkind, Y. Quantitative relationship between phenylalanine ammonia-lyase levels and phenylpropanoid accumulation in transgenic tobacco identifies a rate-determining step in natural product synthesis. *Proc. Natl. Acad. Sci. USA* **1994**, *91*, 7608–7612. [CrossRef] [PubMed]

35. Singh, K.; Kumar, S.; Rani, A.; Gulati, A.; Ahuja, P.S. *Phenylalanine ammonia-lyase* (PAL) and *cinnamate 4-hydroxylase* (C4H) and catechins (flavan-3-ols) accumulation in tea. *Funct. Integr. Genomics* **2009**, *9*, 125–134. [CrossRef] [PubMed]

36. Whetten, R.W.; MacKay, J.J.; Sederoff, R.R. Recent advances in understanding lignin biosynthesis. *Annu. Rev. Plant Biol.* **1998**, *49*, 585–609. [CrossRef] [PubMed]

37. Hoffmann, L.; Maury, S.; Martz, F.; Geoffroy, P.; Legrand, M. Purification, cloning, and properties of an acyltransferase controlling shikimate and quinate ester intermediates in phenylpropanoid metabolism. *J. Biol. Chem.* **2003**, *278*, 95–103. [CrossRef] [PubMed]

38. Hoffmann, L.; Besseau, S.; Geoffroy, P.; Ritzenthaler, C.; Meyer, D.; Lapierre, C.; Pollet, B.; Legrand, M. Silencing of hydroxycinnamoyl-coenzyme a shikimate/quinate hydroxycinnamoyltransferase affects phenylpropanoid biosynthesis. *Plant Cell* **2004**, *16*, 1446–1465. [CrossRef] [PubMed]

39. Besseau, S.; Hoffmann, L.; Geoffroy, P.; Lapierre, C.; Pollet, B.; Legrand, M. Flavonoid accumulation in *arabidopsis* repressed in lignin synthesis affects auxin transport and plant growth. *Plant Cell* **2007**, *19*, 148–162. [CrossRef] [PubMed]

40. Maayan, I.; Shaya, F.; Ratner, K.; Mani, Y.; Lavee, S.; Avidan, B.; Shahak, Y.; Ostersetzer-Biran, O. Photosynthetic activity during olive (*Olea europaea*) leaf development correlates with plastid biogenesis and Rubisco levels. *Physiol. Plant.* **2008**, *134*, 547–558. [CrossRef] [PubMed]

41. Makino, A.; Mae, T.; Ohira, K. Photosynthesis and ribulose 1,5-bisphosphate carboxylase in rice leaves: Changes in photosynthesis and enzymes involved in carbon assimilation from leaf development through senescence. *Plant Physiol.* **1983**, *73*, 1002–1007. [CrossRef] [PubMed]

42. Kennedy, R.; Johnson, D. Changes in photosynthetic characteristics during leaf development in apple. *Photosynth. Res.* **1981**, *2*, 213–223. [CrossRef] [PubMed]

43. Roper, T.R.; Kennedy, R.A. Photosynthetic characteristics during leaf development in bing sweet cherry. *J. Am. Soc. Hortic. Sci.* **1986**, *111*, 938–941.

44. Tu, L.; Lin, Y.; Yang, Z.; Huang, Z.; Zhao, W.; Sun, P. The study of the correlation between the net photosynthetic rate and catechins components of tea plant leaves. *Chin. Agric. Sci. Bull.* **2012**, *28*, 227–233. (In Chinese)

45. Ghasemzadeh, A.; Jaafar, H.Z.; Rahmat, A. Synthesis of phenolics and flavonoids in ginger (*Zingiber officinale* roscoe) and their effects on photosynthesis rate. *Int. J. Mol. Sci.* **2010**, *11*, 4539–4555. [CrossRef] [PubMed]

46. Perl-Treves, R.; Galun, E. The tomato Cu, Zn superoxide dismutase genes are developmentally regulated and respond to light and stress. *Plant Mol. Biol.* **1991**, *17*, 745–760. [CrossRef] [PubMed]

47. Jiang, C.-D.; Li, P.-M.; Gao, H.-Y.; Zou, Q.; Jiang, G.-M.; Li, L.-H. Enhanced photoprotection at the early stages of leaf expansion in field-grown soybean plants. *Plant Sci.* **2005**, *168*, 911–919. [CrossRef]

48. Rodriguez, A.A.; Grunberg, K.A.; Taleisnik, E.L. Reactive oxygen species in the elongation zone of maize leaves are necessary for leaf extension. *Plant Physiol.* **2002**, *129*, 1627–1632. [CrossRef] [PubMed]

49. Turkmen, N.; Sari, F.; Velioglu, Y.S. Effects of extraction solvents on concentration and antioxidant activity of black and black mate tea polyphenols determined by ferrous tartrate and folin–ciocalteu methods. *Food Chem.* **2006**, *99*, 835–841. [CrossRef]

50. Meda, A.; Lamien, C.E.; Romito, M.; Millogo, J.; Nacoulma, O.G. Determination of the total phenolic, flavonoid and proline contents in Burkina Fasan honey, as well as their radical scavenging activity. *Food Chem.* **2005**, *91*, 571–577. [CrossRef]

51. Wang, Z.; Tong, X.; Zhu, S. Quantitative analysis of catechins with HPLC. *J. Tea Sci.* **1991**, *11*, 93–99. (In Chinese)

52. Li, Y.; Li, J.; Gong, X.; Liu, Z. Simultaneous determination of eight catechins, three purine alkaloids and gallic acid in tea by high-performance liquid chromatography. *Food Sci.* **2011**, *32*, 214–217.

53. Tatusov, R.L.; Fedorova, N.D.; Jackson, J.D.; Jacobs, A.R.; Kiryutin, B.; Koonin, E.V.; Krylov, D.M.; Mazumder, R.; Mekhedov, S.L.; Nikolskaya, A.N.; *et al.* The COG database: An updated version includes eukaryotes. *BMC Bioinform.* **2003**, *4*, 41. [CrossRef] [PubMed]

54. Kanehisa, M.; Goto, S.; Sato, Y.; Furumichi, M.; Tanabe, M. Kegg for integration and interpretation of large-scale molecular data sets. *Nucleic Acids Res.* **2012**, *40*, D109–D114. [CrossRef] [PubMed]
55. Livak, K.; Schmittgen, T. Analysis of relative gene expression data using real-time quantitative PCR and the $2^{-\Delta\Delta Ct}$ method. *Methods* **2001**, *25*, 402–408. [CrossRef] [PubMed]

International Journal of
Molecular Sciences

Article

Quantitative Proteomics Analysis of Herbaceous Peony in Response to Paclobutrazol Inhibition of Lateral Branching

Daqiu Zhao, Saijie Gong, Zhaojun Hao, Jiasong Meng and Jun Tao *

Jiangsu Key Laboratory of Crop Genetics and Physiology, College of Horticulture and Plant Protection, Yangzhou University, Yangzhou 225009, China; dqzhao@yzu.edu.cn (D.Z.); gong.sai.jie@163.com (S.G.); haozhaojunko@126.com (Z.H.); jsmeng@yzu.edu.cn (J.M.)

* Correspondence: taojun@yzu.edu.cn; Tel.: +86-514-8799-7219; Fax: +86-514-8734-7537

Academic Editor: Setsuko Komatsu
Received: 25 June 2015; Accepted: 5 October 2015; Published: 14 October 2015

Abstract: Herbaceous peony (*Paeonia lactiflora* Pall.) is an emerging high-grade cut flower worldwide, which is usually used in wedding bouquets and known as the "wedding flower". However, abundant lateral branches appear frequently in some excellent cultivars, and a lack of a method to remove *Paeonia lactiflora* lateral branches other than inefficient artificial methods is an obstacle for improving the quality of its cut flowers. In this study, paclobutrazol (PBZ) application was found to inhibit the growth of lateral branches in *Paeonia lactiflora* for the first time, including 96.82% decreased lateral bud number per branch, 77.79% and 42.31% decreased length and diameter of lateral branches, respectively, declined cell wall materials and changed microstructures. Subsequently, isobaric tag for relative and absolute quantitation (iTRAQ) technology was used for quantitative proteomics analysis of lateral branches under PBZ application and control. The results indicated that 178 differentially expressed proteins (DEPs) successfully obtained, 98 DEPs were up-regulated and 80 DEPs were down-regulated. Thereafter, 34 candidate DEPs associated with the inhibited growth of lateral branches were screened according to their function and classification. These PBZ-stress responsive candidate DEPs were involved in eight biological processes, which played a very important role in the growth and development of lateral branches together with the response to PBZ stress. These results provide a better understanding of the molecular theoretical basis for removing *Paeonia lactiflora* lateral branches using PBZ application.

Keywords: herbaceous peony; paclobutrazol application; proteomics analysis; iTRAQ

1. Introduction

Herbaceous peony (*Paeonia lactiflora* Pall. (*P. lactiflora*)) is an emerging high-grade cut flower worldwide, which is usually used in wedding bouquets and known as the "wedding flower" [1]. As an ideal cut flower, it not only requires excellent flower shape, flower color, flower fragrance, inflorescence stem, leaf and the performance of vase, but also possesses the property of few or no lateral branches. However, the more lateral branches have hampered the cutting flower production of Chinese *P. lactiflora* cultivars. These lateral branches not only consume plant nutrition but also affect the growth and development of the terminal buds, thereby further worsening the quality of cut flowers. Therefore, the lateral branches should be removed as soon as possible in the process of *P. lactiflora* cut flower production, which plays an important role in enhancing the cut flower quality of *P. lactiflora*, such as prolonging the full opening stage of flowers as well as increasing the flower diameters, the thickening rate, and elongation rate of inflorescence stems [2]. However, only artificial methods have been used to remove *P. lactiflora* lateral branches until now, which are inefficient, require a lot of labor and time, and have a negative impact on the large-scale production. Therefore, exploring a method to inhibit the growth of *P. lactiflora* lateral branches is of considerable significance.

Paclobutrazol (PBZ), a triazole compound, is widely used as a growth retardant for controlling vegetative growth in a wide range of horticultural plants [3–6]. In *P. lactiflora*, Wang *et al.* [7] has also studied the effect of PBZ on the growth of five cultivars, their plant heights and crown breadths were reduced, the plant stem diameters were increased, and 100 mg/L concentration PBZ had the best integrated effect. In addition to the above-mentioned effects, we also found that PBZ could significantly inhibit the growth of *P. lactiflora* lateral branches and reduce lateral buds to blossoming out, which was rarely seen in the existing literature. This finding would greatly save on the labor and time needed for removing *P. lactiflora* lateral branches, and reduce the cost and difficulty of the production and management.

Until now, it has been generally believed that the formation of plant lateral branches involves two processes: one is the initiation of the lateral buds, and the other is the growth and development of the lateral buds. After the formation of lateral bud primordiums, they begin with the elongation growth to form lateral buds, and then continue to develop into the lateral branches [8]. Meanwhile, the growth and development of lateral branches is believed to be linked to plant hormones, including auxin (IAA), cytokinin (CK) and strigolactone (SL), and a large number of molecular biology studies have been performed around them [8]. For example, in the biosynthetic pathway of strigolactone, a lot of genes including *dwarf27* (*D27*) [9], *D14* [10], *teosinte branched1* (*TB1*) [11], *branched1* (*BRC1*) [12,13] and *fine culm1* (*FC1*) [14] have been identified as associated with the growth and development of lateral branches. However, the formation and development of plants' characters is often controlled by numerous genes; thus, it is very difficult to fully clarify the inherent mechanism of its development and regulation only from the perspective of hormones. As a powerful technique to perform quantitative proteome analysis, isobaric tag for relative and absolute quantitation (iTRAQ) can shed light on the inherent mechanism of the growth and development of lateral branches [15]. Compared with the two-dimensional-electrophoresis-based approaches, iTRAQ has many advantages including high identification rate of proteins, accurate quantification of different proteins, and high reproducibility [16,17]. This technique has been widely applied in a number of studies, such as fiber differentiation and initiation of cotton [18], and wheat grain development [15]. However, the application of iTRAQ in the growth and development of lateral branches is little known. In this study, to illustrate the molecular metabolism of PBZ inhibiting the growth of *P. lactiflora* lateral branches, proteomics analysis of the lateral branches (except leaves and buds) under control and PBZ application was performed using iTRAQ. A greater understanding of this information can provide a theoretical basis to breed ideal *P. lactiflora* cut flower varieties without lateral branches.

2. Results

2.1. Morphological Indices

The effect of PBZ on the morphology of *P. lactiflora* lateral branches was first investigated in this study (Figure 1). In the full-bloom stage, there were 3.49 lateral buds in a single branch of control, and after PBZ application, the lateral bud number per branch was significantly decreased by 96.82%, the length and diameter of lateral branches were also significantly lower than those of the control with 77.79% and 42.31%, respectively.

Control   PBZ

Figure 1. Effect of PBZ application on morphology of *P. lactiflora* lateral branches: (**A**) Photographs of *P. lactiflora* lateral branches; (**B**) statistics of morphological indices in *P. lactiflora* lateral branches. Three biological replications with each 100 plants were used to perform the statistical analysis. Red circles indicate *P. lactiflora* lateral branches under PBZ application. Different letters indicate highly significant differences ($p < 0.01$). PBZ: paclobutrazol.

2.2. Microstructures and Cell Wall Materials

The microstructures of lateral branches are intuitively shown in Figure 2A–D. Compared with the control, the deformation of sclerenchyma cells was observed and its cell walls had not been thickened under PBZ application. By micrometer measurement, the average cell area and the thickness of the sclerenchyma cell walls in a cross section of the lateral branches under PBZ application were significantly decreased by 19.54% and 60.86%, respectively (Figure 2E).

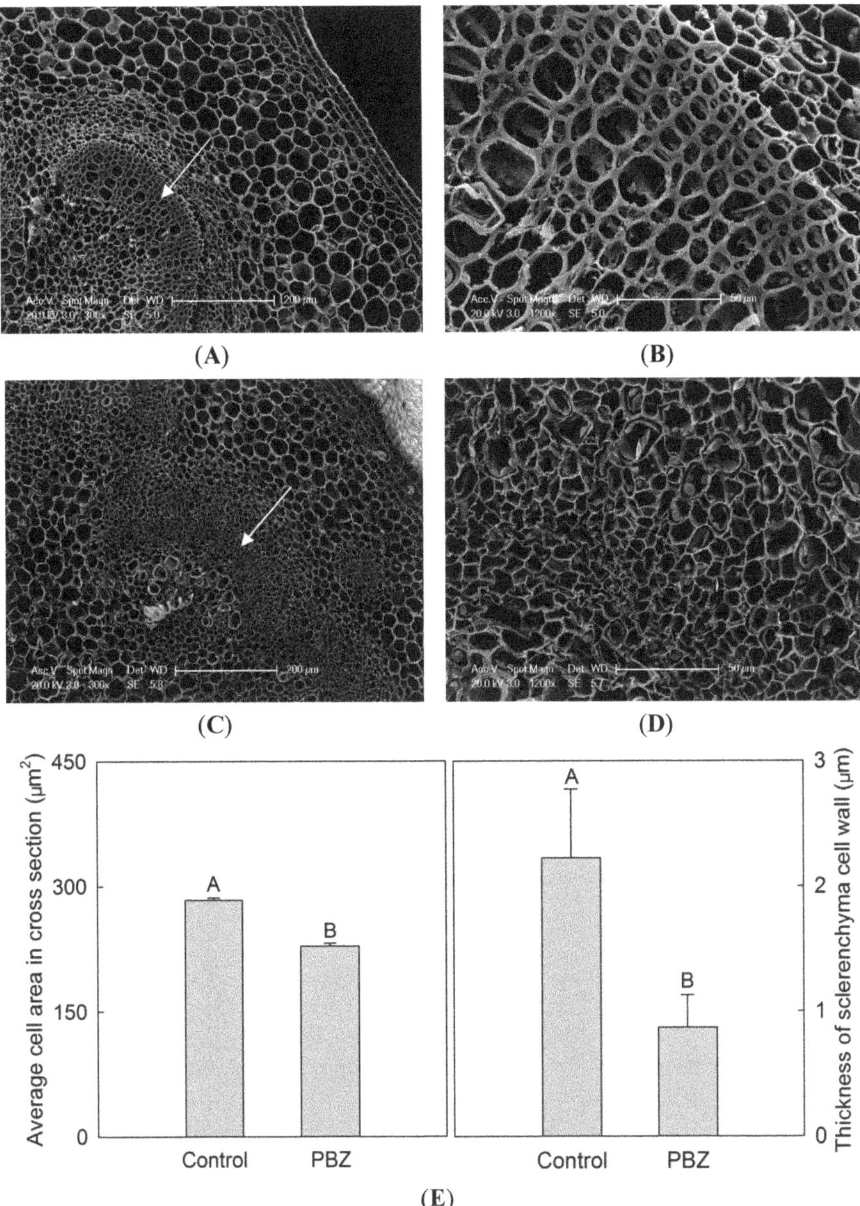

Figure 2. Effect of PBZ application on microstructure of *P. lactiflora* lateral branches: (**A**,**C**) Photographs of *P. lactiflora* lateral branches with a magnification of 200 times (Scale bar = 200 μm); (**B**,**D**) Photographs of partial enlargement in (**A**,**C**) marked by the arrow. Scale bar = 50 μm; and (**E**) Statistics of microstructure in *P. lactiflora* lateral branches. Different letters indicate highly significant differences (*p* < 0.01).

Cellulose and lignin were the major cell wall components, their contents under PBZ application were all lower than those of the control with 13.51% and 42.18%, respectively, and their differences all reached highly significant levels. In addition, the pectin content under PBZ application was also decreased by 3.36%, but the difference was not significant (Figure 3).

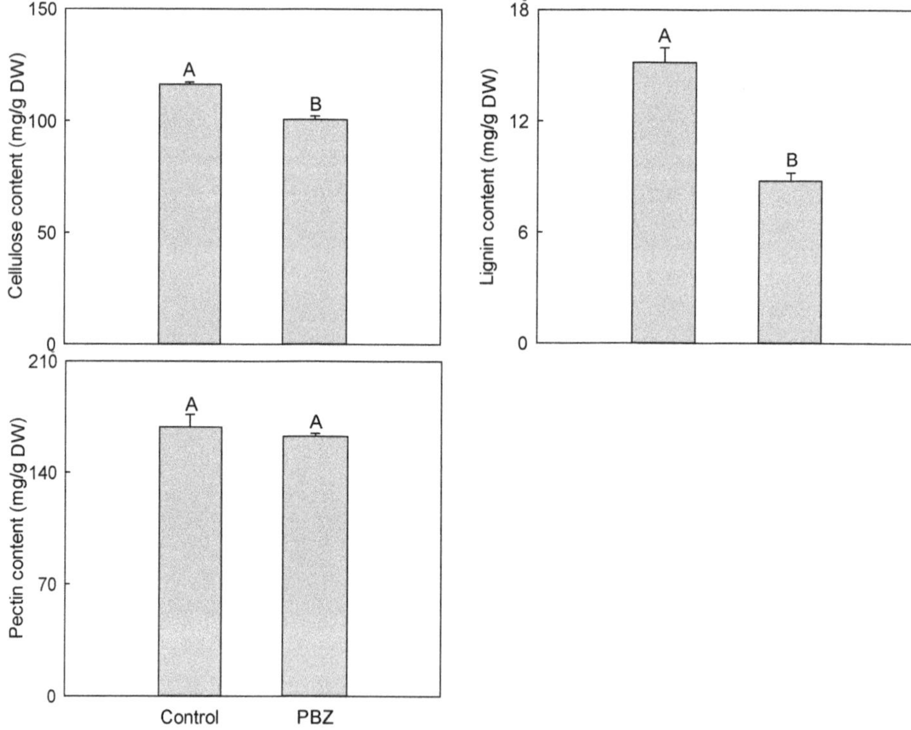

Figure 3. Effect of PBZ application on cell wall materials of *P. lactiflora* lateral branches. Different letters indicate highly significant differences ($p < 0.01$).

2.3. Protein Identification

In order to deeply explore the mechanism of PBZ inhibiting the growth of *P. lactiflora* lateral branches, proteomics analysis was performed by iTRAQ. Firstly, the total proteins of *P. lactiflora* lateral branches under control and PBZ application were extracted. Subsequently, mass spectrometry (MS) was used to identify these proteins according to a Triple TOF5600, and the error distribution of the peptide spectra match quality was adopted to assess its identified quality. Errors over 0.05 Da were defined as false positives. The error distribution of the proteomic identification in *P. lactiflora* lateral branches is shown in Figure S1, which was mainly concentrated between -10 and 10 ppm, indicating that the identified result was relatively good. Furthermore, the basic information of the protein MS according to iTRAQ was analyzed. Overall, 4547 unique peptides and corresponding to 3733 proteins were identified (Figure 4A, Table S1). The distribution of identified protein sequences coverage showed that the number of proteins was basically presented a declining trend with increased coverage (Figure 4B). However, the protein mass distribution was not uniform, and most of the protein molecular weights were concentrated on 20–70 kDa (Figure 4C). In addition, the protein number was basically decreased with the increased number of matching peptides, and the peptide number in most of identified proteins was within 10 (Figure 4D). In addition, the functional annotation and classification of these identified proteins was showed in Figure S2 and Figure S3, Table S2 and Table S3.

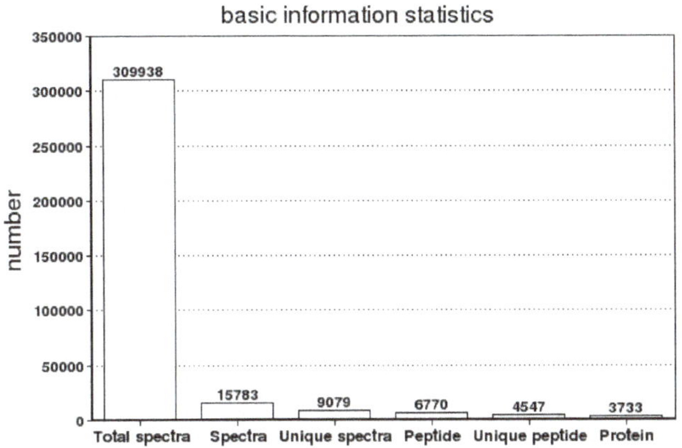

(**A**)

Distribution of Protein's Sequences Coverage

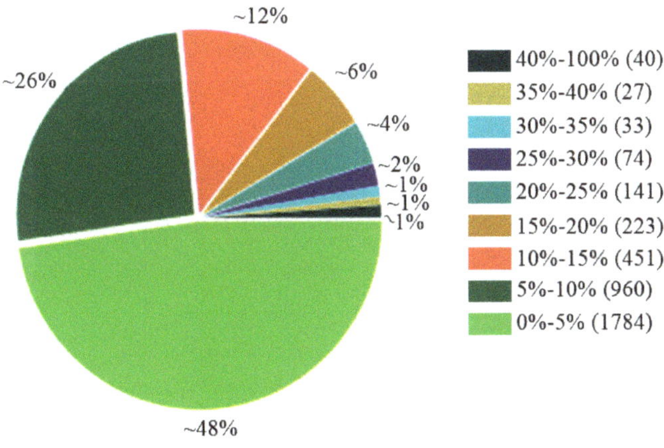

(**B**)

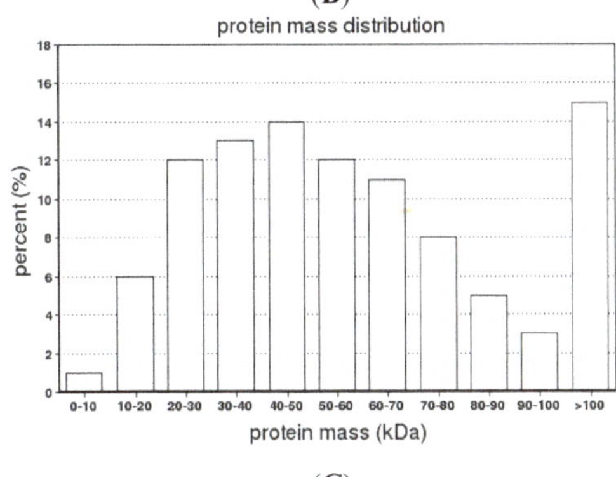

(**C**)

Figure 4. *Cont.*

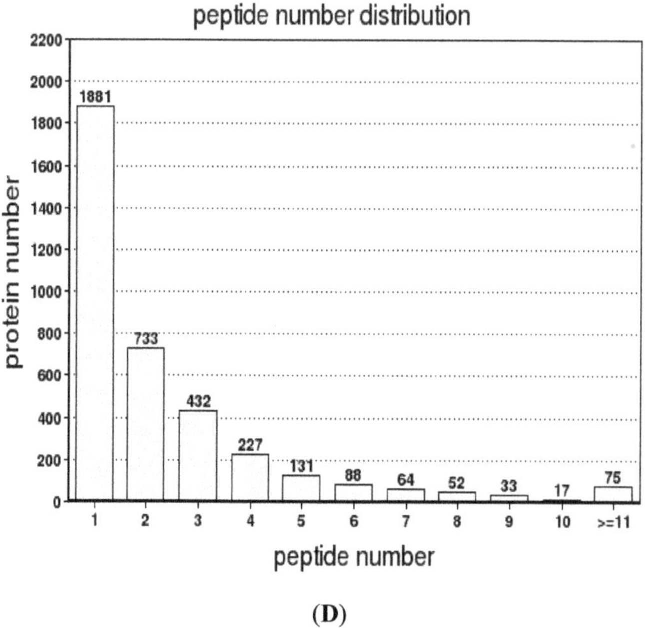

(D)

Figure 4. The basic information of protein mass spectrometry in *P. lactiflora* lateral branches by iTRAQ: (**A**) basic information statistics; (**B**) distribution of protein's sequences coverage; (**C**) protein mass distribution; and (**D**) peptide number distribution. iTRAQ: isobaric tag for relative and absolute quantitation.

2.4. Comparative Analysis of DEPs between Control and PBZ Application

To identify candidate differentially expressed proteins (DEPs) controlling the growth of *P. lactiflora* lateral branches, the distribution of the protein abundance was analyzed, and a 1.2-fold cut-off was used to implicate significant changes in the abundance of DEPs (Figure 5, Table S4 and Table S5). Of the 3733 identified proteins, 178 showed more than 1.2-fold changes with p-value < 0.05 in the protein expression level. Among these 178 proteins, 98 DEPs were up-regulated (red scatters) and 80 DEPs were down-regulated (green scatters). Furthermore, the pathway analysis in Kyoto Encyclopedia of Genes and Genomes database (KEGG) was illustrated to help us understand this result, and 165 DEPs (92.70%) were assigned to 74 KEGG pathways (Table S6). According to the p-value $\leq$ 0.05 and DEPs $\geq$ 5, only nine KEGG pathways were obtained involving large amounts of substance metabolism processes. Among them, metabolic pathways contained the largest number of DEPs (90 DEPs, 54.55%, ko01100), with the biosynthesis of secondary metabolites (55 DEPs, 33.33%, ko01110), phenylpropanoid biosynthesis (14 DEPs, 8.48%, ko00940), ribosome (14 DEPs, 8.48%, ko03010), carbon fixation in photosynthetic organisms (13 DEPs, 7.88%, ko00710) and glyoxylate and dicarboxylate metabolism (12 DEPs, 7.27%, ko00630) following behind, whereas cysteine and methionine metabolism (7 DEPs, 4.24%, ko00270), alanine, aspartate and glutamate metabolism (7 DEPs, 4.24%, ko00250) as well as flavonoid biosynthesis (5 DEPs, 3.03%, ko00941) demonstrated the fewest DEPs.

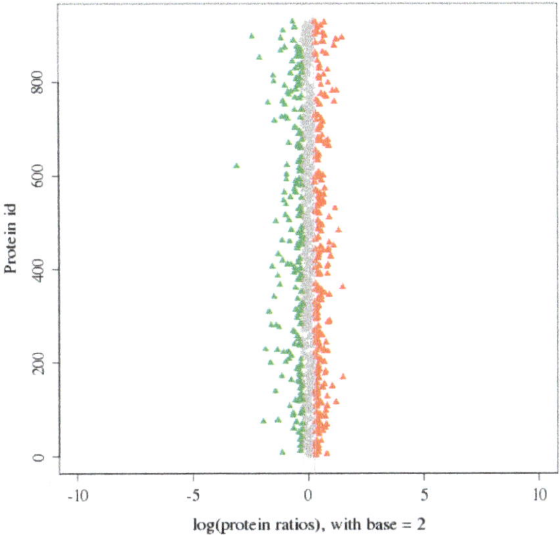

Figure 5. The distribution of protein abundance quantified in *P. lactiflora* lateral branches. The protein would be marked with red or green if protein ratios >1.2; red scatters indicate up-regulated DEPs and green scatters indicate down-regulated DEPs.

2.5. Screening of Candidate DEPs Associated with the Growth Inhibition of Lateral Branches

According to the function and classification of the 178 identified DEPs, 34 candidate DEPs associated with the growth inhibition of *P. lactiflora* lateral branches were obtained, which could be classified into eight categories (Table S7). Among them, cell wall metabolism belonging to biosynthesis of secondary metabolites and phenylpropanoid biosynthesis demonstrated the largest number of DEPs (13 DEPs), followed by cell structure-related proteins (5 DEPs), defense and stress response (4 DEPs), hormone-related proteins (4 DEPs), energy metabolism (3 DEPs), carbohydrate transport and metabolism (2 DEPs) together with protein transport and metabolism (2 DEPs), while lipid transport and metabolism (1 DEPs) contained the fewest DEPs. As expected, the expression levels of DEPs in cell wall metabolism, cell structure-related proteins and energy metabolism all presented downtrends under PBZ application, whereas the opposite trends were observed in defense and stress response, carbohydrate transport and metabolism, lipid transport and metabolism together with protein transport and metabolism in comparison with the control. Of the 34 candidate DEPs, 1-aminocyclopropane-1-carboxylate oxidase (ACO), phenylalanine ammonia-lyase (PAL), and cinnamate 4-hydroxylase (C4H) were blasted to *P. lactiflora*, which had been submitted to NCBI by us, and auxin-repressed protein (ARP) was blasted to tree peony, which was the closest living relative of *P. lactiflora*. Therefore, the gene sequences corresponding to these 4 DEPs were amplified and their changes in the *P. lactiflora* lateral branches under control and PBZ application were verified. Gene expression patterns displayed that *PAL*, *C4H* and *ARP* were expressed lowly under PBZ application, whereas *ACO* presented the opposite trend (Figure 6), which was not consistent with the changes of the corresponding DEPs expression levels due to translational or post-translational regulation [19].

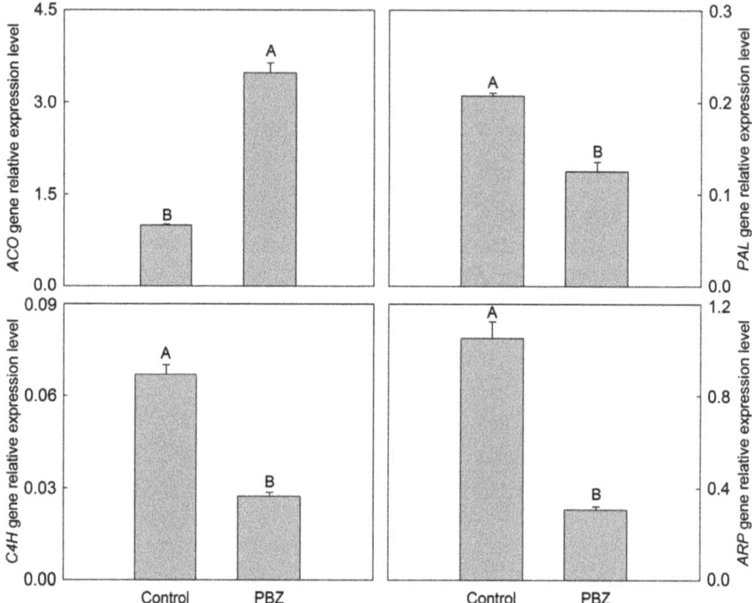

Figure 6. mRNA expression patterns of four representative DEPs in *P. lactiflora* lateral branches. Different letters indicate highly significant differences ($p < 0.01$).

3. Discussion

iTRAQ technology is a powerful means of determining relative protein levels with high repeatability, and it is one of the most sensitive markers in the current studies of proteomics [20,21]. This technology has been widely used in life science research and made great achievements [15,18,19,22], but little information is available about the protein study of *P. lactiflora* using iTRAQ. In the present study, we found PBZ inhibited the growth of *P. lactiflora* lateral branches, including their length, diameter and the number of lateral buds. In order to further clarify the mechanism of PBZ inhibiting the lateral branch growth, iTRAQ technology was firstly used to compare the protein expression differences of *P. lactiflora* lateral branches under control and PBZ application. According to searching of the eudicotyledons database, 3733 proteins were obtained, and 178 DEPs were identified, including 98 up-regulated DEPs and 80 down-regulated DEPs. Thereafter, 34 candidate DEPs associated with the growth inhibition of *P. lactiflora* lateral branches were obtained according to their function and classification. To better understand the molecular mechanism of PBZ inhibiting the growth of lateral branches, these 34 candidate DEPs were classified into eight categories, their functions and regulation in inhibiting the growth of lateral branches were discussed in the below.

3.1. Defense and Stress Response

In this study, four identified DEPs were classified into this category, including lipoxygenase (LOX), peroxiredoxin-2E (PRX2E), catalase (CAT) and heat-shock protein (HSP). Under adversity stress, a plant can perceive the stress signal, and reactive oxygen species (ROS) are produced due to lipid peroxidation catalyzed by LOX, as well as a large amount of H_2O_2 is accumulated due to the enhanced ROS metabolism. Moreover, both ROS and H_2O_2 impair the cells and accelerate cell aging and disintegration [23]. ROS can be scavenged by PRX2E, and H_2O_2 can be reduced to H_2O by CAT [24,25]. In *P. lactiflora* lateral branches, the expression levels of LOX, PRX2E and CAT were all increased under PBZ application, suggesting that *P. lactiflora* suffered from PBZ stress and both PRX2E and CAT were important protective enzymes in plants.

HSP is a type of stress protein in plants induced by high temperature, heavy metal ions and other adverse environmental factors. It is believed that HSP plays a key part in reducing the damage caused by adversity stress [26]. In this study, the expression level of HSP was also enhanced to protect lateral branches under PBZ application, which was consistent with the reports in *Primula* [27].

3.2. Hormone-Related Proteins

The growth and development of lateral branches involves a variety of changes in hormone levels. In this category, ACO, ARP, isopentenyl diphosphate isomerase (IPI) and geranylgeranyl pyrophosphate synthase (GGPS) were identified. ACO catalyzes the synthesis of ethylene, and *ACO* is widely accepted as the rate-limiting gene [28]. When *P. lactiflora* was treated with PBZ, the expression level of ACO was up-regulated, which could increase the content of ethylene inducing the senescence and abscission of lateral branches and finally inhibiting their growth.

The physiological role of PBZ was primarily to accelerate the decomposition of IAA and hinder the biosynthesis of gibberellin (GA), which slowed and inhibited the vegetative growth of plants [29]. Similar results were found in this study. ARP, an auxin-repressed protein, was up-regulated under PBZ application, which revealed that the content of IAA in *P. lactiflora* could be significantly reduced, which enabled the lateral buds to elongate normally, but stop their growth and maintain a dormant state. Similar results were reported in shoot elongation of black locus [30]. Moreover, the down-regulated expression levels of IPI and GGPS involved in the biosynthesis of GA could decrease GA synthesis, which also could not promote the growth of lateral branches.

3.3. Carbohydrate Transport and Metabolism

Aquaporins (AQPs) are a type of membrane proteins that mediate water transport through the membrane and play an important role in maintaining cell osmotic balance [31]. In this category, only two AQPs were identified, one belonged to the plasma membrane intrinsic proteins (PIPs) and the other could not be identified. Liu *et al.* [32] found that the *AQP1* from wintersweet was expressed most heavily in the leaves under high temperature stress. The study of Ahamed *et al.* [33] demonstrated that PIPs controlled rice root water uptake functions in a cold acclimation process. Similarly, the expression levels of our identified AQP and PIP1-2 were higher in *P. lactiflora* lateral branches under PBZ application than those of control, suggesting that *P. lactiflora* lateral branches controlled the expression of AQPs to restore normal growth, which provided water for resuming their growth.

3.4. Lipid Transport and Metabolism

Phospholipase D (PLD), one of the key enzymes of hydrolyzing phospholipids in cell membranes, plays a key part in maintaining the structure, function and stability of cell membranes [34]. Li *et al.* [35] found that UV-B irradiation induced the degradation of membrane lipids, and the degradation of membrane lipids in PLD δ-knockout plants was more severe than that in wild type plants. Under PBZ application, *P. lactiflora* also experienced stress; meantime, the up-regulated PLD in lateral branches could induce the hydrolysis of numerous phospholipids in cell membranes, which damaged the structure and function of the cell membrane, resulting in the inhibited growth of the lateral branches.

3.5. Protein Transport and Metabolism

Proteasome is responsible for the degradation of abnormal or damaged proteins [36]. In this category, two forms, a 20S proteasome subunit α type 7 (PSA7) as well as an ATP-dependent 26S proteasome regulatory subunit (PRS) were obtained, and they were expressed heavily under PBZ application, which was consistent with tea leaves under polyethylene glycol stress [37]. This meant that the increased proteasome activity maintained the growth of the lateral branched by degrading these proteins.

3.6. Energy Metabolism

Embden meyerhof generally existing in the organism can provide energy for the biological activity of an organism [29]. In this study, pyruvate kinase isozyme A (PKIA), phosphoglycerate kinase (PGK) and glyceraldehyde-3-phosphate dehydrogenase (GAPDH) in the embden meyerhof pathway were identified, which were the key enzymes in the process of triose oxidation. These proteins were lowly expressed in *P. lactiflora* lateral branches under PBZ application, which implied that *P. lactiflora* rarely provided energy for the growth of lateral branches.

3.7. Cell Structure-Related Proteins

The cell is the basic unit of organism structure and function. Under PBZ application, lateral branches were significantly affected including the average cell area and thickness of the sclerenchyma cell wall. Moreover, five cell structure-related proteins were identified, including xyloglucan endotransglucosylase (XT), isoflavone reductase (IFR), actin (ACTIN) and actin-3 (ACTIN3). XT catalyzes the transfer of xyloglucan, which has a key part in cell wall modification and the cell elongation process [38]. IFR is associated with the secondary cell wall thickening [39]. The expression levels of XT and IFR in *P. lactiflora* lateral branches under PBZ application were all decreased, which might be closely related to secondary cell wall thinning.

ACTIN plays many important roles in cell shape, cell movement, cell junction, cytoplasmic streaming and cell contraction [40,41]. Two ACTIN were obtained in this study, and they were all down-regulated under PBZ application. These results revealed that the cell shape and cell division were affected, which was consistent with cell shrinkage and a smaller average cell area in cross section (Figure 2).

3.8. Cell Wall Metabolism

Cell wall plays important roles in maintaining cell shape, controlling cell growth, material transportation and information transfer, defense and resistance. In addition, polysaccharides (cellulose, hemi-cellulose, pectin, *etc.*), protein, lignin and other materials are the most important chemical components in the cell wall [29]. In the present study, 13 proteins related to cell wall metabolism were identified. Among of them, bifunctional 3-dehydroquinate dehydratase/shikimate dehydrogenase (DHD/SHD), PAL, C4H, cinnamoyl CoA reductase (CCR), cinnamyl alcohol dehydrogenase (CAD) and cinnamyl alcohol dehydrogenase 2 (CAD2) controlled the synthesis of lignin [42], pectinesterase/pectinesterase inhibitor 35-like (PE/PEI35L) inhibited methyl esterification of cell wall pectin [43], and sucrose synthase 2 (SUS2) was an integral component of the cellulose synthesis machinery [44]. Moreover, UDP-glucose pyrophosphorylase (UGPase) catalyzed the reversible reaction of glucose-1-pyrophosphate and UTP to UDP-glucose and pyrophosphate, and UDPG as the glucose donor was involved in the anabolism of sugar, cellulose, hemi-cellulose and pectin [45]. Under PBZ application, these proteins were all down-regulated, which was consistent with the results of our physiological and biochemical measurements (Figure 3), revealing that PBZ application interfered with the synthesis and accumulation of the main cell wall components and significantly affected the function of the cell wall, eventually inhibited the growth of *P. lactiflora* lateral branches.

4. Experimental Section

4.1. Plant Materials

A *P. lactiflora* cultivar "Zifengyu" with lots of lateral branches was taken as a plant material, which was cultivated in the germplasm repository of Horticulture and Plant Protection College, Yangzhou University, Jiangsu Province, China (32°30′N, 119°25′E). When their buds were exposed to the ground in March, foliar-spraying PBZ with 100 mg/mL was performed once a week until the withering stage of the flowers, whereas the control was treated with deionized water. After counting the number of lateral buds that grew well and could blossom in the full-bloom stage, the length and diameter of the lateral branches under the control and PBZ application were measured. These lateral branches were partly used for microstructure observation which were fixed in 3% glutaraldehyde, and the others were used for physiological measurement and quantitative proteomics analysis, which were immediately frozen in liquid nitrogen and then stored at $-80\,^\circ$C until analysis. Three biological replications with each 100 plants were performed when *P. lactiflora* was treated with PBZ application and control. Furthermore, there were three biological replications in morphological indices, microstructures and cell wall materials. But there were 100 equally mixed samples for the protein identification.

4.2. Microstructures Observation

The fixed top lateral branches of *P. lactiflora* were firstly washed three times with 0.1 mol/L phosphate buffer, and then dehydrated using a gradient ethanol solution (30%, 50%, 70%, 85%, 95% and 100%, 15 min each), treated with the mixture of acetone:anhydrous alcohol (1:1, 2:1, 1:0, v/v) (15 min), the mixture of acetone:isoamyl acetate (1:1, 1:2, v/v) (10 min) and pure isoamyl acetate (30 min). After the critical point drying and spraying gold using ion sputtering equipment (EIKO IB-3, Ibaraki, Japan) for 5 min, the environmental scanning electron microscopy (Philips XL-30 ESEM, Amsterdam, The Netherlands) was used to observe the samples.

4.3. Cell Wall Materials Fractionation and Determination

The cell wall materials were fractioned according to the method of Rose *et al.* [46] with some modifications. Anthrone–H_2SO_4 colorimetry was used to determine the cellulose content [47], the lignin content was measured according to the report of Müsel *et al.* [48] and the measurement of pectin content referred to the reports of Blumenkrantz *et al.* [49] and Majumder *et al.* [50].

4.4. Protein Extraction and Quantification

Lateral branches of *P. lactiflora* were firstly ground into powder, and then their total proteins were extracted from the freeze-dried powder using 500 μL lysis buffer (40 mM Tris-HCl, 7 M Urea, 4% CHAPS, 2 M Thiourea, pH 8.5) including 10 mM DTT, 2 mM EDTA and 1 mM PMSF. Subsequently, the resuspended powder was treated with 200 W ultrasonic for 15 min and 4 $^\circ$C together with 25,000× g centrifugation for 20 min. Moreover, the mixture of supernatant and 5× volume of chilled acetone including 10% (v/v) TCA was performed, and then they were treated with $-20\,^\circ$C incubation for 2 h. After 4 $^\circ$C and 16,000× g centrifugation for 20 min, the precipitate was transferred to another 1.5 mL centrifuge tube washing with lysis buffer including 10 mM DTT, 2 mM EDTA and 1 mM PMSF. After treatment with 200 W ultrasonic for 15 min and 4 $^\circ$C together with 25,000 × g centrifugation for 20 min, in order to reduce the disulfide bonds in proteins and block the cysteines, 10 mM DTT and 55 mM IAM were separately added to the supernatant. Subsequently, the solution was treated with dark incubation for 45 min, and then the mixture of supernatant and 4× volume of chilled acetone was performed to precipitate proteins at $-20\,^\circ$C for 2 h. After 4 $^\circ$C and 25,000× g centrifugation for 20 min as well as 5 min air-drying, 200 μL 0.5 M TEAB was used to dissolve the precipitate which was treated with 200 W ultrasonic for 15 min. Finally, the supernatant was transferred to a new tube after 4 $^\circ$C and 25,000× g centrifugation for 20 min again. The Bradford method was used to quantify total protein and its integrity was assessed by 10% SDS-PAGE. The proteins in the supernatant were sent to the Beijing Genomic Institute (Shenzhen, China) and stored at $-80\,^\circ$C.

4.5. iTRAQ Labeling and Strong Cation Exchange (SCX) Fractionation

A sample of 100 μg total protein was treated with 37 $^\circ$C digestion for 4 h via trypsin (protein/trypsin = 20:1). The digestion using trypsin was repeated at 37 $^\circ$C for 8 h. Subsequently, vacuum centrifugation was used to dry the peptides which was reconstituted in 0.5 M TEAB and iTRAQ labeled by the manufacturer's protocol for 8-plex iTRAQ reagent (Applied Biosystems, Foster, CA, USA). Finally, vacuum centrifugation was used to dry the labeled peptide mixtures again.

Next, a LC-20AB high-performance liquid chromatograph (HPLC) system (Shimazu, Kyoto, Japan) with an Ultremex SCX column (4.6 mm × 250 mm) (Phenomenex Inc., Torrance, CA, USA) was used to fractionate the samples. Four milliliters of buffer A (10 mM KH_2PO_4 in 25% ACN, pH 2.6) were used to reconstitute the iTRAQ-labeled peptide mixtures, subsequently, SCX separation was performed at a flow rate of 1 mL/min using 5% elution buffer B (25 mM NaH_2PO_4, 1 M KCl in 25% ACN, pH 2.7) for 7 min, followed by a linear gradient of 5%–60% buffer B for 20 min, 60%–100% buffer B in 2 min, maintenance at 100% buffer B for 1 min, and finally backed to 5% buffer B for 10 min equilibration. The measurement of the absorbance at 214 nm was used to monitor the elution, and 20 fractions were obtained after screening, a Strata X C18 column (Phenomenex Inc.) was used to desalt them and then vacuum-dried.

4.6. LC-ESI-MS/MS Analysis Based on Triple TOF 5600

Each fraction was resuspended to approximately 0.5 µg/µL using buffer A (5% ACN, 0.1% FA) and centrifuged at $20,000 \times g$ for 10 min to remove the insoluble substances. A 5 µL volume of the supernatant of each fraction was separated by a LC-20AD nanoHPLC (Shimadzu, Japan), and the details were as follows: the samples were loaded onto a C18 trap column in 4 min at 8 µL/min, and then the samples were eluted onto a 10 cm analytical C18 column (inner diameter 75 µm) at 300 nL/min, separated and transferred to an MS system; the eluted system was run starting from 5% buffer B (95% ACN, 0.1% FA) for 5 min, followed by a 35 min linear gradient of 5%–35% buffer B, 35%–60% buffer B for 5 min, 60%–80% buffer B for 2 min, maintenance at 80% B for 2 min, and finally returned to 5% for 1 min and equilibration for 10 min.

A TripleTOF 5600 System (AB SCIEX, Concord, ON, Canada) with an ion spray voltage of 2.5 kV, a curtain gas of 30 psi, a nebulizer gas of 15 psi, and an interface heater temperature of 150 °C was used to acquire data. The MS was operated with an RP of greater than or equal to 30,000 FWHM for the TOF MS scans. For IDA, 250 ms was used to collect survey scans, and up to 30 product ion scans were acquired if exceeding a threshold of 120 counts/s and with a 2+ to 5+ charge-state. Total cycle time was 3.3 s, and Q2 transmission window was 100 Da for 100%. Four time bins with the 40 GHz multichannel TDC detector with a four-anode channel detection were summed for each scan at a pulser frequency value of 11 kHz. All precursor ions for collision-induced dissociation were happened at a sweeping collision energy setting of 35 ± 5 eV with the addition of iTRAQ adjust rolling collision energy. 1/2 of the peak width (15 s) was used for dynamic exclusion, and the precursor was refreshed off the exclusion list.

4.7. Database Search and Quantification

Proteome Discoverer 1.2 (PD 1.2, Thermo Fisher Scientific, San Jose, CA, USA) was used to convert raw data files (sample.wiff) into the mgf format files (sample. mgf), and then Mascot search engine (Matrix Science, London, UK; version 2.3.02) was used to searched against core eudicotyledons database (Available online: http://www.ncbi.nlm.nih.gov/protein/?term=txid91827 (Organism:exp)) including 910,008 sequences. A mass tolerance of 0.05 Da was accepted for intact peptide masses and 0.1 Da for fragmented ions to identify protein, allowing one missed cleavages in the trypsin digests. The considered potential variable modifications contained Gln- > pyro-Glu (N-term Q), Deamidated (NQ) and Oxidation (M), and the fixed modifications contained Carbamidomethyl (C), iTRAQ8plex (K) and iTRAQ8plex (N-term). +2 and +3 were set as the charge states of peptides. Specifically, an automatic decoy database search was conducted in Mascot by choosing the decoy checkbox where a random sequence of database was generated and tested for raw spectra together with the real database. In order to decrease the probability of false peptide identification, only the significance scores (≥ 20) at the 99% confidence interval peptides were considered as the identified protein. Moreover, each confident protein contained at least one unique peptide. In order to quantify protein, at least two independent peptides were included in a protein. The median ratio in Mascot was used to weigh and normalize the quantified protein ratios, and differentially expressed proteins were only defined as those with fold changes >1.2 and *p*-values < 0.05.

4.8. Bioinformatics and Annotations

Blast2GO program [51] was used to perform the functional annotations of proteins against the non-redundant protein database (NR; NCBI, Bethesda, MD, USA; Available online: http://www.ncbi.nlm.nih.gov), and these identified proteins were also classified and grouped by matching the COG database (NCBI, Bethesda, MD, USA; Available online: http://www.ncbi.nlm.nih.gov/COG/). Moreover, the GO enrichment and KEGG pathway enrichment analysis were conducted.

4.9. Quantitative RT-PCR (Q-PCR) for Identified Proteins

Gene expression pattern was detected by Q-PCR technology via a BIO-RAD CFX96™ Real-Time System (Bio-Rad, Hercules, CA, USA). CTAB extraction protocol was modified to extract total RNA [52], and then 1% agarose gel electrophoresis and a spectrophotometer (Eppendorf, Hamburg, Germany) were used to confirm the integrity of RNA. Next, 1 µg RNA from each sample was used as templates to synthesize the cDNA with PrimeScript® RT reagent Kit With gDNA Eraser (TaKaRa, Kyoto, Japan). In this study, the internal control was *P. lactiflora Actin* (JN105299) [53], and Table S8 listed all gene-specific primers. SYBR® Premix Ex Taq™ (TaKaRa) was used to conduct the Q-PCR, and the $2^{-\Delta\Delta Ct}$ comparative threshold cycle (C_t) method was calculated the levels of gene relative expression [54], meanwhile, the *PAL* expression level of control was used as the control. A Bio-Rad CFX Manager V1.6.541.1028 software (Bio-Rad, Hercules, CA, USA) was used to gather the C_t values of three triplicate reactions.

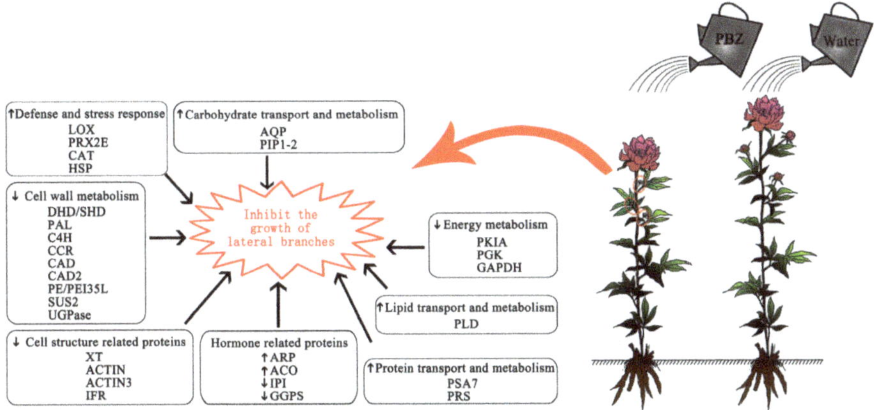

Figure 7. An overview of *P. lactiflora* lateral branches response to PBZ application. LOX: lipoxygenase; PRX2E: peroxiredoxin-2E; CAT: catalase; HSP: heat-shock protein; AQP: Aquaporin; PIP: plasma membrane intrinsic protein; DHD/SHD: dehydratase/shikimate dehydrogenase; PAL: phenylalanine ammonia-lyase; C4H: cinnamate 4-hydroxylase; CCR: cinnamoyl CoA reductase; CAD: cinnamyl alcohol dehydrogenase; PE/PEI35L: pectinesterase/pectinesterase inhibitor 35-like; SUS2: sucrose synthase 2; UGPase: UDP-glucose; PKIA: pyruvate kinase isozyme A; PGK: phosphoglycerate kinase; GAPDH: glyceraldehyde-3-phosphate dehydrogenase; PLD: Phospholipase D; XT: xyloglucan endotransglucosylase; ACTIN3: actin-3; IFR: isoflavone reductase; ARP: auxin-repressed protein; ACO: 1-aminocyclopropane-1-carboxylate oxidase; IPI: isopentenyl diphosphate isomerase; GGPS: geranylgeranyl pyrophosphate synthase; PSA7: proteasome subunit α type 7; PRS: proteasome regulatory subunit.

5. Conclusions

In conclusion, differential proteomic analysis of *P. lactiflora* lateral branches under PBZ application and control displayed significant changes in the biological processes. Figure 7 summarizes the major findings. When PBZ was applied in *P. lactiflora*, defense and stress response began to reduce the damage quickly, but the excessive accumulation of abnormal proteins made lateral branches degrade the damaged proteins by increasing the PSA7 and PRS expression levels in protein transport and metabolism, and water was transported to lateral branches by enhancing the AQP and PIP1-2 expression levels in carbohydrate transport and metabolism, which made the lateral branches restore normal growth. However, the increased PLD expression level in the lipid transport and metabolism and the decreased cell structure-related proteins (XT, ACTIN, ACTIN3 and IFR) expression levels caused the cell membrane and other cellular structures to be destroyed. Furthermore, a significant reduction of energy in the embden meyerhof pathway (decreased PKIA PGK and GAPDH expression levels) was produced, and the synthesis of the cell wall materials was blocked (decreased DHD/SHD, PAL, C4H, CCR, CAD, CAD2, PE/PEI35L, SUS2 and UGPase expression levels). In addition, the contents of IAA (increased ARP expression level) and GA (decreased IPI and GGPS expression levels) promoting the growth of lateral branches were reduced, and ethylene (increased ACO level) causing the lateral branches to age was produced in abundance, eventually inhibiting the growth of lateral branches. These results provide a theoretical basis for removing *P. lactiflora* lateral branches using PBZ.

Supplementary Materials: Supplementary materials can be found at http://www.mdpi.com/1422-0067/16/10/24332/s1.

Acknowledgments: This work was supported by the Agricultural Science & Technology Independent Innovation Fund of Jiangsu Province (CX[14]2023), the Scientific Research Foundation for high level talents of Yangzhou University (137010420) and the Priority Academic Program Development from Jiangsu Government.

Author Contributions: Daqiu Zhao and Jun Tao conceived the study and designed the experiments; Daqiu Zhao, Saijie Gong, Zhaojun Hao and Jiasong Meng performed the experiments; Daqiu Zhao analyzed the data and wrote the manuscript with suggestions by Jun Tao. All authors read and approved the final manuscript.

Conflicts of Interest: The authors declare no conflict of interest.

References

1. Stevens, S.; Stevens, A.B.; Gast, K.L.B.; O'Mara, J.A.; Tisserat, N.A.; Bauernfeind, R. Commercial specialty cut flower production, peonies. Cooperative Extension Service, Kansas State University: Manhattan, KS, USA, 1993. Available online: http://www.oznet.ksu.edu/library/hort2/mf1083.pdf (accessed on 1 March 2012).

2. Me, Z. Effects of removing lateral buds and spraying calcium on qualities of peony cut flowers. Master Thesis, Chinese Academy of Forestry, Beijing, June 2012. Available online: http://cdmd.cnki.com.cn/Article/CDMD-82201-1012423671.htm (accessed on 1 January 2015).

3. Baninasab, B.; Ghobadi, C. Influence of paclobutrazol and application methods on high-temperature stress injury in cucumber seedlings. *J. Plant Growth Regul.* **2011**, *30*, 213–219. [CrossRef]

4. Carver, S.T.; Arnold, M.A.; Byrne, D.H.; Armitage, A.R.; Lineberger, R.D.; King, A.R. Growth and flowering responses of sea marigold to daminozide, paclobutrazol, or uniconazole applied as drenches or sprays. *J. Plant Growth Regul.* **2014**, *33*, 626–631. [CrossRef]

5. Saxena, P.; Singh, V.K.; Pathak, N. Antioxidative enzymes and biochemical changes in paclobutrazol induced flowering in mango (*M. indica*) cultivars. *J. Environ. Biol.* **2014**, *35*, 1061–1066. [PubMed]

6. Xu, G.; Luo, R.; Yao, Y. Paclobutrazol improved the reproductive growth and the quality of seed oil of *Jatropha curcas*. *J. Plant Growth Regul.* **2013**, *32*, 875–883. [CrossRef]

7. Wang, N.; Han, Y.; Zhang, W. Effects of paclobutrazol of different concentration on growth of *Paeonia lactiflora*. *For. Sci. Technol. Inf.* **2014**, *46*, 16–17.

8. Chen, X.Y.; Xue, W.H. Plant growth and regulation. In *Plant Physiology and Molecular Biology*, 3rd ed.; Chen, X.Y., Xue, W.H., Eds.; Higher Education Press: Beijing, China, 2012; pp. 406–523.

9. Lin, H.; Wang, R.; Qian, Q.; Yan, M.X.; Meng, X.B.; Fu, Z.M.; Yan, C.; Jiang, B.; Su, Z.; Li, J.; *et al.* DWARF27, an iron-containing protein required for the biosynthesis of strigolactones, regulates rice tiller bud outgrowth. *Plant Cell* **2009**, *21*, 1512–1525. [CrossRef] [PubMed]

10. Arite, T.; Umehara, M.; Ishikawa, S.; Hanada, A.; Maekawa, M.; Yamaguchi, S.; Kyozuka, J. *d14*, a strigolactone-insensitive mutant of rice, shows an accelerated outgrowth of tillers. *Plant Cell Physiol.* **2009**, *50*, 1416–1424. [CrossRef] [PubMed]

11. Wang, R.L.; Stec, A.; Hey, J.; Lukens, L.; Doebley, J. The limits of selection during maize domestication. *Nature* **1999**, *398*, 253–279.

12. Chen, X.; Zhou, X.; Xi, L.; Li, J.; Zhao, R. Roles of *DgBRC1* in regulation of lateral branching in chrysanthemum (*Dendranthema* × *grandiflora* cv. Jinba). *PLoS ONE* **2013**, *8*, e61717. [CrossRef] [PubMed]

13. Aguilar-Martínez, J.A.; Poza-Carrió, C.; Cubas, P. *Arabidopsis BRANCHED1* acts as an integrator of branching signals within axillary buds. *Plant Cell* **2007**, *19*, 458–472. [CrossRef] [PubMed]

14. Minakuchi, K.; Kameoka, H.; Yasuno, N.; Umehara, M.; Luo, L.; Kobayashi, K.; Hanada, A.; Ueno, K.; Asami, T.; Yamaguchi, S.; *et al.* *FINE CULM1* (*FC1*) works downstream of strigolactones to inhibit the outgrowth of axillare buds in rice. *Plant Cell Physiol.* **2010**, *51*, 1127–1135. [CrossRef] [PubMed]

15. Ma, C.; Zhou, J.; Chen, G.; Bian, Y.; Lv, D.; Li, X.; Wang, Z.; Yan, Y. iTRAQ-based quantitative proteome and phosphoprotein characterization reveals the central metabolism changes involved in wheat grain development. *BMC Genom.* **2014**, *15*. [CrossRef] [PubMed]

16. Ge, P.; Hao, P.; Cao, M.; Guo, G.; Lv, D.; Subburaj, S.; Li, X.; Yan, X.; Xiao, J.; Ma, W.; *et al.* iTRAQ-based quantitative proteomic analysis reveals new metabolic pathways of wheat seedling growth under hydrogen peroxide stress. *Proteomics* **2013**, *13*, 3046–3058. [CrossRef] [PubMed]

17. Molloy, M.P.; Herbert, B.R.; Walsh, B.J.; Tyler, M.I.; Traini, M.; Sanchez, J.C.; Hochstrasser, D.F.; Williams, K.L.; Gooley, A.A. Extraction of membrane proteins by differential solubilization for separation using two-dimensional gel electrophoresis. *Electrophoresis* **1998**, *19*, 837–844. [CrossRef] [PubMed]

18. Ma, Q.; Wu, M.; Pei, W.; Li, H.; Li, X.; Zhang, J.; Yu, J.; Yu, S. Quantitative phosphoproteomic profiling of fiber differentiation and initiation in a fiberless mutant of cotton. *BMC Genom.* **2014**, *15*. [CrossRef] [PubMed]

19. Liu, R.; Chen, S.; Jiang, J.; Zhu, L.; Zheng, C.; Han, S.; Gu, J.; Sun, J.; Li, H.; Wang, H.; *et al.* Proteomic changes in the base of chrysanthemum cuttings during adventitious root formation. *BMC Genom.* **2013**, *14*. [CrossRef] [PubMed]

20. Wiese, S.; Reidegeld, K.A.; Meyer, H.E.; Warscheid, B. Protein labeling by iTRAQ: A new tool for quantitative mass spectrometry in proteome research. *Proteomics* **2007**, *7*, 340–350. [CrossRef] [PubMed]

21. Zieske, L.R. A perspective on the use of iTRAQ™ reagent technology for protein complex and profiling studies. *J. Exp. Bot.* **2006**, *57*, 1501–1508. [CrossRef] [PubMed]

22. Ali, A.; Alexandersson, E.; Sandin, M.; Resjö, S.; Lenman, M.; Hedley, P.; Levander, F.; Andreasson, E. Quantitative proteomics and transcriptomics of potato in response to *Phytophthora infestans* in compatible and incompatible interactions. *BMC Genom.* **2014**, *15*. [CrossRef] [PubMed]

23. Cao, S.X.; Zhang, C.; Tang, Y.F.; Qi, H.Y. Protein characteristic of the plant lipoxygenase and the function on fruit ripening and senescence and adversity stress. *Plant Physiol. J.* **2014**, *50*, 1096–1108.

24. Bhaduri, A.M.; Fulekar, M.H. Antioxidant enzyme responses of plants to heavy metal stress. *Rev. Environ. Sci. Biotechnol.* **2012**, *11*, 55–69. [CrossRef]

25. Jin, H.; Huang, F.; Cheng, H.; Song, H.; Yu, D. Overexpression of the *GmNAC2* gene, an NAC transcription factor, reduces abiotic stress tolerance in tobacco. *Plant Mol. Biol. Rep.* **2013**, *31*, 435–442. [CrossRef]

26. Morimoto, R.I.; Santoro, M.G. Stress-inducible responses and heat shock proteins: New pharmacologic targets for cytoprotection. *Nat. Biotechnol.* **1998**, *16*, 833–838. [CrossRef] [PubMed]

27. Zhang, L.; Zhang, Q.; Gao, Y.; Pan, H.; Shi, S.; Wang, Y. Overexpression of heat shock protein gene *PfHSP21.4* in *Arabidopsis thaliana* enhances heat tolerance. *Acta Physiol. Plant.* **2014**, *36*, 1555–1564. [CrossRef]

28. Yang, S.F.; Hoffman, N.E. Ethylene biosynthesis and its regulation in higher plants. *Annu. Rev. Plant Physiol.* **1984**, *35*, 155–189. [CrossRef]

29. Wang, Z. Plant growth substances, the structure and function of plant cells. In *Plant Physiology*, 3rd ed.; Wang, Z., Ed.; China Agriculture Publishing House: Beijing, China, 2000; pp. 11–219.

30. Park, S.; Han, K.H. An auxin-repressed gene (*RpARP*) from black locus (*Robinia pseudoacacia*) is posttranscriptionally regulated and negatively associated with shoot elongation. *Tree Physiol.* **2003**, *23*, 815–823. [CrossRef] [PubMed]

31. Maurel, C.; Verdoucq, L.; Luu, D.T.; Véronique, S. Plant aquaporins: Membrane channels with multiple integrated functions. *Annu. Rev. Plant Biol.* **2008**, *59*, 595–624. [CrossRef] [PubMed]

32. Liu, X.; Zhao, Z.X.; Wang, F.; Qin, C.; Sui, S.Z.; Li, M.Y. Changes in transcription level of the aquaporin gene *CpAQP1* from *Chimonanthus praecox* and the cloning of its promoter. *J. Southwest Univ.* **2014**, *36*, 48–55.

33. Ahamed, A.; Murai-Hatano, M.; Ishikawa-Sakurai, J.; Hayashi, H.; Kawamura, Y.; Uemura, M. Cold stress-induced acclimation in rice is mediated by root-specific aquaporins. *Plant Cell Physiol.* **2012**, *53*, 1445–1456. [CrossRef] [PubMed]

34. Zhao, P.; Wang, D.L. Study on the effect of phospholipase D on the freezing tolerance of *Arabidopsis thaliana*. *Chin. Agric. Sci. Bull.* **2011**, *27*, 107–111.

35. Li, Y.; Han, B.; Li, W.Q. Suppression of phospholipase Dδ enhances the membrane damage induced by UV-B irradiation. *Plant Divers. Resour.* **2011**, *33*, 27–31.

36. Smalle, J.; Vierstra, R.D. The ubiquitin 26S proteasome proteolytic pathway. *Annu. Rev. Plant Biol.* **2004**, *55*, 555–590. [CrossRef] [PubMed]

37. Guo, C.F.; Sun, Y.; Lai, C.C.; Zhang, M.Q. Proteomic analysis of tea leaf under polyethylene glycol stress. *J. Tea Sci.* **2009**, *29*, 79–88.

38. Jan, A.; Yang, G.; Nakamura, H.; Ichikawa, H.; Kitano, H.; Matsuoka, M.; Matsumoto, H.; Komatsu, S. Characterization of a xyloglucan endotransglucosylase gene that is up-regulated by gibberellin in rice. *Plant Physiol.* **2004**, *136*, 3670–3681. [CrossRef] [PubMed]

39. Qin, Z.X.; Yang, Y.M.; Zhang, C.H.; Xu, C.N.; Zhai, Z.X. Cloning and analysis of cDNA related to the genes of secondary wall thickening of cotton (*Gossypium hirsutum* L.) fiber. *Acta Agron. Sin.* **2003**, *29*, 860–866.

40. Pantaloni, D.; Clainche, C.L.; Carlier, M.F. Mechanism of actin-based motility. *Science* **2001**, *292*, 1502–1506. [CrossRef] [PubMed]

41. Wang, Z.; Wu, Z.; Jian, J.; Lu, Y. Cloning and expression of an actin gene in the haemocytes of pearl oyster (*Pinctada fucata*, Gould 1850). *Mar. Genom.* **2008**, *1*, 63–67. [CrossRef] [PubMed]

42. Humphreys, J.M.; Chapple, C. Rewriting the lignin roadmap. *Curr. Opin. Plant Biol.* **2002**, *5*, 224–229. [CrossRef]

43. Mei, X.H.; Hao, Y.L.; Zhu, H.L.; Gao, H.Y.; Luo, Y.B. Cloning of pectin methylesterase inhibitor from kiwi fruit and its high expression in *Pichia pastoris*. *Enzym. Microb. Technol.* **2007**, *4*, 1001–1005. [CrossRef]

44. Fujii, S.; Hayashi, T.; Mizuno, K. Sucrose synthase is an integral component of the cellulose synthesis machinery. *Plant Cell Physiol.* **2010**, *51*, 294–301. [CrossRef] [PubMed]

45. Zhang, G.; Qi, J.; Xu, J.; Niu, X.; Zhang, Y.; Tao, A.; Zhang, L.; Fang, P.; Lin, L. Overexpression of UDP-glucose pyrophosphorylase gene could increase cellulose content in Jute (*Corchorus capsularis* L.). *Biochem. Biophys. Res. Commun.* **2013**, *442*, 153–158. [CrossRef] [PubMed]

46. Rose, J.K.; Hadfield, K.A.; Labavitch, J.M.; Bennet, A.B. Temporal sequence of cell wall disassembly in rapidly ripening melon fruit. *Plant Physiol.* **1998**, *117*, 345–361. [CrossRef] [PubMed]

47. Updegraff, D.M. Semimicro determination of cellulose inbiological materials. *Anal. Biochem.* **1969**, *32*, 420–424. [CrossRef]

48. Müsel, G.; Schindler, T.; Bergfeld, R.; Ruel, K.; Jacquet, G.; Lapierre, C.; Speth, V.; Schopfer, P. Structure and distribution of lignin in primary and secondary cell walls of maize coleoptiles analyzed by chemical and immunological probes. *Planta* **1997**, *201*, 146–159. [CrossRef]

49. Blumenkrantz, N.; Asboe-Hansen, G. New method for quantitative determination of uronic acids. *Anal. Biochem.* **1973**, *54*, 484–489. [CrossRef]

50. Majumder, K.; Mazumdar, B.C. Changes of pectic substances in developing fruits of cape-gooseberry (*Physalis peruviana* L.) in relation to the enzyme activity and evolution of ethylene. *Sci. Hortic.* **2002**, *96*, 91–101. [CrossRef]

51. Conesa, A.; Gotz, S.; Garcia-Gomez, J.M.; Terol, J.; Talon, M.; Robles, M. Blast2GO: A universal tool for annotation, visualization and analysis in functional genomics research. *Bioinformatics* **2005**, *21*, 3674–3676. [CrossRef] [PubMed]

52. Zhao, D.Q.; Zhou, C.H.; Tao, J. Carotenoid accumulation and carotenogenic genes expression during two types of persimmon fruit (*Diospyros kaki* L.) development. *Plant Mol. Biol. Rep.* **2011**, *29*, 646–654. [CrossRef]

53. Zhao, D.Q.; Tao, J.; Han, C.X.; Ge, J.T. An *actin* gene as the internal control for gene expression analysis in herbaceous peony (*Paeonia lactiflora* Pall.). *Afr. J. Agric. Res.* **2012**, *7*, 2153–2159.

54. Schmittgen, T.D.; Livak, K.J. Analyzing real-time PCR data by the comparative C_T method. *Nat. Protoc.* **2008**, *36*, 1101–1108. [CrossRef]

International Journal of
Molecular Sciences

Review

Abiotic Stresses: Insight into Gene Regulation and Protein Expression in Photosynthetic Pathways of Plants

Mohammad-Zaman Nouri [1,*], Ali Moumeni [1] and Setsuko Komatsu [2,*]

[1] Rice Research Institute of Iran, Mazandaran Branch, Agricultural Research, Education and Extension Organization (AREEO), Amol 46191-91951, Iran; amoumeni@areo.ir

[2] National Institute of Crop Science, National Agriculture and Food Research Organization, Tsukuba 305-8518, Japan

* Correspondence: m.nouri@areo.ir (M.-Z.N.); skomatsu@affrc.go.jp (S.K.); Tel.: +98-11-4325-3912 (M.-Z.N.); +81-298-38-8693 (S.K.); Fax: +98-11-4325-3995 (M.-Z.N.); +81-298-38-8694 (S.K.)

Academic Editor: Jianhua Zhu

Received: 25 June 2015; Accepted: 21 August 2015; Published: 28 August 2015

Abstract: Global warming and climate change intensified the occurrence and severity of abiotic stresses that seriously affect the growth and development of plants, especially, plant photosynthesis. The direct impact of abiotic stress on the activity of photosynthesis is disruption of all photosynthesis components such as photosystem I and II, electron transport, carbon fixation, ATP generating system and stomatal conductance. The photosynthetic system of plants reacts to the stress differently, according to the plant type, photosynthetic systems (C_3 or C_4), type of the stress, time and duration of the occurrence and several other factors. The plant responds to the stresses by a coordinate chloroplast and nuclear gene expression. Chloroplast, thylakoid membrane, and nucleus are the main targets of regulated proteins and metabolites associated with photosynthetic pathways. Rapid responses of plant cell metabolism and adaptation to photosynthetic machinery are key factors for survival of plants in a fluctuating environment. This review gives a comprehensive view of photosynthesis-related alterations at the gene and protein levels for plant adaptation or reaction in response to abiotic stress.

Keywords: abiotic stress; photosynthesis; gene regulation; protein expression

---

1. Introduction

Abiotic stresses are major constraints to all living organisms with more challenges to the plants, as they cannot move as other organisms [1]. Several reports indicated that the structure of living macromolecules such as lipids, proteins, and nucleic acids are disposed to damage and/or degradation under severe abiotic stress conditions [2]. From an agricultural context, abiotic stresses are ultimately defined in terms of their effects on crop yield as final economic output [3]. Despite that plant growth is controlled by a variety of physiological, biochemical, and molecular processes, photosynthesis is a key mechanism, which provides to a large extent energy as well as organic molecules for plant growth and development [4]. Generally in the plant kingdom and specifically in higher plants, leaves serve as a highly specialized part that is basically appointed in the photosynthetic process [5]. Photosynthesis represents one of the most important photo-chemical reactions in plants, since energy from sunlight is trapped and converted into biological energy. Hence, improving the efficiency of photosynthesis could have a huge beneficial impact [6]. Photosynthesis is a consequence of a multi-step and complicated process that involves several biological pathways. The Pathways are photosynthetic electron transport system (PETs), in which the light energy is altered into ATP and NADPH; The Calvin–Benson cycle that is also known as a photosynthetic carbon fixation cycle in which CO_2 is fixed into carbohydrates, as well as assimilation, transport, and utilization of photoassimilates as the organic products of photosynthesis [7–9].

The two important steps, PETs and the Calvin–Benson cycle, are under the control of many genes/gene products encoded from chloroplast as well as nuclear genomes. While the products of genes involved in photosynthesis have obvious functions, they operate together within the framework of an extensively coordinated photosynthetic network of genes, regulatory components, signaling factors, and metabolic processes. The expression of genes in both cellular organelles is highly variable and affected by a diverse range of environmental factors [10]. Many environmental stresses such as drought, salinity, flooding, light, unfavorable temperatures, and its rapid fluctuations adversely affect the process of photosynthetic carbon metabolism in plants. It may alter the ultrastructure of the organelles, change the concentration of various pigments and metabolites as well as stomatal regulation [3,11]. Several reports indicate that photosynthesis cascades are highly correlated with the accumulation of some important proteins such as ribulose-1,5-bisphosphate carboxylase/oxygenase (RuBisCO) and other photosynthesis-related proteins [12,13].

To get insight into the photosynthetic gene expression and regulation under abiotic stresses, OMICS technologies such as genomics, transcriptomics, proteomics, and metabolomics can provide detailed information which can be later applied to improve plant yield potentials. In response to various abiotic stresses plants continuously need to adjust their transcriptome profile [1]. In recent decades, transcriptomic and proteomic approaches have emerged as powerful tools to analyze genome expression at the transcription and translational levels, respectively [14]. These high-throughput technologies have been extensively accepted to study the expression of certain genes and proteins in response to different abiotic stresses [15]. Proteomics, as one of the cutting edge molecular techniques, efficiently deals with the functional molecular studies. Recently, improvement of techniques for isolation and purification of cell organelles and compartments gave new insights into organelle proteomics [16].

Photosynthesis in plants is under the control of a complex network of proteins. Four major multisubunit protein complexes, photosystem (PS) I, PSII, the ATP synthase complex and cytochrome b_6/f complex are involved in the process [17]. These proteins are greatly affected under abiotic stress conditions. This review paper provides an overview of the effect of abiotic stresses on gene regulation and protein expression involved in photosynthesis in plants with emphasis on the data reported through transcriptome and proteome technologies. It describes molecular mechanisms that determine how these different classes of genes and proteins are regulated in response to abiotic stress conditions.

2. Photosynthesis in C$_3$ and C$_4$ Plants in Response to Elevated CO$_2$ Concentration

2.1. C$_3$ Plants

Plants with the metabolic pathways of C_3 for carbon fixation are distributed worldwide. They represent over 95% of the earth's plant species, especially in cold and wet climates, usually with low light intensity. In C_3 plants, the photosynthetic Carbon Reduction or Calvin–Benson cycle for CO_2 fixation produces a three-carbon compound, phosphoglycerate. Therefore, plants utilizing this pathway are often named as C_3 species [18]. According to a systems biology analysis, the photosynthetic metabolism of C_3 plants has a highly cooperative regulation in changing environments [19]. Effects of environmental changes and abiotic stresses on photosynthesis system of many C_3 plants, from stomatal conductance to carbon assimilation and from gene regulation to protein expression are well documented [3,20]. Various components are involved in the mechanism of photosynthesis in response to environmental stresses, including photosynthetic pigments and photosystems, the electron transport system, and CO_2 reduction pathways.

Changes in CO_2 level of atmosphere is an environmental factor with the most direct and instant effect on photosynthesis. Global atmospheric CO_2 concentration of the earth is 380 μL/L which is 40%

more than pre-industrial times. Values are predicted to reach between 530 and 970 μL/L by the end of this century [21]. In theory, elevated CO_2 will directly affect the balance between photosynthetic carbon fixation and photorespiration. However, plant response to high CO_2 is under the influence of several factors, including plant carbon fixation pathways. Foyer *et al.* [9] reviewed the literature related to the C_3 and C_4 plant responses to elevated CO_2 concentration compared with those grown with ambient CO_2 [9]. Exposing C_3 leaves to high CO_2, immediately increases net photosynthesis because of decreased photorespiration [22,23] and enhances the expression of genes associated with cyclic electron flow pathways. However, long-term elevated CO_2 often decreases photosynthetic capacity, RuBisCO activity and CO_2 fixation [9].

2.2. C_4 Plants

C_4 plants are named for the four-carbon organic acids produced in the first product of carbon fixation. C_4 plants have an improved photosynthetic efficiency with minimized water loss in hot and dry environments. Generally, these kind of species are native to the tropics and warmer climates with high light intensity exhibiting a higher photosynthetic and growth rate due to gains in the water, carbon and nitrogen efficiency uses [24]. Maize (*Zea mays*), sugar cane (*Saccharum officinarum*) and sorghum (*Sorghum bicolor*) are among the most productive crops with C_4 photosynthesis pathway [25]. Although, C_3 and C_4 plants are alike in the basic photosynthetic pathways such as Calvin–Benson cycle and electron transport chain components, significant differences exist in their response to environmental changes.

Response of C_4 plants to elevated CO_2 concentration is not similar to those in C_3 plants. C_4 species have greater rates of CO_2 assimilation for a given leaf nitrogen [25]. The association of photosynthesis rate and intercellular CO_2 concentration was compared in soybean (C_3) and corn (C_4). The CO_2 concentration of 384 μmol/mol as the ambient level of 2009 was compared with 700 μmol/mol for the predicted concentration at the end of this century. According to the results, while photosynthesis was stimulated by 39% in soybean, there was no change in the photosynthesis rate of corn under elevated CO_2 concentration [26]. Ghannoum [27] reviewed the C_4 photosynthesis response to water stress and in interaction with CO_2 concentration and emphasized that elevated CO_2 concentration alleviates the deleterious effect of drought on plant productivity. It is well known that abiotic stresses such as drought, reduces stomatal conductance, CO_2 assimilation rate, and intercellular CO_2 [27,28]. Therefore, saturating CO_2 concentration keeps the photosynthetic capacity unchanged.

3. Impact of Abiotic Stress on the Photosynthetic System of Plants

Abiotic stresses such as drought, cold, salinity, high temperature and so on can adversely affect growth and productivity of plants. Hence, an overview of the effect of these stresses is presented on the profile of gene expression/protein abundance of photosynthesis related pathways and their regulation networks in plants.

3.1. Drought

Drought has been recognized as a primary constraint in limiting the growth and development of plants. It usually causes loss of water content, reduced leaf water potential, stomatal conductance, and transpiration rate [18]. Stomatal closure is the earliest response to drought causing a decrease in mesophyll CO_2 diffusion and reduction in the photosynthesis rate [29]. In crop plants, decrease in carbon gain through photosynthesis is the major reason for loss of yield under drought. Drought stress has been shown to inhibit photosynthesis in plants within a few days of limiting water supply, thereby causing a significant reduction in CO_2 assimilation rate [30]. Decrease in photosynthesis under stress, reduces utilization of absorbed light energy in chloroplasts and the excess light energy could lead to photoinhibition. Photoinhibition, reduces quantum yield of PSII and induces photorespiration and H_2O_2 production [31,32]. In this situation, plants can increase utilization of absorbed light energy by improving CO_2 fixation to minimize photoinhibition [33]. Prolonged and severe drought stress will result in interruption of the energy production process and metabolism and ultimately cell death.

Gene expression profiles of model plants from mocots and dicots includingrice [34–38], and Arabidopsis [39,40] in response to drought stress were extensively studied using heterogenous genotypes [41–46] or near isogenic lines [36] at both vegetative and reproductive stages through transcriptome analysis with single or multiple stress treatments. Results from these experiments indicated that most genes involved in photosynthesis are down-regulated in response to drought stress treatments. Chlorophyll a/b-binding protein CP24, PSI reaction center subunit V, protochlorophyllide reductase A, peptidyl-prolyl cis-trans isomerase, and others functioning in the photosynthetic pathways are examples of down-regulated genes in rice leaf tissues when subjected to drought stress [37]. Gene expression profiling of physic nut (*Jatropha curcas* L.) seedlings exposed to drought indicated that the expression of genes involved in PSI, PSII and Calvin cycle components such as light-harvesting complex proteins, and genes encoding key enzymes in the Calvin cycle, RuBisCO small subunit, phosphoglycerate kinase and phosphoribulokinase were significantly down-regulated. However, several genes encoding glycolysis and the TCA cycle, including 6-phosphofructokinase, aconitate hydratase, and dihydrolipoamide succinyltransferase were up-regulated [47]. The evidence from these reports showed that inhibition of photosynthesis is the major consequence of drought reaction in rice leaf.

Gene expression analyses in tolerant genotypes of C_3 and C_4 plants were already reported. Genes encoding components of PSI, PSII as well as several genes related to the Calvin–Benson cycle, such as triosephosphate isomerase, fructose-1,6-bisphosphatase, RuBisCO small subunit and RuBisCO activase were repressed in tolerantrice genotypes, such as a C_3 plant [36,41] and in tolerant maize genotypes, as a C_4 plant [48] in response to drought stress. There were several common genes related to photosynthesis, which were down-regulated in tolerant and susceptible genotypes in response to drought stress. Therefore, it can be assumed that plants avoid photo-oxidation of the photosynthetic machinery and the creation of free radicals that are destructive for the cell.

In the meantime, activation of several key genes at different cycles of the photosynthetic pathways were reported in rice [36], maize [48] and Arabiodopsis [49]. Some of the activated gene transcripts were PSII (P680 chlorophyll a, LOC_Os07g01480) from photosystem and electron transport, phosphoglycolate phosphatases (LOC_Os03g24070) from photorespiration, and RuBisCO (LOC_Os03g09090) together with eight more gene transcripts from Calvin–Benson cycle which was reported in two near-isogenic lines (NILs) of rice against water-deficit treatments (Table 1). Reduction in photosynthetic activity in response to drought stress is due to a decline in stomatal conductance as well as RuBisCO activities resulting in lower carbon fixation followed by the over-reduction of components of the electron transport system and production of reactive oxygen species.

Table 1. Significant differentially expressed genes involved in photosynthesis pathway in rice genotypes including two pairs of near-isogenic lines (NILs) and their susceptible parent (IR64) in response to severe drought stress [42].

Photosynthesis Cycle	Locus_ID	Reaction ID	Gene Expression, $\log_2$ratio *				
			IR64 †	NIL10	NIL13	NIL11	NIL18
	LOC_Os03g56869	ribose-5-phosphate isomerase	−3.465	−3.211	−3.139	−3.229	−2.624
	LOC_Os07g08030	ribose-5-phosphate isomerase	−3.844	−3.463	−3.263	−3.015	−2.710
	LOC_Os04g50880	uridine kinase	−3.140	−3.140	−2.952	−2.185	−2.102
	LOC_Os02g47020	uridine kinase	−2.574	−2.495	−3.062	−2.185	−1.861
	LOC_Os03g07300	orotidine-5′-phosphate decarboxylase	−2.570	−2.449	−2.164	−1.826	−1.507
	LOC_Os06g04270	Transketolase	−1.844	−1.333	−1.390	−1.394	−1.293
	LOC_Os04g19740	Transketolase	3.525	3.068	2.905	3.630	3.341
	LOC_Os06g04270	Transketolase	−1.844	−1.333	−1.390	−1.394	−1.293
	LOC_Os04g19740	Transketolase	3.525	3.068	2.905	3.630	3.341
	LOC_Os03g16050	phosphoric ester hydrolase	−2.582	−2.252	−2.539	−1.504	−1.300
	LOC_Os01g64660	phosphoric ester hydrolase	−3.897	−3.802	−4.389	−3.268	−2.591
	LOC_Os11g07020	fructose-bisphospate aldolase isozyme	−3.383	−2.558	−2.670	−2.045	−2.522
	LOC_Os06g40640	fructose-bisphosphate aldolase	−2.820	−2.738	−3.224	−2.088	−1.927
	LOC_Os01g02880	fructose-bisphosphate aldolase	1.500	1.333	1.412	1.480	1.342
	LOC_Os10g08960	pyridoxin biosynthesis protein ER1, putative	−1.946	−1.623	−1.534	−1.371	−1.176
	LOC_Os10g30550	tRNA methyltransferase, putative, expressed	1.974	2.427	1.890	2.048	2.217
	LOC_Os06g45710	phosphoglycerate kinase	3.423	3.643	3.260	3.662	3.066
Calvin cycle	LOC_Os05g41640	phosphoglycerate kinase	−3.760	−3.652	−4.003	−3.513	−3.312
	LOC_Os03g19240	AMP-binding enzyme, putative, expressed	−4.206	−4.022	−3.931	−3.537	−3.428
	LOC_Os03g51740	tyrosine transaminase	−3.771	−5.202	−5.041	−4.316	−4.475
	LOC_Os11g02760	ribulose-bisphosphate carboxylase	−3.068	−3.428	−2.905	−3.534	−2.015
	LOC_Os03g09090	ribulose-bisphosphate carboxylase	1.578	1.350	1.501	1.521	1.820
	LOC_Os12g17600	ribulose-bisphosphate carboxylase	−5.512	−5.732	−6.115	−4.999	−5.034
	LOC_Os12g19381	ribulose-bisphosphate carboxylase	−3.091	−3.135	−3.337	−2.661	−3.216
	LOC_Os12g19394	ribulose-bisphosphate carboxylase	−4.124	−4.089	−4.936	−3.427	−3.177
	LOC_Os12g19470	ribulose-bisphosphate carboxylase	−3.197	−2.578	−3.097	−2.215	−1.978
	LOC_Os11g32770	ribulose bisphosphate carboxylase large chain precursor	−2.279	−2.726	−2.444	−1.093	−1.093
	LOC_Os11g47970	ribulose bisphosphate carboxylase/oxygenase activase, chloroplast precursor	−4.174	−3.598	−3.109	−2.294	−2.931
	LOC_Os10g21280	ribulose bisphosphate carboxylase large chain precursor	−2.597	−2.746	−2.203	−1.288	−1.083
	LOC_Os03g52840	glycine hydroxymethyltransferase	−3.369	−3.170	−3.199	−2.666	−2.509
	LOC_Os08g37940	phosphoglycolate phosphatase	−3.938	−3.938	−4.637	−4.789	−4.146
Photorespiration	LOC_Os04g41340	phosphoglycolate phosphatase	−3.807	−3.432	−3.621	−2.600	−2.386
	LOC_Os03g36750	phosphoglycolate phosphatase	−2.712	−2.462	−2.400	−2.516	−2.228
	LOC_Os03g19760	phosphoglycolate phosphatase	−1.144	−1.076	−1.134	−1.253	−1.167
	LOC_Os02g57100	Hydrolase	−2.129	−1.749	−1.795	−2.283	−1.990
	LOC_Os01g41710		−6.434	−6.434	−6.409	−5.622	−6.342
	LOC_Os01g52240		−6.346	−6.441	−5.927	−6.023	−5.902
	LOC_Os02g10390	Light-harvesting chlorophyll-protein complex	−1.972	−1.823	−2.190	−1.020	−1.958
	LOC_Os02g52650	Light-harvesting chlorophyll-protein complex	−2.998	−2.967	−3.451	−3.194	−2.554
	LOC_Os03g39610		−3.766	−3.797	−4.078	−3.380	−3.969
	LOC_Os04g38410		−5.128	−4.972	−5.069	−5.043	−5.131
	LOC_Os06g21590	Light-harvesting chlorophyll-protein complex	−2.223	−2.249	−2.761	−1.919	−2.098
Antenna proteins	LOC_Os07g38960	Light-harvesting chlorophyll-protein complex	−3.290	−3.182	−3.243	−2.598	−2.337
	LOC_Os08g33820	Light-harvesting chlorophyll-protein complex	−2.651	−2.651	−2.999	−2.582	−2.504
	LOC_Os07g37240	Light-harvesting chlorophyll-protein complex	−5.463	−5.674	−5.660	−4.875	−4.614
	LOC_Os09g17740		−2.365	−1.729	−1.918	−1.438	−1.943
	LOC_Os07g37550	Light-harvesting chlorophyll-protein complex	−2.005	−1.726	−2.079	−1.022	−1.637
	LOC_Os11g13890	Light-harvesting chlorophyll-protein complex	−4.353	−4.353	−4.858	−3.783	−3.739

Table 1. *Cont.*

Photosynthesis Cycle	Locus_ID	Reaction ID	Gene Expression, $\log_2$ratio *				
			IR64 †	NIL10	NIL13	NIL11	NIL18
Photosystem and electron transport system	LOC_Os02g51470	F-type ATPase	−1.708	−1.475	−1.876	−1.781	−1.645
	LOC_Os07g32880	F-type ATPase	−2.999	−2.940	−2.441	−2.189	−2.222
	LOC_Os02g01340	Photosynthetic electron transport	−1.446	−1.317	−1.640	−1.011	−1.184
	LOC_Os03g48040	Photosynthetic electron transport	−2.058	−1.945	−2.137	−1.937	−1.935
	LOC_Os06g01210	Photosynthetic electron transport	−3.000	−2.915	−3.315	−2.440	−3.101
	LOC_Os06g01850	Photosynthetic electron transport	−2.085	−1.803	−2.114	−1.728	−1.800
	LOC_Os08g01380	Photosynthetic electron transport	−3.417	−2.474	−2.750	−2.516	−2.461
	LOC_Os03g56670	Photosystem I (P700 chlorophyll a)	−2.801	−2.714	−2.990	−2.314	−2.535
	LOC_Os05g48630	Photosystem I (P700 chlorophyll a)	−1.841	−1.810	−1.876	−1.990	−1.596
	LOC_Os07g05480	Photosystem I (P700 chlorophyll a)	−4.478	−4.499	−5.141	−4.467	−4.418
	LOC_Os07g25430	Photosystem I (P700 chlorophyll a)	−2.573	−2.470	−2.949	−2.652	−2.617
	LOC_Os08g44680	Photosystem I (P700 chlorophyll a)	−2.151	−2.231	−2.468	−1.649	−2.432
	LOC_Os09g30340	Photosystem I (P700 chlorophyll a)	−4.405	−4.523	−4.857	−3.808	−3.681
	LOC_Os12g08770	Photosystem I (P700 chlorophyll a)	−4.552	−4.540	−4.975	−3.700	−4.317
	LOC_Os12g23200	Photosystem I (P700 chlorophyll a)	−2.255	−1.965	−2.449	−1.349	−1.924
	LOC_Os01g31690	Photosystem II (P680 chlorophyll a)	−3.108	−3.089	−2.917	−2.665	−2.833
	LOC_Os01g56680	Photosystem II (P680 chlorophyll a)	−3.117	−3.034	−3.817	−3.196	−3.728
	LOC_Os01g64960	Photosystem II (P680 chlorophyll a)	−2.733	−2.504	−3.035	−2.133	−2.002
	LOC_Os01g71190	Photosystem II (P680 chlorophyll a)	−2.982	−2.887	−3.073	−2.683	−2.349
	LOC_Os03g21560	Photosystem II (P680 chlorophyll a)	−2.878	−2.814	−3.727	−3.547	−3.263
	LOC_Os07g04840	Photosystem II (P680 chlorophyll a)	−3.009	−2.943	−3.494	−2.656	−2.836
	LOC_Os07g36080	Photosystem II (P680 chlorophyll a)	−5.421	−5.449	−5.847	−4.821	−4.372
	LOC_Os08g02630	Photosystem II (P680 chlorophyll a)	−3.645	−3.537	−3.488	−2.570	−2.621
Xanthophyll_Cycle	LOC_Os04g31040	violaxanthin de-epoxidase	−2.836	−2.562	−2.680	−2.452	−2.284
	LOC_Os01g51860	violaxanthin de-epoxidase	−1.326	−1.326	−1.479	−1.290	−1.198
chlorophyll a biosynthesis II	LOC_Os02g51080	geranylgeranyl reductase	−2.447	−2.447	−2.803	−1.883	−2.140
	LOC_Os02g51080	geranylgeranyl reductase	−2.447	−2.447	−2.803	−1.883	−2.140

† Rice NILs: NIL10 = IR77298-14-1-2-B-10; NIL13 = IR77298-14-1-2-B-13, NIL18 = IR77298-5-6-B-18; NIL11 = IR77298-5-6-B-11; * Signal intensities of gene expression from microarray data (4 × 44 K, agilenttechnologies) which converted to log2; Negative log2ratio are down-regulated genes and positive (yellow) $\log_2$ratio are up-regulated genes; Genes with no change in their expression are not shown; Severe drought stress: water deficit treatments with a fraction of transpirable soil water (FTSW) of 20 percent.

Drought stress in various plants has also been studied using proteomics techniques. Most of the leaf proteome analysis in response to water stress clearly confirmed the regulation of the proteins related to photosynthesis pathway. The activity of enzymes related to photoelectron transport and carbon reduction cycle, including the key enzyme RuBisCO are reduced under drought stress [50–52]. Regulation of this protein is highly affected by the duration and severity of the stress as well as plant type. Summary of selected proteins involved in photosynthesis pathway in response to abiotic stress is represented in Table 2. Up-regulation of chloroplast ATP synthase, both the CF1α and CF1β [52,53], cytochrome b_6/f complex, chloroplast oxygen-evolving enhancer protein 1 [54,55], two key enzymes involved in sucrose utilisation, invertase and sucrose synthase [56] were reported in response to drought stress. However, the activity of the key enzyme in sucrose synthesis, sucrose-phosphate synthase, is down-regulated by water stress [56].

Table 2. Summary of selected differentially expressed proteins involved in photosynthesis pathway in response to abiotic stress.

Stress	Plant Species	Protein Description	Expression *	Ref.
Drought	Sugar cane (*Saccharum officinarum* L.)	phosphoenolpyruvate carboxylase; NADP malic enzyme; pyruvate orthophosphate dikinase	−	[4]
	Maize (*Zea mays* L.)	phosphoenolpyruvate carboxylase	−	[4,27]
	Arabidopsis (*Arabidopsis thaliana*)	fructose-1,6-bisphosphatase; genes related to ATP synthesis, PSI and PSII	−	[20]
	Norway spruce (*Piceaabies*)Wheat (*Triticum aestivum* L.)	oxygen-evolving enhancer protein 2 RuBisCO LS	+	[55,57]
	Cotton (*Gossypium herbaceum*)	RuBisCO subunit binding protein	−	[52]
	Cotton (*Gossypium herbaceum*)	chloroplast ATP synthase	+	[52]
	Wheat (*Triticum aestivum* L.)	RuBisCOactivase; RuBisCO LS	+	
	Wild watermelon (*Citrullus lanatus*)	ATP synthase; chloroplast Rieske ISP; RuBisCO SS; PSI subunit D; oxygen-evolving enhancer protein 2	+	
Salinity	Common bean (*Phaseolus vulgaris* L.); Sunflower (*Helianthus annuus* L.); Wheat (*Triticum aestivum* L.); Maize (*Zea mays* L.)	fructose-1,6-bisphosphatase; RuBisCO; phosphoenolpyruvate carboxylase; ATP synthase	−	[4,54,58]
	Potato (*Solanum tuberosum* L.); Rice (*Oryza sativa* L.)	fructose-1,6-bisphosphatase	+	[4]
	Arabidopsis (*Arabidopsis thaliana*)	fructose-1,6-bisphosphatase; genes related to ATP synthesis, PSI and PSII	−	[20]
	Wheat (*Triticum aestivum* L.)	oxygen-evolving enhancer protein 2; RuBisCOactivase;	+	[54]
	Maize (*Zea mays* L.)	23 kDa polypeptides of PSII; ferredoxin NADPH1; oxidoreductase; chlorophyll a/b binding protein	+	[58]
Heat	Maize (*Zea mays* L.)	RuBisCOactivase	−	[4]
	Cotton (*Gossypium hirsutum* L.); Tobacco (*Nicotiana tabacum* L.)	RuBisCO	−	[4]
	Wheat (*Triticum aestivum* L.)	RuBisCO; phosphoenolpyruvate carboxylase	−	[4]
	Populus euphratica	PSII stability/assembly factor	−	[59]
	Populus euphratica	plastid ATP synthase CF1 a chain	+	[59]
	Rice (*Oryza sativa* L.)	glyceraldehyde-3-phosphate dehydrogenase; RuBisCO LS; 23 kDa polypeptide of PSII; Oxygen-evolving complex protein 1; oxygen-evolving protein of PSII;	−	[60]
	Rice (*Oryza sativa* L.)	RuBisCOactivase; glutamine synthetase; glyceraldehyde-3-phosphate dehydrogenase;	+	[60]
Flooding	Cacao (*Theobroma cacao*)	oxygen-evolving enhancer protein; light harvesting chlorophyll a/b-binding protein; light-harvesting complex II protein Lhcb2; light-harvesting complex I chlorophyll a/b binding protein 3; phosphate dikinase 1	+	[61]
Cold	*Thellungiella halophila*	glyceraldehyde-3-phosphate dehydrogenase; oxygen-evolving enhancer 33; RuBisCO SS	+	[62]
	Maize (*Zea mays* L.)	cytochrome b561/ferric reductase	−	[62]

* Up- and down-regulation are represented as + and −, respectively.

Not only the leaf, but also the root proteome is also highly affected by drought stress which subsequently impairs photosynthesis in plants. Two key enzymes of carbohydrate metabolism, UDP-glucose pyrophosphorylase and 2,3-bisphosphoglycerate independent phosphoglycerate mutase, were down-regulated in soybean root upon exposure to drought [63]. The levels of abundance of both proteins tended to revert to that of the control plants when watering was restored. Because the shift in carbon partitioning under drought stress is an adaptive response, a decrease in the expression of glycolytic enzymes in response to drought stress might be a consequence of reduced growth. Furthermore, it is a mechanism for accumulating sugars as an energy source for recovery and rapid growth once water is available. It has been shown that the expression of *S*-adenosylmethionine synthetase in soybean root decreases upon exposure to drought stress [63]. Down-regulation of this enzyme under drought is consistent with the inhibition of photosynthetic activity as a general feature of abiotic stresses. It is postulated that whole plant should be considered to study of photosynthesis activity in response to drought stress.

3.2. Salinity

Salt stress derived from the high accumulation of salts near the root zone causes accumulation of saline ions in plant tissues. Osmotic effect and ion toxicity are two main reasons of growth reduction, when plants are exposed to salt stress. Although several reports did not separate these two effects, Munns [64] proposed a two-phase growth model in the plant response to salt. According to the model, water deficit is the primary effect of salt stress (Phase 1) with equal effect on leaf expansion rate of a given plant species regardless of the degree of tolerance. The effect of ions (Phase 2) causes the leaves of a sensitive variety to die faster. Restriction of CO_2 diffusion into the chloroplast and reduction of carbon metabolism are among the physiological changes in leaves under salt stress [20]. Intensity and duration of the stress, leaf age as well as plant species are the main determining factors in plant response to the stress.

It is crucial to determine the molecular basis of the variation in important traits such as photosynthesis rate in plants in response to salinity stress. Photosynthesis is a complex pathway and many genes are involved in this system, hence when plants are exposed to salinity stress, expression of several photosynthesis related genes may change. Chaves *et al.* [20] summarized the number of affected genes and proteins of some model plants under drought and salt stress. They showed that several genes related to ATP synthase, PSI and PSII were down-regulated by salt and drought stress. In-depth study of the mitochondrial proteome during salt stress induced programmed cell death in rice was performed by Chen *et al.* [65]. *S*-adenosylmethionine synthetase was among the four down-regulated proteins in response to salt stress. This protein was already discussed to be down-regulated in response to drought stress.

Photosynthesis-related proteins were down-regulated in soybean seedling leaf under salt stress [66]. Sobhanian *et al.* [67], reviewed the effects of salt stress on several plants, including rice, soybean, wheat, potato and *Aleuropus lagopoides*. They concluded that reducing photosynthesis activity under salt stress was the only common response in the plants. Using *in vivo* hydroponic rice seedling culture system, proteome of rice leaves under salt stress was evaluated. Among the photosynthesis related proteins, oxygen evolution proteins, a protein related to PSII, was up-regulated in response to salt stress [68]. Salt stress alters the expression of proteins even after hours of stress exposure. It has been shown that in the initial phase of moderate salt stress (up to 4 h), sodium ions accumulate quickly and excessively in chloroplast of maize. This could enhance the expression of polypeptides of PSII, ferredoxin $NADPH^+$ oxidoreductase, ATP synthase and chlorophyll a/bbinding protein [58]. Since the water potential of the leaves remained unchanged, it can be assumed that the rapid response of plant to salinity, is a reflecting mechanism to alleviate the detrimental effects of sodium ions on the photosynthetic machinery.

3.3. Cold

Cold climate significantly reduces crop productivity and when the temperature drops to the freezing point, the damage is more severe. The key change in plant cells when exposed to cold stress is the fluidity of the membrane and by reduction of the fluidity, the plant cell senses cold stress [69]. Changes in carbohydrate metabolism, secondary metabolism and photosynthesis are other common responses of plant cells under cold stress. It has been shown that cold stress significantly altered the maximum quantum yield of PSII (Fv/Fm), the maximum photo-oxidizable P700 (Pm), the energy distribution in PSII and the redox state of P700 in seedlings of three promising oilseed crops originating from tropical regions [70]. Although PSI is the main target of stress under cold conditions, it has been shown that PSII is more sensitive to low temperature than PSI [70,71]. A decrease of photosynthetic activity, may cause photodamage which ultimately causes generation of ROS. Aggregation of ROS may inhibit protein synthesis, necessary for the repair of photodamage. The D1 protein, required for the repair of PSII, is one of the proteins suppressed by ROS [70].

In rice, as a model C_3 plant, several reports indicated that cold stress prevents chlorophyll synthesis and chloroplast formation in leaf tissues. Therefore, a reduction in chlorophyll content can be a sign of low temperature effect on rice genotypes [72]. Previous study showed that when rice seedling were treated at 10 °C for 72 h, a large number of genes including those involved in photosynthesis were highly down-regulated (Figure 1). Proteins related to the photosynthetic pathway are widely affected by low temperature stress. In C_4 plants, however, cold stress is one of the main limiting factors for growth and development. Although there are examples of C_4 species with cold adaptation, they cannot compete with C_3 plants in cold climates. Sage and McKown [73] noted possible reasons of poor C_4 photosynthesis rate at cold climate which are: declinein activity of the C_4-cycle enzymes phosphoenolpyruvate carboxylase and pyruvate phosphate dikinase; lower maximum quantum yield of C_4 photosynthesis compared to C_3 species in low temperature environment; and limitation in RuBisCO capacity [73]. Thus, the effect of cold on the photosynthesis of C_4 plants is more severe than C_3 plants.

Gao *et al.* [62] analyzed the proteome of *Thellungiella halophyila*, a chilling-tolerant plant, under cold stress and reported that 28% of the regulated proteins were photosynthesis-related proteins. Glyceraldehydes-3-phosphate dehydrogenase B, chloroplast precursor, RuBisCO small and large subunits, chloroplast carbonic anhydrase precursor andpastocyanin, oxygen-evolving enhancer, cytochrome b_6/f complex iron- sulfur subunit, and alanine-2-oxoglutarate aminotransferase were among the regulated proteins. Identification of a large number of chloroplast-related proteins (nearly half of the regulated proteins) supports the idea that cold stress tolerance of *T. halophila*is achieved, at least partly, by regulation of chloroplast function.

No	GO Term	Onto	Number	Description	Z-score	Mean	FDRnbsp;
1	GO:0009607	P	207	response to biotic stimulus	-2	-1.4	0.043
2	GO:0006091	P	46	generation of precursor metabolites and energy	-2.1	-1.5	0.04
3	GO:0019748	P	139	secondary metabolic process	-2.3	-1.4	0.022
4	GO:0006575	P	12	cellular amino acid derivative metabolic process	-2.3	-1.8	0.02
5	GO:0006950	P	494	response to stress	-2.4	-1.4	0.018
6	GO:0050896	P	706	response to stimulus	-2.5	-1.4	0.012
7	GO:0008152	P	1417	metabolic process	-2.9	-1.3	0.0041
8	GO:0019684	P	12	photosynthesis, light reaction	-3.3	-2	0.001
9	GO:0009765	P	12	photosynthesis, light harvesting	-3.3	-2	0.001
10	GO:0009719	P	309	response to endogenous stimulus	-3.3	-1.4	0.00088
11	GO:0015979	P	41	photosynthesis	-4.3	-1.8	1.6e-05
12	GO:0016747	F	45	transferase activity, transferring acyl group	-2	-1.5	0.05
13	GO:0016741	F	30	transferase activity, transferring one-carbon groups	-2.1	-1.6	0.04
14	GO:0046914	F	250	transition metal ion binding	-2.1	-1.4	0.039
15	GO:0004091	F	15	carboxylesterase activity	-2.1	-1.7	0.039
16	GO:0070008	F	13	serine-type exopeptidase activity	-2.1	-1.7	0.035
17	GO:0004185	F	13	serine-type carboxypeptidase activity	-2.1	-1.7	0.035
18	GO:0008168	F	29	methyltransferase activity	-2.1	-1.6	0.034
19	GO:0016798	F	93	hydrolase activity, acting on glycosyl bonds	-2.4	-1.5	0.016
20	GO:0004553	F	89	hydrolase activity, hydrolyzing O-glycosyl compounds	-2.7	-1.5	0.0075
21	GO:0046906	F	88	tetrapyrrole binding	-2.7	-1.5	0.0063
22	GO:0020037	F	88	heme binding	-2.7	-1.5	0.0063
23	GO:0043169	F	379	cation binding	-3.1	-1.4	0.002
24	GO:0043167	F	379	ion binding	-3.1	-1.4	0.002
25	GO:0016491	F	287	oxidoreductase activity	-3.2	-1.4	0.0015
26	GO:0009055	F	135	electron carrier activity	-3.3	-1.5	0.00093
27	GO:0031072	F	19	heat shock protein binding	-3.5	-1.9	0.00053
28	GO:0005506	F	109	iron ion binding	-3.5	-1.5	0.00048
29	GO:0005622	C	1054	intracellular	-2	-1.3	0.05
30	GO:0005618	C	207	cell wall	-2	-1.4	0.043
31	GO:0043229	C	914	intracellular organelle	-2.1	-1.3	0.04
32	GO:0043226	C	914	organelle	-2.1	-1.3	0.04
33	GO:0005635	C	11	nuclear envelope	-2.1	-1.7	0.036
34	GO:0030312	C	209	external encapsulating structure	-2.1	-1.4	0.035
35	GO:0005829	C	32	cytosol	-2.2	-1.6	0.027
36	GO:0044424	C	1010	intracellular part	-2.2	-1.3	0.025
37	GO:0044436	C	14	thylakoid part	-2.3	-1.7	0.019
38	GO:0005576	C	79	extracellular region	-2.4	-1.5	0.016
39	GO:0043231	C	868	intracellular membrane-bounded organelle	-2.5	-1.3	0.011
40	GO:0005634	C	306	nucleus	-2.6	-1.4	0.0082
41	GO:0048046	C	20	apoplast	-2.7	-1.7	0.0078
42	GO:0043227	C	871	membrane-bounded organelle	-2.7	-1.4	0.007
43	GO:0009523	C	10	photosystem II	-3	-2	0.0027
44	GO:0009536	C	257	plastid	-3.2	-1.4	0.0014
45	GO:0034357	C	20	photosynthetic membrane	-3.8	-1.9	0.00017
46	GO:0009521	C	16	photosystem	-4.6	-2.1	5.4e-06
47	GO:0009579	C	174	thylakoid	-5.7	-1.6	1.2e-08

Figure 1. Gene ontology (GO) analysis of the down-regulated differentially expressed genes of rice in response to cold treatment. This figure shows a colorful model of the PAGE analysis of gene expression data under the cold treatment after 72 h. The information includes the following: GO terms, (including 3 GO categories: biological process (P), molecular function (F) and cellular component (C)), number of annotated genes for each GO term, GO description, a simple colorful model in which the red color system indicates up-regulation and blue color indicates down-regulation, and different statistical parameters such as z-scores, means and adjusted P values (FDR) in the different rice genotypes.

3.4. Light

The photosynthesis system in plants is directly related to both quality and quantity of light. Changes in light intensity leads to imbalance in light captue by the photosystems. To ensure optimal photosynthesis efficiency, plants adjust the relative abundance of PSI and PSII according to the light quality [9]. When plants are exposed to high light intensity, a rapid repression may happen in genes encoding light-harvesting complex components, PSI and PSII reaction centre subunits [74,75]. While PSII is highly susceptible to photodamage, PSI is efficiently protected against photodamage. However, photoinhibition of PSI has been reported similar to PSII in Arabidopsis when plants are exposed to low temperature [76]. Foyer *et al.* [9], explained the mechanisms that regulate reactions in the photosynthetic electron transport chain so that the rate of production of ATP and NADPH is coordinated with the rate of their utilization in metabolism. This mechanism optimizes light use efficiency at low irradiance or dissipates excess excitation energy as heat at high light condition. The energy absorbed by plants under high irradiance exceeds the capacity of light utilization in photosynthesis and this cause photoinhibition. Although PSII is a primary site of inhibition, there is evidence that under certain circumstances, PSI can be photoinhibited even faster than PSII [77]. According to a review of the literature, most of the studies focused on high light stress rather than low light irradiance, because of the deleterious effects of photoinhibition and photodamage on plants [76,77].

It has been shown that transcripts encoding proteins involved in photoprotection such as the PSII-S and early light inducible protein 2 (ELIP2) are enhanced in high light [74,75]. ELIPs are nuclear-encoded proteins belonging to the chlorophyll a/b-binding protein family located in thylakoid membranes. The proposed function of the protein is a transient binding to the released chlorophylls under high light stress and prevent the formation of free radicals [78]. Recently, a proteome analysis of *Arabidopsis* in response to increased light conditions could identify proteins related to photosynthesis, carbon metabolism and plastid mRNA processing. The results confirmed the participation of the EXECUTER proteins in signalling and control of chloroplast metabolism, and in the regulation of plant response to environmental changes [79]. A proteome analysis of *Arabidopsis* exposed to high light conditions revealed that 35 out of 64 identified proteins were related to photosynthesis [80]. Therefore, it can be assumed that the deletorius effects of light stress on the photosynthetic pathway is higher than otherabiotic stresses.

3.5. Flooding

Flooding is a complex abiotic stress that affects the growth and development of plants and significantly decreases the productivity of crops. Higher plants are aerobic organisms that die when oxygen availability is limited due to soil flooding [81]. Plants may encounter oxygen depletion as a preliminary stress signal, when the soil water content rises above field capacity [69]. The response of plant to flooding stress is highly correlated with the level of tolerance or susceptibility to the stress. It has been shown that reduction in gas exchange parameters was varied among the citrus genotypes, and the tolerant genotypes maintain CO_2 assimilation rate and carboxylative efficiency at control levels for a longer time than sensitive genotypes under continuously flooded conditions [82]. Thus, the ability to maintain gas exchange parameters may be presumed as the main indication of tolerance to flooding. Mutava *et al.* [83] used four contrasting soybean genotypes for tolerance or susceptibility to flooding and drought. They reported different mechanisms contributing to the reduction of net photosynthesis under drought and flooding. Under drought stress, ABA and stomatal conductance were responsible for reduced photosynthetic rate; while under flooding stress, accumulation of starch granules played a major role.

Transcriptional responses to flooding stress in soybean seedlings have been studied by Nanjo *et al.* [84]. Using a soybean microarray chip, genome-wide changes in gene expression were analyzed in which photosynthsis related genes were up-regulated [84]. A comparison of soybean tolerant and susceptible genotypes showed that seven fibrillin proteins were up-regulated in the tolerant genotype, but down-regulated in the susceptible genotype [83]. Photosynthesis related proteins can be identified not only in the leaf, but also in other green organs. A comparative proteomic analysis of cotyledon of soybean under flooding stress using gel-free quantitative techniques resulted in the identification of 165 proteins which are commonly observed in both control and flooding-stressed plants. Photosynthesis related proteins were not among the main group of the regulated proteins. However, the role of ferritin in protecting plant cells against oxidative damage under flooding conditions was highlighted [85]. Study of gene expression and protein profiles of clonal genotypes of cocoa in response to flooding stress indicated that photosynthesis-associated proteins such asoxygen-evolving enhancer protein, light harvesting chlorophyll a/b binding protein, RuBisCO activase, light harvesting complex I and complex II proteins were highly up-regulated [61]. These proteins improve the plant's ability to maintain glycolysis and induce fermentation against anoxia and may also serve to distinguish tolerant and susceptible genotypes.

3.6. Heat

Physiological studies have indicated that short and long-term exposure to heat stress in temperate and tropical crops reduced the net photosynthesis rate and the PSII activity [86]. High temperature can significantly reducethe maximal quantum yield of PSII in plants [87] such as rice [60,88], *Populus* [59] and tomato [89]. Chloroplast and thylakoid membrane are highly susceptible to heat stress. Carbon metabolism of the stroma, photochemical reactions in thylakoid lamellae and changes instructural organization of thylakoids are considered as the primary sites of injury in chloroplast at heat stress [87,90]. It also caused reduction in electron transport, damage to photosystems and activation of the glycolate pathway and generation of H_2O_2 in poplar [91].

Hasanuzzaman *et al.* [87] reviewed the molecular mechanisms of high temperature stress in different crop species. Reduction inthe amount of photosynthetic pigments as a result of lipid peroxidation of chloroplast and thylakoid membranes, decrease in gas exchange and CO_2 assimilation rates, leaf water potential, leaf stomatal conductance, intercellular CO_2 concentration, total chlorophyll content and leaf area were major effects of heat stress in plants. Heat stress reduces starch and sucrose synthesis, soluble proteins, RuBisCO binding proteins and large and small subunits. It also decreases the activity of sucrose phosphate synthase, ADP-glucose pyrophosphorylase, and invertase [87].

Gene ontology analysis under moderately high temperatures (30 °C) revealed a significant increase in the expression of transcripts related to photosynthesis and those encoding polypeptides associated with PSII, ferredoxins, subunits of RuBisCO, RuBisCO activase and a number of Calvin cycle enzymes [92]. It seems that under moderate high temperature, plants may improve the photosynthesis mechanisms aimed at reaching the best possible performance in the new situation. However, when plant encounter heat stress, the expression of a wide range of proteins are down-regulated. Proteins associated with primary carbon assimilation, Calvin cycles, PSI, PSII, RuBisCO subunits, carbonic anhydrase, electron transport proteins such as oxygen-evolving enhancer protein and ferredoxin-NADP reductase are down-regulated following exposure to heat [60,86]. The opposite reaction of plants under acute heat stress, indicates that the photosynthesis apparatus is impaired by the stress.

4. Nucleus and Chloroplat Genes/Proteins Control Photosynthesis under Abiotic Stress

4.1. Coordination of Nucleus and Chloroplast

Photosynthesis is a complex mechanism in green plants as a result of a coordination between chloroplast and other organelles and compartments of the cell. Chloroplasts are the main site for light- and dark-dependent reactions of photosynthesis. Besides photoassimilation of carbohydrates, chloroplasts are involved in the biosynthesis of lipids, aromatic amino acids, hormones, vitamins, and secondary metabolites [93]. This organelle is highly sensitive to various environmental stresses and it is one of the initial sites of the stress response in plant cells [3]. Under stress conditions, electron balance and redox homeostasis are highly important. Therefore, proteins involved in electron transport play a major role in the chloroplast. Ferredoxins are small and soluble proteins that play a key role in electron distribution in all types of plastids [94]. Down-regulation of ferredoxin, shortage of $NADP^+$and over-reduction of the photosynthetic electron transport chain forms superoxide radicals and singlet oxygen in the chloroplast.

In the process of photosynthesis, the chloroplast and nucleus regulates almost all of the required genes and proteins in photosynthesis. The number of proteins in chloroplasts which are encoded in the nucleus is significantly higher than that in plastid. It has been reported that chloroplasts contain 3000–4000 different proteins which are mostly encoded in the nucleus, with only a small number encoded in the plastid genome [95]. According to an estimation, in comparison with about 2300 nuclear-encoded proteins in the chloroplast of Arabidopsis, only 87 proteins are plastid-encoded [96]. Therefore, a high degree of coordination should exist between nucleus and chloroplast to express the required genes. The mechanisms of coordination and bilateral information exchange between nucleus and chloroplast are explained by Fey *et al.* [95], where the role of redox signals as well as photosynthetic products like sugars and ROS as photosynthetic by-products in plastid-to-nucleus signaling are highlighted. The signaling components such as genomes uncoupled 1, cryptochrome1 and chloroplastic EXECUTER proteins which mediate signaling processes to the nucleus were explained by Kangasjärvi *et al.* [93]. Recently, various mechanisms involved in the signals coming from the chloroplasts to the nucleus via retrograde signaling were explained [97]; the nucleus-plastid coordination and signaling mechanisms are more important when a plant is exposed to abiotic stress since this coordination should lead to the alleviation of damage to the photosynthesis system of plants.

4.2. Overexpression of Photosynthesis-Related Proteins

The performance of photosynthetic pathways is improved in transgenic plants overexpressing the genes encoding photosynthesis-related proteins under abiotic stress. It has been shown that co-expression of the two soluble flavoproteins in the chloroplast stroma reduced ROS accumulation and improved the tolerance to stress [98]. Overexpression of MYB transcription factor in *Arabidopsis* [99]; glycine betaine in rice and tomato [100,101]; betaine aldehyde dehydrogenase in sweet potato [102]; plastidal protein synthesis elongation factor in wheat [103]; chloroplast small heat shock protein in tomato [104]; sucrose non-fermenting1-related protein kinase 2 in *Arabidopsis* [105]; Na^+/H^+ antiporter in *Arabidopsis* and cotton [106,107] and aquaporin in tobacco [108] are already reported. Transgenic plants with the overexpressed genes performed at least one of the following characteristics under abiotic stress conditions: enhance cell membrane stability, CO_2 fixation rate, PSII activity and photosynthetic rate, and reduce heat injury to thylakoids, electrolyte leakage and destruction of chlorophyll. To have efficient protein overexpression, understanding the type of the stress and its effects on plant is necessary. A survey on the localization of the overexpressed proteins related to photosynthesis pathway indicated that not all of the proteins are localized in the chloroplast [16]. Abundance of proteinin other organelles indicates the complexity of the pathway and highlights the role of signaling especially under abiotic stress.

4.3. Role of Thylakoid Membrane in Photosynthesis

A thylakoid, a membrane-bound compartment inside chloroplasts, is a place for primary reactions of photosynthesis. The thylakoid membrane contains the four major multisubunit protein complexes, PSI, PSII, ATP synthase complex and cytochrome b_6/f complex. Around 100 proteins are controlling the reactions in thylakoid membranes. The proteins are mainly involved in the conversion of light energy to chemical energy, but several other proteins have a function in assembly, maintenance, and regulation of the four multiprotein complexes [17]. Isolation and analysis of proteins located in thylakoid membrane can give an overview of the complex function of the membrane especially under stress conditions. Thylakoid membrane proteins were isolated from wild-type and mutant strains of *Chlamydomonas reinhardtii* and analyzed using proteomic techniques. More than 30 different spots were identified as light-harvesting complex proteins. The function of this protein in response to abiotic stress was reported in plants by Kono *et al.* [109] and Muneer *et al.* [110].

Environmental stress impairs the activity of the thylakoid membrane by disruption of the membrane, thereby, inhibiting the activities of membrane-associated electron carriers and enzymes and resulting in reduction of the PSI and PSII and photosynthesis rate [3]. Fluctuation of light intensity affects the architecture and protein distribution of thylakoids. It has been shown that chloroplasts in the shade havea higher density of thylakoids per chloroplast sectional area and more extensive grana stacks, and thereby more granal thylakoids than chloroplasts in the sun [109] indicating variation in the capacity of photosynthetic electron transport.

5. Phosphorylation of the Photosynthesis-Related Proteins under Abiotic Stress

Protein posphorylation at specific serine, threonine and tyrosine residues is able to change many properties of proteins such as interaction with other proteins, stability, localization and activity [111]. Reversible phosphorylation plays an important role in the regulation of cellular mechanism, signaling pathways and several developmental processes of plant such as cell growth, differentiation, migration, metabolism, apoptosis and stress responses [112,113]. Under abiotic stress, reversible protein phosphorylation is a powerful tool to alleviate damage to plant cell and specifically photosynthesis system. Phosphorylation of the thylakoid proteins in response to drought, high light, cold, heat and nutrient deficiency was reported; phosphorylation sites in thylakoid proteins from the green alga exposed to different environmental conditions indicated that 31 *in vivo* protein phosphorylation sites affect the photosynthetic machinery in the alga [111]. A survey of the photosynthetic pathways indicated that several critical functions of photosynthesis-related proteins are under control of reversible phosphorylation. The reversible phosphorylation of chlorophyll a/b binding proteins is part of the light-harvesting complex under stress for balancing the excitation energy between the PSI and PSII [109,114]. PSII core protein, the D1, D2, and CP43 phosphorylation in the photoinhibition-repair cycle [115] are known to be involved in photosynthesis pathways. Although, phosphorylation and dephosphorylation of the photosynthesis-related proteins occur under ambient conditions, the role of the modification in reduction of damage is of greater significance under stress conditions.

6. Conclusions and Future Perspectives

Environmental stresses are great challenges for the growth and development of plants. When plants are exposed to abiotic stress, photosynthetic pathways are highly affected. In crops, reduction in photosynthesis rate, significantly decrease assimilates and ultimately reduce the yield. Response to the specific stress is highly dependent on the level of tolerance or susceptibility of plants to the stress which is mostly controlled by the expression of nuclear genes and proteins. Nuclear gene expression under stress condition is controlled by retrograde signaling pathways. The signals can regulate the expression of genes which leads to the expression of proteins. There are four major multi-subunit protein complexes in the process of photosynthesis. Although all of the complexes are affected by abiotic stress, review of the regulated proteins under stresses highlights a significant role for PSII in the thylakoid membrane. Therefore, more attention should be paid to the photosynthesis system, especially to PSII, when the target is the production of an abiotic stress tolerant plant. Furthermore, because of the coordinated mechanisms in photosynthetic pathways, the whole plant should be considered to alleviate the deleterious effects of the abiotic stress.

Acknowledgments: The authors gratefully acknowledge financial support from the National Institute of Crop Science, Tsukuba, Japan.

Author Contributions: Mohammad-Zaman Nouri wrote and edited the paper and generated Table 2; Ali Moumeni wrote and edited the paper and generated Figure 1 and Table 1; Setsuko Komatsu advised the review process and edited the manuscript.

Conflicts of Interest: The authors declare no conflict of interest.

References

1. Gupta, B.; Sengupta, A.; Saha, J.; Gupta, K. Plant abiotic stress: 'Omics' approach. *J. Plant Biochem. Physiol.* **2013**, *1*, e108. [CrossRef]

2. Miranda, H. Stress Response in Cyanobacterium *Synechocystis* sp. PCC 6803. Ph.D. Thesis, Umeå University, Umeå, Sweden, December 2011.

3. Pinheiro, C.; Chaves, M.M. Photosynthesis and drought: Can we make metabolic connections from available data? *J. Exp. Bot.* **2011**, *62*, 869–882. [CrossRef] [PubMed]

4. Ashraf, M.; Harris, P.J.C. Photosynthesis under stressful environments: An overview. *Photosynthetica* **2013**, *51*, 163–190. [CrossRef]

5. Komatsu, S.; Hossain, Z. Organ-specific proteome analysis for identification of abiotic stress response mechanism in crop. *Front. Plant Sci.* **2013**, *4*, 71. [CrossRef] [PubMed]

6. Gunawardana, D. Supercharging the rice engine. *Rice Today* **2008**, *7*, 20–21.

7. Pego, J.V.; Kortstee, A.J.; Huijser, C.; Smeekens, S.C.M. Photosynthesis, sugars and the regulation of gene expression. *J. Exp. Bot.* **2000**, *51*, 407–416. [CrossRef] [PubMed]

8. Eberhard, S.; Finazzi, G.; Wollman, F.A. The dynamics of photosynthesis. *Annu. Rev. Genet.* **2008**, *42*, 463–515. [CrossRef] [PubMed]

9. Foyer, C.H.; Neukermans, J.; Queval, G.; Noctor, G.; Harbinson, J. Photosynthetic control of electron transport and the regulation of gene expression. *J. Exp. Bot.* **2012**, *63*, 1637–1661. [CrossRef] [PubMed]

10. Berry, J.O.; Yerramsetty, P.; Zielinski, A.M.; Mure, C.M. Photosynthetic gene expression in higher plants. *Photosynth. Res.* **2013**. [CrossRef] [PubMed]

11. Ambavaram, M.M.R.; Basu, S.; Krishnan, A.; Ramegowda, V.; Batlang, U.; Rahman, L.; Baisakh, N.; Pereira, A. Coordinated regulation of photosynthesis in rice increases yield and tolerance to environmental stress. *Nat. Commun.* **2014**. [CrossRef]

12. Maayan, I.; Shaya, F.; Ratner, K.; Mani, Y.; Lavee, S.; Avidan, B.; Shahak, Y.; Ostersetzer-Biran, O. Photosynthetic activity during olive (*Olea europaea*) leaf development correlates with plastid biogenesis and RuBisCO levels. *Physiol. Plant.* **2008**, *134*, 547–558. [CrossRef] [PubMed]

13. Urban, O.; Sprtova, M.; Kosvancova, M.; Tomaskova, I.; Lichtenthaler, H.K.; Marek, M.V. Comparison of photosynthetic induction and transient limitations during the induction phase in young and mature leaves from three poplar clones. *Tree Physiol.* **2008**, *28*, 1189–1197. [CrossRef] [PubMed]

14. Woodson, J.D.; Chory, J. Coordination of gene expression between organellar and nuclear genomes. *Nat. Rev. Genet.* **2008**, *9*, 383–395. [CrossRef] [PubMed]

15. Blankenburg, M.; Haberland, L.; Elvers, H.D.; Tannert, C.; Jandrig, B. High-throughput omics technologies: Potential tools for the investigation of influences of EMF on biological systems. *Curr. Genom.* **2009**, *10*, 86–92. [CrossRef] [PubMed]

16. Nouri, M.Z.; Komatsu, S. Subcellular protein overexpression to develop abiotic stress tolerant plants. *Front. Plant Sci.* **2013**, *4*, 2. [CrossRef] [PubMed]

17. Hippler, M.; Klein, J.; Fink, A.; Allinger, T.; Hoerth, P. Towards functional proteomics of membrane protein complexes: Analysis of thylakoid membranes from *Chlamydomonas reinhardtii*. *Plant J.* **2001**, *28*, 595–606. [CrossRef] [PubMed]

18. Taiz, L.; Zeiger, E. *Plant Physiology*, 5th ed.; Sinauer Associates: Sunderland, MA, USA, 2010.

19. Ahuja, I.; de Vos, R.C.; Bones, A.M.; Hall, R.D. Plant molecular stress responses face climate change. *Trends Plant Sci.* **2010**, *15*, 664–674.

20. Chaves, M.M.; Flexas, J.; Pinheiro, C. Photosynthesis under drought and salt stress: regulation mechanisms from whole plant to cell. *Ann. Bot.* **2009**, *103*, 551–560. [CrossRef] [PubMed]

21. Prins, A.; Mukubi, J.M.; Pellny, T.K.; Verrier, P.J.; Beyene, G.; Lopes, M.S.; Emami, K.; Treumann, A.; Lelarge-Trouverie, C.; Noctor, G.; *et al.* Acclimation to high CO_2 in maize is related to water status and dependent on leaf rank. *Plant Cell Environ.* **2011**, *34*, 314–331. [CrossRef] [PubMed]

22. Kramer, P.J. Carbon-dioxide concentration, photosynthesis, and dry-matter production. *Bioscience* **1981**, *31*, 29–33. [CrossRef]

23. Bowes, G. Growth at elevated CO_2: Photosynthesis responses mediated through RuBisCO. *Plant Cell Environ.* **1991**, *14*, 795–806. [CrossRef]

24. Sage, R.F.; Monson, R.K. C_4 *Plant Biology*; Academic Press: San Diego, CA, USA, 1999.

25. Lara, M.V.; Andreo, C.S. C_4 plants adaptation to high levels of CO_2 and to drought environments. In *Abiotic Stress in Plants-Mechanisms and Adaptations*; Shanker, A., Ed.; InTech: Hampshire, UK, 2011; pp. 415–428.

26. Leakey, A.D.B.; Ainsworth, E.A.; Bernacchi, C.J.; Rogers, A.; Long, S.P.; Ort, D.R. Elevated CO_2 effects on plant carbon, nitrogen and water relations: six important lessons from FACE. *J. Exp. Bot.* **2009**, *60*, 2859–2876. [CrossRef] [PubMed]

27. Ghannoum, O. C_4 photosynthesis and water stress. *Ann. Bot.* **2009**, *103*, 635–644. [CrossRef] [PubMed]

28. Ripley, B.S.; Gilbert, M.E.; Ibrahim, D.G.; Osborne, C.P. Drought constraints on C_4 photosynthesis: Stomatal and metabolic limitations in C_3 and C_4 subspecies of *Alloteropsis semialata*. *J. Exp. Bot.* **2007**, *58*, 1351–1363. [CrossRef] [PubMed]

29. Chaves, M.M.; Maroco, J.P.; Pereira, J.S. Understanding plant responses to drought-from genes to the whole plant. *Funct. Plant Biol.* **2003**, *30*, 239–264. [CrossRef]

30. Ribas-Carbo, M.; Taylor, N.L.; Giles, L.; Busquets, S.; Finnegan, P.M.; Day, D.A.; Lambers, H.; Medrano, H.; Berry, J.A.; Flexas, J. Effects of water stress on respiration in soybean leaves. *Plant Physiol.* **2005**, *139*, 466–473. [CrossRef] [PubMed]

31. Ort, D.R.; Baker, N.R. A photoprotective for O_2 as an alternative electron sink in photosynthesis. *Curr. Opin. Plant Biol.* **2002**, *5*, 193–198. [CrossRef]

32. Sanda, S.; Yoshida, K.; Kuwano, M.; Kawamura, T.; Munekage, Y.N.; Akashi, K.; Yokota, A. Responses of the photosynthetic electron transport system to excess light energy caused by water deficit in wild watermelon. *Physiol. Plant.* **2011**, *142*, 247–264. [CrossRef] [PubMed]

33. Flexas, J.; Medrano, H. Drought-inhibition of photosynthesis in C_3 plants: Stomatal and non-stomatal limitations revisited. *Ann. Bot.* **2002**, *89*, 183–189. [CrossRef] [PubMed]

34. Wang, H.; Zhang, H.; Li, Z. Analysis of gene expression profile induced by water stress in upland rice (*Oryza sativa* L. var. IRAT109) seedlings using subtractive expressed sequence tags library. *J. Integr. Plant Biol.* **2007**, *49*, 1455–1463. [CrossRef]

35. Rabello, A.R.; Guimarães, C.M.; Rangel, P.H.N.; Silva, F.R.; Seixas, D.; Souza, E.; Brasileiro, A.C.M.; Spehar, C.R.; Ferreira, M.E.; Mehta, A. Identification of drought-responsive genes in roots of upland rice (*Oryza sativa* L.). *BMC Genom.* **2008**, *9*, 485. [CrossRef] [PubMed]

36. Moumeni, A.; Satoh, K.; Kondoh, H.; Asano, T.; Hosaka, A.; Venuprasad, R.; Serraj, R.; Kumar, A.; Leung, H.; Kikuchi, S. Comparative analysis of root transcriptome profiles of two pairs of drought-tolerant and susceptible rice near-isogenic lines under different drought stress. *BMC Plant Biol.* **2011**, *11*, 174. [CrossRef] [PubMed]

37. Wang, D.; Pan, Y.; Zhao, X.; Zhu, L.; Fu, B.; Li, Z. Genome-wide temporal-spatial gene expression profiling of drought responsiveness in rice. *BMC Genom.* **2011**, *12*, 149. [CrossRef] [PubMed]

38. Hazen, S.P.; Pathan, M.S.; Sanchez, A.; Baxter, I.; Dunn, M.; Estes, B.; Chang, H.S.; Zhu, T.; Kreps, J.A.; Nguyen, H.T. Expression profiling of rice segregating for drought tolerance QTLs using a rice genome array. *Funct. Integr. Genom.* **2005**, *5*, 104–116. [CrossRef] [PubMed]

39. Bray, E.A. Genes commonly regulated by water-deficit stress in *Arabidopsis thaliana*. *J. Exp. Bot.* **2004**, *55*, 2331–2341. [CrossRef] [PubMed]

40. Dinneny, J.R.; Long, T.A.; Wang, J.Y.; Jung, J.W.; Mace, D.; Pointer, S.; Barron, C.; Brady, S.M.; Schiefelbein, J.; Benfey, P.N. Cell identity mediates the response of Arabidopsis roots to abiotic stress. *Science* **2008**, *320*, 942–945. [CrossRef] [PubMed]

41. Degenkolbe, T.; Do, P.T.; Zuther, E.; Repsilber, D.; Walther, D.; Hincha, D.K.; Köhl, K.I. Expression profiling of rice cultivars differing in their tolerance to long-term drought stress. *Plant Mol. Biol.* **2009**, *69*, 133–153. [CrossRef] [PubMed]

42. Gorantla, M.; Babu, P.R.; Lachagari, V.B.R.; Reddy, A.M.M.; Wusirika, R.; Jeffrey, L.; Bennetzen, J.L.; Reddy, A.R. Identification of stress-responsive genes in an indica rice (*Oryza sativa* L.) using ESTs generated from drought-stressed seedlings. *J. Exp. Bot.* **2007**, *58*, 253–265. [CrossRef] [PubMed]

43. Reddy, A.R.; Ramakrishna, W.; Sekhar, A.C.; Ithal, N.; Babu, P.R.; Bonaldo, M.F.; Soares, M.B.; Bennetzen, J.L. Novel genes are enriched in normalized cDNA libraries from drought-stressed seedlings of rice (*Oryza sativa* L. subsp. indica cv. Nagina 22). *Genome* **2002**, *45*, 204–211. [CrossRef] [PubMed]

44. Yang, L.; Zheng, B.; Mao, C.; Qi, X.; Liu, F.; Wu, P. Analysis of transcripts that are differentially expressed in three sectors of the rice root system under water deficit. *Mol. Gen. Genom.* **2004**, *272*, 433–442. [CrossRef] [PubMed]

45. Zhou, J.; Wang, X.; Jiao, Y.; Qin, Y.; Liu, X.; He, K.; Chen, C.; Ma, L.; Wang, J.; Xiong, L.; *et al.* Global genome expression analysis of rice in response to drought and high-salinity stresses in shoot, flag leaf, and panicle. *Plant Mol. Biol.* **2007**, *63*, 591–608. [CrossRef] [PubMed]

46. Ji, K.; Wanga, Y.; Sunb, W.; Louc, Q.; Meic, H.; Shena, S.; Chen, H. Drought-responsive mechanisms in rice genotypes with contrasting drought tolerance during reproductive stage. *J. Plant Physiol.* **2012**, *169*, 336–344. [CrossRef] [PubMed]

47. Zhang, C.; Zhang, L.; Zhang, S.; Zhu, S.; Wu, P.; Chen, Y.; Li, M.; Jiang, H.; Wu, G. Global analysis of gene expression profiles in physic nut (*Jatropha curcas* L.) seedlings exposed to drought stress. *BMC Plant Biol.* **2015**, *15*, 17. [CrossRef] [PubMed]

48. Hayano-Kanashiro, C.; Calderon-Vazquez, C.; Ibarra-Laclette, E.; Herrera-Estrella, L.; Simpson, J. Analysis of gene expression and physiological responses in three Mexican maize landraces under drought stress and recovery irrigation. *PLoS ONE* **2009**, *4*, e7531. [CrossRef] [PubMed]

49. Osakabe, Y.; Osakabe, K.; Shinozaki, S.; Tran, L.-S.P. Response of plants to water stress. *Front. Plant Sci.* **2014**. [CrossRef] [PubMed]

50. Ramachandra Reddy, A.; Chaitanva, K.V.; Vivekanandan, M. Drought-induced responses of photosynthesis and antioxidant metabolism in higher palnts. *J. Plant Physiol.* **2004**, *161*, 1189–1202. [CrossRef] [PubMed]

51. Ali, G.H.; Komatsu, S. Proteomic analysis of rice leaf sheath during drought stress. *J. Proteome Res.* **2006**, *5*, 396–403.

52. Deeba, F.; Pandey, A.K.; Ranjan, S.; Mishra, A.; Singh, R.; Sharma, Y.K.; Shirke, P.A.; Pandey, V. Physiological and proteomic responses of cotton (*Gossypium herbaceum* L.) to drought stress. *Plant Physiol. Biochem.* **2012**, *53*, 6–18. [CrossRef] [PubMed]

53. Kosmala, A.; Perlikowski, D.; Pawłowicz, I.; Rapacz, M. Changes in the chloroplast proteome following water deficit and subsequent watering in a high- and a low-drought-tolerant genotype of *Festuca arundinacea*. *J. Exp. Bot.* **2012**, *63*, 6161–6172. [CrossRef] [PubMed]

54. Kamal, A.H.; Cho, K.; Choi, J.-S.; Jin, Y.; Park, C.-S.; Lee, J.S.; Woo, S.H. Patterns of protein expression in water-stressed wheat chloroplasts. *Biol. Plant.* **2013**, *57*, 305–331. [CrossRef]

55. Blödner, C.; Majcherczyk, A.; Kües, U.; Polle, A. Early drought-induced changes to the needle proteome of Norway spruce. *Tree Physiol.* **2007**, *27*, 1423–1431. [CrossRef] [PubMed]

56. Praxedes, S.C.; DaMatta, F.M.; Loureiro, M.E.; Ferrão, M.A.G.; Cordeiro, A.T. Effects of long-term soil drought on photosynthesis and carbohydrate metabolism in mature robusta coffee (*Coffea canephora* Pierre var. *kouillou*) leaves. *Environ. Exp. Bot.* **2006**, *56*, 263–273. [CrossRef]

57. Faghani, E.; Gharechahi, J.; Komatsu, S.; Mirzaei, M.; Khavarinejad, R.A.; Najafi, F.; Farsad, L.K.; Salekdeh, G.H. Comparative physiology and proteomic analysis of two wheat genotypes contrasting in drought tolerance. *J. Proteom.* **2015**, *114*, 1–15. [CrossRef] [PubMed]

58. Zörb, C.; Herbst, R.; Forreiter, C.; Schubert, S. Short-term effects of salt exposure on the maize chloroplast protein pattern. *Proteomics* **2009**, *9*, 4209–4220. [CrossRef] [PubMed]

59. Ferreira, S.; Hjernø, K.; Larsen, M.; Wingsle, G.; Larsen, P.; Fey, S.; Roepstorff, P.; Salomé Pais, M. Proteome profiling of *Populus euphratica* Oliv. upon heat stress. *Ann. Bot.* **2006**, *98*, 361–377. [CrossRef] [PubMed]

60. Han, F.; Chen, H.; Li, X.J.; Yang, M.F.; Liu, G.S.; Shen, S.H. A comparative proteomic analysis of rice seedlings under various high-temperature stresses. *Biochim. Biophys. Acta* **2009**, *1794*, 1625–1634. [CrossRef] [PubMed]

61. Bertolde, F.Z.; Almeida, A.A.; Pirovani, C.P. Analysis of gene expression and proteomic profiles of clonal genotypes from *Theobroma cacao* subjected to soil flooding. *PLoS ONE* **2014**, *7*, e108705. [CrossRef] [PubMed]

62. Gao, F.; Zhou, Y.; Zhu, W.; Li, X.; Fan, L.; Zhang, G. Proteomic analysis of cold stress-responsive proteins in *Thellungiella rosette* leaves. *Planta* **2009**, *230*, 1033–1046. [CrossRef] [PubMed]

63. Alam, I.; Sharmin, S.A.; Kim, K.H.; Yang, J.K.; Choi, M.S.; Lee, B.H. Proteome analysis of soybean roots subjected to short-term drought stress. *Plant Soil* **2010**, *333*, 491–505. [CrossRef]

64. Munns, R. Physiological processes limiting plant growth in saline soils: Some dogmas and hypotheses. *Plant Cell Environ.* **1993**, *16*, 15–24. [CrossRef]

65. Chen, X.; Wang, Y.; Li, J.; Jiang, A.; Cheng, Y.; Zhang, W. Mitochondrial proteome during salt stress-induced programmed cell death in rice. *Plant Physiol. Biochem.* **2009**, *47*, 407–415. [CrossRef] [PubMed]

66. Sobhanian, H.; Razavizadeh, R.; Nanjo, Y.; Ehsanpour, A.A.; RastgarJazii, F.; Motamed, N.; Komatsu, S. Proteome analysis of soybean leaves, hypocotyls and roots under salt stress. *Proteome Sci.* **2010**, *8*, 1–15. [CrossRef] [PubMed]

67. Sobhanian, H.; Aghaei, K.; Komatsu, S. Changes in the plant proteome resulting from salt stress: Toward the creation of salt-tolerant crops? *J. Proteom.* **2011**, *74*, 1323–1337. [CrossRef] [PubMed]
68. Kim, D.W.; Rakwal, R.; Agrawal, G.K.; Jung, Y.H.; Shibato, J.; Jwa, N.S. A hydroponic rice seedling culture model system for investigating proteome of salt stress in rice leaf. *Electrophoresis* **2005**, *26*, 4521–4539. [CrossRef] [PubMed]
69. Arbona, V.; Manzi, M.; Ollas, C.; Gómez-Cadenas, A. Metabolomics as a tool to investigate abiotic stress tolerance in plants. *Int. J. Mol. Sci.* **2013**, *14*, 4885–4911.
70. Lei, Y.; Zheng, Y.; Dai, K.; Duan, B.; Cai, Z. Different responses of photosystem I and photosystem II in three tropical oilseed crops exposed to chilling stress and subsequent recovery. *Trees* **2014**, *28*, 923–933. [CrossRef]
71. Huang, W.; Zhang, S.B.; Cao, K.F. Stimulation of cyclic electron flow during recovery after chilling-induced photoinhibition of PSII. *Plant Cell Physiol.* **2010**, *51*, 1922–1928. [CrossRef] [PubMed]
72. Sharma, P.; Sharma, N.; Deswal, R. The molecular biology of the low temperature response in plants. *Bioessays* **2005**, *27*, 1048–1059. [CrossRef] [PubMed]
73. Sage, R.F.; McKown, A.D. Is C_4 photosynthesis less phenotypically plastic than C_3 photosynthesis? *J. Exp. Bot.* **2006**, *57*, 303–317. [CrossRef] [PubMed]
74. Kimura, M.; Yamamoto, Y.Y.; Seki, M.; Sakurai, T.; Sato, M.; Abe, T.; Yoshida, S.; Manabe, K.; Shinozaki, K.; Matsui, M. Identification of Arabidopsis genes regulated by high light-stress using cDNA microarray. *Photochem. Photobiol.* **2003**, *77*, 226–233. [PubMed]
75. Murchie, E.H.; Hubbart, S.; Peng, S.; Horton, P. Acclimation of photosynthesis to high irradiance in rice: Gene expression and interactions with leaf development. *J. Exp. Bot.* **2005**, *56*, 449–460. [CrossRef] [PubMed]
76. Zhang, S.; Scheller, H.V. Photoinhibition of photosystem I at chilling temperature and subsequent recovery in *Arabidopsis thaliana*. *Plant Cell Physiol.* **2004**, *45*, 1595–1602. [CrossRef] [PubMed]
77. Barth, C.; Krause, G.H.; Winter, K. Responses of photosystem I compared with photosystem II to high-light stress in tropical shade and sun leaves. *Plant Cell Environ.* **2001**, *24*, 163–176. [CrossRef]
78. Wang, X.; Peng, Y.; Singer, J.W.; Fessehaie, A.; Krebs, S.L.; Rajeev Arora, R. Seasonal changes in photosynthesis, antioxidant systems and ELIP expression in a thermonastic and non-thermonastic Rhododendron species: A comparison of photoprotective strategies in overwintering plants. *Plant Sci.* **2009**, *177*, 607–617. [CrossRef]
79. Uberegui, E.; Hall, M.; Lorenzo, Ó.; Schröder, W.P.; Balsera, M. An Arabidopsis soluble chloroplast proteomic analysis reveals the participation of the Executer pathway in response to increased light conditions. *J. Exp. Bot.* **2015**, *66*, 2067–2077. [CrossRef] [PubMed]
80. Phee, B.K.; Cho, J.H.; Park, S.; Jung, J.H.; Lee, Y.H.; Jeon, J.S.; Bhoo, S.H.; Hahn, T.R. Proteomic analysis of the response of Arabidopsis chloroplast proteins to high light stress. *Proteomics* **2004**, *4*, 3560–3568. [CrossRef] [PubMed]
81. Voesenek, L.A.; Colmer, T.D.; Pierik, R.; Millenaar, F.F.; Peeters, A.J. How plants cope with complete submergence. *New Phytol.* **2006**, *170*, 213–226. [CrossRef] [PubMed]
82. Hossain, Z.; López-Climent, M.F.; Arbona, V.; Pérez-Clemente, R.M.; Gómez-Cadenas, A. Modulation of the antioxidant systemin citrus under waterlogging and subsequent drainage. *J. Plant Physiol.* **2009**, *166*, 1391–1404. [CrossRef] [PubMed]
83. Mutava, R.N.; Prince, S.J.; Syed, N.H.; Song, L.; Valliyodan, B.; Chen, W.; Nguyen, H.T. Understanding abiotic stress tolerance mechanisms in soybean: A comparative evaluation of soybean response to drought and flooding stress. *Plant Physiol. Biochem.* **2015**, *86*, 109–120. [CrossRef] [PubMed]
84. Nanjo, Y.; Maruyama, K.; Yasue, H.; Yamaguchi-Shinozaki, K.; Shinozaki, K.; Komatsu, S. Transcriptional responses to flooding stress in roots including hypocotyl of soybean seedlings. *Plant Mol. Biol.* **2011**, *77*, 129–144. [CrossRef] [PubMed]
85. Kamal, A.H.; Rashid, H.; Sakata, K.; Komatsu, S. Gel-free quantitative proteomic approach to identify cotyledon proteins in soybean under flooding stress. *J. Proteom.* **2015**, *112*, 1–13. [CrossRef] [PubMed]
86. Ahsan, N.; Donnart, T.; Nouri, M.Z.; Komatsu, S. Tissue-specific defense and thermo-adaptive mechanisms of soybean seedlings under heat stress revealed by proteomic approach. *J. Proteome Res.* **2010**, *9*, 4189–4204. [CrossRef] [PubMed]
87. Hasanuzzaman, M.; Nahar, K.; Alam, M.M.; Roychowdhury, R.; Fujita, M. Physiological, biochemical, and molecular mechanisms of heat stress tolerance in plants. *Int. J. Mol. Sci.* **2013**, *14*, 9643–9684. [CrossRef] [PubMed]

88. Vani, B.; PardhaSaradhi, P.; Mohanty, P. Alteration in chloroplast structure and thylakoid membrane composition due to *in vivo* heat treatment of rice seedlings: correlation with the functional changes. *J. Plant Physiol.* **2001**, *158*, 583–592. [CrossRef]

89. Morales, D.; Rodriguez, P.; Dellamico, J.; Nicolas, E.; Torrecillas, A.; Sanchez-Blanco, M.J. High-temperature preconditioning and thermal shock imposition affects water relations, gas exchange and root hydraulic conductivity in tomato. *Biol. Plant.* **2003**, *47*, 203–208. [CrossRef]

90. Wang, J.Z.; Cui, L.J.; Wang, Y.; Li, J.L. Growth, lipid peroxidation and photosynthesis in two tall fescue cultivars differing in heat tolerance. *Biol. Plant.* **2009**, *53*, 247–242. [CrossRef]

91. Song, Y.; Chen, Q.; Ci, D.; Shao, X.; Zhang, D. Effects of high temperature on photosynthesis and related gene expression in poplar. *BMC Plant Biol.* **2014**, *14*, 111. [CrossRef] [PubMed]

92. Hancock, R.D.; Morris, W.L.; Ducreux, L.J.; Morris, J.A.; Usman, M.; Verrall, S.R.; Fuller, J.; Simpson, C.G.; Zhang, R.; Hedley, P.E.; *et al.* Physiological, biochemical and molecular responses of the potato (*Solanum tuberosum* L.) plant to moderately elevated temperature. *Plant Cell Environ.* **2014**, *37*, 439–450. [CrossRef] [PubMed]

93. Kangasjärvi, S.; Nurmi, M.; Tikkanen, M.; Aro, E.M. Cell-specific mechanisms and systemic signalling as emerging themes in light acclimation of C_3 plants. *Plant Cell Environ.* **2009**, *32*, 1230–1240. [CrossRef] [PubMed]

94. Hase, T.; Schürmann, P.; Knaff, D.B. The Light-Driven Plastocyanin: Ferredoxin Oxidoreductase. In *Photosystem I*; Golbeck, J.H., Ed.; Springer: Dordrecht, The Netherlands, 2006; pp. 477–498.

95. Fey, V.; Wagner, R.; Bräutigam, K.; Pfannschmidt, T. Photosynthetic redox control of nuclear gene expression. *J. Exp. Bot.* **2005**, *56*, 1491–1498. [CrossRef] [PubMed]

96. Abdallah, F.; Salamini, F.; Leister, D. A prediction of the size and evolutionary origin of the proteome of chloroplasts of Arabidopsis. *Trends Plant Sci.* **2000**, *5*, 141–142.

97. Singh, R.; Singh, S.; Parihar, P.; Singh, V.P.; Prasad, S.M. Retrograde signaling between plastid and nucleus: A review. *J. Plant Physiol.* **2015**, *181*, 55–66. [CrossRef] [PubMed]

98. Giró, M.; Ceccoli, R.D.; Poli, H.O.; Carrillo, N.; Lodeyro, A.F. An *in vivo* system involving co-expression of cyanobacterial flavodoxin and ferredoxin-NADP⁺ reductase confers increased tolerance to oxidative stress in plants. *FEBS Open Biol.* **2011**, *1*, 7–13. [CrossRef] [PubMed]

99. Mao, X.; Jia, D.; Li, A.; Zhang, H.; Tian, S.; Zhang, X.; Jia, J.; Jing, R. Transgenic expression of TaMYB2A confers enhanced tolerance to multiple abiotic stresses in Arabidopsis. *Funct. Integr. Genom.* **2011**, *11*, 445–465. [CrossRef] [PubMed]

100. Su, J.; Hirji, R.; Zhang, L.; He, C.; Selvaraj, G.; Wu, R. Evaluation of the stress-inducible production of choline oxidase in transgenic rice as a strategy for producing the stress-protectant glycine betaine. *J. Exp. Bot.* **2006**, *57*, 1129–1135. [CrossRef] [PubMed]

101. Park, E.J.; Jeknić, Z.; Pino, M.T.; Murata, N.; Chen, T.H. Glycinebetaine accumulation is more effective in chloroplasts than in the cytosol for protecting transgenic tomato plants against abiotic stress. *Plant Cell Environ.* **2007**, *30*, 994–1005. [CrossRef] [PubMed]

102. Fan, W.; Zhang, M.; Zhang, H.; Zhang, P. Improved tolerance to various abiotic stresses in transgenic sweet potato (*Ipomoea batatas*) expressing spinach betaine aldehyde dehydrogenase. *PLoS ONE* **2012**, *7*, e37344. [CrossRef] [PubMed]

103. Fu, J.; Momcilović, I.; Clemente, T.E.; Nersesian, N.; Trick, H.N.; Ristic, Z. Heterologous expression of a plastid EF-Tu reduces protein thermal aggregation and enhances CO_2 fixation in wheat (*Triticum aestivum*) following heat stress. *Plant Mol. Biol.* **2008**, *68*, 277–288. [CrossRef] [PubMed]

104. Wang, L.; Zhao, C.M.; Wang, Y.J.; Liu, J. Overexpression of chloroplast-localized small molecular heat-shock protein enhances chilling tolerance in tomato plant. *J. Plant Physiol. Mol. Biol.* **2005**, *31*, 167–174.

105. Zhang, H.; Mao, X.; Wang, C.; Jing, R. Overexpression of a common wheat gene TaSnRK2.8 enhances tolerance to drought, salt and low temperature in Arabidopsis. *PLoS ONE* **2010**, *30*, e16041. [CrossRef] [PubMed]

106. Brini, F.; Hanin, M.; Mezghani, I.; Berkowitz, G.A.; Masmoudi, K. Overexpression of wheat Na⁺/H⁺ antiporter TNHX1 and H⁺-pyrophosphatase TVP1 improve salt- and drought-stress tolerance in *Arabidopsis thaliana* plants. *J. Exp. Bot.* **2007**, *58*, 301–308. [CrossRef] [PubMed]

107. He, C.; Yan, J.; Shen, G.; Fu, L.; Holaday, A.S.; Auld, D.; Blumwald, E.; Zhang, H. Expression of an Arabidopsis vacuolar sodium/proton antiporter gene in cotton improves photosynthetic performance under salt conditions and increases fiber yield in the field. *Plant Cell Physiol.* **2005**, *46*, 1848–1854. [CrossRef] [PubMed]

108. Aharon, R.; Shahak, Y.; Wininger, S.; Bendov, R.; Kapulnik, Y.; Galili, G. Overexpression of a plasma membrane aquaporin in transgenic tobacco improves plant vigor under favorable growth conditions but not under drought or salt stress. *Plant Cell* **2003**, *15*, 439–447. [CrossRef] [PubMed]

109. Kono, M.; Terashima, I. Long-term and short-term responses of the photosynthetic electron transport to fluctuating light. *J. Photochem. Photobiol. B* **2014**, *137*, 89–99. [CrossRef] [PubMed]

110. Muneer, S.; Park, Y.G.; Manivannan, A.; Soundararajan, P.; Jeong, B.R. Physiological and proteomic analysis in chloroplasts of *Solanum lycopersicum* L. under silicon efficiency and salinity stress. *Int. J. Mol. Sci.* **2014**, *15*, 21803–21824. [CrossRef] [PubMed]

111. Turkina, M.V.; Kargul, J.; Blanco-Rivero, A.; Villarejo, A.; Barber, J.; Vener, A.V. Environmentally modulated phosphoproteome of photosynthetic membranes in the green alga *Chlamydomonas reinhardtii*. *Mol. Cell. Proteom.* **2006**, *5*, 1412–1425. [CrossRef] [PubMed]

112. Baena-Gonzalez, E.; Rolland, F.; Thevelein, J.M.; Sheen, J. A central integrator of transcription networks in plant stress and energy signalling. *Nature* **2007**, *448*, 938–942. [CrossRef] [PubMed]

113. Huber, S.C. Exploring the role of protein phosphorylation in plants: from signalling to metabolism. *Biochem. Soc. Trans.* **2007**, *35*, 28–32. [CrossRef] [PubMed]

114. Liu, X.D.; Shen, Y.G. NaCl-Induced phosphorylation of light harvesting chlorophyll a/b proteins in thylakoid membranes from the halotolerant green alga, *Dunaliella salina*. *FEBS Lett.* **2004**, *569*, 337–340. [CrossRef] [PubMed]

115. Aro, E.M.; Suorsa, M.; Rokka, A.; Allahverdiyeva, Y.; Paakkarinen, V.; Saleem, A.; Battchikova, N.; Rintamäki, E. Dynamics of photosystem II: A proteomic approach to thylakoid protein complexes. *J. Exp. Bot.* **2005**, *56*, 347–356. [CrossRef] [PubMed]

 International Journal of
Molecular Sciences

Review

Plant Responses to Nanoparticle Stress

Zahed Hossain [1],*, Ghazala Mustafa [2] and Setsuko Komatsu [2],*

[1] Department of Botany, University of Kalyani, Kalyani 741235, West Bengal, India
[2] National Institute of Crop Science, National Agriculture and Food Research Organization, Tsukuba 305-8518, Japan; ghazalamustafa@affrc.go.jp
* Correspondence: zahed_kly@yahoo.com (Z.H.); skomatsu@affrc.go.jp (S.K.); Tel.: +91-33-2582-8750 (ext. 217) (Z.H.); +81-29-838-8693 (S.K.); Fax: +91-33-2582-8282 (Z.H.); +81-29-838-8694 (S.K.)

Academic Editor: Jianhua Zhu
Received: 14 September 2015; Accepted: 23 October 2015; Published: 6 November 2015

Abstract: With the rapid advancement in nanotechnology, release of nanoscale materials into the environment is inevitable. Such contamination may negatively influence the functioning of the ecosystems. Many manufactured nanoparticles (NPs) contain heavy metals, which can cause soil and water contamination. Proteomic techniques have contributed substantially in understanding the molecular mechanisms of plant responses against various stresses by providing a link between gene expression and cell metabolism. As the coding regions of genome are responsible for plant adaptation to adverse conditions, protein signatures provide insights into the phytotoxicity of NPs at proteome level. This review summarizes the recent contributions of plant proteomic research to elaborate the complex molecular pathways of plant response to NPs stress.

Keywords: nanoparticles; oxidative stress; proteomics

---

1. Introduction

Nanotechnology is an emerging multidisciplinary field with a wide range of applications in cancer therapy, targeted drug delivery, electronics, cosmetic industry, and biosensors [1]. Nevertheless, unspecified release of metal-based nanoparticles (NPs) into the ecosystem has raised global concern about their potential phytotoxic effects. The NPs are extremely fine particles with lengths between 1 and 100 nm in at least two of their dimensions [2]. These are in fact intermediate in size between molecules and bulk materials. Division of bulk materials into smaller and smaller pieces gives them very unique physical and chemical properties [3]. Moreover, a high surface-to-volume ratio renders these nanoscale materials highly reactive or catalytic.

Bulk production of NPs often leads to their indiscriminate release in nature through industrial waste-waters [4,5]. A majority of the manufactured NPs contain heavy metals. Thus, soil and water contamination with metallic NPs has become an important environmental issue. Nanoparticles interact with the plants and results in the uptake and accumulation that affect their fate and transport in the ecosystem. Moreover, NPs could remain attached to the plant surface and impart physical and chemical damage to the plant organs. Usually, NPs enter the plant root system through the lateral root junctions and reach the xylem through the cortex and the pericycle [6]. Notably, NPs' entry into the plant can be stopped by the cell wall. The specific properties of cell wall allowing the transport of NPs could be attributed to the pore size of cell wall [7]. The NPs that are in the size range within the cell wall pore size could effectively cross the cell wall and reach the plasma membrane [8]. The rate of entry depends on the size and surface properties of NPs. Indeed, the smaller NPs can enter into plant cells easily. In contrast, larger NPs, being unable to enter the cells, cannot affect the cell metabolic pathways [9]. Larger NPs can only penetrate through the hydathodes, flower stigmas, and stomata. Mechanism of interaction between NPs and plants could be chemical or physical. Chemical interactions involve the production of reactive oxygen species [1], disturbance of ion cell membrane transport activity [10], oxidative damage [11], and lipid peroxidation [12]. Following entry into the plant cells, NPs after mixing behave as metal ions and react with sulfhydryl, carboxyl groups and ultimately alter the protein activity. However, while conducting engineered nanomaterials (ENMs) mediated ecotoxicity study, much attention needs to be paid towards various artifacts which often lead to misinterpretations of results [13]. These potential factors include toxic impurities in ENM materials, their proper storage and dispersion in testing medium. Moreover, ENMs exert indirect toxicity which affects plant growth and development through nutrient depletion with passage of time, and estimation of ENM dispersal in organisms. In addition, ENMs face different changes (viz. settling, dissolution, agglomeration, *etc.*) during the exposure period, which is difficult to measure accurately. Due to increased surface area and properties, ENMs readily adsorb organic molecules and inorganic ions from the nutrient medium resulting indirect toxicity symptoms including chlorosis and wilting [14,15]. Moreover, during ENM exposure, organic acid in plant root exudates decreases the pH of the media, thus altering nutrient supply and ENM properties [16]. Inefficiency to explore the influence of these factors can direct to an inappropriate explanation of phytotoxicity and ultimately a fabricated impact of ENMs [13].

2. Plant Response to Nanoparticle Stress

NPs with different composition, size, and concentration, physical/chemical properties have been reported to influence growth and development of various plant species with both positive and negative effects [17]. Khodakovskaya *et al.* [18] reported that multi-walled carbon nanotubes markedly influenced tomato seed germination and seedling growth by up-regulating stress-related gene expression. In *Arabidopsis*, Al_2O_3-NPs were reported to be least toxic compared to zinc oxide, iron oxide, and silicon oxide nanoparticles [19]. Previous study highlighted the toxic effects of NPs on algae [20]. NPs like titanium oxide, zinc oxide, cerium oxide, and silver NPs were deposited on the surface of cell as well as in the organelles, which resulted in oxidative stress to the cell through the induction of oxidative stress signaling [21]. In *Cucurbita pepo*, the effect of silver, copper (Cu), zinc oxide, and silicon nanoparticles indicated that seed germination was unaffected by these NPs and their counterpart bulk materials; however, Cu nanoparticles reduced root length compared to the control and plants treated with the bulk Cu powder [22]. In rice, ZnO NPs, but not titanium oxide cause deleterious effects on the root length at early growth stages [23]. Riahi-Madvar *et al.* [24] indicated that the root growth of *Triticum aestivum* was affected by different concentrations of the alumina-nanoparticles; however, NPs did not affect the seed germination, shoot length, and dry biomass. In rice seedlings, nano-CuO treatment led to an increase in activity of antioxidant enzymes and elevated MDA concentration [25]. A similar experiment on the nano-CuO modulated photosynthetic performance and antioxidative defense system in *Hordeum vulgare* demonstrated restriction in root and shoot growth with decreased photosynthetic performance index [26]. Moreover, nano-CuO mediated DNA damage and plant growth restriction were reported in radish (*Raphanus sativus*) and ryegrass (*Lolium perenne* and *Lolium rigidum*) [27]. Changes in enzyme activities, ascorbate and free thiol levels resulting in higher membrane damage and photosynthetic stress have been documented in shoots of germinating rice seedlings on exposure to very high concentration of cerium oxide NPs [28]. Generation of ROS and reactive nitrogen species and H_2O_2 upon exposure to Ag and ZnO engineered NPs on the duckweed (*Spirodela punctuta*) suggest that toxicity of Ag and ZnO-NPs predominantly caused by both the particulates and ionic forms [29].

Among the various metal NPs, much attention has been paid to Ag-NP owing to their characteristic physiochemical and biological properties compared to the massive bulk material [30]. The Ag-NPs have wide applications as an essential component in different products like household, food, and industries because of their bactericidal and fungicidal properties [31]. Compared to the silver-based compounds, Ag-NPs, with increased surface area available for microbe interaction, are reported to be more toxic to bacteria, fungi, and viruses. Like other metal ions, Ag-NPs can also induce oxidative stress in bacteria, animals, algae as well as higher plants [32]. However, the impact of Ag-NPs on plants largely depends on various factors such as plant species, growth stage of plant, composition and concentration of the nanoparticles, and the experimental setup (temperature, treatment period, media composition, and method of exposure, *etc.*) [33]. Nano-Ag is one of the most extensively studied NPs whose toxicology has been examined in various crops [22,32,34]. Although Ag-NPs exposure is reported to be detrimental for plant growth, some studies have demonstrated the growth-enhancing properties of Ag-NPs in *Brassica juncea* [35], *Eruca sativa* [33], wetland plants [36], and *Phaseolus vulgaris* and *Zea mays* [37]. An investigation by Kumari *et al.* [34] revealed chromotoxic effects of Ag-NPs on the mitotic cell division in root-tip cells of *Allium cepa*. Moreover, Ag-NPs interact with the membrane proteins and activate signaling pathways, that leads to inhibition of cell proliferation [38,39].

Perusal of all these nanotoxicity studies over the past decade reveals that plant response to NPs stress has been evaluated extensively in various crops largely *at physiological and biochemical levels*. Rather less focus has been given to the study of plant-NPs interface at transcript level (Table 1). Microarray-based gene expression analysis of *Arabidopsis thaliana* roots on exposure to ZnO-NPs, TiO$_2$-NPs, and fullerene soot indicates that the underlying mechanisms of phytotoxicity are highly specific to the nanoparticle [40]. Khodakovskaya *et al.* [41] designed an advanced method by amalgamating genetic, photothermal, and photoacoustic strategies for highly sensitive detection of NPs in different parts of tomato plants, most importantly the reproductive organs. Total gene expression analysis of tomato leaves and roots exposed to carbon nanotubes (CNTs) revealed up-regulation in the stress and water channel-related genes. A separate study demonstrated selective root growth in maize upon exposure to single-walled carbon nanotubes (SWCNTs) [42]. Transcriptional analysis suggests that nanoparticle-root cell interaction selectively modulates gene expression in seminal roots, thus affecting relative root growth and development. Similar to transcriptome analysis, only limited numbers of studies have emphasized the effects of nanoparticles stress on plants at proteome level.

3. Modulation of Proteome Composition under Nanoparticle Stress

Over the past decade, phytotoxicity of Ag-NPs has been evaluated extensively in various crops, largely at morphological, physiological, and biochemical levels. However, only limited studies have emphasized the effects of Ag-NPs stress on plants at proteome level (Table 2). Recently, Mirzajani *et al.* [43] performed a gel-based proteomic study to understand the effects of Ag-NPs toxicity on *Oryza sativa*. The root proteome study revealed that Ag-NPs-responsive proteins were primarily associated with oxidative stress response pathway, Ca^{2+} regulation and signaling, transcription, protein degradation, cell wall synthesis, cell division, and apoptosis. Increased abundance of defense-related proteins including superoxide dismutase, L-ascorbate peroxidase, glutathione-S-transferase implies accelerated production of ROS under Ag-NPs treatment. It has been hypothesized that Ag-NPs or released ions impede cell metabolism by binding to second messenger calcium ion receptors, calcium ion channels, and Ca^{2+}/Na^{+}–ATPases.

Proteomic study on *Eruca sativa* roots exposed to Ag-NPs and AgNO$_3$ revealed that both forms of Ag caused changes in the proteins related to redox regulation, disrupting cellular homeostasis (Figure 1) [33]. However, the Ag-NPs alone were responsible to alter the ER and vacuolar proteins, thus indicating these organelles as target sites of Ag-NPs. These findings suggest that phytotoxicity of Ag-NPs is primarily due to their characteristic physiochemical properties, and not by releasing the Ag^{+} [33]. We also studied the toxicity mechanisms of Ag-NPs on early-stage-soybean growth under flooding stress [44]. In total, three different particle sizes (2, 15, and 50–80 nm) and concentrations (0.2, 2, and 20 ppm) were screened. The Ag-NPs of 15 nm facilitated the soybean growth under flooding, compared to the larger and smaller nanoparticles. The changed proteins under Ag-NPs exposure were mainly related to stress, signaling, and cell metabolism.

Table 1. Summary of gene expression analyses in response to nanoparticle stress.

Plant (Cultivar)	Tissue/Organ	Growth Stage	Nanoparticles (Particle Size, Dose)	Treatment Period	Technique	Major Findings	Ref.
Arabidopsis thaliana accession Columbia-0	Root	3-week-old	Ag (10–80 nm), TiO_2 (10–40 nm), MWNT	7 days	Microarray analysis	Exposure to NPs repressed expression of phosphate-starvation and root-development genes.	[45]
Arabidopsis thaliana (ecotype, Columbia) Wild, *tir1*, *abi5* mutants	Seedling	4-day-old	Ag-NPs: triangular (47 nm), spherical (8 nm), decahedral (45 nm)	3 days	qRT-PCR	Ag-NPs induced ROS accumulation; interfered with ethylene biosynthesis; promoted root growth; triggered gene expression involved in cellular events-cell proliferation, metabolism, and hormone signaling pathways.	[46]
Arabidopsis thaliana (Col-0/Redei-L211497)	Seedling	Germinating	Ag-NPs (20 nm; 5 ppm)	10 days	Microarray analysis	Up-regulated genes primarily associated with metals and oxidative stress, while down-regulated genes linked to biotic and hormonal stimuli.	[47]
Chlamydomonas reinhardtii (wild-type strain C137)	–	–	Ag (20 nm; 1 ppm), TiO_2(5 nm; 1 ppm), ZnO (20 nm; 1 ppm), QDs (6–10 nm, 0.12 ppm)	2 h	RNA-seq analysis	Genes associated with photosynthesis were markedly decreased on exposure to TiO_2. Ag-NPs exposure led to the elevation of transcripts encoding components of cell wall and flagella.	[48]
Arabidopsis thaliana (wild type, cv. Columbia)	Root	3-week-old	ZnO (<100 nm; 100 ppm), TiO_2 (<150 nM; 100 ppm), FS (100 ppm)	7 days	Microarray analysis	Both abiotic (oxidative, salt, water deprivation) and biotic (wounding and defense to pathogens) stress responsive genes were up-regulated under ZnO and FS; while cell organization and biogenesis associated genes were down-regulated upon ZnO-NPs.	[40]

Abbreviation: FS, fullerene soot; MWNT, multi-wall nanotubes; QDs, CdTe/CdS quantum dots.

Table 2. Summary of proteomic analyses in response to nanoparticle stress.

Plant (Cultivar)	Organ	Growth Stage	Nanoparticles (Particle Size, Dose)	Treatment Period	Technique	Major Findings	Ref.
Soybean (*Glycine max* L. cv. Enrei)	Root including hypocotyl	2-day-old	Ag, ZnO, Al$_2$O$_3$-NPs with flooding stress (0.5–500 ppm)	1–3 days	Gel-free (nanoLC MS/MS)	Al$_2$O$_3$-NPs (50 ppm) promote seedling growth under flooding stress by regulating energy metabolism and cell death.	[49]
Soybean (*Glycine max* L. cv. Enrei)	Root, cotyledon	2-day-old	Ag-NPs with flooding stress (2, 15, 50–80 nm; 0.2, 2, 20 ppm)	1–4 days	Gel-free (nanoLC MS/MS)	Ag-NPs (15 nm: 2 ppm) treatments facilitate seedling growth under flooding stress. Decreased abundance of glyoxalase II 3 and fermentation-related proteins: pyruvate decarboxylase 2 and alcohol dehydrogenase 1, indicating metabolic shift from fermentative pathways towards normal cellular processes.	[44]
Rice (*Oryza sativa* L. cv. IR651)	Root	10-day-old	Ag-NPs (30, 60 ppm)	20 days	Gel-based (2-DE, nanoLC/FT-ICR MS)	Increased abundance of proteins related to oxidative stress response pathway, Ca^{2+} regulation signaling, transcription, protein degradation, cell wall synthesis, cell division and apoptosis.	[43]
Eruca sativa (common name: rocket)	Root	Germinating seeds	Ag-NPs or AgNO$_3$ (0.1, 1, 10, 20, 100 ppm)	5 days	Gel-based (2-DE, nanoLC-nESI-MS/MS)	Alteration of some proteins related to the ER and vacuole indicating these two organelles as targets of the Ag-NPs action. Effects of Ag-NPs are not simply due to the release of Ag$^+$.	[33]

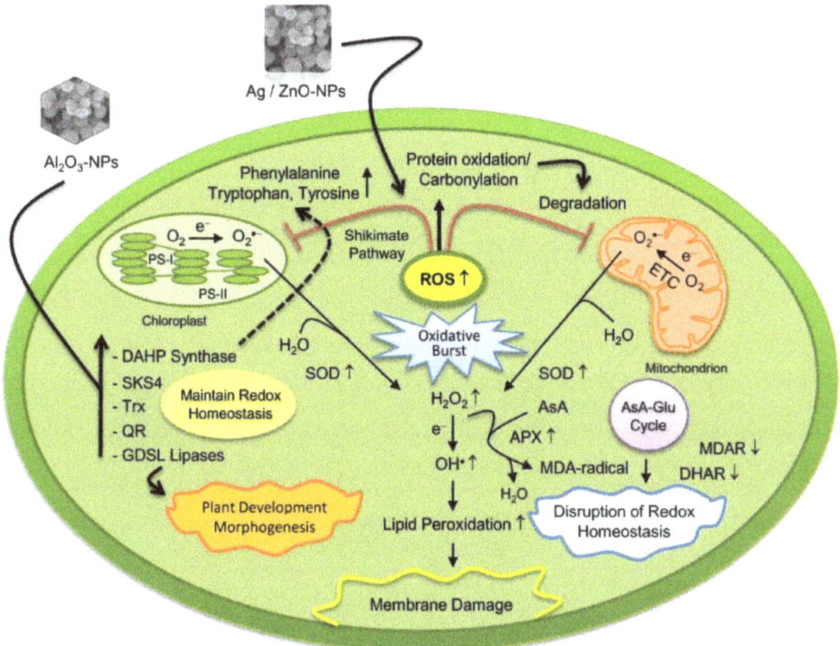

Figure 1. Cellular toxicity induced by nanoparticles (NPs). Exposure to NPs potentially leads to toxic side effects such as enhanced ROS generation, disruption of redox homeostasis, lipid peroxidation, impaired mitochondrial function, and membrane damage. Upward arrows indicate increased and downward arrows indicate decreased protein abundance in response to NPs stress, respectively. Dotted arrow represents shikimate pathway, a common biosynthetic route for the synthesis of aromatic amino acids. Abbreviations: APX, ascorbate peroxidase; AsA, reduced ascorbate; DAHP, 3-deoxy-*D*-arabino-heptulosonate-7-phosphate; DHAR, dehydroascorbate reductase; ETC, electron transport chain; H_2O_2, hydrogen peroxide; MDA, malondialdehyde; MDAR, monodehydroascorbate reductase; PS, photosystem; QR, quinone reductase; ROS, reactive oxygen species; SKS4, SKU5 similar 4 protein; SOD, superoxide dismutase; Trx, thioredoxin.

In roots and cotyledons the abundance of glyoxalase II 3, an important enzyme of glyoxalase detoxification pathway, was increased in a time-course manner under flooding stress; however, it declined in response to Ag-NPs. Furthermore, Ag-NPs treatment caused a metabolic shift from fermentative pathways towards normal cellular processes. The results suggested that the Ag-NPs (15 nm at 2 ppm) treated soybeans experienced less deprivation of oxygen, which acts as an important factor for enhanced growth of soybeans under Ag-NPs treatment with flooding stress. In contrast, high concentration of Ag-NPs (20 ppm, 15 nm particle size) was lethal to soybean seedlings [44].Very recently, we compared the effects of Ag-, ZnO- and Al₂O₃-NPs on two-day-old soybean under flooding stress [49]. Interestingly, enhanced soybean growth was observed in 50 ppm Al₂O₃-NPs treatment. The Al₂O₃-NPs-responsive proteins were predominantly related to protein synthesis/degradation, glycolysis, and lipid metabolism. Moreover, 5-fold enhanced abundance of NmrA-like negative transcriptional regulator family protein was recorded under Al₂O₃-NPs treatment. In summary, proteomic findings suggest that regulation of energy metabolism and reduced root cell death might promote soybean growth under flooding stress.

4. Conclusions and Future Prospects

By summarizing proteomic contributions, efforts have been made in the present review to delineate the molecular basis of acquisition of nanoparticles stress response mechanism. However, only limited numbers of proteomic studies have so far been conducted in the plant system. Most of the studies carried out so far primarily deal with the overall plant response towards a specific NPs stress showing differential abundance of proteins involved in oxidation-reduction, ROS detoxification, stress signaling, and hormonal pathways. Proteomic studies on Ag-NPs induced phytotoxicity revealed that the size of the nanoparticle is the key factor in determining the type and magnitude of the cellular response. Future initiatives need to be taken to find out whether the metallic nanoparticles exert their toxicity solely due to their unique properties or to the released metal ions. Moreover, research aimed at identifying and characterizing subcellular organelle proteins are expected for exploring the precise alterations in the protein signature of cell to withstand the NPs stress. In addition to proteomics, other "omics" based high-throughput techniques such as transcriptomics and metabolomics have immense potential to evaluate the effects and toxicity of nanoscale materials [45,50]. Metabolomics allows fast screening for biomarkers of oxidative stress following the application of NPs. Moreover, combination of NMR- and LC/MS-based metabolomics approach is being exploited to investigate the specific pathways of interest including those related to oxidative stress, an inevitable consequence of nanoparticle exposure [50]. MALDI MS imaging technique, a powerful tool for nanotoxicology study, provides a snapshot of how NPs are distributed in tissues, which is important for characterizing and understanding nanomaterial-based toxicity [51]. Furthermore, the plant's response to combined NPs would be another topic for future "omics" based research that could highlight the possible interaction between stress signaling pathways. All these valuable information would further provide us an extensive and elaborated picture about the response mechanism of NPs stress in plants.

Acknowledgments: The authors gratefully acknowledge financial support from the National Institute of Crop Science, Tsukuba, Japan.

Author Contributions: Zahed Hossain wrote and edited, Ghazala Mustafa wrote, Setsuko Komatsu advised the review process and edited the manuscript. All authors read and approved the final version of this review article.

Conflicts of Interest: The authors declare no conflict of interest.

References

1. Nel, A.; Xia, T.; Mädler, L.; Li, N. Toxic potential of materials at the nanolevel. *Science* **2006**, *311*, 622–627. [CrossRef] [PubMed]
2. Nowack, B.; Bucheli, T.D. Occurrence, behavior and effects of nanoparticles in the environment. *Environ. Pollut.* **2007**, *150*, 5–22. [CrossRef] [PubMed]
3. Jefferson, D.A. The surface activity of ultrafine particles. *Philos. Trans. R. Soc. Lond. Ser. A* **2000**, *358*, 2683–2692. [CrossRef]
4. Brunner, T.J.; Wick, P.; Manser, P.; Spohn, P.; Grass, R.N.; Limbach, L.K.; Bruinink, A.; Stark, W.J. *In vitro* cytotoxicity of oxide nanoparticles: Comparison to asbestos, silica, and the effect of particle solubility. *Environ. Sci. Technol.* **2006**, *40*, 4374–4381. [CrossRef] [PubMed]
5. Owen, R.; Handy, R. Formulating the problems for environmental risk assessment of nanomaterials. *Environ. Sci. Technol.* **2007**, *41*, 5582–5588. [PubMed]
6. Dietz, K.J.; Herth, S. Plant nanotoxicology. *Trends Plant Sci.* **2011**, *16*, 582–589. [CrossRef] [PubMed]
7. Fleischer, A.; O'Neill, M.A.; Ehwald, R. The pore size of non-graminaceous plant cell walls is rapidly decreased by borate ester cross-linking of the pectic polysaccharide rhamnogalacturonan II. *Plant Physiol.* **1999**, *121*, 829–838. [CrossRef] [PubMed]
8. Navarro, E.; Baun, A.; Behra, R.; Hartmann, N.B.; Filser, J.; Miao, A.J.; Quigg, A.; Santschi, P.H.; Sigg, L. Environmental behavior and ecotoxicity of engineered nanoparticles to algae, plants, and fungi. *Ecotoxicology* **2008**, *17*, 372–386. [CrossRef] [PubMed]

9. Verano-Braga, T.; Miethling-Graff, R.; Wojdyla, K.; Rogowska-Wrzesinska, A.; Brewer, J.R.; Erdmann, H.; Kjeldsen, F. Insights into the cellular response triggered by silver nanoparticles using quantitative proteomics. *ACS Nano* **2014**, *8*, 2161–2175. [CrossRef] [PubMed]

10. Auffan, M.; Achouak, W.; Rose, J.; Roncato, M.A.; Chanéac, C.; Waite, D.T.; Masion, A.; Woicik, J.C.; Wiesner, M.R.; Bottero, J.Y. Relation between the redox state of iron-based nanoparticles and their cytotoxicity toward *Escherichia coli*. *Environ. Sci. Technol.* **2008**, *42*, 6730–6735. [CrossRef] [PubMed]

11. Foley, S.; Crowley, C.; Smaihi, M.; Bonfils, C.; Erlanger, B.F.; Seta, P.; Larroque, C. Cellular localisation of a water-soluble fullerene derivative. *Biochem. Biophys. Res. Commun.* **2002**, *294*, 116–119. [CrossRef]

12. Kamat, J.P.; Devasagayam, T.P.; Priyadarsini, K.I.; Mohan, H. Reactive oxygen species mediated membrane damage induced by fullerene derivatives and its possible biological implications. *Toxicology* **2000**, *155*, 55–61. [CrossRef]

13. Petersen, E.J.; Henry, T.B.; Zhao, J.; MacCuspie, R.I.; Kirschling, T.L.; Dobrovolskaia, M.A.; Hackley, V.; Xing, B.; White, J.C. Identification and avoidance of potential artifacts and misinterpretations in nanomaterial ecotoxicity measurements. *Environ. Sci. Technol.* **2014**, *48*, 4226–4246. [CrossRef] [PubMed]

14. Slomberg, D.L.; Schoenfisch, M.H. Silica nanoparticle phytotoxicity to *Arabidopsis thaliana*. *Environ. Sci. Technol.* **2012**, *46*, 10247–10254. [PubMed]

15. Begum, P.; Fugetsu, B. Phytotoxicity of multi-walled carbon nanotubes on red spinach (*Amaranthus tricolor* L) and the role of ascorbic acid as an antioxidant. *J. Hazard. Mater.* **2012**, *243*, 212–222. [CrossRef] [PubMed]

16. Marschner, H. *Mineral. Nutrition of Higher Plants*, 2nd ed.; Academic Press: San Diego, CA, USA, 1995.

17. Ma, X.; Geisler-Lee, J.; Deng, Y.; Kolmakov, A. Interactions between engineered nanoparticles (ENPs) and plants: Phytotoxicity, uptake and accumulation. *Sci. Total Environ.* **2010**, *408*, 3053–3061. [CrossRef] [PubMed]

18. Khodakovskaya, M.; Dervishi, E.; Mahmood, M.; Xu, Y.; Li, Z.; Watanabe, F.; Biris, A.S. Carbon nanotubes are able to penetrate plant seed coat and dramatically affect seed germination and plant growth. *ACS Nano* **2009**, *3*, 3221–3227. [CrossRef] [PubMed]

19. Lee, C.W.; Mahendra, S.; Zodrow, K.; Li, D.; Tsai, Y.C.; Braam, J.; Alvarez, P.J. Developmental phytotoxicity of metal oxide nanoparticles to *Arabidopsis thaliana*. *Environ. Toxicol. Chem.* **2010**, *29*, 669–675. [CrossRef] [PubMed]

20. Arouja, V.; Dubourguier, H.C.; Kasemets, K.; Kahru, A. Toxicity of nanoparticles of CuO, ZnO, and TiO_2 to microalgae *Pseudokirchneriella subcapitata*. *Sci. Total Environ.* **2009**, *407*, 1461–1468. [CrossRef] [PubMed]

21. Buzea, C.; Pacheco, I.I.; Robbie, K. Nanomaterials and nanoparticles: sources and toxicity. *Biointerphases* **2007**, *2*, 17–71. [CrossRef]

22. Stampoulis, D.; Sinha, S.K.; White, J.C. Assay-dependent phytotoxicity of nanoparticles to plants. *Environ. Sci. Technol.* **2009**, *43*, 9473–9479. [CrossRef] [PubMed]

23. Boonyanitipong, P.; Kositsup, B.; Kumar, P.; Baruah, S.; Dutta, J. Toxicity of ZnO and TiO_2 nanoparticles on germinating rice seed *Oryza sativa* L. *Int. J. Biosci. Biochem. Bioninform.* **2011**, *1*, 282–285. [CrossRef]

24. Riahi-Madvar, A.; Rezaee, F.; Jalili, V. Effects of alumina nanoparticles on morphological properties and antioxidant system of *Triticum aestivum*. *Iran. J. Plant Physiol.* **2012**, *3*, 595–603.

25. Shaw, A.K.; Hossain, Z. Impact of nano-CuO stress on rice (*Oryza sativa* L.) seedlings. *Chemosphere* **2013**, *93*, 906–915. [CrossRef] [PubMed]

26. Shaw, A.K.; Ghosh, S.; Kalaji, H.M.; Bosa, K.; Brestic, M.; Zivcak, M.; Hossain, Z. Nano-CuO stress induced modulation of antioxidative defense and photosynthetic performance of syrian barley (*Hordeum vulgare* L.). *Environ. Exp. Bot.* **2014**, *102*, 37–47. [CrossRef]

27. Atha, D.H.; Wang, H.; Petersen, E.J.; Cleveland, D.; Holbrook, R.D.; Jaruga, P.; Dizdaroglu, M.; Xing, B.; Nelson, B.C. Copper oxide nanoparticle mediated DNA damage in terrestrial plant models. *Environ. Sci. Technol.* **2012**, *46*, 1819–1827. [CrossRef] [PubMed]

28. Rico, C.M.; Hong, J.; Morales, M.I.; Zhao, L.; Barrios, A.C.; Zhang, J.Y.; Peralta-Videa, J.R.; Gardea-Torresdey, J.L. Effect of cerium oxide nanoparticles on rice: A study involving the antioxidant defense system and *in vivo* fluorescence imaging. *Environ. Sci. Technol.* **2013**, *47*, 5635–5642. [CrossRef] [PubMed]

29. Thwala, M.; Musee, N.; Sikhwivhilu, L.; Wepener, V. The oxidative toxicity of Ag and ZnO nanoparticles towards the aquatic plant *Spirodela punctata* and the role of testing media parameters. *Environ. Sci. Process. Impacts* **2013**, *15*, 1830–1843. [CrossRef] [PubMed]

30. Sharma, V.K.; Yngard, R.A.; Lin, Y. Silver nanoparticles: Green synthesis and their antimicrobial activities. *Adv. Colloid. Interface Sci.* **2009**, *145*, 83–96. [CrossRef] [PubMed]
31. Tran, Q.H.; Nguyen, V.Q.; Le, A.T. Silver nanoparticles: synthesis, properties, toxicology, applications and perspectives. *Adv. Nat. Sci.* **2013**, *4*, 033001. [CrossRef]
32. Jiang, H.S.; Li, M.; Chang, F.Y.; Li, W.; Yin, L.Y. Physiological analysis of silver nanoparticles and AgNO$_3$ toxicity to *Spirodela polyrrhiza*. *Environ.Toxicol. Chem.* **2012**, *31*, 1880–1886. [CrossRef] [PubMed]
33. Vannini, C.; Domingo, G.; Onelli, E.; Prinsi, B.; Marsoni, M.; Espen, L.; Bracale, M. Morphological and proteomic responses of *Eruca sativa* exposed to silver nanoparticles or silver nitrate. *PLoS ONE* **2013**, *8*, e68752. [CrossRef] [PubMed]
34. Kumari, M.; Mukherjee, A.; Chandrasekaran, N. Genotoxicity of silver nanoparticles in *Allium cepa*. *Sci. Total Environ.* **2009**, *407*, 5243–5246. [CrossRef] [PubMed]
35. Sharma, P.; Bhatt, D.; Zaidi, M.G.; Saradhi, P.P.; Khanna, P.K.; Arora, S. Silver naoparticle-mediated enhancement in growth and antioxidant status of *Brassica juncea*. *Appl. Biochem. Biotechnol.* **2012**, *167*, 2225–2233. [CrossRef] [PubMed]
36. Yin, L.; Colman, B.P.; McGill, B.M.; Wright, J.P.; Bernhardt, E.S. Effects of silver nanoparticle exposure on germination and early growth of eleven wetland plants. *PLoS ONE* **2012**, *7*, e47674. [CrossRef] [PubMed]
37. Salama, H.M.H. Effects of silver nanoparticles in some crop plants, common bean (*Phaseolus vulgaris* L.) and corn (*Zea mays* L.). *Int. Res. J. Biotechnol.* **2012**, *3*, 190–197.
38. Roh, J.Y.; Eom, H.J.; Choi, J. Involvement of *Caenorhabditis elegans* MAPK signaling pathways in oxidative stress response induced by silver nanoparticles exposure. *Toxicol. Res.* **2012**, *28*, 19–24. [CrossRef] [PubMed]
39. Gopinath, P.; Gogoi, S.K.; Sanpui, P.; Paul, A.; Chattopadhyay, A.; Ghosh, S.S. Signaling gene cascade in silver nanoparticle induced apoptosis. *Colloids Surf. B* **2010**, *77*, 240–245. [CrossRef] [PubMed]
40. Landa, P.; Vankova, R.; Andrlova, J.; Hodek, J.; Marsik, P.; Storchova, H.; White, J.C.; Vanek, T. Nanoparticle-specific changes in *Arabidopsis thaliana* gene expression after exposure to ZnO, TiO$_2$, and fullerene soot. *J. Hazard. Mater.* **2012**, *241–242*, 55–62. [CrossRef] [PubMed]
41. Khodakovskaya, M.V.; de Silva, K.; Nedosekin, D.A.; Dervishi, E.; Biris, A.S.; Shashkov, E.V.; Galanzha, E.I.; Zharov, V.P. Complex genetic, photothermal, and photoacoustic analysis of nanoparticle-plant interactions. *Proc. Natl. Acad. Sci. USA* **2011**, *108*, 1028–1033. [CrossRef] [PubMed]
42. Yan, S.; Zhao, L.; Li, H.; Zhang, Q.; Tan, J.; Huang, M.; He, S.; Li, L. Single-walled carbon nanotubes selectively influence maize root tissue development accompanied by the change in the related gene expression. *J. Hazard. Mater.* **2013**, *246–247*, 110–118. [CrossRef] [PubMed]
43. Mirzajani, F.; Askari, H.; Hamzelou, S.; Schober, Y.; Römpp, A.; Ghassempour, A.; Spengler, B. Proteomics study of silver nanoparticles toxicity on *Oryza sativa* L. *Ecotoxicol. Environ. Saf.* **2014**, *108*, 335–339. [CrossRef] [PubMed]
44. Mustafa, G.; Sakata, K.; Hossain, Z.; Komatsu, S. Proteomic study on the effects of silver nanoparticles on soybean under flooding stress. *J. Proteom.* **2015**, *122*, 100–118. [CrossRef] [PubMed]
45. García-Sánchez, S.; Bernales, I.; Cristobal, S. Early response to nanoparticles in the *Arabidopsis* transcriptome compromises plant defence and root-hair development through salicylic acid signalling. *BMC Genom.* **2015**, *16*, 341–356. [CrossRef] [PubMed]
46. Syu, Y.Y.; Hung, J.H.; Chen, J.C.; Chuang, H.W. Impacts of size and shape of silver nanoparticles on *Arabidopsis* plant growth and gene expression. *Plant Physiol. Biochem.* **2014**, *83*, 57–64. [CrossRef] [PubMed]
47. Kaveh, R.; Li, Y.S.; Ranjbar, S.; Tehrani, R.; Brueck, C.L.; van Aken, B. Changes in *Arabidopsis thaliana* gene expression in response to silver nanoparticles and silver ions. *Environ. Sci. Technol.* **2013**, *47*, 10637–10644. [PubMed]
48. Simon, D.F.; Domingos, R.F.; Hauser, C.; Hutchins, C.M.; Zerges, W.; Wilkinson, K.J. Transcriptome sequencing (RNA-seq) analysis of the effects of metal nanoparticle exposure on the transcriptome of *Chlamydomonas reinhardtii*. *Appl. Environ. Microbiol.* **2013**, *79*, 4774–4785. [CrossRef] [PubMed]
49. Mustafa, G.; Sakata, K.; Komatsu, S. Proteomic analysis of flooded soybean root exposed to aluminum oxide nanoparticles. *J. Proteom.* **2015**, *128*, 280–297. [CrossRef] [PubMed]
50. Schnackenberg, L.K.; Sun, J.; Beger, R.D. Metabolomics techniques in nanotoxicology studies. *Methods Mol. Biol.* **2012**, *926*, 141–156. [PubMed]

51. Burnum, K.E.; Frappier, S.L.; Caprioli, R.M. Matrix-assisted laser desorption/ionization imaging mass spectrometry for the investigation of proteins and peptides. *Annu. Rev. Anal. Chem.* **2008**, *1*, 689–705. [CrossRef] [PubMed]

 International Journal of
Molecular Sciences

Review

Maize-Pathogen Interactions: An Ongoing Combat from a Proteomics Perspective

Olga Pechanova [1] and Tibor Pechan [2,*]

[1] Mississippi State Chemical Laboratory, Mississippi State University, Mississippi State, MS 39762, USA; opechanova@mscl.msstate.edu

[2] Institute for Genomics, Biocomputing and Biotechnology, Mississippi State University, Mississippi State, MS 39762, USA

* Correspondence: pechan@ra.msstate.edu; Tel.: +1-662-325-8278; Fax: +1-662-325-7692

Academic Editors: Setsuko Komatsu and Zahed Hossain
Received: 1 October 2015; Accepted: 18 November 2015; Published: 30 November 2015

Abstract: Maize (*Zea mays* L.) is a host to numerous pathogenic species that impose serious diseases to its ear and foliage, negatively affecting the yield and the quality of the maize crop. A considerable amount of research has been carried out to elucidate mechanisms of maize-pathogen interactions with a major goal to identify defense-associated proteins. In this review, we summarize interactions of maize with its agriculturally important pathogens that were assessed at the proteome level. Employing differential analyses, such as the comparison of pathogen-resistant and susceptible maize varieties, as well as changes in maize proteomes after pathogen challenge, numerous proteins were identified as possible candidates in maize resistance. We describe findings of various research groups that used mainly mass spectrometry-based, high through-put proteomic tools to investigate maize interactions with fungal pathogens *Aspergillus flavus*, *Fusarium* spp., and *Curvularia lunata*, and viral agents Rice Black-streaked Dwarf Virus and Sugarcane Mosaic Virus.

Keywords: proteomics; maize; pathogen; interaction; defense; resistance; mycotoxin; pathogenesis-related; *Aspergillus flavus*; *Fusarium* spp.; *Curvularia lunata*; plant virus

1. Introduction

Maize is the most grown cereal crop in the world (990.64 million tons per year according to USDA WASDE report, May 2015) valued for its nutritional properties as food and feed, and as a source material for many industrial uses. Diseases inflicted on this important plant species by naturally occurring pathogens are detrimental to the economy and threaten the food supplies around the globe, especially in the regions where maize is a staple food commodity. In addition, fungal infection and consequent contamination with mycotoxins pose serious health hazards to both humans and animals, which can be directly infected by fungus or can be affected by consumption of contaminated food and feed. Among the mycotoxins, the aflatoxin B1 (product of *Aspergillus flavus*) is the most carcinogenic compound found in nature [1,2], and it represents the most dangerous agent the maize producers and consumers might potentially face.

For decades, researches around the world have been investigating the processes underlying the mechanism of maize pathogen recognition and the launch of effective defenses. Engaging the most advanced system biology approaches, including "omics" disciplines, such as genomics, transcriptomics, and proteomics, they have moved towards a better understanding of the many processes occurring during maize interactions with its pathogens. A thorough knowledge of molecular mechanisms of maize host response to pathogens is critical for the elucidation of a genetic basis of host resistance.

The advent of mass spectrometry (MS) application in proteomic research caused a true paradigm shift, allowing for many scientists to focus on the maize proteome. Continuous advancements of high-throughput technologies, availability of protein, EST and genomic databases (B73 genotype [3]) further accelerated the maize proteomics field.

In this review, we present a summary of proteome-based research of maize interaction with, and defense against, diverse pathogens. Various proteomic techniques were used, including both gel-based and gel-free approaches. Numerous potential protein and gene candidates in maize disease resistance were identified. Of special interest, these proteins could serve as selective markers for the development of resistant elite varieties, either through conventional breeding or via genetic engineering.

2. Maize-Pathogen Interactions from a Proteomics View

2.1. Maize vs. Aspergillus Flavus and Aflatoxins

Aspergillus flavus is a globally found fungal pathogen that causes disease in many agricultural crops and contaminates them with aflatoxins, very toxic metabolites that are produced via secondary metabolism [4,5]. Aspergillus ear rot caused by *A. flavus* is the major maize disease worldwide, and grains contaminated with aflatoxins present immense agronomical problems leading to more than one billion of dollars lost annually, according to The American Phytopathological Society [6]. Disease development can occur pre- and post-harvest and is most serious in the Southern United States, where hot weather and frequent droughts trigger aflatoxin production. If not controlled, aflatoxins might be present in a wide range of maize-based foods and feeds, as well as in dairy products. They pose serious health hazards to both humans and animals, if digested via contaminated food and feed. In humans, aflatoxins have been directly linked to hepatocellular carcinoma, since they are metabolized in the liver [7].

2.1.1. Kernel Resistance

A notable number of proteomic investigations has been performed on maize kernel tissues [8–17]. Chen and colleagues examined numerous maize genotypes with dissimilar levels of resistance to aflatoxins accumulation. In one of their earliest studies using one-dimensional electrophoresis (1DE) and N-terminal sequencing, they identified a 14-kDa trypsin inhibitor that was significantly more abundant in kernel protein extracts from seven aflatoxin-resistant inbred lines, while not detectable or less abundant in kernels of six susceptible genotypes [13]. High constitutive levels of this protein are believed to be contributing to the maize kernels' resistance as demonstrated by antifungal assays. Trypsin inhibitor purified from corn kernels caused spore rupture and abnormal hyphal growth of *A. flavus*. It also inhibited *A. flavus* growth via inhibition of fungal α-amylase, under certain experimental conditions [15]. Including other kernel tissues (endosperm and embryo), and applying more advanced proteomic tools such as two-dimensional electrophoresis (2DE) and tandem mass spectrometry (MS/MS), additional constitutive resistance-related proteins were identified [10–12]. Both embryo and endosperm of resistant lines (MP420, Mp313E, GT-MAS:gk, CI2, MI82, and T115) exhibited similar defense patterns and unique and/or higher expression (at least 5-fold) of several interesting proteins. These proteins can be grouped into the following categories: (i) storage proteins, which include globulin 1 and 2, and late embryogenesis abundant proteins (LEA3, LEA 14); (ii) stress-related proteins, such as aldose reductase (ALD), osmotic stress-related proteins WSI18, peroxredoxin antioxidant (PER1), cold regulated protein, anionic peroxidase, glyoxalase I protein (GLX I) and several small heat shock proteins (HSP); and (iii) antifungal proteins, such as trypsin inhibitor and pathogenesis-related (PR) protein 10 (PR-10). Antifungal PR-10 protein undoubtedly plays a role in host kernel resistance. Overexpression of *ZmPR-10* gene was found to be inhibitory to both *A. flavus* hyphal growth and conidial germination [14]. Furthermore, using RNAi gene silencing approach, fungal colonization and aflatoxin accumulation were more extensive in *pr10*-silenced transgenic kernels, while *pr10*-silenced callus lines displayed enhanced sensitivity to heat stress treatment, and significantly reduced *pr10* transcription.

Based on their observations, the authors proposed the existence of an association between stress tolerance and disease resistance [8]. LEA proteins, for example, are known to protect higher plants from dehydration caused by environmental stresses, especially drought [18]. In maize kernels, LEA proteins probably assist in aflatoxin resistance as well. Direct and active involvement of GLX1 in aflatoxin resistance was validated through the control of its substrate, methylglyoxal (MG), which induced aflatoxin production in infected kernels [11]. Few susceptible genotypes did exhibit significant increases in MG content when infected. The evidence of association between stress tolerance and disease resistance can be exploited in the future as a novel strategy to enhance maize resistance to fungal diseases.

2.1.2. Rachis Resistance

Contamination of grains is of the highest concern for maize marketing and consumption. Therefore, a majority of the research is being devoted to this tissue. However, infection of other parts of the maize plant is equally important, as these organs may contribute to the resistance or susceptibility by affecting the pathogen's behavior within a maize ear and/or the whole plant. For example, maize rachis (cob) plays a role in the delivery of nutrients to developing kernels, and it was reported that *A. flavus* uses it as a conduit for its spread within a maize ear [19,20]. When developing ears of aflatoxin-susceptible and aflatoxin-resistant hybrids were inoculated with GUS-tagged *A. flavus* strain, fungus moved freely throughout the maize cob and to the kernels in susceptible hybrids. In resistant hybrids, however, fungal spread was halted in rachis preventing it from entering the kernels [19]. The same phenomenon was observed in maize inbreds [20]. To elucidate the possible role of rachis in aflatoxin resistance, a comprehensive large-scale study was carried out by Pechanova *et al.* [21,22]. The two-dimensional Difference In Gel Electrophoresis (2D-DIGE) comparison of rachis from 21-day-old maize of four genetically unrelated maize inbreds (aflatoxin-resistant Mp313E and Mp420, and aflatoxin-susceptible SC212m and B73) revealed that resistant rachis contained 44 proteins of significantly higher abundance, mainly abiotic stress-related proteins such as heat shock proteins, antioxidant enzymes (superoxide dismutase (SOD), ascorbate peroxidase (APX), thioredoxin, *etc.*), and other diverse stress-related proteins. Proteins from the phenylpropanoid pathway, such as phenylalanine ammonia lyase (PAL), caffeoyl-CoA-3-*O*-methyltransferase1 and chalcone flavonone isomerase, were also present at considerably higher levels. Susceptible rachis, on the other hand, had significantly higher expression of 26 proteins, the majority of which were biotic stress-associated PR proteins, such as chitinases, basic endochitinase C and germin-like proteins (GLP) subfamily 1 member 17.

In regards to a fungal challenge, tremendous differences between resistant and susceptible genotypes were observed in response to *A. flavus*, especially after a long-term exposure. Thirty-five days post-infection, genotypes Mp313E and SC212m reacted with differential expression of 30 and 48 proteins, respectively. However, unlike resistant Mp313E, susceptible inbred SC212m exhibited a much more robust response via strong upregulation (3–10-fold) of fifteen different PR proteins including chitinases, glucanases, protein P21, permatin, PR-1, PR-5, PRm3, and PRm6b. The resistant genotype did not induce strong defenses as these proteins were already constitutively accumulated in its rachis, as shown in proteomic maturation studies (young *vs.* mature rachis). The most remarkable difference was observed for PRm3 (class III chitinase) that accumulated 40-fold during maturation of resistant rachis, as opposed to a 7.3-fold increase in the susceptible line. These findings indicated that rachis from resistant and susceptible genotypes differ in their defense mechanisms. While susceptible rachis relies on inducible defenses via action of antifungal proteins, resistant rachis is almost exclusively dependent on constitutive defenses. Abiotic stress-related proteins appear to be vital for young rachis to control oxidative stress caused by heat and drought, whereas biotic stress-responsive proteins, such as PR proteins, become more important as the ear matures. The subset of defense-related proteins including several isoforms of chitinases, PRm 6b, β-1.3-glucanase, GLP subfamily 1 member 17, catalase 3, Asr protein, abscisic stress ripening protein 1, auxin-binding protein 1, abscisic acid (ABA)-responsive protein, caffeoyl-CoA 3-*O*-methyltransferase 1, and remorin might be especially important for rachis resistance, as these proteins mapped near quantitative trait locus (QTL) for resistance to aflatoxin accumulation present on several chromosomes [22].

Similarly to kernels [13], association between abiotic stress tolerance and aflatoxin-resistance seem to be a crucial factor in rachis defense. This was the most obvious for two small HSPs that were almost 11-fold more abundant in resistant *vs.* susceptible rachis [22]. Unlike in susceptible inbreds, resistant rachis is able to cope better with heat and drought challenges via high expression of abiotic stress-related proteins, which simultaneously provide a good protection against *A. flavus*/aflatoxins [22].

2.1.3. Silk Resistance

Comparative protein profiles of silk from maize genotypes (resistant Mp313E, Mp420, and susceptible SC212m, Mp339) were also investigated via 2DE [23]. Silk seems to play an essential role in pre-harvest contamination of maize, as it is easily accessible to the pathogenic agents, thus serving as a main point of entry and conduit for microbial infection. Once external silk of field-grown maize is colonized with *A. flavus*, fungal growth down the silk into the cob interior is rapid, and followed by subsequent colonization of kernel surfaces [24]. At the same time, silk very likely serves as a first line of defense against microbial infection, and its antifungal properties have previously been reported [25]. Three PR proteins could play a major role in *A. flavus*/aflatoxin resistance. Hydrolytic enzymes chitinase A, PRm3 chitinase, and chitinase I were constitutively expressed at higher levels in silks of both resistant maize varieties when compared to their susceptible counterparts [23]. These findings were supported by antifungal assay showing that extracts from resistant inbreds had higher antifungal activity than those from susceptible lines.

2.1.4. Summary of Maize Protein-Based Defenses against *A. flavus*

Maize resistance to *A. flavus* infection and aflatoxin accumulation is a polygenic, quantitatively inherited trait strongly influenced by the environment [26–28].

As indicated from proteomic studies reviewed above, maize ear defense is correspondingly a very complicated mechanism consisting of diverse proteins, metabolites, and signaling molecules. It involves antifungal activity, host cells detoxification, cell wall reinforcement, primary and secondary metabolism, signaling, *etc.* Constitutive proteins seem to serve as major resistance factors, but inducible defenses are also essential. Figure 1 summarizes protein-based defenses of maize plant in tissues that are most instrumental to both infection and resistance by/against the *A. flavus*.

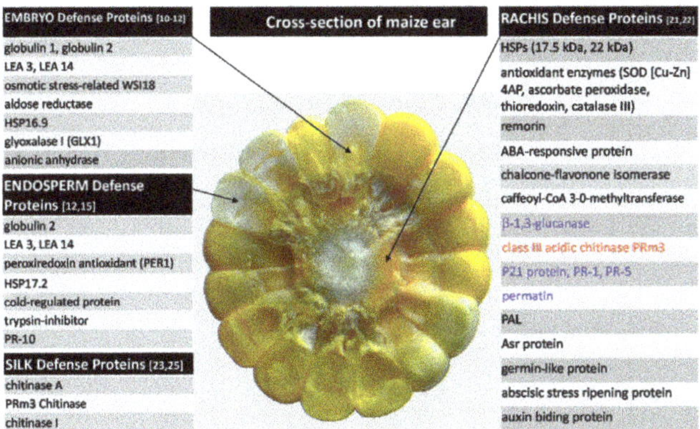

Figure 1. Selected proteins associated with maize defense against *Aspergillus flavus*. References are given in parenthesis next to the tissue-specific protein groups. Proteins in black and blue color represent constitutive and induced proteins, respectively. Protein in red was constitutively more abundant in resistant line, and induced in both resistant and susceptible lines (Modified from [29]).

2.2. Maize vs. Fusarium Pathogens

Fusarium species represent another major group of fungal pathogens associated with maize diseases. This ubiquitous soil-borne genus contains several toxigenic species, with *F. graminearum* and *F. verticillioides* being the most common pathogens of maize. They can infect different parts of the maize plant in all stages of its development, causing significant reductions in maize yield and quality. Most importantly, under certain favorable environmental conditions during both pre- and post-harvest processing, these fungi produce harmful mycotoxins further diminishing maize crops [30,31]. The most common mycotoxins produced by toxigenic Fusaria include fumonisin, deoxynivalenol, and zearalenone. They contaminate maize grains, and therefore represent similar health hazard like aflatoxins.

2.2.1. *F. graminearum* and Gibberella Ear Rot

Gibberella ear rot is caused by *F. graminearum* and results in moldy, mycotoxin-contaminated maize kernels. Infection is more prevalent in cool and wet weather, at the beginning stages of silking, and usually originates at the tip of an ear, followed by spreading towards the base [32]. Deoxynivalenol and zearalenone produced by this fungus are harmful to both humans and farm animals.

To evaluate defense processes occurring early upon encountering this pathogen, protein profiles of developing maize kernels of two inbred lines were examined 48 h post-inoculation [33]. In this large-scale study conducted over three growing seasons, 878 proteins were identified and quantified via iTRAQ MS/MS as differentially expressed between the mock- and fungal-treated kernels, and/or between the two inbred lines. Both genotypes, resistant (CO441) and susceptible (B73), exhibited changes in proteomes during both mock inoculation and pathogen challenge. Ninety-six proteins were differentially abundant in at least one of the treatments. They consisted of several major groups of defense proteins including PR proteins PR-10, chitinases, xylanase inhibitors (XIP), thaumatin-like protein, zeamatin precursor, GLPs, and peroxidase. Not only were these proteins among the most strongly induced ones, but also they were often at significantly higher levels in the resistant *vs.* susceptible genotype. Proteins PR-10, PRm3 chitinase, and PR-5 thaumatin, were also identified via 2DE approach in *Fusarium*-treated silks of B73 inbred, 48 h post-infection, playing role in silk defense [33]. Differential expression was also obvious for several important enzymes from secondary metabolism, especially the phenylpropanoid pathway. PAL, the protein catalyzing the first and committed step of the pathway, was more significantly induced in susceptible inbred (B73) when compared to CO441. Cinnamyl alcohol dehydrogenase, 4-coumarate-CoA ligase, and phenolic *O*-methyltransferase did not show significant induction in any of the inbreds, but they were more abundant in the resistant genotype CO441. From the terpenoid biosynthetic pathway, hydroxymethylbutenyl 4-diphosphate synthase (HDS) of the isoprenoid methylerythritol phosphate (MEP) pathway was found among the most *F. graminearum*-responsive proteins in both genotypes. HDS was also more abundant in resistant CO441. Finally, of the two chalcone flavonone isomerases that are involved in flavonoid biosynthesis, one was induced by fungal infection in both genotypes, while another was more abundant in CO441. Overall, resistant genotype CO441 exhibited higher levels of many defense-associated proteins than susceptible B73, under both mock- and fungus-treated conditions. However, B73 displayed stronger response to the *F. graminearum* than the resistant inbred.

2.2.2. *F. verticillioides* and Fusarium Ear Rot

F. verticillioides causes Fusarium ear and stalk rot. Disease is prevalent, especially during hot and dry weather, both before and after harvest. Fumonisins produced by this species appear to be one of the most common maize-based food and feed contaminants, therefore they are of high importance to the producers [30,31]. Fumonisins are highly toxic and have been associated with several livestock and humans diseases including cancer [34,35].

Protection of seeds during germination is critical for plant propagation and ultimately, for survival of plant species. Defense responses during this vulnerable developmental phase were examined in 20-hour-old germinating maize embryos challenged with *F. verticillioides* [36]. Twenty-four hours after infection, proteins from fungus-infected *vs.* sterile embryos were compared via 2DE, using neutral and acidic extraction conditions. Several groups of proteins have shown changes in their expression. Detoxifying enzymes catalase 2, SOD, and glutathione-*S*-transferase (GST) were almost completely absent in extracts from sterile embryos, while present in fungus-infected embryos, likely in order to protect them from oxidative damage and xenobiotics. Proteins involved in protein synthesis, folding and stabilization such as small HSP 17.2, peptidylprolyl *cis*-trans isomerase, and cyclophilin were also more abundant in fungus-challenged embryos. Another up-regulated protein belonging to this group, eukaryotic translation initiation factor 5A (eIF-5A), might be in demand to facilitate synthesis of proteins needed for defense. Protein extracts from sterile and fungus-infected embryos were also compared using immunological reactions, which confirmed induction of P23 protein, and constitutive expression of β-1,3-glucanases and chitinases. Finally, carbohydrate metabolism in embryos also appeared to be altered by *F. verticillioides*. As seen from the expression of two glyceraldehyde-3-phosphate dehydrogenases (GAPDH) (cytosolic 1 and 2), and fructose-bisphosphate aldolase, fungus seems to repress glycolysis and activate gluconeogenesis.

2.2.3. *F. verticillioides* and Extracellular Matrix

Besides its other biological functions, extracellular matrix (ECM) serves as a first line of defense during the plant-pathogen interaction. As such, ECM is critical for plant survival. Actions taken in this compartment trigger a complex series of events leading to the rapid defense responses at the infection site and throughout the plant. To gain an insight into this earliest host-pathogen encounter, a pathogen attack was simulated by treating cell cultures with elicitor molecules found in an attenuated fungal pathogen [37]. After a 6-hour-long treatment, they responded by activating three major events. The first event was characterized by elicitor-induced secretion of six putative XIP isoforms into the ECM culture medium. These defensive enzymes inhibit xylanases that fungi secrete during infection to hydrolyze the host's cell walls [38]. The second event included rapid changes in the phosphorylation status of several apoplastic peroxidases, implying that the phosphorylation regulatory mechanism is very likely controlling this host-pathogen contact. More specifically, two peroxidases were consistently and significantly dephosphorylated within 10 min of *Fusarium* elicitor treatment, although their relative abundances remained unchanged. Finally, cell wall-enriched protein fractions contained considerably increased levels of cytosolic GAPDH and HSPs suggesting that during elicitor-induced ECM-mediated defenses these classical cytosolic proteins are specifically targeted to the cell walls. Furthermore, the three major elicitor-induced occurrences were accompanied by increased production of H_2O_2 as determined by histochemical staining. Accumulation of H_2O_2 is consistent with the onset of elicitor-induced oxidative burst, a central event in the establishment of an effective resistance response. Also observed was a disappearance of β-*N*-acetylglucosaminidase (β-NGase) protein spots. This protein hydrolyses bacterial cell wall peptidoglycan [39], and the authors speculate that the disappearance of β-NGase is likely the result of its oxidative cross-linking to the cell walls. Fungal elicitor-induced changes in maize ECM proteome thus strongly imply that this important anatomical structure plays a complex role in maize defense machinery.

2.2.4. Summary of Maize Protein-Based Defenses against *Fusarium* spp.

Studies reviewed above strongly indicate that maize resistance to *Fusarium* spp. and *A. flavus* are comparable. Similar groups of proteins were found to be responsive against diseases caused by both types of fungi [22,33,36]. Suggestion of a common resistance mechanism between *Fusarium* ear rot/fumonisins and *Aspergillus* ear rot/aflatoxins was previously made from QTL mapping, which showed that some of the genes implicated in resistance to ear rots and mycotoxin accumulation were identical or genetically linked [40]. Furthermore, in very recent study [41], selected defense *pr* genesand proteins (catalase, superoxide dismutase, peroxidases) were monitored in developing kernels of resistant and susceptible maize genotypes inoculated with *F. proliferatum*, *F. subglutinans*, and *A. flavus*. Similarly to rachis tissue [22], higher gene transcription and enzymatic activities in uninoculated kernels of resistant line testify about constitutive defense against the pathogens. On the other hand, susceptible lines responded with induction of reactive oxygen species-scavenging proteins. In addition, maize 9-lipoxygenases (LOX) ZmLOX12 and ZmLOX3 suppressed contamination by *F. verticillioides* and *A. flavus*, respectively [42,43]. Hence, breeding for Aspergillus and Fusarium resistances may produce parallel outcomes. Figure 2 summarizes protein-based defenses of maize plant in tissues that are of interest in regards to infection by *Fusarium* spp.

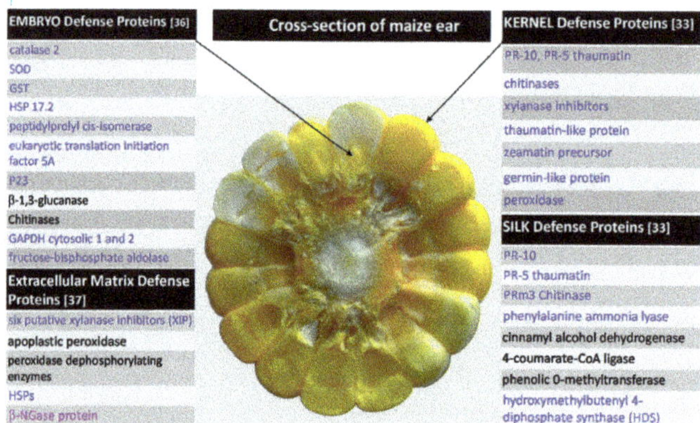

Figure 2. Selected proteins associated with maize defense against *Fusarium* spp. References are given in parenthesis next to the tissue-specific protein groups. Proteins in black and blue color represent constitutive and induced proteins, respectively. Protein in violet was not detected after the pathogen challenge.

2.3. Maize vs. Curvularia lunata

Curvularia lunata is a foliar fungal pathogen and a causal agent of Curvularia leaf spot of maize. It can cause damage to the maize yields, especially in hot and humid climates [44]. It has been reported that *C. lunata* produces a furanoid type toxin, both *in vitro* and *in planta*, which can possibly lead to leaf lesions [45]. Unlike Aspergilli and Fusaria, not much is known about *C. lunata* pathogenesis in maize at molecular level. Huang *et al.* attempted to unravel the pattern of resistance by comparing four maize inbreds ranging from highly resistant to susceptible to *C. lunata* [46,47]. Leaf proteomes analyzed 24 h post-infection revealed that resistant lines were much more responsive to infection than susceptible lines [47]. From 100 differentially expressed spots, only 8 were identified from 10 selected and analyzed spots. The expression levels ofGLP, translation initiation factor eIF-5A, oxygen-evolving complex (OEC), and two oxygen-evolving enhancer proteins (OEE 1, OEE 2) were found to be significantly more abundant in highly resistant variety Luyuan 92, when compared to the three other genotypes. GLPs, for instance, are ubiquitous developmentally regulated cell wall glycoproteins that possess a wide range of functions [48], including defense against pathogens [49,50]. Several GLPs have enzymatic activities of SOD and oxalate oxidase (OXO) producing hydrogen peroxide (H_2O_2), which can serve as a signaling molecule for a range of defense reactions, including cell death, and as a cofactor for cell wall strengthening by cross-linking [50,51]. In addition, GLPs often bind non-covalently to cell walls [52,53], potentially contributing to structural reinforcement. Consequently, GLPs might be able to modulate host resistance against pathogen by signaling via H_2O_2, and/or cell wall strengthening. This also may be the case for Luyuan 92 resistance to *C. lunata* but it remains to be confirmed in further studies. Similarly, Photosystem II proteins OEC and OEEs were previously reported as responsive to abiotic and biotic stresses [54,55], therefore they might act in the maize-*C. lunata* interaction as well. When longer viral exposures (24, 36, 48, 60 and 72 h) were examined, more proteins were identified as differentially expressed, especially in the resistant genotype 78599-1 [46]. The majority of these proteins were, nevertheless, common for both types of inbreds. Twenty-seven proteins that changed at least 3-fold in abundance were mainly related to photosynthesis, respiration, oxidative and drought stress, and signal transduction. Among them, putative glutathione peroxidase (GPX), APX and 22 kDa drought-inducible protein are the most common facilitators of plant tolerance to environmental challenges, as shown for maize-*A. flavus* interactions. By analogy, these proteins are also possible mediators between maize stress tolerance and resistance to *C. lunata*.

2.4. Maize vs. Viral Pathogens

Viral diseases are also fairly common in maize. They occur throughout the maize-growing regions around the world, and they can cause sporadic but very damaging epidemics. Proteomics of maize-virus interactions is a relatively new research area, and unlike for fungal pathogenesis, not much is known about the virus impact on the biology of a maize plant at the molecular, especially proteomic, level. Several recent studies [56–58] addressing the effect of viruses on maize protein expressions will be reviewed in the following paragraphs.

2.4.1. Rice Black-Streaked Dwarf Virus

Li and colleagues scrutinized the long-term adjustment of the maize elite, but very susceptible line (Ye478), to Rice Black-streaked Dwarf Virus (RBSDV) [58]. This insect-transmitted pathogen is considered a major causal agent of maize rough dwarf disease (MRDD) in China, and it is responsible for dwarfed and severely damaged maize plants. Fifty days post-infection, virus-infected leaves responded with significant changes in abundance for 123 proteins, as revealed via 2DE and MS/MS. Functional variety of 91 identified proteins suggested an engagement of a large arsenal of biological processes and pathways, most likely due to severe morphological changes taking place upon viral encounter. For example, the accumulation of detoxifying enzymes peroxidase 39, APx2-cytosolic APX, and catalase isozyme 3 implies an onset of oxidative stress caused by RBSDV, and a subsequent need for regulation of cellular redox state. Higher levels of lipoxygenases and small GTP-binding proteins in virus-infected plants suggest that these proteins modulate the triggering of downstream defenses via signaling pathways.

UDP-glucosyltransferase BX9 catalyzes conversion of benzoxazinoids to their inactive glucoside form for their deposition to vacuoles. Benzoxazinoids are major grass phytotoxins that offer protection against a wide range of herbivores, fungi and bacteria [59]. Up-regulation of UDP-glucosyltransferase BX9 in RBSDV-infected plants indicates that these natural insecticides are very likely a part of maize defenses against RBSDV.

Consistent with severely altered leaf morphology, the upregulation of cinnamyl alcohol dehydrogenase and caffeic acid 3-*O*-methyltransferase may have led to modified cell walls in infected plants. These plants had rough leaves with enations on the veins. MRDD establishment is also followed by profound adaptation in carbohydrate metabolism, as numerous proteins from glycolysis, TCA cycle, glycogenesis, and pentose phosphate metabolism, were extensively altered in expression. Particularly evident was the decrease in transketolase levels in diseased plants, leading to a severely weakened influx of proteins affecting plant growth and development. On the other hand, elevated amounts of starch granules that were observed in MRDD-diseased leaves were very likely the result of enhanced expression of ADP-glucose phospohorylase small subunit in virus-infected plants. Finally, photosynthesis, carbon fixation and assimilation as well as starch synthesis also appeared intensified in virus-infected leaves [58].

Overall, the invasion by RBSDV causes tremendous changes in maize's metabolism, which subsequently leads to severe morphological differences when compared to normal plants. During MRDD disease, numerous fundamental biochemical pathways become seriously compromised, which is accompanied by significant changes in expression of proteins, particularly those associated with plant growth. The proteomic study reviewed above clearly showed that the occurrence of MRDD results in very complex responses involving a number of proteins, as well as defensive and signaling molecules.

2.4.2. Sugarcane Mosaic Virus

Sugarcane Mosaic Virus (SCMV) is another viral pathogen and causal agent of mosaic disease in maize and other graminaceous plants. SCMV occurs around the globe especially in regions where susceptible varieties are grown. Two proteomic gel-based investigations were conducted by Wu *et al.*, in which authors compared leaf protein profiles from resistant (Siyi) and susceptible (Mo17) inbred lines, 6 and 12 days post-infection (dpi) with SCMV, respectively [56,57]. With regard to functional classifications, similar groups of responsive proteins were observed in both genotypes for both time points. Ninety-six (6 dpi) and 93 (12 dpi) differentially expressed and identified proteins were predominantly from energy and metabolism, stress and defense, and photosynthesis, followed by protein synthesis and folding, signal transduction/transcription, and carbon fixation. However, a larger proportion of stress/defense-related and signal transduction/transcription-associated proteins were upregulated in the resistant genotype relative to the susceptible line, especially after longer viral exposure. In contrast, photosynthesis-associated proteins were down-regulated in the resistant inbred. After longer exposure to the virus, (12 dpi) enzymes from the major energy-producing glycolysis and gluconeogenesis pathways exhibited an increase in expression in the resistant line, while they were considerably lower in the susceptible maize genotype. Nineteen (6 dpi) and 17 (12 dpi) differentially expressed proteins were novel proteins that have not been previously known as virus-responsive. Few examples for 6 dpi treatment include remorin, glutamate dehydrogenase, calcium-dependent protein kinase, ATP synthase CF1 alpha subunit, and bZIP transcription factor ABI5. Abscisic stress ripening protein, serine/threonine-protein kinase, ferredoxin-NADP reductase, nucleolar RNA helicase 2, and 30S ribosomal protein S5 were among 17 newly identified proteins during 12 dpi experiment. More importantly, these proteins broaden the repertoire of possible virus-defensive maize proteins and may serve as good candidates for future resistance-related studies. Although similar biological processes in both lines were affected by infection, only 17 proteins were common for both inbreds. They include aconitase, transketolase, fructose-bisphosphate aldolase, nucleoside diphosphate kinase, thioredoxin, ABA stress ripening protein, GST, chaperonin, T-complex protein, Ribulose-1,5-bisphosphate carboxylase/oxygenase (RuBisCo) large subunit, RuBisCo subunit binding protein beta subunit, ferredoxin-NADP reductase, remorin, histidine triad nucleotide binding protein, cysteine synthase, β-D-glucosidase precursor and electron transporter protein. Similarly to RBSDV infection, there were dramatic changes in expression of proteins involved in carbohydrates metabolism in both genotypes, suggesting a significant shift of this metabolic pathway. Viral proteins were also found and identified in the susceptible line at 6 dpi confirming the presence of infection. In addition, resistance to SCMV seems to be also instituted via phytohormones-mediated defenses. Evidence exists that plant pathogens evolved a mechanism to modify the host's hormone levels to establish pathogenicity [60–62]. Indeed, quantitative differences in concentrations of salicylic acid (SA), ABA, ethylene, jasmonic acid (JA) and azelaic acid (AZA) were observed between mock-inoculated and SCMV-inoculated plants, as well as between the resistant and susceptible plants.

Both studies brought valuable input into previously not much known maize–SCMV interactions. Identified proteins are subject for further research and possible resistance candidates. Summary of maize leaf proteins implicated in its defense against viral pathogens is given in Table 1.

Table 1. Examples of proteins associated with maize defense against viral pathogens.

Pathogen	Maize Genotype	Induced Proteins	Repressed Proteins	Reference
Black-Streaked Dwarf Virus	Ye478 Susceptible	Peroxidase 39, APx2-cytosolic ascorbate peroxidase, catalase 3, oxygenase, caffeic acid 3-*O*-methyltransferase, cinnamyl alcohol degydrogenase, UDP-glucose pyrophosphorylase, GAPDH, ADP-glucose phosphorylase, lipoxygenases, GTP-binding proteins, enzymes of starch synthesis, hotosynthesis, carbon fixation and assimilation	Enzymes of glycolysis, TCA cycle, glycogenesis, and pentose phosphate metabolism, particularly transketolase	[58]
Sugarcane Mosaic Virus	Siyi Resistant	Enzymes of glycolysis and gluconeogenesis, remorin, cysteine synthase, glutamate dehydrogenase, calcium-dependent protein kinase, bZIP transcription factor ABI5, serine/threonine-protein kinase, nucleolar RNA helicase 2, 30S ribosomal protein S5		[56,57]
Sugarcane Mosaic Virus	Mo17 Susceptible		Enzymes of glycolysis and gluconeogenesis, remorin, cysteine synthase, ATP synthase CF1 α subunit	[56,57]
Sugarcane Mosaic Virus	Siyi and Mo17	Stress/defense-related and signal transduction/transcription-associated proteins,aconitase, transketolase,nucleoside diphosphate kinase, abscisic stress ripening protein, chaperonin, T-complex protein, ferredoxin-NADP reductase, histidine triad nucleotide binding protein, β-D-glucosidase precursor, electron transporter protein	Photosynthesis-associated proteins, thioredoxin, glutathione *S*-transferase, RuBisCo large subunit, RuBisCo subunit binding protein α subunit	[56,57]

3. Commonly Responsive Proteins in Maize-Pathogen Interaction

Several groups of proteins were found commonly associated with the maize response to fungal and/or viral pathogens (Table 2). Their main functions will be discussed briefly in the following paragraphs.

3.1. Pathogenesis–Related Proteins

PR proteins represented the largest and most diverse group of proteins associated with maize resistance. Generally, PRs are inducible by a variety of pathogens via salicylic acid, jasmonic acid, or ethylene signaling, which leads into their accumulation at the site of the attack but also systemically at distant parts of the plant [63]. Many are expressed constitutively in healthy plants during growth and development. This large and diverse group of proteins classified into 17 families possesses a broad range of enzymatic activities rendering them indispensable for plant survival upon pathogen attack. An arsenal of PR proteins was found differentially expressed in maize genotypes associated with fungal diseases imposed by *F. verticillioides* [36], *F. graminearum* [33], and *A. flavus* [14,22,23], including chitinase and β-1,3-glucanases. These antifungal hydrolytic enzymes are perhaps the most important among PR proteins. They are ubiquitous in the plant kingdom, can be preformed or inducible, and often work synergistically. The mode of their action is degradation of chitin and glucans, major structural polysaccharides of fungal cell walls and exoskeletons of arthropods and nematodes [64–67].

Table 2. Commonly responsive protein groups during maize interactions with fungal and/or viral pathogens. Selected proteins are listed.

Protein Group	Elicited by Pathogen	Tissue	Proteins	References
Pathogen related proteins	*Af, Fv, Fg*	Embryo, silk, rachis, kernels	Chitinase, glucanase, trypsin and amylase inhibitor, peroxidase	[14,22,23,33,36]
Detoxifying enzymes	*Af, Fv, Fg, Cl,* RBSDV, SCMV	Leaves, foliage, embryo, silk, rachis, kernels	SOD, catalase, PER, thioredoxin, glutaredoxin, glutathione reductase, GST, dehydroascorbate reductase	[12,22,33,36,46,56,57,68]
Proteins involved in secondary metabolism	*Af, Fv, Fg,* RBSDV, SCMV	Rachis, leaves, ear	PAL, caffeoyl-CoA 3-*O*-methyltransferase, chalcone-flavonone isomerase, cinnamyl alcohol dehydrogenase, hydroxymethyl-butenyl 4-diphosphate synthase	[22,56,58]
Proteins involved in energy producing pathways	*Af, Fv,* RBSDV, SCMV	Rachis, leaves	GAPDH, ADP-glucose pyro-phosphorylase, UDP-glucose pyrophosphorylase, glucose phosphate isomerase, fructose-bisphosphate aldolase, transketolase	[22,33,36,37,47,56–58]
Proteins involved in protein synthesis, folding and stabilization	*Af, Fv, Fg*	Embryo, rachis, kernel	elF-5A/initiation of translation; HSPs, chaperonins, peptidylprolyl *cis*-trans isomerase, cyclophilin	[22,33,36,37,46,47,56]

Af—Aspergillus flavus; Fv—Fusarium verticillioides; Fg—Fusarium graminearum; Cl—Curvularia lunata; RBSVD—Rice Black-streaked Dwarf Virus; SCMV—*Sugarcane mosaicvirus.*

Other PR proteins important in maize fungal resistance included zeamatin and proteins P21 and P23. These are the members of PR-5 family that have a close resemblance to a sweet-tasting protein thaumatin and are also known as thaumatin-like proteins (TLs). They permeabilize fungal cell walls causing a rapid cell lysis and killing the pathogen [69]. Others have bifunctional trypsin/α-amylase inhibitor activities that impair fungal germination [70]. Highly basic PR-5 protein P23 was induced in maize embryo infected by *F. verticillioides* [36] and highly acidic protein P21 in rachis infected by *A. flavus* [22]. Finally, both silks and kernels infected with *F. graminearum* responded with up-regulation of zeamatin and thaumatin, respectively [33].

A class III peroxidase belonging to the PR-9 group was also found as fungus responsive. ZmPrx16 increased in abundance in response to *F. graminearum* in kernels of both resistant and susceptible inbred, and it was also constitutively expressed at higher levels in resistant kernels [33]. Class III peroxidases are well-known defense-associated enzymes that are located in plant cell walls and vacuoles [71,72]. They create a physical barrier in host tissues to halt pathogen invasion by catalyzing the H_2O_2-dependentcross-linking of cell wall components resulting in enhanced lignification or suberization.

3.2. Detoxifying Enzymes

One of the most critical events upon plant infection is pathogen recognition followed by rapid apoplastic production of reactive oxygen species (ROS) via a process known as the oxidative burst [73,74]. The most common reactive ROS include superoxide, hydroxyl radical and hydrogen peroxide, and their induction is imperative for the outcome of host-pathogen interactions as they mediate the launch of diverse defensive systems. ROS serve as substrates for strengthening of host cell walls via oxidative cross-linking of glycoproteins and as secondary messengers for activation of molecular and physiological responses in plant cells. Although their induction is transient, ROS are extremely reactive, and if not controlled they can cause detrimental oxidative damage to the host cells that can lead to their death. Detoxification of ROS is therefore paramount to the cell survival. For this purpose, plants are equipped with effective ROS-scavenging systems consisting of detoxifying enzymes that are responsible for the maintenance of the steady state levels of ROS in different cell compartments. Key detoxifying enzymes found in plant cells include APX [22,74–76], glutathione peroxidase (GPX) [46,74,77], SOD [22,36,74,76–78], catalase [25,36,41,58,68,77], peroxiredoxin (PER) [10–12,57,74,79], thioredoxin [22,56,57,74], glutaredoxin [74,80], and glutathione reductase [74,76,81], and dehydroascorbate reductase [58,82], and their inducibility or repression was a common characteristic feature following exposure of maize to multiple stressors such as *F. verticillioides* [36], *F. graminearum* [33], *A. flavus* [22,68], *C. lunata* [46]. RBSDV [58], and SMV [56]. Catalase 2 and SOD, for example, were two main enzymes protecting maize embryo from oxidative damage during Fusarium ear rot establishment [36], and combined action of GPX and APX seems to be required for supporting maize defense response against *C. lunata* in both resistant and susceptible varieties [46]. Catalase activity was also significantly higher in immature embryos of maize genotypes resistant to *A. flavus* infection, while the challenge by the pathogen caused further increased in catalase activity [68]. Antioxidant enzymes SOD [Cu–Zn] 4AP, peroxiredoxin, and thioredoxin-like protein 5 were also found constitutively expressed at higher levels in multiple tissues of aflatoxin-resistant maize varieties [12,22], likely contributing to the maize tolerance to the heat and drought which renders them better protection against aflatoxin production by *A. flavus*.

With regard to viral pathogens, APx2-cytosolic ascorbate peroxidase and catalase isozyme 3 were accumulated in RBSDV-infected leaves suggesting that these proteins protect cells from toxicity of ROS [58]. Finally, distinct antioxidant activities were observed in SCMV-infected leaves. SOD was down-regulated in susceptible maize while ascorbate peroxidase was up-regulated resistant line [56].

In addition to ROS, plants also detoxify other contaminants by conjugating them to diverse small non-protein antioxidants such as glutathione. These reactions are catalyzed by GSTs, a family of metabolic isozymes that are known to be highly expressed in major cereal crops representing up to 2% of their total foliage proteome [83]. They have been implicated in maize detoxification of herbicides [84, 85], and during maize infection with *F. verticillioides* [36] and *A. flavus* [22]. The xenobiotic detoxifier GST14 was the most upregulated protein (6.9-fold) in maize rachis after a 35 days exposure to *A. flavus* [22].

3.3. Proteins Involved in Secondary Metabolism

Higher plants harbor a wide spectrum of organic compounds that perform a myriad of cellular roles including protection against predators and pathogenic microorganisms. They are produced via secondary metabolism, predominantly the phenylpropanoid, the isoprenoid and the alkaloid pathways [86]. The phenylpropanoid pathway in particular supplies plant cells with a variety of phenolics with a broad range of defensive properties. Some act as toxins, for example phytoalexins and isoflavonoids with antimicrobial and antifungal properties, respectively.

Maize 9-LOX converts linolenic and linoleic acid into 10-oxo-11-phytodienoic acid (10-OPDA), which acts as a phytoalexin, inhibiting growth of *A. flavus* and *F. verticillioides* [87]. Direct toxicity of phenolics to *A. flavus* is also probable. It was previously reported against *A. flavus* [88] and

Fusarium spp. [89,90]. Other phenolics are precursors of lignins, major components of plant secondary cell walls where they are responsible for wall's mechanical rigidity [91].

Although proteomic studies on maize-pathogen interaction did not aspire to identify stress-inducible secondary metabolites, proteins responsible for their production were of great interest as their induction indicates involvement of chemical defenses. For example, PAL, a key regulatory enzyme of the phenylapropanoid pathway, together with caffeoyl-CoA 3-*O*-methyltransferase1 and chalcone–flavonone isomerase were up to 5-fold higher in rachis from aflatoxin-resistant maize genotypes [22]. This observation implies that the impediment of *A. flavus* spread within an ear of resistant genotypes may be likely a consequence of increased cell wall lignification [19,20]. Cinnamyl alcohol dehydrogenase is another example of a cell wall reinforcing protein, catalyzing the final step in a branch of phenylpropanoid synthesis that leads to the production of lignin monomers. The enzyme was inducible upon viral infection where it was preferentially accumulated in the leaves of plants diseased with RBSDV [58] as well as with SCMV [56]. Thus, fortification of cell walls via enhanced lignin biosynthesis appears to be a common mechanism of resistance against fungal and viral disease.

3.4. Proteins Involved in Energy-Producing Carbohydrate Metabolic Pathways

Recognition of pathogen and subsequent deployment of downstream defenses is a complex process requiring immense reprogramming within plant cells that involves generation of signaling molecules, cell wall reinforcements, cytoskeleton rearrangement, induced synthesis of proteins, secondary metabolites, *etc.* Needlessly to say, all these processes have high demand for energy influx that is predominantly provided by primary metabolic pathways [92–95]. In maize-pathogen interactions, carbohydrate metabolism was one of the maize primary metabolic processes impacted the most by pathogen infection. Response-associated proteins included enzymes from glycolysis, gluconeogenesis, pentose phosphate pathway, TCA cycle, as well as associated mitochondrial electron transport and ATP biosynthesis [22,33,36,37,46,56–58]. Whether these pathways serve solely as the energy providers for costly maize defenses or they also participate in resistance, will need to be further inspected. There is evidence demonstrating that plants also use carbohydrate metabolism as a source of signaling molecules to directly or indirectly activate downstream defenses [94]. Plant GAPDHs, for example, have been known for some time to have non-traditional activities that are beyond the catalysis of energy production by glycolysis. In *Arabidopsis*, cytosolic GAPDH can be reversibly inactivated by H_2O_2 followed by reactivation with reduced glutathione, thus mediating signaling via ROS [96,97]. In maize, GADPH was upregulated in response to *A. flavus* [22] while it was lowered in expression after *F. verticillioides* infection [36]. The most robust impact on carbohydrate metabolism was seen in maize infected with viral agents. ADP-glucose pyrophosphorylase, UDP-glucose pyrophosphorylase, glucose phosphate isomerase, GAPDH, fructose-bisphosphate aldolase, and transketolase were all dramatically altered upon infection with RBSDV [58], the agent that causes immense changes in maize physiology. Similarly, major energy-producing pathways of carbohydrate metabolism were significantly compromised in maize infected by sugarcane mosaic virus, especially after long viral exposure of a susceptible line [56,57].

Given that simple sugars are actively involved in signaling [98] and gene regulation [99], their concentration in plant cells under a biotic stressor may directly or indirectly correlate with a plant's capability to trigger adequate defenses. Nonetheless, functional studies are necessary to fully clarify how alterations in carbohydrate metabolism impact pathogen establishment, disease development and maize resistance.

3.5. Proteins Involved in Protein Synthesis, Folding and Stabilization

This group of proteins implicated in maize-pathogen interaction most commonly consisted of eukaryotic translation initiation factor 5A (eIF-5A), multiple HSPs, chaperonins, peptidylprolyl *cis*-trans isomerases, and cyclophilin [22,33,36,37,46,47,56]. Eukaryotic translation initiation factor 5A is best known for the initiation of eukaryotic cellular protein biosynthesis, but in recent years, it has also been found as pathogen-responsive. Its exact role in maize-pathogen interaction remains to be solved, but in other plant species, the function of eIF-5A is slowly becoming better defined. As demonstrated in *Arabidopsis*, eIF-5A can serve as a key element in regulating the induction of programmed cell death caused by infection with a virulent pathogen [100]. HSPs, on the other hand, are known stress proteins inducible under heat and other environmental stimuli [101]. As intra-molecular chaperones HSPs prevent the aggregation of newly synthesized proteins and the accumulation of proteins damaged by stresses [102]. Hence, they are responsible for maintaining proteins' functional conformation essential for their cellular activities and ultimately for cell survival under stress [103]. These proteins seem valuable for maize resistance to *A. flavus* and aflatoxins, especially in southern states of the USA whose climate conditions facilitate aflatoxin production. HSPs were one of the most differentially expressed proteins with significantly higher levels (11-fold) in rachis tissues from aflatoxin-resistant maize genotypes [22] thus helping resistant lines to alleviate heat stress and subsequent control of aflatoxin production. Peptidylprolyl *cis*-trans isomerases (PPIs) and cyclophilins have similar roles as HSPs. They are upregulated under stress to accelerate folding or refolding of nascent or damaged proteins, respectively [104–107], showing that they are critical for cell adaptation under stress conditions.

4. Concluding Remarks

The presented review outlines the impact of several fungal and viral pathogens on the biology of maize, the world's leading cereal crop. As seen from the compiled proteomic-based evidences, maize interaction with pathogens is a remarkably complex system engaging multiple levels of defense occurring in multiple tissues. This is anticipated since maize resistance to the majority of diseases is a polygenic quantitatively controlled trait that is strongly influenced by environmental conditions [108]. Therefore, the full comprehension of events underlying maize-pathogen interaction is a difficult task requiring a system biology approach. Future functional studies are still needed to fully unravel this intricate and complex defensive network that will help geneticists and breeders to develop resistant maize varieties.

Acknowledgments: The authors would like to express their gratitude to Linda Breazeale for grammatical review of the manuscript. This work was funded in part by grant NIH 2P20GM103476-12 (MS-INBRE) and 2015 Mississippi Agriculture and Forestry Experiment Station Strategic Research Initiative award.

Author Contributions: Authors contributed equally to writing of this manuscript.

Conflicts of Interest: The authors declare no conflict of interest.

References

1. Castegnaro, M.; McGregor, D. Carcinogenic risk assessment of mycotoxins. *Rev. Med. Vet.* **1998**, *149*, 671–678.
2. McKean, C.; Tang, L.; Billam, M.; Tang, M.; Theodorakis, C.W.; Kendall, R.J.; Wang, J.S. Comparative acute and combinative toxicity of aflatoxin B1 and T-2 toxin in animals and immortalized human cell lines. *J. Appl. Toxicol.* **2006**, *26*, 139–147. [CrossRef] [PubMed]
3. Schnable, P.S.; Ware, D.; Fulton, R.S.; Stein, J.C.; Wei, F.; Pasternak, S.; Liang, C.; Zhang, J.; Fulton, L.; Graves, T.A.; *et al.* The B73 maize genome: Complexity, diversity, and dynamics. *Science* **2009**, *326*, 1112–1115. [CrossRef] [PubMed]
4. Diener, U.L.; Davis, N.D. Aflatoxin formation by *Aspergillus flavus*. In *Aflatoxin: Scientific Background, Control and Implications*; Goldblatt, L.A., Ed.; Academic Press: New York, NY, USA, 1969; pp. 360–391.
5. Payne, G.A. Aflatoxins in maize. *Crit. Rev. Plant. Sci.* **1992**, *10*, 423–440. [CrossRef]
6. Schmale, D.G.; Munkvold, G.P. Mycotoxins in Crops: A Threat to Human and Domestic Animal Health. Available online: http://www.apsnet.org/edcenter/intropp/topics/Mycotoxins/Pages/default.aspx (accessed on 11 November 2015).
7. Neal, G.E. Participation of animal biotransformation in mycotoxin toxicity. *Rev. Med. Vet.* **1998**, *149*, 555–560.
8. Chen, Z.Y.; Brown, R.L.; Cleveland, T.E. Evidence for an association in corn between stress tolerance and resistance to *Aspergillus flavus* infection and aflatoxin contamination. *Afr. J. Biotechnol.* **2004**, *3*, 693–699.

9. Chen, Z.Y.; Brown, R.L.; Cleveland, T.E.; Damann, K.E.; Russin, J.S. Comparison of constitutive and inducible maize kernel proteins of genotypes resistant or susceptible to aflatoxin production. *J. Food Prot.* **2001**, *64*, 1785–1792. [PubMed]

10. Chen, Z.Y.; Brown, R.L.; Damann, K.E.; Cleveland, T.E. Identification of unique or elevated levels of kernel proteins in aflatoxin-resistant maize genotypes through proteome analysis. *Phytopathology* **2002**, *92*, 1084–1094. [CrossRef] [PubMed]

11. Chen, Z.Y.; Brown, R.L.; Damann, K.E.; Cleveland, T.E. Identification of a maize kernel stress-related protein and its effect on aflatoxin accumulation. *Phytopathology* **2004**, *94*, 938–945. [CrossRef] [PubMed]

12. Chen, Z.Y.; Brown, R.L.; Damann, K.E.; Cleveland, T.E. Identification of maize kernel endosperm proteins associated with resistance to aflatoxin contamination by *Aspergillus flavus*. *Phytopathology* **2007**, *97*, 1094–1103. [CrossRef] [PubMed]

13. Chen, Z.Y.; Brown, R.L.; Lax, A.R.; Guo, B.Z.; Cleveland, T.E.; Russin, J.S. Resistance to *Aspergillus flavus* in corn kernels is associated with a 14-kDa protein. *Phytopathology* **1998**, *88*, 276–281. [CrossRef] [PubMed]

14. Chen, Z.Y.; Brown, R.L.; Rajasekaran, K.; Damann, K.E.; Cleveland, T.E. Identification of a maize kernel pathogenesis-related protein and evidence for its involvement in resistance to *Aspergillus flavus* infection and aflatoxin production. *Phytopathology* **2006**, *96*, 87–95. [CrossRef] [PubMed]

15. Chen, Z.Y.; Brown, R.L.; Russin, J.S.; Lax, A.R.; Cleveland, T.E. A corn trypsin inhibitor with antifungal activity inhibits *Aspergillus flavus* α-amylase. *Phytopathology* **1999**, *89*, 902–907. [CrossRef] [PubMed]

16. Guo, B.Z.; Chen, Z.Y.; Brown, R.L.; Lax, A.R.; Cleveland, T.E.; Russin, J.S.; Mehta, A.D.; Selitrennikoff, C.P.; Widstrom, N.W. Germination induces accumulation of specific proteins and antifungal activities in corn kernels. *Phytopathology* **1997**, *87*, 1174–1178. [CrossRef] [PubMed]

17. Lozovaya, V.V.; Waranyuwat, A.; Widholm, J.M. B-1,3-glucanase and resistance to *Aspergillus flavus* infection in maize. *Crop Sci.* **1998**, *38*, 1255–1260. [CrossRef]

18. Hong-Bo, S.; Zong-Suo, L.; Ming-An, S. Lea proteins in higher plants: Structure, function, gene expression and regulation. *Colloids Surf.* **2005**, *45*, 131–135. [CrossRef] [PubMed]

19. Alfaro, Y. Response of Resistant and Susceptible Maize Genotypes to Inoculation with Transformed *Aspergillus flavus* Isolates. Ph.D. Thesis, Mississippi State University, Mississippi State, MS, USA, 1999.

20. Magbanua, Z.V.; Williams, W.P.; Luthe, D.S. The maize rachis affects *Aspergillus flavus* spread during ear development. *Maydica* **2013**, *58*, 182–188.

21. Pechanova, O.; Pechan, T.; Ozkan, S.; McCarthy, F.M.; Williams, W.P.; Luthe, D.S. Proteome profile of the developing maize (*Zea mays* L.) rachis. *Proteomics* **2010**, *10*, 3051–3055. [CrossRef] [PubMed]

22. Pechanova, O.; Pechan, T.; Williams, W.P.; Luthe, D.S. Proteomic analysis of the maize rachis: Potential roles of constitutive and induced proteins in resistance to *Aspergillus flavus* infection and aflatoxin accumulation. *Proteomics* **2011**, *11*, 114–127. [CrossRef] [PubMed]

23. Peethambaran, B.; Hawkins, L.; Windham, G.L.; Williams, W.P.; Luthe, D.S. Anti-fungal activity of maize silk proteins and role of chitinase in *Aspergillus flavus* resistance. *Toxin Rev.* **2010**, *29*, 27–39. [CrossRef]

24. Marsh, S.F.; Payne, G.A. Preharvest infection of corn silks and kernels by *Aspergillus flavus*. *Phytopathology* **1984**, *74*, 1284–1289. [CrossRef]

25. Neucere, J.N. Inhibition of *Aspergillus flavus* growth by silk extracts of resistant and susceptible corn. *J. Agric. Food Chem.* **1996**, *44*, 1982–1983. [CrossRef]

26. Naidoo, G.; Forbes, A.M.; White, C.P.; Rocheford, T.R. Resistance to *Aspergillus flavus* ear rot and aflatoxin accumulation in maize f1 hybrids. *Crop Sci.* **2002**, *42*, 360–364. [CrossRef]

27. Paul, C.; Naidoo, G.; Forbes, A.; Mikkilineni, V.; White, D.; Rocheford, T. Quantitative trait loci for low aflatoxin production in two related maize populations. *TAG Theor. Appl. Genet.* **2003**, *107*, 263–270. [CrossRef] [PubMed]

28. Walker, R.D.; White, D.G. Inheritance of resistance to Aspergillus ear rot and aflatoxin production of corn from ci2. *Plant Dis.* **2001**, *85*, 322–327. [CrossRef]

29. Pechanova, O.; Takac, T.; Samaj, J.; Pechan, T. Maize proteomics: An insight into the biology of an important cereal crop. *Proteomics* **2013**, *13*, 637–662. [CrossRef] [PubMed]

30. Fandohan, P.; Hell, K.; Marasas, W.F.O.; Wingfield, M.J. Infection of maize by Fusarium species and contamination with fumonisin in Africa. *Afr. J. Biotechnol.* **2003**, *2*, 570–579.

31. Munkvold, G.P. Epidemiology of Fusarium diseases and their mycotoxins in maize ears. *Eur. J. Plant Pathol.* **2003**, *109*, 705–713. [CrossRef]

32. Woloshuk, C.; Wise, K. Diseases of Corn: Gibberella Ear Rot. Available online: http//www.extension.purdue.edu/extmedia/BP/BP-77-W.pdf (accessed on 3 September 2015).

33. Mohammadi, M.; Anoop, V.; Gleddie, S.; Harris, L.J. Proteomic profiling of two maize inbreds during early gibberella ear rot infection. *Proteomics* **2011**, *11*, 3675–3684. [CrossRef] [PubMed]

34. Marasas, W.F. Fumonisins: Their implications for human and animal health. *Nat. Toxins* **1995**, *3*, 193–198. [CrossRef] [PubMed]

35. Stockmann-Juvala, H.; Savolainen, K. A review of the toxic effects and mechanisms of action of fumonisin B1. *Hum. Exp. Toxicol.* **2008**, *27*, 799–809. [CrossRef] [PubMed]

36. Campo, S.; Carrascal, M.; Coca, M.; Abian, J.; San Segundo, B. The defense response of germinating maize embryos against fungal infection: A proteomics approach. *Proteomics* **2004**, *4*, 383–396. [CrossRef] [PubMed]

37. Chivasa, S.; Simon, W.J.; Yu, X.L.; Yalpani, N.; Slabas, A.R. Pathogen elicitor-induced changes in the maize extracellular matrix proteome. *Proteomics* **2005**, *5*, 4894–4904. [CrossRef] [PubMed]

38. Juge, N.; Payan, F.; Williamson, G. XIP-I, a xylanase inhibitor protein from wheat: A novel protein function. *Biochim. Biophys. Acta* **2004**, *1696*, 203–211. [CrossRef] [PubMed]

39. Ortiz, J.M. Mutant of bacillus subtilis lacking Exo-β-*N*-acetylglucosaminidase activity. *J. Bacteriol.* **1974**, *117*, 909–910. [PubMed]

40. Robertson-Hoyt, L.A.; Betran, J.; Payne, G.A.; White, D.G.; Isakeit, T.; Maragos, C.M.; Molnar, T.L.; Holland, J.B. Relationships among resistances to *Fusarium* and *Aspergillus* ear rots and contamination by fumonisin and aflatoxin in maize. *Phytopathology* **2007**, *97*, 311–317. [CrossRef] [PubMed]

41. Lanubile, A.; Maschietto, V.; de Leonardis, S.; Battilani, P.; Paciolla, C.; Marocco, A. Defense responses to mycotoxin-producing fungi *Fusarium proliferatum*, *F. subglutinans*, and *Aspergillus flavus* in kernels of susceptible and resistant maize genotypes. *Mol. Plant Microbe Interact.* **2015**, *28*, 546–557. [CrossRef] [PubMed]

42. Christensen, S.A.; Nemchenko, A.; Park, Y.S.; Borrego, E.; Huang, P.C.; Schmelz, E.A.; Kunze, S.; Feussner, I.; Yalpani, N.; Meeley, R.; *et al.* The novel monocot-specific 9-lipoxygenase ZmLOX12 is required to mount an effective jasmonate-mediated defense against *Fusarium verticillioides* in maize. *Mol. Plant Microbe Interact.* **2014**, *27*, 1263–1276. [CrossRef] [PubMed]

43. Gao, X.; Brodhagen, M.; Isakeit, T.; Brown, S.H.; Gobel, C.; Betran, J.; Feussner, I.; Keller, N.P.; Kolomiets, M.V. Inactivation of the lipoxygenase ZmLOX3 increases susceptibility of maize to *Aspergillus* spp. *Mol. Plant Microbe Interact.* **2009**, *22*, 222–231. [CrossRef] [PubMed]

44. Ito, M.F.; Paradela, F.; Soave, J.; Sugimori, M.H. Leaf spot caused in maize (*Zea mays* L.) by *Culvularia lunata* (wakker) boedijn. *Summa Phytopathol.* **1979**, *5*, 181–184.

45. Liu, T.; Liu, L.X.; Huang, X.L.; Jiang, X.; Zhou, P.H. A new furanoid toxin produced by *Culvularia lunata*, the causal agent of maize Curvularia leaf spot. *Can. J. Plant Pathol.* **2009**, *31*, 22–27. [CrossRef]

46. Huang, X.; Liu, L.; Chen, J.; Zhai, Y. Comparative proteomic analysis of the response in resistant and susceptible maize inbred lines to infection by *Culvularia lunata*. *Prog. Nat. Sci.* **2009**, *19*, 845–850. [CrossRef]

47. Huang, X.; Liu, L.; Zhai, Y.; Liu, T.; Chen, C. Proteomic comparison of four maize inbred lines with different levels of resistance to *Culvularia lunata* (wakker) boed infection. *Prog. Nat. Sci.* **2009**, *19*, 353–358. [CrossRef]

48. Bernier, F.; Berna, A. Germins and germin-like proteins: Plant do-all proteins. But what do they do exactly? *Plant Physiol. Biochem.* **2001**, *39*, 545–554. [CrossRef]

49. Manosalva, P.M.; Davidson, R.M.; Liu, B.; Zhu, X.; Hulbert, S.H.; Leung, H.; Leach, J.E. A germin-like protein gene family functions as a complex quantitative trait locus conferring broad-spectrum disease resistance in rice. *Plant Physiol.* **2009**, *149*, 286–296. [CrossRef] [PubMed]

50. Zimmermann, G.; Baumlein, H.; Mock, H.P.; Himmelbach, A.; Schweizer, P. The multigene family encoding germin-like proteins of barley. Regulation and function in basal host resistance. *Plant Physiol.* **2006**, *142*, 181–192. [CrossRef] [PubMed]

51. Christensen, A.B.; Thordal-Christensen, H.; Zimmermann, G.; Gjetting, T.; Lyngkjaer, M.F.; Dudler, R.; Schweizer, P. The germinlike protein GLP4 exhibits superoxide dismutase activity and is an important component of quantitative resistance in wheat and barley. *Mol. Plant Microbe Interact.* **2004**, *17*, 109–117. [CrossRef] [PubMed]

52. Lane, B.G. Oxalate, germins, and higher-plant pathogens. *IUBMB Life* **2002**, *53*, 67–75. [CrossRef] [PubMed]

53. Vallelian-Bindschedler, L.; Mosinger, E.; Metraux, J.P.; Schweizer, P. Structure, expression and localization of a germin-like protein in barley (*Hordeum vulgare* L.) that is insolubilized in stressed leaves. *Plant Mol. Biol.* **1998**, *37*, 297–308. [CrossRef] [PubMed]

54. Jones, A.M.; Thomas, V.; Bennett, M.H.; Mansfield, J.; Grant, M. Modifications to the *Arabidopsis* defense proteome occur prior to significant transcriptional change in response to inoculation with pseudomonas syringae. *Plant Physiol.* **2006**, *142*, 1603–1620. [CrossRef] [PubMed]

55. Sugihara, K.; Hanagata, N.; Dubinsky, Z.; Baba, S.; Karube, I. Molecular characterization of cDNA encoding oxygen evolving enhancer protein 1 increased by salt treatment in the mangrove *Bruguiera gymnorrhiza*. *Plant Cell Physiol.* **2000**, *41*, 1279–1285. [CrossRef] [PubMed]

56. Wu, L.; Han, Z.; Wang, S.; Wang, X.; Sun, A.; Zu, X.; Chen, Y. Comparative proteomic analysis of the plant-virus interaction in resistant and susceptible ecotypes of maize infected with sugarcane mosaic virus. *J. Proteom.* **2013**, *89*, 124–140. [CrossRef] [PubMed]

57. Wu, L.; Wang, S.; Chen, X.; Wang, X.; Wu, L.; Zu, X.; Chen, Y. Proteomic and phytohormone analysis of the response of maize (*Zea mays* L.) seedlings to sugarcane mosaic virus. *PLoS ONE* **2013**, *8*, e70295. [CrossRef] [PubMed]

58. Li, K.; Xu, C.; Zhang, J. Proteome profile of maize (*Zea mays* L.) leaf tissue at the flowering stage after long-term adjustment to rice black-streaked dwarf virus infection. *Gene* **2011**, *485*, 106–113. [CrossRef] [PubMed]

59. Von Rad, U.; Huttl, R.; Lottspeich, F.; Gierl, A.; Frey, M. Two glucosyltransferases are involved in detoxification of benzoxazinoids in maize. *Plant J.* **2001**, *28*, 633–642. [CrossRef] [PubMed]

60. Block, A.; Schmelz, E.; O'Donnell, P.J.; Jones, J.B.; Klee, H.J. Systemic acquired tolerance to virulent bacterial pathogens in tomato. *Plant Physiol.* **2005**, *138*, 1481–1490. [CrossRef] [PubMed]

61. DebRoy, S.; Thilmony, R.; Kwack, Y.B.; Nomura, K.; He, S.Y. A family of conserved bacterial effectors inhibits salicylic acid-mediated basal immunity and promotes disease necrosis in plants. *Proc. Natl. Acad. Sci. USA* **2004**, *101*, 9927–9932. [CrossRef] [PubMed]

62. Nomura, K.; Melotto, M.; He, S.Y. Suppression of host defense in compatible plant-pseudomonas syringae interactions. *Curr. Opin. Plant Biol.* **2005**, *8*, 361–368. [CrossRef] [PubMed]

63. Van Loon, L.C.; Rep, M.; Pieterse, C.M. Significance of inducible defense-related proteins in infected plants. *Annu. Rev. Phytopathol.* **2006**, *44*, 135–162. [CrossRef] [PubMed]

64. Bowles, D.J. Defense-related proteins in higher plants. *Annu. Rev. Biochem.* **1992**, *59*, 873–907. [CrossRef] [PubMed]

65. Kasprzewska, A. Plant chitinases—Regulation and function. *Cell. Mol. Biol.* **2003**, *8*, 809–824.

66. Leah, R.; Tommerup, H.; Svendsen, I.; Mundy, J. Biochemical and molecular characterization of three barley seed proteins with antifungal properties. *J. Biol. Chem.* **1991**, *266*, 1564–1573.

67. Schlumbaum, A.; Mauch, F.; Vogeli, U.; Boller, T. Plant chitinases are potent inhibitors of fungal growth. *Nature* **1986**, *324*, 365–367. [CrossRef]

68. Magbanua, Z.V.; de Moraes, C.M.; Brooks, T.D.; Williams, W.P.; Luthe, D.S. Is catalase activity one of the factors associated with maize resistance to *Aspergillus flavus*? *Mol. Plant Microbe Interact.* **2007**, *20*, 697–706. [CrossRef] [PubMed]

69. Vigers, A.; Roberts, W.; Selitrennikof, C. A new family of plant antifungal proteins. *Mol. Plant Microbe Interact.* **1991**, *4*, 315–323. [CrossRef] [PubMed]

70. Richardson, M.; Valdes-Rodrigues, S.; Blanco-Labra, A. A possible function for thaumatin and TMV-induced protein suggested by homology to a maize inhibitor. *Nature* **1987**, *327*, 432–434. [CrossRef]

71. Passardi, F.; Cosio, C.; Penel, C.; Dunand, C. Peroxidases have more functions than a swiss army knife. *Plant Cell Rep.* **2005**, *24*, 255–265. [CrossRef] [PubMed]

72. Almagro, L.; Gomez Ros, L.V.; Belchi-Navarro, S.; Bru, R.; Ros Barcelo, A.; Pedreno, M.A. Class III peroxidases in plant defence reactions. *J. Exp. Bot.* **2009**, *60*, 377–390. [CrossRef] [PubMed]

73. Bolwell, G.P.; Wojtaszek, P. Mechanisms for the generation of reactive oxygen species in plant defence—A broad perspective. *Physiol. Mol. Plant Pathol.* **1997**, *51*, 347–366. [CrossRef]

74. Mittler, R.; Vanderauwera, S.; Gollery, M.; van Breusegem, F. Reactive oxygen gene network of plants. *Trends Plant Sci.* **2004**, *9*, 490–498. [CrossRef] [PubMed]

75. Hiraga, S.; Sasaki, K.; Ito, H.; Ohashi, Y.; Matsui, H. A large family of class III plant peroxidases. *Plant Cell Physiol.* **2001**, *42*, 462–468. [CrossRef] [PubMed]

76. Jiang, M.; Zhang, J. Effect of abscisic acid on active oxygen species, antioxidative defence system and oxidative damage in leaves of maize seedlings. *Plant Cell Physiol.* **2001**, *42*, 1265–1273. [CrossRef] [PubMed]

77. Kurama, E.E.; Fenille, R.C.; Rosa, V.E., Jr.; Rosa, D.D.; Ulian, E.C. Mining the enzymes involved in the detoxification of reactive oxygen species (ROS) in sugarcane. *Mol. Plant Pathol.* **2002**, *3*, 251–259. [CrossRef] [PubMed]

78. Guan, L.; Scandalios, J.G. Two structurally similar maize cytosolic superoxide dismutase genes, *Sod4* and *Sod4A*, respond differentially to abscisic acid and high osmoticum. *Plant Physiol.* **1998**, *117*, 217–224. [CrossRef] [PubMed]

79. Dietz, K.J.; Jacob, S.; Oelze, M.L.; Laxa, M.; Tognetti, V.; de Miranda, S.M.; Baier, M.; Finkemeier, I. The function of peroxiredoxins in plant organelle redox metabolism. *J. Exp. Bot.* **2006**, *57*, 1697–1709. [CrossRef] [PubMed]

80. Gutsche, N.; Thurow, C.; Zachgo, S.; Gatz, C. Plant-specific CC-type glutaredoxins: Functions in developmental processes and stress responses. *Biol. Chem.* **2015**, *396*, 495–509. [CrossRef] [PubMed]

81. Chugh, V.; Kaur, N.; Grewal, M.S.; Gupta, A.K. Differential antioxidative response of tolerant and sensitive maize (*Zea mays* L.) genotypes to drought stress at reproductive stage. *Indian J. Biochem. Biophys.* **2013**, *50*, 150–158. [PubMed]

82. Mittler, R. Oxidative stress, antioxidants and stress tolerance. *Trends Plant Sci.* **2002**, *7*, 405–410. [CrossRef]

83. Dixon, D.P.; Lapthorn, A.; Edwards, R. Plant glutathione transferases. *Genome Biol.* **2002**, *3*. REVIEWS3004. [CrossRef] [PubMed]

84. Sari-Gorla, M.; Ferrario, S.; Rossini, L.; Frova, C.; Villa, M. Developmental expression of glutathione *S*-transferase in maize and its possible connection with herbicide tolerance. *Euphytica* **1993**, *67*, 221–230. [CrossRef]

85. Timmerman, K.P. Molecular characterization of corn glutathione *S*-transferase isozymes involved in herbicide detoxication. *Physiol. Plant* **1989**, *77*, 465–471. [CrossRef]

86. Iriti, M.; Faoro, F. Chemical diversity and defence metabolism: How plants cope with pathogens and ozone pollution. *Int. J. Mol. Sci.* **2009**, *10*, 3371–3399. [CrossRef] [PubMed]

87. Christensen, S.A.; Huffaker, A.; Kaplan, F.; Sims, J.; Ziemann, S.; Doehlemann, G.; Ji, L.; Schmitz, R.J.; Kolomiets, M.V.; Alborn, H.T.; *et al.* Maize death acids, 9-lipoxygenase-derived cyclopente(a)nones, display activity as cytotoxic phytoalexins and transcriptional mediators. *Proc. Natl. Acad. Sci. USA* **2015**, *112*, 11407–11412. [CrossRef] [PubMed]

88. Kim, J.H.; Campbell, B.C.; Mahoney, N.E.; Chan, K.L.; Molyneux, R.J. Identification of phenolics for control of *Aspergillus flavus* using saccharomyces cerevisae in a model target-gene bioassay. *J. Agric. Food Chem.* **2004**, *52*, 7814–7821. [CrossRef] [PubMed]

89. Beekrum, S.; Govinden, R.; Padayachee, T.; Odhav, B. Naturally occurring phenols: A detoxification strategy for fumonism B$_1$. *Food Addit. Contam.* **2003**, *20*, 490–493. [CrossRef] [PubMed]

90. Curir, P.; Dolci, M.; Dolci, P.; Lanzotti, V.; de Cooman, L. Fungitoxic phenols from carnation (*Dianthus caryophyllus*) effective against *Fusarium oxysporum*, f. sp. dianthi. *Phytochem. Anal.* **2003**, *14*, 8–12. [CrossRef] [PubMed]

91. Boerjan, W.; Ralph, J.; Baucher, M. Lignin biosynthesis. *Annu. Rev. Plant Biol.* **2003**, *54*, 519–546. [CrossRef] [PubMed]

92. Berger, S.; Sinha, A.K.; Roitsch, T. Plant physiology meets phytopathology: Plant primary metabolism and plant-pathogen interactions. *J. Exp. Bot.* **2007**, *58*, 4019–4026. [CrossRef] [PubMed]

93. Bolton, M.D. Primary metabolism and plant defense—Fuel for the fire. *Mol. Plant Microbe Interact.* **2009**, *22*, 487–497. [CrossRef] [PubMed]

94. Rojas, C.M.; Senthil-Kumar, M.; Tzin, V.; Mysore, K.S. Regulation of primary plant metabolism during plant-pathogen interactions and its contribution to plant defense. *Front. Plant Sci.* **2014**, *5*, 17. [CrossRef] [PubMed]

95. Van Loon, L.C. Disease induction by plant viruses. *Adv. Virus Res.* **1987**, *33*, 205–255. [PubMed]

96. Hancock, J.T.; Henson, D.; Nyirenda, M.; Desikan, R.; Harrison, J.; Lewis, M.; Hughes, J.; Neill, S.J. Proteomic identification of glyceraldehyde 3-phosphate dehydrogenase as an inhibitory target of hydrogen peroxide in arabidopsis. *Plant Physiol. Biochem.* **2005**, *43*, 828–835. [CrossRef] [PubMed]

97. Henry, E.; Fung, N.; Liu, J.; Drakakaki, G.; Coaker, G. Beyond glycolysis: GAPDHs are multi-functional enzymes involved in regulation of ROS, autophagy, and plant immune responses. *PLoS Genet.* **2015**, *11*, e1005199. [CrossRef] [PubMed]

98. Jang, J.C.; Sheen, J. Sugar sensing in higher plants. *Trends Plant Sci.* **1997**, *2*, 208–214. [CrossRef]

99. Tsukaya, H.; Ohshima, T.; Naito, S.; Chino, M.; Komeda, Y. Sugar-dependent expression of the *CHS-A* gene for chalcone synthase from petunia in transgenic *Arabidopsis. Plant Physiol.* **1991**, *97*, 1414–1421. [CrossRef] [PubMed]

100. Hopkins, M.T.; Lampi, Y.; Wang, T.W.; Liu, Z.; Thompson, J.E. Eukaryotic translation initiation factor 5A is involved in pathogen-induced cell death and development of disease symptoms in *Arabidopsis. Plant Physiol.* **2008**, *148*, 479–489. [CrossRef] [PubMed]

101. Andreeva, L.; Heads, R.; Green, C.J. Cyclophilins and their possible role in the stress response. *Int. J. Exp. Pathol.* **1999**, *80*, 305–315. [CrossRef] [PubMed]

102. Hendrich, J.P.; Hartl, F.U. Molecular chaperone functions of heat shock proteins. *Annu. Rev. Biochem.* **1993**, *62*, 349–384. [CrossRef] [PubMed]

103. Sun, W.; van Montagu, M.; Verbruggen, N. Small heat shock proteins and stress tolerance in plants. *Biochim. Biophys. Acta* **2002**, *1577*, 1–9. [CrossRef]

104. Kim, S.K.; You, Y.N.; Park, J.C.; Joung, Y.; Kim, B.G.; Ahn, J.C.; Cho, H.S. The rice thylakoid lumenal cyclophilin OsCYP20-2 confers enhanced environmental stress tolerance in tobacco and *Arabidopsis. Plant Cell Rep.* **2012**, *31*, 417–426. [CrossRef] [PubMed]

105. Sekhar, K.; Priyanka, B.; Reddy, V.D.; Rao, K.V. Isolation and characterization of a pigeonpea cyclophilin (*CcCYP*) gene, and its over-expression in *Arabidopsis* confers multiple abiotic stress tolerance. *Plant Cell Environ.* **2010**, *33*, 1324–1338. [CrossRef] [PubMed]

106. Sharma, A.D.; Singh, P. Effect of water stress on expression of a 20 kD cyclophilin-like protein in drought susceptible and tolerant cultivars of sorghum. *J. Plant Biochem. Biotechnol.* **2003**, *12*, 77–80. [CrossRef]

107. Trivedi, D.K.; Ansari, M.W.; Tuteja, N. Multiple abiotic stress responsive rice cyclophilin: (OsCYP-25) mediates a wide range of cellular responses. *Commun. Integr. Biol.* **2013**, *6*, e25260. [CrossRef] [PubMed]

108. Wisser, R.J.; Balint-Kurti, P.J.; Nelson, R.J. The genetic architecture of disease resistance in maize: A synthesis of published studies. *Phytopathology* **2006**, *96*, 120–129. [CrossRef] [PubMed]

 International Journal of
Molecular Sciences

Article

Microtubule-Associated Protein SBgLR Facilitates Storage Protein Deposition and Its Expression Leads to Lysine Content Increase in Transgenic Maize Endosperm

Chen Liu [1,2], Shixue Li [1], Jing Yue [1], Wenhan Xiao [1], Qian Zhao [1], Dengyun Zhu [1] and Jingjuan Yu [1,*]

[1] State Key Laboratory for Agro-Biotechnology, College of Biological Sciences, China Agricultural University, No. 2 Yuanmingyuan West Road, Beijing 100193, China; liuchen@iae.ac.cn (C.L.); lishixue19890105@126.com (S.L.); yuejinglove@126.com (J.Y.); zb1106052@cau.edu.cn (W.X.); zhaoqian@cau.edu.cn (Q.Z.); zhudy@cau.edu.cn (D.Z.)
[2] Key Laboratory of Pollution Ecology and Environmental Engineering, Institute of Applied Ecology, Chinese Academy of Sciences, No. 72 Wenhua Road, Shenyang 110016, China
* Correspondence: yujj@cau.edu.cn; Tel.: +86-10-6273-3462

Academic Editors: Setsuko Komatsu and Zahed Hossain
Received: 26 October 2015; Accepted: 7 December 2015; Published: 12 December 2015

Abstract: Maize (*Zea mays*) seed is deficient in protein and lysine content. Many studies have been made to improve the nutritional quality of maize seeds. Previously, we reported the role of a natural lysine-rich protein gene *SBgLR* in increasing protein and lysine content. However, how the SBgLR improves lysine and protein content remains unclear. Here, the reasons and possible mechanism for SBgLR in protein and lysine improvement have been analyzed and discussed. Through seed-specific expression of *SBgLR*, we obtained transgenic maize with the simultaneously increased lysine and protein contents. High-protein and high-lysine characters were stably inherited across generations. The expression of *SBgLR* in maize kernels increased the accumulation of both zeins and non-zein proteins. Transmission electron microscopy showed that the number of protein bodies (PBs) was increased obviously in SBgLR transgenic immature endosperms with the morphology and structure of PBs unchanged. The proteinaceous matrix was more abundant in transgenic mature endosperms under scanning electron microscopy. The stabilities of zein and lysine-rich non-zein genes were also increased in transgenic endosperms. Finally, the potential application of SBgLR in maize nutrient improvement was evaluated. This study shows that a cytoskeleton-associated protein has potential applicable value in crop nutrient improving, and provided a feasible strategy for improvement of maize grain quality.

Keywords: SBgLR; microtubule-associated protein; protein body formation; maize nutrient improvement

1. Introduction

Maize is one of the most important cereals for the food and feed industry. It is an important source of proteins for humans and livestock, especially in developing countries [1,2]. Zein, the most abundant storage protein in maize kernels, is deficient in lysine and tryptophan, leading to a poor quality of protein for monogastric animals. Many approaches have been used to improve the nutritional quality of maize. *Opaque 2* (*o2*) is a naturally-occurring high-lysine mutant, which has twice the normal level of lysine and tryptophan [3]. However, this mutant has a soft endosperm [2]. The kernel softness is further overcome by several *o2* modifiers, and quality protein maize (QPM) has been developed by this approach [4]. In addition, a bacterial gene encoding a lysine feedback-insensitive dihydrodipicolinate synthase (DHDPS) has been introduced into maize to increase free lysine level [5]. Unfortunately, no lysine increase was observed when the gene was specifically expressed in starchy endosperm, even though more than a two-fold level of lysine accumulation was detected when expressed in the embryo and aleurone [5]. More recently, high-lysine maize has been obtained by decreasing lysine catabolism through RNA interference technique [6], by reducing α-zein mRNA [7,8], by expressing a milk protein α-lactalbumin [9], and by expressing a lysine-rich protein gene *GhLRP* in maize endosperm [10]. Except for some *GhLRP* transgenic lines, these high lysine corn plants contained normal or less level protein content in the seeds. However, seed-specific expression of lysine-rich protein gene *SB401* or its homolog *SBgLR* significantly increased both lysine and total protein contents in maize seeds [11–13]. Functional analysis showed that SB401 and SBgLR can bind and interact with the cytoskeleton and function in microtubule (MT) reorganizations [14,15]. But how these two genes improve protein and lysine content of transgenic maize seeds remains unclear.

It is known that zein accumulates in protein bodies (PBs) in the endosperm cells. However, the mechanism of PB formation, especially the role of the cytoskeleton system, is incompletely understood. Previous studies demonstrated that the cytoskeleton plays an important role in the spatial and temporal distribution of materials and information by affecting targeting, tethering, transporting and translation of mRNA [16–18]. In maize endosperm cells, extensive evidence showed that zeins are synthesized on cytoskeleton-bound polysomes [19,20] and their mRNAs were localized through binding to a specific site on the tubulin and actin cytoskeleton [21]. These data indicated that the cytoskeleton serves as an attachment scaffold for mRNA translation and protein synthesis [22,23]. eEF1α, a lysine-rich cytoskeleton-associated protein, plays an important role in protein synthesis and actin-bundling in maize endosperms [24–26]. PBs in endosperm cells are distributed between starch granules, where eEF-1α and cytoskeletal elements, such as actin filaments and microtubules (MTs), are abundant [27]. The eEF1α concentration was found to be a good predictor of the endosperm lysine content [28,29]. However, the roles of microtubule-associated proteins, and the relationship between the cytoskeleton system and storage protein accumulation are not fully understood.

The *SBgLR*, we cloned from *Solanum berthaultii*in previously [30], encodes a pollen-specific protein with high lysine content (18.93%, w/w) [12]. SBgLR contains five imperfectly repeated motifs of V-V-E-K-K-N/E-E, which resemble the repetitive domain responsible for MT-binding activity in murine and plants cells [15,31]. Considering these characteristics, we further investigated the activities of SBgLR on MT regulation. We found that the recombinant SBgLR binds to both tubulin and MTs *in vitro*. SBgLR overexpressing tobacco showed curving and right-handed twisting root growth; and abnormal directional expansion of cotyledon pavement cells, suggesting that SBgLR interacts with MTs and regulates their organization [15]. Our previous work showed that overexpression of *SBgLR* can increase the lysine and protein content in maize seeds [12,13]. In this research, we obtained SBgLR transgenic maize with the simultaneously increased lysine and protein contents in seeds and analyzed heredity of high-lysine and high-protein characters. In addition, through analyzing the SBgLR transgenic maize seeds by transmission electron microscopy (TEM) and scanning electron microscopy (SEM), we found that the expression of *SBgLR* promotes PB formation and proteinaceous matrix deposition. And transcription analysis showed increasing stabilities of storage protein genes and non-zein genes. The potential application of SBgLR transgenic maize lines was evaluated and the probable mechanism of SBgLR in improving the protein and lysine contents of maize seeds was discussed.

2. Results and Discussion

2.1. High-Protein and High-Lysine Characters are Stably Inherited in SBgLR Transgenic Lines

SBgLR was previously reported to increase lysine and protein content in T_1 transgenic maize seeds [8,9]. In order to analyze the heredity of the high-protein and high-lysine characters and to obtain the SBgLR transgenic maize with the selection marker free, we introduced *SBgLR* gene into maize hybrid line Hi3027 (08 × 178) under the control of the seed-specific promoter *pF128* with the double "T-DNA" *Agrobacterium*-binary vector pSB130-SBgLR (Figure 1A) [32]. Finally, a total of 23 independent transgenic events were identified by polymerase chain reaction (PCR) amplification using *SBgLR*-specific primer pair S1/S2 (Table S1). Kernels of these plants were obtained by self-pollination. T_1 plant genotyping showed that 14 transgenic lines exhibited segregation at a ratio of approximately 3:1, suggesting the presence of a single-site integration of the *SBgLR* gene. The expressions of the *SBgLR* gene in transgenic maize endosperm was detected by reverse transcription-polymerase chain reaction (RT-PCR). A 241-bp fragment was amplified from the coding region of the *SBgLR* gene in T_2 immature kernels at 15 days after pollination (DAP) of each transgenic line, but not from wild type (WT) control (Figure 1B). This indicated that *SBgLR* was transcribed and spliced accurately in transgenic kernels as reported by Lang *et al.* [30]. Furthermore, Western blot was performed to investigate the accumulation of SBgLR protein in maize seeds. The result showed that two bands with different staining signals (weak and dark staining) were detected from transgenic seeds protein, but not from WT control (Figure 1C). Sequence analysis revealed that predicted glycosylation and phosphorylation sites existed in the SBgLR amino acid sequence (Figure S1). The weak staining band might be a post-translational modification form of SBgLR protein.

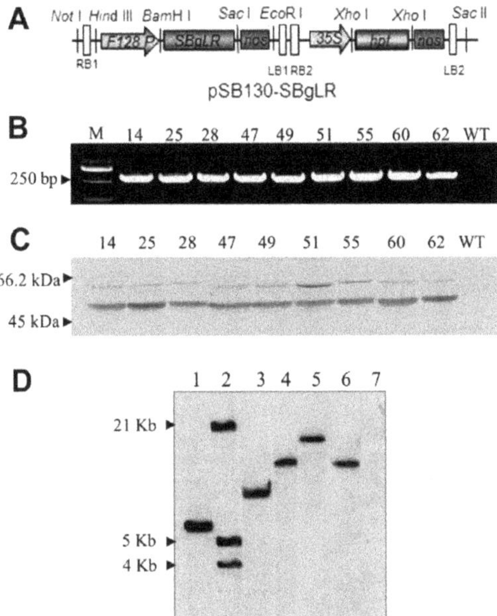

Figure 1. The diagram of binary vector and molecular analysis of transgenic plants. (**A**), Diagram of pSB130-SBgLR; (**B**), RT-PCR analysis of *SBgLR* transcript in transgenic line 14, 25, 28, 47, 49, 51, 55, 60, 62 and WT kernels. M, DNA marker; (**C**), Western blot analysis of SBgLR protein in transgenic line 14, 25, 28, 47, 49, 51, 55, 60, 62 and WT kernels. Two specific bands were detected from transgenic seeds protein, but not from WT protein; (**D**), Southern blot analysis of transgenic Q14 and Q51 plants. Lane 1, positive control. Lane 2, DNA marker, from bottom to top: 4 Kb, 5 Kb and 21 Kb respectively. Lane 3 and lane 4, Q14 genomic DNA digested with *Hind* III and *Bam*H I respectively. Lane 5 and 6, Q51 genomic DNA digested with *Hind* III and *Bam*H I respectively. Line 7, WT genomic DNA digested with *Hind* III as a negative control.

The protein and lysine content were analyzed in the seeds of T_1 plants from 14 transgenic lines. Seeds of Hi3027 segregation progenies were used as a reference. The protein and lysine content of line

14, 25, 28, 47, 49, 51, 55, 60 and 62 were markedly increased (more than 20%) and were selected to sow in T_2 generation (Table 1). We monitored the protein and lysine content of these transgenic lines in the continuous generations, the results showed that the high-lysine and high-protein characters could be stably inherited across generations (Table 1). Among all the transgenic lines, the protein content in T_2 seeds of line 14 and 51 was 15.17 and 14.16 g/100 g dry weight, respectively, while it was 10.35 g/100 g dry weight in WT control (Table 1). Meanwhile, the lysine content in the seeds of these two lines was 0.49 and 0.44 g/100 g dry weight, respectively, and it was 0.28 g/100 g dry weight in WT control (Table 1). Southern blot further confirmed that a single copy of *SBgLR* expression cassette insertion in these two lines (Figure 1D) and the marker gene was not detected by PCR and Southern blot in these two lines (Data not shown). So these two lines named Q14 and Q51 were selected as materials for further study on how SBgLR increase protein and lysine content in transgenic maize seeds.

Table 1. Protein and lysine content in different transgenic lines.

Line	Protein Content (g/100 g) [$]			Lysine Content (g/100 g) [$]		
	T_1	T_2	T_3	T_1	T_2	T_3
14	14.21 ± 0.07 **	15.17 ± 0.84 **	15.20 ± 0.37 **	0.42 ± 0.03 **	0.49 ± 0.06 **	0.49 ± 0.03 **
25	12.78 ± 0.12 **	13.17 ± 0.96 **	13.05 ± 0.89 **	0.39 ± 0.02 **	0.38 ± 0.02 **	0.39 ± 0.02 **
28	14.26 ± 1.10 **	13.19 ± 1.48 **	13.67 ± 0.99 **	0.37 ± 0.02 **	0.34 ± 0.01 *	0.36 ± 0.01 **
47	12.72 ± 0.14 **	13.19 ± 1.04 **	13.34 ± 1.17 **	0.39 ± 0.01 **	0.39 ± 0.02 **	0.38 ± 0.03 **
49	12.61 ± 0.09 **	12.22 ± 0.10 **	12.99 ± 0.76 **	0.39 ± 0.01 **	0.41 ± 0.03 **	0.40 ± 0.01 **
51	14.29 ± 0.11 **	14.16 ± 1.07 **	14.37 ± 0.92 **	0.41 ± 0.03 **	0.44 ± 0.02 **	0.43 ± 0.01 **
55	12.60 ± 0.19 **	12.40 ± 0.28 **	13.94 ± 0.39 **	0.36 ± 0.03 **	0.35 ± 0.02 **	0.37 ± 0.01 **
60	12.59 ± 0.17 **	13.77 ± 0.46 **	13.17 ± 0.56 **	0.40 ± 0.02 **	0.42 ± 0.01 **	0.41 ± 0.02 **
62	12.46 ± 0.56 **	12.74 ± 0.79 **	13.41 ± 1.07 **	0.38 ± 0.01 **	0.39 ± 0.02 **	0.39 ± 0.01 **
CT [$$]	10.54 ± 0.40	10.30 ± 0.88	10.78 ± 0.53	0.26 ± 0.01	0.28 ± 0.01	0.25 ± 0.01

Values shown are grams of protein or lysine per 100 g of kernel dry weight. Data were mean ± SD. [$], T_1, T_2 and T_3 represents the plant generation. Data were collected from the seeds of T_1, T_2 and T_3 plants. [$$], Hi3027 segregation progenies with the same generation as transgenic lines were used as control (CT). The data of protein and lysine content of control (CT) were the average value of twenty Hi3027 segregation progenies. Data were analyzed by One-way ANOVA ($p < 0.05$) and followed by Least Significant Difference (*LSD*) post test (* $p < 0.05$; ** $p < 0.01$).

2.2. Zeins And Non-Zein Proteins Accumulation in SBgLR Transgenic Maize Seeds

Maize storage proteins are mainly divided into two groups: zein and non-zein according to the differences of solubility. To explore which fractions of storage proteins were changed in transgenic kernels, we extracted and analyzed zein, non-zein and total protein extracts of the seeds of transgenic T_2 and Hi3027 F_2 plants by sodium dodecyl sulfate-polyacrylamide gel electrophoresis (SDS-PAGE) and protein content quantification assays. The results are shown in Figure 2. Total zein content of Q14 and Q51 mature seeds was 4.62 mg/50 mg and 4.75 mg/50 mg flour respectively, whereas it was 3.18 mg/50 mg flour of WT (Figure 2A). Except 10-kDa δ-zein, the contents of 15-kDa β-zein, 16-kDa γ-zein, 19-kDa α-zein, 22-kDa α-zein, 27-kDa γ-zein and 50-kDa γ-zein were all increased in transgenic mature seeds (Figure 2D). The non-zein proteins that contribute the most to maize lysine content were also increased in Q14 and Q51 mature seeds on SDS-PAGE staining gel (Figure 2E). They were increased to 2.06 mg/50 mg flour and 1.98 mg/50 mg flour in protein content quantification assay (Figure 2B). Total protein contents of WT, Q14 and Q51 were 4.27 mg/50 mg, 6.16 mg/50 mg and 5.38 mg/50 mg flour, respectively (Figure 2C,F). These results suggested that SBgLR is involved in both zeins and non-zein proteins enhancement in transgenic seeds.

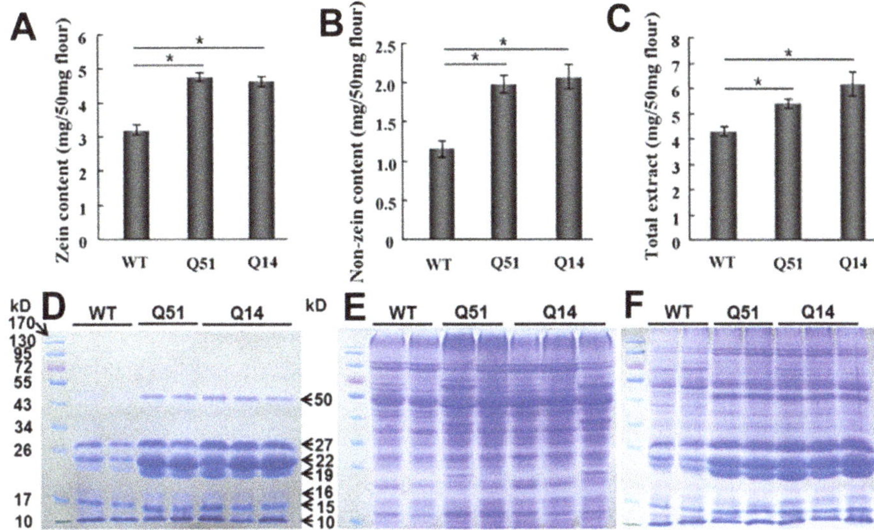

Figure 2. Analysis of protein accumulation in transgenic maize seeds. (**A–C**), Quantification assay of zein (**A**), non-zein (**B**) and total extract (**C**); (**D–F**), SDS-PAGE analysis of zein (**D**), non-zein (**E**) and total extract (**F**). The significance of difference was evaluated by Student's *t*-test (* $p < 0.05$).

2.3. Transmission Electron Microscopy of SBgLR-Overexpression Endosperm

PBs are the major storage form of different types of zeins in maize endosperm. The composition of zeins in PBs has a decisive effect on PB morphology [33]. We have shown that zeins were markedly increased in transgenic endosperm cells. But how are these proteins stored in the endosperm cells? Two possibilities exist. One is that the increased zeins deposited into PBs form larger ones; and the other is that these zeins form more PBs in endosperm cells. To investigate the storage form of the increased zeins in SBgLR transgenic endosperm, several cell layers of Q14 and Q51 endosperm from T_2 plants at 15 DAP and 20 DAP were analyzed with TEM. Endosperm samples from WT kernels (Hi3027 F_2 segregation progeny) were used as references. Compared with WT (Figure 3A), the PBs were more abundant in Q14 and Q51 endosperms at 15 DAP and 20 DAP (Figure 3B,C). Quantitative analysis showed that the number of PBs in Q14 and Q15 endosperms was significantly greater than that in WT control (Student's *t*-test, $p < 0.01$). In WT endosperm cells at 15 DAP, the average number of PBs per 60 μm^2 micrograph was 379.54 ± 18.41, while it was 1115.20 ± 39.18 and 952.37 ± 32.67, respectively, in Q14 and Q51 endosperm cells (Figure 3D). Similarly, the number of PBs was increased significantly in Q14 (Figure 3F) and Q51 (Figure 3G) endosperm cells at 20 DAP. They were 2.16 and 1.91 times of that in WT control, respectively (Figure 3H). Morphology observation showed that the size and shape of PBs were the same in Q14 (Figure 3I) and Q51 (Figure 3J) endosperm cells as in WT references (Figure 3K) at 15 DAP. As the seed develops, PBs increase in size in WT endosperm cells (Figure 3N) whereas almost not changed in Q14 (Figure 3L) and Q51 (Figure 3M) endosperm cells. These results indicated that the expression of SBgLR increases the number of PBs in transgenic endosperm cells.

As the localization of zeins within PBs is important in the morphology of PBs [34], we conducted immunogold staining assays using antibody against 19-kDa α-zein; it was mainly, but not entirely, distributed over the light-staining region of PBs in WT (Figure 4A) cells. A similar distribution of 19-kDa α-zein were observed in Q14 endosperm cells (Figure 4B), suggesting a normal localization of zeins within PBs. No particle was detected when primary antibody was omitted in WT control (Figure 4C). These results indicate that SBgLR does not affect the localization of zeins in PBs.

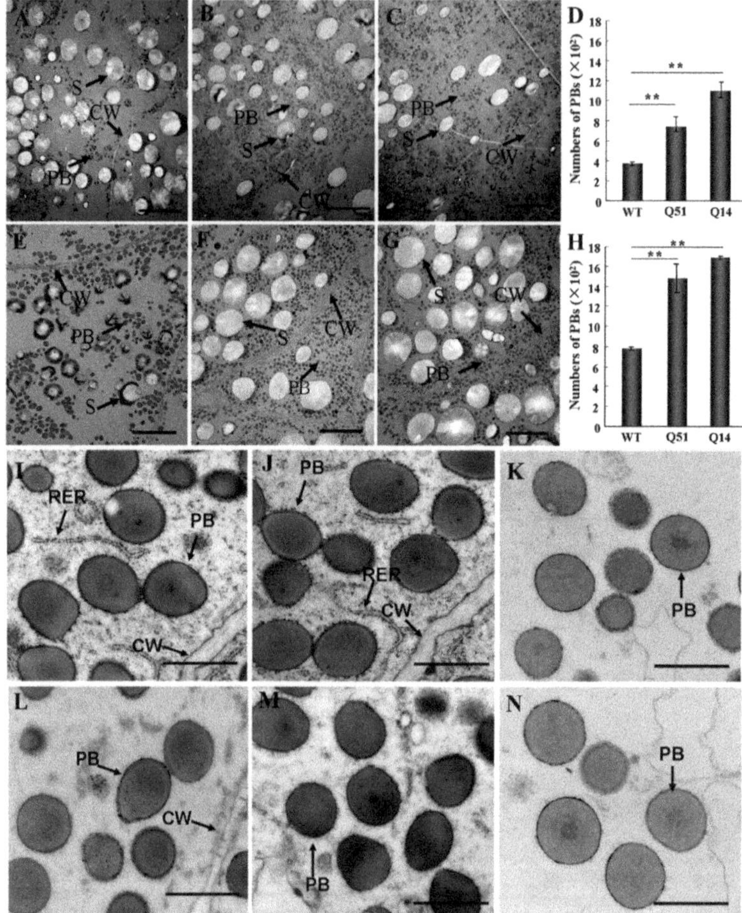

Figure 3. Transmission electron microscopy of 15 and 20 DAP endosperm cells. (**A–C**), Transmission electron microscopy of WT (**A**), Q14 (**B**) and Q51 (**C**) 15 DAP endosperm. Bar = 20 μm; (**E–G**), Transmission electron microscopy of WT (**E**), Q14 (**F**) and Q51 (**G**) 20 DAP endosperm. Bar = 20 μm; (**D,H**), Statistic of PB numbers per 60 μm² micrograph in WT, Q14 and Q51 at 15 and 20 DAP endosperms. Data were mean ± SD, (** $p < 0.01$, Student's *t*-test); (**I–K**), PB morphology observation of Q14 (**I**), Q51 (**J**) and WT (**K**) at 15 DAP endosperm; (**L–N**), PB morphology observation of Q14 (**L**), Q51 (**M**) and WT (**N**) at 20 DAP endosperm. Bar = 200 nm in (**I–N**). RER, rough endoplasmic reticulum; CW, cell wall; S, starch granule.

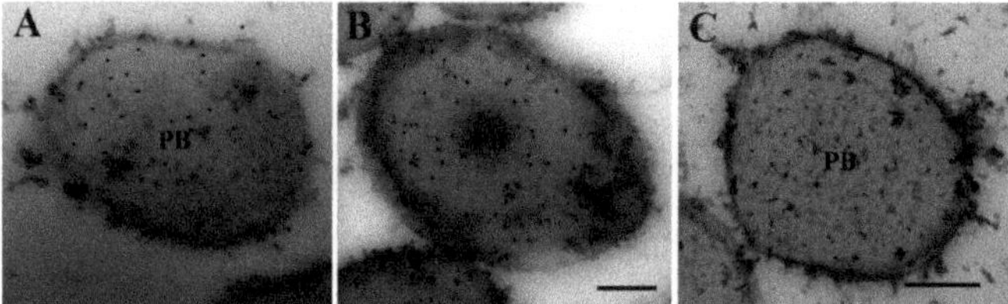

Figure 4. Immunogold labeling of 19-kDa α-zein. (**A**), WT; (**B**), Q14 20 DAP endosperm; (**C**), Primary antibody was omitted as negative control. Bar = 200 nm in (**A–C**). PB, protein body.

2.4. Scanning Electron Microscopy of Transgenic Mature Endosperm

TEM results showed that the number of PBs was increased in SBgLR transgenic immature endosperm cells. We further observed the proteinaceous matrix deposition of SBgLR transgenic mature seeds under SEM. The proteinaceous matrix mainly consisted of zein and non-zein proteins, and was located between starch granules. We examined the central region of the mature starchy endosperm where the deposition of proteinaceous matrix between starchy grains was easily observed under SEM. In WT mature kernels, endosperm cells in the central region of the starchy endosperm contained smooth starch grains with little proteinaceous matrix (Figure 5A,D). However, greater amounts of proteinaceous matrix were accumulated surrounding the spheroidal starch granules in Q14 (Figure 5B,E) and Q51 (Figure 5C,F) mature starchy endosperms. Observation at higher magnification revealed that proteinaceous matrix surrounding starch grains was more dense compared to that in WT endosperm (white arrow in Figure 5D), and distributed consecutively in Q14 (white arrow in Figure 5E) and Q51 endosperms (white arrow in Figure 5F) .

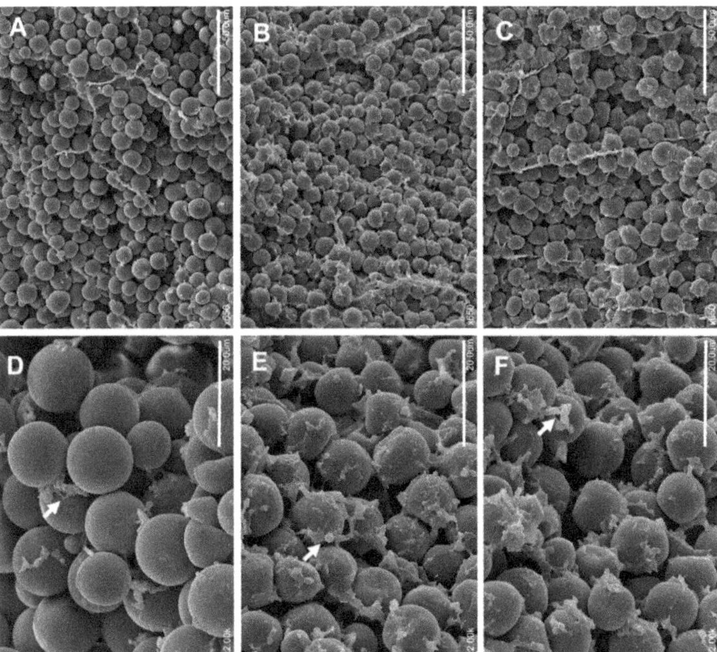

Figure 5. Scanning electron microscopy of mature seeds starchy endosperm (central region). (**A,D**), WT mature endosperm; (**B,E**), Transgenic Q14 mature endosperm; (**C,F**), Transgenic Q51 mature endosperm. Bar = 50 μm in (**A–C**); Bar = 20 μm in (**D–F**). White arrows indicate proteinous matrix.

2.5. mRNA Stability Analysis of Zein and Lysine-Rich Non-Zein Genes in Transgenic Endosperm Cells

We have shown that zein and non-zein protein accumulation were enhanced in transgenic endosperm cells. In order to investigate the differences of mRNA levels between transgenic and WT endosperm cells, a gene chip assay was performed using WT and Q51 20 DAP endosperm. The result showed that, compared to WT reference, the mRNA levels of some zein and lysine-rich non-zein protein genes were higher in Q51 immature endosperm cells. This primary result was further confirmed by quantitative RT-PCR. The expression level of maize 19-kDa α *zein* (AF371268), 22-kDa α *zein* (AF371277.1) and δ *zein storage protein gene* (AF371265.1) in transgenic endosperms were

12.01, 16.08, and 269.86 times higher, respectively, than in WT endosperms (Figure 6). Meanwhile, the expression level of five non-zein genes, *histone H2B.3* (BM080549), *cylicin-1* (CF627896), *ER lumen protein retaining receptor C28H8.4* (CF602623), *fructose-bisphosphatealdolase* (CF624216) and *60S ribosomal protein L22-2* (BI543125), which encode lysine-rich proteins, were 2.68, 2.21, 3.71, 1.95 and 2.28 times higher, respectively, than in WT endosperms (Figure 6). This result indicated that the mRNA stabilities of storage protein genes were also increased in SBgLR transgenic endosperm cells.

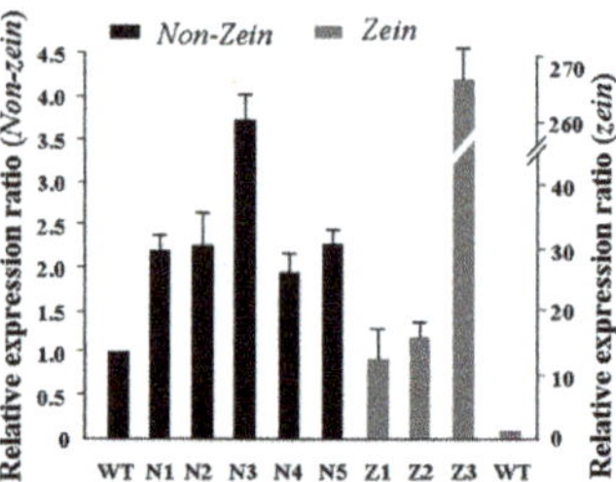

Figure 6. Quantitative analysis of some *zein* and lysine-rich *non-zein* transcripts in 20 DAP endosperm. N1, *histone H2B.3* (BM080549); N2, *cylicin-1* (CF627896); N3, *ER lumen protein retaining receptor C28H8.4* (CF602623); N4, *fructose-bisphosphatealdolase* (CF624216); N5, *60S ribosomal protein L22-2* (BI543125); Z1, 19 kDa α *zein* (AF371268); Z2, 22 kDa α *zein* (AF371277.1); Z3, δ *zein storage protein gene* (AF371265.1). Maize *actin* (LOC100280540) was used as reference gene.

2.6. Potential Application of Transgenic Maize Lines in Nutrient Maize Breeding

High protein and high lysine maize inbred lines are valuable in maize nutrient improvement. To evaluate the potential application of SBgLR maize, kernel quality and agronomic traits were analyzed. Kernel hardness, a critical agronomic trait, is very important in maize breeding. Under incandescent and transmitted light, the kernels of T_5 generations of Q14 and Q51 showed the vitreous phenotypes as well as WT (Figure 7). No obvious differences of seed weight and kernel density were found (Figure S2A,B). Agronomic characters such as plant height, ear height, ear length, kernel weight per ear and ear rows were not affected in different generations of these transgenic lines (Tables S2–S4). Other nutrients such as starch and oil are also important. We analyzed the starch and oil contents of the seeds of T_1, T_4 and T_5 transgenic plants. The data shown in Table 2 indicate that starch and oil content were not changed significantly in these transgenic lines (Table 2). Furthermore, data from our field test showed that the germination rates of T_2 seeds of these two lines was also similar to that of the WT (Figure S2C). These results indicate that the increase of protein and lysine contents in transgenic lines did not affect the agronomic characters and kernel qualities and these transgenic maize lines are potentially promising for corn breeding.

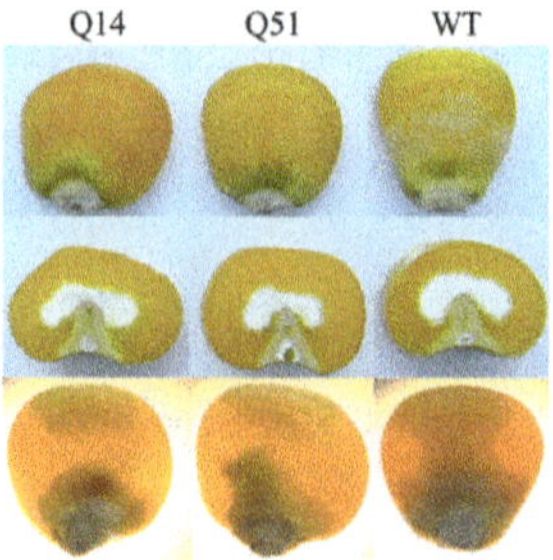

Figure 7. Kernel phenotype for Q14, Q51 and WT. Photographs of kernels were taken under incandescent light (**Top** and **Middle**) or with transmitted light (**Bottom**).

Table 2. Starch and oil content of the seeds of T_1, T_4 and T_5 plants.

Line	Oil Content (g/100 g)			Starch Content (g/100 g)		
	T_1	T_4	T_5	T_1	T_4	T_5
14	4.66 ± 0.04	4.89 ± 0.59	4.49 ± 0.69	65.76 ± 0.40	64.34 ± 1.01	64.02 ± 1.60
25	3.94 ± 0.32	4.35 ± 0.20	4.13 ± 0.47	66.55 ± 0.22	65.72 ± 0.40	65.43 ± 1.09
28	4.02 ± 0.37	4.75 ± 0.86	4.66 ± 0.83	66.25 ± 0.61	68.10 ± 2.25	69.61 ± 0.72
47	5.45 ± 0.26	4.95 ± 0.38	4.60 ± 0.15	67.72 ± 0.92	67.85 ± 0.76	66.18 ± 0.67
49	4.65 ± 0.30	4.20 ± 0.32	4.42 ± 0.82	67.27 ± 0.70	68.51 ± 0.97	68.81 ± 1.01
51	5.87 ± 0.31	5.26 ± 0.26	5.13 ± 0.64	65.50 ± 0.27	67.41 ± 0.65	68.41 ± 2.65
55	5.22 ± 0.78	5.12 ± 0.28	5.42 ± 0.41	65.45 ± 0.84	66.45 ± 0.94	66.95 ± 1.17
60	5.08 ± 0.34	5.03 ± 0.14	5.13 ± 0.33	65.26 ± 0.68	66.39 ± 1.17	67.57 ± 0.54
62	5.27 ± 0.50	5.06 ± 0.30	4.92 ± 0.34	63.09 ± 0.76	67.28 ± 1.16	67.35 ± 0.64
CT	4.15 ± 0.58	4.34 ± 0.31	4.41 ± 0.60	65.60 ± 0.94	67.12 ± 0.54	67.32 ± 0.91

CT, Hi3027 segregation progenies. Data are mean ± SD.

2.7. Discussion

Protein and lysine content are the most important nutritional characters in maize. Overexpression of *SBgLR* can increase the lysine and protein content in maize seeds [12,13]. Here, we analyzed the protein and lysine content of different generations of *SBgLR* transgenic maize seeds and showed that the high-lysine and high-protein content characters could be stably inherited across generations (Table 1). Additionally, kernel quality and agronomic traits of transgenic lines were as normal as WT control (Figure 7, Table 2 and Tables S2–S4). This indicated that SBgLR transgenic maize had potential use in maize breeding for nutritional improvement.

Further, we analyzed the reasons why SBgLR increased protein and lysine content. The result showed that the expression of SBgLR increased both zein and non-zein proteins in transgenic seeds (Figure 2). Further observations through TEM and SEM indicated that the number of PBs and proteinous matrix was significantly increased in SBgLR transgenic endosperm (Figures 3 and 5), while, the morphologies of PBs and zein localizations in PBs were not affected (Figure 3I–N and Figure 4).

The cytoskeleton plays an important role in storage protein accumulation and PB formation. Zeins are synthesized on cytoskeleton-bound polysomes and their mRNA localize by binding to a specific site on the tubulin and actin [19,20]. PBs in developing maize endosperms are surrounded by an extensive cytoskeletal network through which zeins are deposited into them [35]. In the *o2* mutant, the morphology of PBs is abnormal, and the cytoskeleton matrix is loose and unstable simultaneously [33]. Therefore, storage protein synthesis and PB formation are highly correlated and dependent on cytoskeleton activity. Many studies have demonstrated that the activity and organization of the cytoskeleton is regulated by cytoskeleton-associated protein [36]. For example, elongation factor 1α (EF-1α), a lysine-rich protein, bundles microfilaments and co-localizes with actin surrounding PBs to facilitate zein peptide elongation [27,37]. Our previous work showed that SBgLR binds and bundles MTs *in vitro*, and has an effect on MT organization, indicating SBgLR is a microtubule-associated protein [15]. The expression of SBgLR in tobacco root epidermal cells leads to curving and twisting root growth, abnormal cell expansion and cell layer arrangement [15]. In this study, a preliminary immunogold labeling assay showed that SBgLR was mainly detected at the periphery of PBs (white arrows in Figure S3A) and cytoskeleton-like structures surrounding PBs in transgenic maize endosperm (white arrows in Figure S3B). This result suggested the possible role of SBgLR in MT organization in endosperm cells, and that increased protein and lysine content is correlated with its cytoskeleton regulation activity. When storage proteins begin to accumulate, MTs and microfilaments organize into a meshwork and serve as attachment sites for protein synthesis and PB formation [20–22]. We hypothesize, in transgenic endosperm, SBgLR interacts with MTs and regulates their organization, and that a more stable and more compact cytoskeleton system compared with the WT is established, which promotes mRNA stability; this may explain why *zein* and *non-zein* transcripts were increased in transgenic endosperm (Figure 6). Consequently, the translation efficiency is increased, leading to more zein and non-zein protein accumulation in transgenic endosperm (Figure 2). The increased zeins were further deposited into PBs, increasing the number of PBs in transgenic endosperm (Figure 3A–H). Though the efficiency of storage protein synthesis increases in transgenic endosperm, the accumulation of zein and non-zein proteins cannot increase to unlimited levels, possibly because the soil nutrients, especially nitrogen supplementation, are relatively limited. This is perhaps the most reasonable explanation why the number of PBs was increased, but the size of PBs was almost unchanged in transgenic endosperm (Figure 3I,L,J,M). Some evidence indicates that cytoskeleton-associated non-zein proteins, such as carbohydrate-metabolizing enzymes (CHMEs) and EF-1α, make a significant contribution to the lysine content of corn [29,38]. Therefore, non-zein proteins together with the accumulation of SBgLR should contribute to the increased lysine content in transgenic maize. The discussion above provides a possible explanation for the role of SBgLR in increasing protein and lysine content in an indirect manner. However, further studies are needed to confirm the function and clarify the mechanism of SBgLR in transgenic maize seeds.

3. Experimental Section

3.1. Plasmid Construction and Maize Transformation

To construct the binary vector pSB130-SBgLR for maize transformation, the *SBgLR* (GenBank: AY377987.1) expressing cassette was sub-cloned into *Hind* III and *Eco*RI sites of pSB130, which was kindly provided by Siwen Xin (Chinese University of Hong Kong, Hong Kong, China) (Figure 1A).

To obtain transgenic maize lines, two elite maize inbred lines 08 and 178 were pollinated as female and male for the production of hybrid Hi3027. Immature embryos of the maize hybrid Hi3027 (08 × 178) were used as explants, and *Agrobacterium*-mediated transformation was performed as in a previously described method [39].

3.2. PCR Amplification and Southern Blot Hybridization

The presence of transgene in regenerated plants was confirmed by PCR amplification. The PCR was carried out using 50 ng DNA and the primer sequences (S1/S2) used in the reactions are listed in Table S1. The amplification parameters were set as follows: 95 °C for 5 min, 94 °C for 1 min, 56 °C for 1 min, 72 °C for 1 min, repeat for 30 cycles and a final extension for 10 min at 72 °C. Reactions were carried out with Taq DNA polymerase (Tiangen, Beijing, China) in a Biometra Thermo cycler (TG-96; Biometra, Horsham, PA, USA), and the PCR products were analyzed on 0.8% agarose gels.

For Southern blot hybridization, genomic DNA was isolated from fresh leaves using a cetyltrimethyl ammonium bromide (CTAB) method [40]. A total of 30 µg genomic DNA was digested with *Hind* III or *Bam*H I restriction enzyme (TaKaRa, Tokyo, Japan) and electrophoresed on 0.8% agarose gel and blotted onto Hybond TM-XL membrane (Amersham, Piscataway, NJ, USA). Hybridization was carried out according to the manufacturer's protocol of Dig High Primer DNA Labeling and Detection Starter Kit I (Roche, Penzberg, Germany).

3.3. RT-PCR and Western Blot Detection

Immature kernels of T$_2$ plants of each line at 20 DAP were randomly selected for genotyping. Embryos were separated for DNA extraction, and kernel genotype was determined by PCR using primer pair S1/S2 (Table S1). Each transgenic endosperm was cut into two halves, one for RNA extraction and the other for protein extraction, which were used for RT-PCR and Western blot analysis respectively.

Total RNA of immature endosperm at 20 DAP was extracted using TRIzol reagent (CWBIO, Beijing, China). The cDNA first strand synthesis was carried out according to the manufacturer's protocol of Reverse Transcription System (Promega A3500, Fitchburg, WI, USA). RT-PCR amplification was performed using SBgLR (ORF) specific primers (SBRT) (Table S1) and carried out according to the procedure described above.

For Western bolt analysis, half of the 20 DAP endosperm was ground in 500 µL extraction buffer (20 mM Tris-HCl, pH 8.0, 5 mM EDTA, 0.05% SDS, 10 mM DTT, 1 mM PMSF) and vortexed at 4 °C for 1 h, then centrifuged at 4 °C for 20 min and the suspension was used for Western blot detection. In this study, 20 µL samples were separated on 12% SDS-PAGE and transferred to PVDF membranes (Millipore Corporation, Billerica, MA, USA). The membranes were blocked (3% BSA in TBST, 50 mM Tris-HCl, 150 mM NaCl, 0.05% Tween-20, pH 7.5) at 4 °C overnight and incubated with anti-SBgLR antibody (1:1000) for 1 h. Alkaline phosphatase-conjunct goat anti-rabbit IgG antibody was added (1:5000) after washing the membrane 15 min for three times. The hybridization signal was detected by using NBT/BCIP reaction kit (Promega, S380C, S381C, Fitchburg, WI, USA).

3.4. Agronomic Quality Measurement

To analyze the kernel density, 100-kernel-weight and volume was measured. Germination rate was calculated according to the statistical data obtained from our field test. Forty-five seeds were sowed in the field test and the germination rate was the ratio of germinated seeds to total seeds. The starch and oil content of mature seeds were analyzed by Fourier transform near-infrared (FT-NIR) spectrometer VECTOR22/N (BRUKER, Bremen, Germany).

3.5. Protein, Lysine, Zein and Non-Zein Contents Analysis

About 5 g of mature kernels were dehydrated in electric thermostatic drying oven at 80 °C for 48 h. The dehydrated kernels were ground into powder. Total nitrogen of kernels were determined by a nitrogen detection apparatus based on the principle of Kjeldahl determination using national standard GB2905-82 at Beijing Academy of Agriculture and Forestry Science (Beijing, China). The nitrogen content was converted to protein content by multiplying a conversion factor of 6.25.

Lysine content was analyzed based on the principle of ninhydrin reaction. Briefly, about 10 mg of defatted powder, 1 mL ddH$_2$O and 2 mL ninhydrin reaction reagent were added into a 30-mL tube, and then the mixture was thoroughly vortexed. Subsequently, the tube was incubated in boiling water for 20 min, and 3 mL 50% ethanol was added to the cooled sample, followed by a centrifugation at $13,400 \times g$ for 10 min. Supernatant was collected, and the absorbance at 570 nm was recorded using a spectrophotometer (TECHCOMP UV2300, Beijing, China). Standard curve was conducted using a series concentration of leucine solution (from 0 to 50 µg/mL). Lysine content of each sample was calculated using the equation as follows: Lys content (g/100 g dry weight) = (measured lysine content × hydrolysis volume)/sample weight.

Zein and non-zein proteins were extracted from 50 mg endosperm flour according to the previously described method [41]. A total of 200 µL ddH$_2$O was added to each fraction for SDS-PAGE and protein content analysis. Quantification of total extract, zein and non-zein fractions were performed using a BCA protein assay kit (Pierce, Appleton, WI, USA). For SDS-PAGE, 10 µL samples were loaded and electrophoresed in 12% polyacrylamide gel, and the gel was stained with Coomassie brilliant blue R250 (Amresco, Solon, OH, USA).

For statistical analyses, One-way ANOVA ($p < 0.05$) was performed to evaluate the difference between the measured traits. The significance of difference between WT and the transgenic lines was evaluated by *LSD* post-test (* $p < 0.05$; ** $p < 0.01$). All statistical analysis were made using the "Data Analysis" function in Microsoft Excel 2010 (Microsoft, Redmond, WA, USA).

3.6. Quantitative RT-PCR Analysis

For quantitative RT-PCR, embryos taken from single 20 DAP kernels were used for genotyping. Total RNA was extracted from the endosperm harboring the transgene using TRIzol reagent (CWBIO). The cDNA first strand synthesis was carried out according to the manufacturer's protocol of Reverse Transcription System (Promega A3500, Fitchburg, WI, USA). Quantitative RT-PCR was performed in a

20-µL reaction system containing SYBGREEN reaction mixture (Takara, Japan), 10 µM of each primer and 50 ng cDNA. The sequences of the primers, δ-zein for *δ storage protein zein gene* (AF371265.1), 19-zein for 19 kDa *α zein* (AF371268), 22-zein for 22 kDa *α zein* (AF371277.1), histone H2B for *histone H2B.3* (BM080549), cylicin for *cylicin-1* (CF627896), ER protein for *ER lumen protein retaining receptor C28H8.4* (CF602623), FBA for *fructose-bisphosphatealdolase* (CF624216) and Ribosomal protein for *60S ribosomal protein L22-2* (BI543125), are listed in Table S1. Amplification was carried out on a Bio-Rad Real-Time System CFX96TM C1000 thermal cycler (Bio-Rad, Waltham, MA, USA) using the following conditions: 95 °C for 30 s, 35 cycles of 95 °C for 10 s, 60 °C for 10 s and 72 °C for 10 s. Maize *actin* (Table S1) was used as reference gene. Expression levels of these genes were determined as the Ct values [42]. The Ct values were then converted into relative quantities.

3.7. Transmission Electron Microscopy (TEM)

The presence of a transgene was detected by PCR using the primer pair S1/S2 as described previously. The transgenic endosperms with the thickness of 2-mm from T_2 plants were perpendicularly sectioned to the pericarp, including the aleurone and 10 to 20 cell layers of endosperm. The slices were fixed in freshly prepared potassium phosphate buffer (50 mM, pH 6.8) containing 1% glutaraldehyde, 4% paraformaldehyde and 5 mM EGTA at 4 °C overnight. The fixed slices were rinsed at room temperature for 2 h, post-fixed with fixation buffer containing 1% osmium tetroxide at 4 °C overnight and dehydrated in gradient ethanol. The tissues were embedded in Spurr and LR White resin for transmission electron microscopy (TEM) or immunogold labeling assay, respectively. Plastic blocks were sectioned into 90 nm including six to eight cell layers, and collected on formvar-coated nickel grids. Non-transgenic endosperm segregated from transgenic line was used as negative control.

For TEM, grids were stained with 2% uranyl acetate and 2.66% lead citrate, followed by three rinses in ddH$_2$O. The stained grids were air-dried and observed under a Hitachi 7500 electron microscope (Hitachi, Japan) operated at 80 kV. Pictures were taken using ITEM (Olympus, Tokyo, Japan). For PBs numbers quantification, data were collected from the measurement of at least 30 cells from two different kernels from two individual ears. Stutent's *t*-test was performed between WT and two transgenic lines respectively to evaluate the significance of differences (* $p < 0.05$; ** $p < 0.01$).

For immunogold labeling assay, grids were blocked in blocking buffer (3% BSA, 0.05% Tween-20 in 50 mM potassium phosphate buffer, pH 6.9) and incubated with primary antibody (1:200) (the anti-19 kDa α-zein antibody was kindly provided by Brian A. Larkins) at 4 °C overnight. Subsequently, the grids were incubated with gold-conjugated secondary antibody at room temperature for 1 h and rinsed in ddH$_2$O for three times. The grids were then stained as described above. The stained grids were examined with a Hitachi 7500 transmission electron microscope (Hitachi, Tokyo, Japan). Pictures were taken using ITEM (OSIS, Germany).

3.8. Scanning Electron Microscopy (SEM)

Mature seeds of T_2 transgenic and Hi3027 F_2 plants were dissected from the crown to the base and mounted on the surface of a brass disk using double-sided adhesive silver-tape. They were then coated with gold/palladium using an ion coater (EIKO IB.3, Tokyo, Japan), and the central region of starchy endosperm was examined using SEM (JEOL, Tokyo, Japan).

4. Conclusions

We analyzed SBgLR transgenic maize with high-lysine and high-protein content in seeds. Both PB formation and proteinaceous matrix deposition were promoted. The increased accumulation of zeins and lysine-rich non-zein proteins were the reason for protein content increment. Further, the increased deposition of non-zein protein, as well as the accumulation of SBgLR, contributes to the lysine increment in transgenic endosperm. The high-lysine and high-protein traits are stably inherited across generations. SBgLR can be considered for use in improving nutrient quality of maize seeds.

This work indicates that a cytoskeleton-associated protein has potential applicable value in improving crop seed nutrient quality and provides a feasible strategy for maize grain quality improvement.

Supplementary Materials: Supplementary materials can be found at http://www.mdpi.com/1422-0067/16/12/26199/s1.

Acknowledgments: We are grateful to Junzhen Jia and Haihong Liu at the "985" Technology Platform (China Agricultural University, Beijing, China) for their assistance on electron microscopy, Brian A. Larkins (Arizona University, Tucson, AZ, USA) for his advice on this manuscript and providing the anti-19 kDa α-zein antibody, and Siwen Xin (Chinese University of Hong Kong, Hong Kong, China) for providing twin T-DNA vector pSB130. This work was supported by the National Transgenic Major Program of China (Grant no. 2014ZX003-002, 2013 ZX 003-002, and 2011ZX003-002).

Author Contributions: Jingjuan Yu and Chen Liu conceived and designed the research; Chen Liu prepared the samples, designed and performed the experiments, and wrote the original manuscript; Shixue Li, Jing Yue, Wenhan Xiao and Qian Zhao managed the experimental condition; Dengyun Zhu managed the plant materials; Jingjuan Yu revised thoroughly the manuscript and finalized the manuscript. All authors read and approved the final manuscript.

Conflicts of Interest: The authors declare that they have no competing interests.

References

1. Wu, Y.; Messing, J. Proteome balancing of the maize seed for higher nutritional value. *Front. Plant Sci.* **2014**, *5*, 240. [CrossRef] [PubMed]

2. Azevedo, R.A.; Arruda, P. High-lysine maize: The key discoveries that have made it possible. *Amino Acids* **2010**, *39*, 979–989. [CrossRef] [PubMed]

3. Mertz, E.T.; Bates, L.S.; Nelson, O.E. Mutant gene that changes protein composition and increases lysine content of maize endosperm. *Science* **1964**, *145*, 279–280. [CrossRef] [PubMed]

4. Gibbon, B.C.; Larkins, B.A. Molecular genetic approaches to developing quality protein maize. *Trends Genet.* **2005**, *21*, 227–233. [CrossRef] [PubMed]

5. Mazur, B.; Krebbers, E.; Tingey, S. Gene discovery and product development for grain quality traits. *Science* **1999**, *285*, 372–375. [CrossRef] [PubMed]

6. Houmard, N.M.; Mainville, J.L.; Bonin, C.P.; Huang, S.; Luethy, M.H.; Malvar, T.M. High-lysine corn generated by endosperm-specific suppression of lysine catabolism using RNAi. *Plant Biotechnol. J.* **2007**, *5*, 605–614. [CrossRef] [PubMed]

7. Wu, Y.; Messing, J. RNA interference can rebalance the nitrogen sink of maize seeds without losing hard endosperm. *PLoS ONE* **2012**, *7*, e32850. [CrossRef] [PubMed]

8. Wu, Y.; Messing, J. Novel genetic selection system for quantitative trait loci of quality protein maize. *Genetics* **2011**, *188*, 1019–1022. [CrossRef] [PubMed]

9. Bicar, E.H.; Woodman-Clikeman, W.; Sangtong, V.; Peterson, J.M.; Yang, S.S.; Lee, M.; Scott, M.P. Transgenic maize endosperm containing a milk protein has improved amino acid balance. *Transgenic Res.* **2008**, *17*, 59–71. [CrossRef] [PubMed]

10. Yue, J.; Li, C.; Zhao, Q.; Zhu, D.; Yu, J. Seed-specific expression of a lysine-rich protein gene, *GhLRP*, from cotton significantly increases the lysine content in maize seeds. *Int. J. Mol. Sci.* **2014**, *15*, 5350–5365. [CrossRef] [PubMed]

11. Yu, J.J.; Peng, P.; Zhang, X.J.; Zhao, Q.; Zhu, D.Y.; Sun, X.H.; Liu, J.Q.; Ao, G.M. Seed-specific expression of a lysine rich protein *sb401* gene significantly increases both lysine and total protein content in maize seeds. *Mol. Breed.* **2004**, *14*, 1–7. [CrossRef]

12. Lang, Z.H.; Yu, J.J.; Zhu, D.Y.; Zhao, Q.; Ao, G.M. Cloning of a lysine-rich gene SBgLR and the effect of improving the protein and lysine content in transgenic maize seed. *Chin. J. Agric. Biotechnol.* **2004**, *12*, 487–492.

13. Wang, M.; Liu, C.; Li, S.; Zhu, D.; Zhao, Q.; Yu, J. Improved nutritive quality and salt resistance in transgenic maize by simultaneously overexpression of a natural lysine-rich protein gene, *SBgLR*, and an ERF transcription factor gene, *TSRF1*. *Int. J. Mol. Sci.* **2013**, *14*, 9459–9474. [CrossRef] [PubMed]

14. Huang, S.L.; Jin, L.F.; Du, J.Z.; Li, H.; Zhao, Q.; Ou, G.S.; Ao, G.M.; Yuan, M. SB401, a pollen-specific protein from *Solanum berthaultii*, binds to and bundles microtubules and F-actin. *Plant J.* **2007**, *51*, 406–418. [CrossRef] [PubMed]

15. Liu, C.; Qi, X.; Zhao, Q.; Yu, J. Characterization and functional analysis of the potato pollen-specific microtubule associated protein SBgLR in tobacco. *PLoS ONE* **2013**, *8*, e60543. [CrossRef] [PubMed]

16. Muench, D.G.; Park, N.-I. Messages on the move: The role of the cytoskeleton in mRNA localization and translation in plant cells. *Can. J. Bot.* **2006**, *84*, 572–580. [CrossRef]

17. Hesketh, J.E.; Pryme, I.F. Association of mRNAs and polyribosomes with the cytoskeleton: Potential roles in transport and compartmentalization of mRNAs. In *The Cytoskeleton*; JAI Press: Stamford, CT, USA, 1996; pp. 31–58.

18. Johnston, D.S. The intracellular localization of messenger RNAs. *Cell* **1995**, *81*, 161–170. [CrossRef]

19. Stankoviċ, B.; Abe, S.; Davies, E. Co-localization of polysomes, cytoskeleton, and membranes with protein bodies from corn endosperm. *Protoplasma* **1993**, *177*, 66–72. [CrossRef]

20. Stankovic, B.; Clore, A.; Larkins, B.A.; Abe, S.; Davies, E. Actin in protein body synthesis and protein body formation. In *Actin: A Dynamic Framework for Multiple Plant Cell Functions*; Baluska, F., Barlow, P., Staiger, C.J., Vokmann, D., Eds.; Kluwer: Alphen aan den Rijn, The Netherlands, 2000; pp. 129–143.

21. Muench, D.G.; Wu, Y.J.; Coughlan, S.J.; Okita, T.W. Evidence for a cytoskeleton-associated binding site involvd in prolamine mRNA localization to the protein bodies in rice endosperm tissue. *J. Plant Physiol.* **1998**, *116*, 559–569. [CrossRef]

22. Davies, E.; Comer, E.C.; Lionberger, J.M.; Stankoviċ, B.; Abe, S. Cytoskeleton-bound polysomes in plants. III. Polysomes-cytoskeleton-membrane interactions in maize endosperm. *Cell Biol. Int.* **1993**, *17*, 331–340. [CrossRef]

23. Zak, E.A.; Bocharova, M.A.; Sokolov, I.A.; Klyachio, N.L. Plant polysomes bound to the cytoskeleton. *Dokl. Biol. Sci.* **1995**, *344*, 549–551.

24. Lopez-Valenzuela, J.A. Cytoskeleton proteins are coordinately increased in maize genotypes with high levels of eEF1A. *Plant Physiolol.* **2004**, *135*, 1784–1797. [CrossRef] [PubMed]

25. Mateyak, M.K.; Kinzy, T.G. eEF1A: Thinking outside the ribosome. *J. Biol. Chem.* **2010**, *285*, 21209–21213. [CrossRef] [PubMed]

26. Gross, S.R.; Kinzy, T.G. Translation elongation factor 1A is essential for regulation of the actin cytoskeleton and cell morphology. *Nat. Struct. Mol. Biol.* **2005**, *12*, 772–778. [CrossRef] [PubMed]

27. Clore, A.M.; Dannenhoffer, J.M.; Larkins, B.A. EF-1α is associated with a cytoskeletal network surrounding protein bodies in maize endosperm cells. *Plant Cell* **1996**, *8*, 2003–2014. [PubMed]

28. Habben, J.E.; Kirleis, A.W.; Larkins, B.A. The origin of lysine-containing proteins in *opaque-2* maize endosperm. *Plant Mol. Biol.* **1993**, *23*, 825–838. [CrossRef] [PubMed]

29. Habben, J.E.; Moro, G.L.; Hunter, B.G.; Hamaker, B.R.; Larkins, B.A. Elongation factor 1α concentration is highly correlated with the lysine content of maize endosperm. *Proc. Natl. Acad. Sci. USA* **1995**, *92*, 8640–8644. [CrossRef] [PubMed]

30. Lang, Z.H.; Zhao, Q.; Yu, J.J.; Zhu, D.Y.; Ao, G.M. Cloning of potato *SBgLR* gene and its intron splicing in transgenic maize. *Plant Sci.* **2004**, *166*, 1227–1233. [CrossRef]

31. Wang, X.; Zhu, L.; Liu, B.Q.; Wang, C.; Jin, L.F.; Zhao, Q.; Yuan, M. *Arabidopsis* microtubule-associated protein18 functions in directional cell growth by destabilizing cortical microtubules. *Plant Cell* **2007**, *19*, 877–889. [CrossRef] [PubMed]

32. Pan, Y.; Ma, X.; Liang, H.; Zhao, Q.; Zhu, D.; Yu, J. Spatial and temporal activity of the foxtail millet (*Setaria italica*) seed-specific promoter *pF128*. *Planta* **2015**. [CrossRef] [PubMed]

33. Wu, Y.; Holding, D.R.; Messing, J. γ-Zeins are essential for endosperm modification in quality protein maize. *Proc. Natl. Acad. Sci. USA* **2010**, *107*, 12810–12815. [CrossRef] [PubMed]

34. Holding, D.R.; Otegui, M.S.; Li, B.; Meeley, R.B.; Dam, T.; Hunter, B.G.; Jung, R.; Larkins, B.A. The maize *floury1* gene encodes a novel endoplasmic reticulum protein involved in zein protein body formation. *Plant Cell* **2007**, *19*, 2569–2582. [CrossRef] [PubMed]

35. Larkins, B.A.; Hurkman, W.J. Synthesis and deposition of zein in protein bodies of maize endosperm. *J. Plant Physiol.* **1978**, *62*, 256–263. [CrossRef]

36. Wasteneys, G.O.; Ambrose, J.C. Spatial organization of plant cortical microtubules: Close encounters of the 2D kind. *Trends Cell Biol.* **2009**, *19*, 62–71. [CrossRef] [PubMed]

37. Durso, N.A.; Cyr, R.J. Beyond translation: Elongation factor-1α and the cytoskeleton. *Protoplasma* **1994**, *180*, 99–105. [CrossRef]

38. Kishu, A.; Shunnosuke, A.; Hideki, S. Lysine-containing proteins in maize endosperm: A major contribution from cytoskeleton-associated carbohydrate-metabolizing enzymes. *Planta* **2003**, *217*, 628–638.
39. Frame, B.R.; Shou, H.; Chikwamba, R.K.; Zhang, Z.; Xiang, C.; Fonger, T.M.; Pegg, S.E.K.; Li, B.; Nettleton, D.S.; Pei, D.; *et al. Agrobacterium*-mediated transformation of maize embryos using a standard binary vector system. *J. Plant Physiol.* **2002**, *129*, 13–22. [CrossRef] [PubMed]
40. Doyle, J.J.; Doyle, J.L. A rapid total DNA preparation procedure for fresh plant tissue. *Focus* **1990**, *12*, 13–15.
41. Wallace, J.C.; Lopes, M.A.; Paiva, E.; Larkins, B.A. New methods for extraction and quantitation of zeins reveal a high content of γ-zein in modified *opaque-2* maize. *J. Plant Physiol.* **1990**, *92*, 191–196. [CrossRef]
42. Walker, N.J. A technique whose time has come. *Science* **2002**, *296*, 557–559. [CrossRef] [PubMed]

International Journal of
Molecular Sciences

Article

Stable Expression of Basic Fibroblast Growth Factor in Chloroplasts of Tobacco

Yun-Peng Wang [1], Zheng-Yi Wei [1], Xiao-Fang Zhong [1], Chun-Jing Lin [1], Yu-Hong Cai [2], Jian Ma [3], Yu-Ying Zhang [1,4], Yan-Zhi Liu [1] and Shao-Chen Xing [1,*]

[1] Jilin Provincial Key Laboratory of Agricultural Biotechnology, Agro-Biotechnology Research Institute, Jilin Academy of Agricultural Sciences, No. 1363, Shengtai st., Changchun 130033, China; wangypbio@163.com (Y.-P.W.); weizy@cjaas.com (Z.-Y.W.); zhongxf@cjaas.com (X.-F.Z.); lincj@cjaas.com (C.-J.L.); yuying0609@126.com (Y.-Y.Z.); liuyz_g@126.com (Y.-Z.L.)

[2] Institute of Agricultural Quality Standard and Testing Technology, Jilin Academy of Agricultural Sciences, No. 1363, Shengtai st., Changchun 130033, China; yhcai64@163.com

[3] Faculty of Agronomy, Jilin Agricultural University, No. 2888, Xincheng st., Changchun 130118, China; majian19790106@163.com

[4] College of Biological Sciences, China Agricultural University, No. 2 West Yuanmingyuan Road, Beijing 100094, China

* Correspondence: xingsc@cjaas.com; Tel.: +86-431-8706-3121; Fax: +86-431-8706-3080

Academic Editors: Setsuko Komatsu and Marcello Iriti
Received: 4 November 2015; Accepted: 15 December 2015; Published: 23 December 2015

Abstract: Basic fibroblast growth factor (bFGF) is a multifunctional factor in acceleration of cell proliferation, differentiation and transference, and therefore widely used in clinical applications. In this study, expression vector pWX-Nt03 harboring a codon-optimized *bFGF* gene was constructed and introduced into the tobacco chloroplasts by particle bombardment. After four rounds of selection, *bFGF* was proved to integrate into the chloroplast genome of regenerated plants and two of four transgenic plants were confirmed to be homoplastomic by PCR and Southern hybridization. ELISA assay indicated that *bFGF* represented approximately 0.1% of total soluble protein in the leaves of transplastomic tobacco plants. This is the first report of *bFGF* expression via chloroplast transformation in model plant, providing an additional option for the production of chloroplast-produced therapeutic proteins.

Keywords: tobacco; chloroplast; genetic transformation; basic fibroblast growth factor; green fluorescent protein

1. Introduction

Fibroblast growth factors (FGF) are a large family of heparin-binding protein mitogens, with potential therapeutic utilities due to their broad target cells specificity. Twenty three different FGFs have been described thus far [1]. One FGF, basic fibroblast growth factor (bFGF or FGF2), a single-subunit protein with a molecular weight of ~17 kDa and a pI of 9.6, was first cloned from the bovine brain and pituitary in 1978 [2].

bFGF acts to accelerate cellular proliferation, differentiation and transference [3–5]. Recombinant bFGF is highly desirable in the marketing due to its critical role in the growth of blood vessels, wound healing, tissue regeneration and nerve system [6–8] and many attempts have been made to meet the great demand over the past few decades. Currently bFGF is produced via genetic engineering technology in several systems, from conventional microbial expression systems like *Escherichia coli* (*E. coli*) [9,10], *Bacillus subtilis* [11] and yeast [12,13], to insect cells [14]. More importantly, two main crops, soybean and rice, were reported to express functional recombinant bFGF specifically in seeds by *Agrobacterium*-mediated transformation [15,16]. However, the insolubility and biological activity of target proteins would likely be an intractable obstacle for microbial [17] and plant-produced [16] platform to lag the further application when high level of protein expression was achieved. Plant-based production of therapeutic proteins or antigens can be cost-effective, as, unlike microbial systems, it does not require sterile laboratories or production equipment [18,19]. But, thus far the efficiency of plant nuclear transformation with many genes remains low, producing low protein yields complicated by gene silencing and unstable inheritance [20]. Moreover, pollen-mediated transgene flow has drawn significant attention due to its potential impact on biodiversity [21]. Chloroplast transformation will be likely to overcome those disadvantages due to the features of higher expression, generally soluble proteins and mostly maternal inheritance, although some challenges such as protein degradation, pleiotropic effects, time-consuming homoplasmic selection are still the main hurdles for most of plants and need to be addressed [22,23]. Since tobacco was firstly engineered via chloroplast transformation, plastid transformation has drawn more attention, achieving high-level expression, gene-stacking, no gene silencing, and even facilitating successful transfer of whole metabolic pathways to plastids [24].

In this study, we engineered tobacco plants to express bFGF via plastid transformation. To our knowledge, this is the first report of bFGF expressed via plastid transformation and therefore provides a novel alternative of bFGF expression.

2. Results

2.1. Construction of Expression Vector and Functional Verification

An expression vector named pWXL-Nt03 for tobacco chloroplast transformation was constructed based on previously published protocol used in our laboratory [25] with two modifications (Figure 1). The homologous fragment from alfalfa was replaced with one from the tobacco plastid genome, and the codon-optimized *bFGF* gene was inserted into the vector multiple cloning site of the vector. The vector was verified by sequencing.

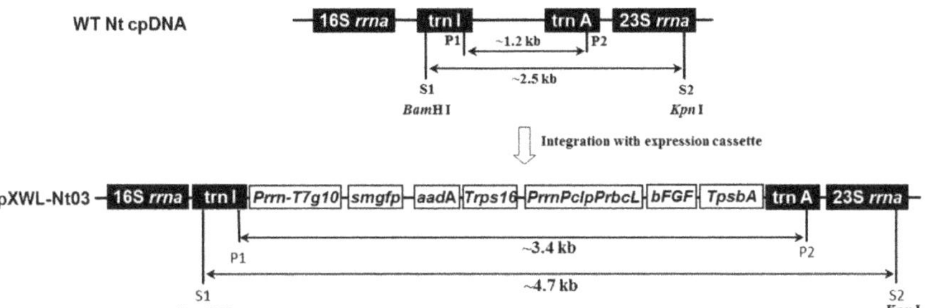

Figure 1. Construction of chloroplast transformation vector pXWL-Nt03. pXWL-Nt03 vector flanked with chloroplast regions of *16S-trnI* and *trnA-23S* comprised a synthetic expression cassette containing the *smgfp* reporter gene and *aadA* selectable marker gene. These two genes were jointly driven by the Prrn-T7g10 promoter and the terminator was 3′ untranslated regions (UTR) rps16 (*Trps16*). The target basic fibroblast growth factor (*bFGF*) gene was driven by the fusion promoter *PrrnPclpPrbcL* previously described [26] and *psbA* 3′ UTR as terminator (*TpsbA*). P1/P2 and S1/S2 indicated the primers used for PCR and digestion sites for Southern blot analysis, respectively. A ~3.4 kb fragment was amplified for transplastomic plants, but a ~1.2 kb fragment in wild-type tobacco chloroplast DNA (WT Nt cpDNA) without expression cassette. Similarly, total DNA was digested by *BamH* I and *Kpn* I for Southern blot analysis. A ~4.7 kb signal was hybridized for transplastomic plants, but only a ~2.5 kb for untransformed plants.

The chloroplast originated from a prokaryote and thus contains prokaryotic protein synthesis machinery. The specific expression vector for chloroplast transformation was therefore generally transferred into prokaryotic system like *E. coli* to test its genetic construction and function [27]. After the MACH 1, an *E. coli* strain, was transformed with this vector following the heat shock protocol, the overnight-cultured cells were collected by centrifugation. The pellet was observed to appear green in color, indicating the successful expression of green fluorescent protein (GFP) in transformed bacterial cells. This observation suggested that the other genes within the same expression cassette will play their roles in prokaryotic-like chloroplast expression system.

2.2. Tobacco Chloroplast Transformation and Homoplastomic Selection

Young leaves were placed on the Murashige and Skoog (MS) medium with adaxial side down, cultured for 4 h in the dark and bombarded. The bombarded leaves were cultured in the dark at 26 °C for 3 days before being cut into 5 mm × 5 mm pieces and transferred into the selection media containing 500 mg/L of spectinomycin. Selection media was replaced every 3 weeks. Forty-three resistant transformants were regenerated in selection media, and integration of the expression cassette was verified by PCR. Young leaves from PCR-positive shoots were cut into pieces and placed abaxial side down in selective media and cultured for two to three additional regeneration cycles to obtain homoplastomic plants (Figure 2). Seeds were harvested from homoplastomic tobacco plants and sowed in soil until to T_3 generation.

Figure 2. Procedure for homoplastomic selection after chloroplast transformation of tobacco. (**A**) Bombarded leaves were cut into pieces for selection; (**B**) Resistant shoots appear while most leave pieces bleached on antibiotic medium; (**C**) Surviving antibiotic resistant shoots; (**D**) Regenerated plantlet after first round of selection; (**E**) GFP expression in T_0 transplastomic tobacco with green color and chlorophyll autofluorescence with red color under ultraviolet (UV) lamp; (**F**) GFP expression in T_3 homoplastomic tobacco seedlings, line 1 and line 4, and wild-type control under UV lamp. Scale bars = 1 cm.

2.3. Molecular Characterization of T_0 Transplastomic Plants

After four rounds of selection, antibiotic resistant shoots were obtained and transferred to new media without phytohormone for rooting. Integration of the foreign expression cassette into the plastid genome of transformants was verified by PCR. As shown in Figure 3A, two putative transgenic plants were homoplastomic with the single band of ~3.4 kb, but an additional ~1.2 kb band in lanes 2 and 3 indicated the heteroplastomic status of other two plants. Both hetero-/homoplastomic plants displayed normal phenotypes compared with wild-type plants under greenhouse condition.

Total DNA was isolated from the leaves of T_0 transgenic and wild-type plants and integration of the *bFGF* gene into the tobacco plastid genome was confirmed by Southern blot. As indicated in Figure 3B, the single ~4.7 kb band in lane 1 and 4 implied homoplastomic status, whereas lanes 2 and 3 displayed an additional ~2.5 kb band, identically amplified from wild-type plants, indicating heteroplastomic status.

Northern blot was carried out to investigate the transcript level of *gfp* and *bFGF*. As shown in Figure 3C, no strong differences of hybridization signal intensity was observed between *gfp* and *bFGF* genes, indicating the similar accumulation of mRNAs before translation into proteins unless the hybridizations are obviously probe sequence dependent.

Western blot revealed proteins of approximately 17 kDa in lanes 1 to 4, representing bFGF expression in the four T_0 transgenic lines (Figure 3D) of identical size to commercially produced bFGF. As expected, no signal was observed for wild-type plant.

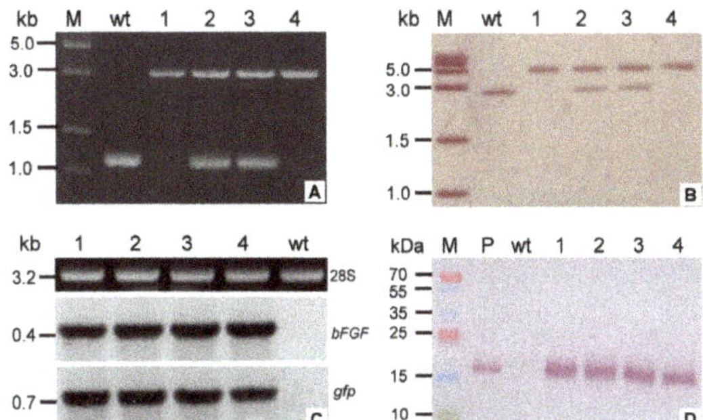

Figure 3. Molecular testing of T_0 transplastomic plants. (**A**) PCR analysis of integration of the expression cassette into transgenic plants. M, DNA molecular marker; Lanes 1–4: T_0 putative transgenic plants; (**B**) Southern blot analysis. M, DNA molecular marker; Lanes 1–4: the leaves of T_0 putative transgenic plants. The single 4.7 kb band in lanes 1 to 4 indicated homoplastomic status due to the insertion of foreign expression cassette, whereas lanes 2 and 3 with an additional ~2.5 kb band, amplified from wild-type plants, indicating heteroplastomic status; (**C**) Northern blot analysis for *bFGF* (middle) and *gfp* (bottom) mRNAs. The ethidium bromide-stained gel of 28S band of total RNA is shown in the top as a loading quantitative control. wt is untransformed tobacco as negative control; (**D**) Western blot for determining the expression of bFGF in T_0 transplastomic plants. Expression of bFGF (17 kDa) is observed in the leaves of four transgenic plants (lanes 1–4), but absent in the leaves of a wild-type plant (wt). Commercial bFGF protein (P) served as positive control. M is protein marker.

2.4. Measurement of bFGF and GFP Expression in T_0 Transformed Plants by ELISA

Total soluble protein (TSP) was extracted from the leaves of T_0 transformed plants to quantify the expression of bFGF and GFP by ELISA. In the four transgenic plants bFGF represented between 0.095% and 0.106% of TSP from mature leaves (Figure 4A) and much lower expression in young and old leaves (data not shown), indicating a far lower level of bFGF expression than GFP, which the maximum average expression level can reach up to 22.343% of TSP in mature leaves. The average expression in young and old leaves is 13.441% and 1.947% of TSP, relatively lower than mature leaves (Figure 4B).

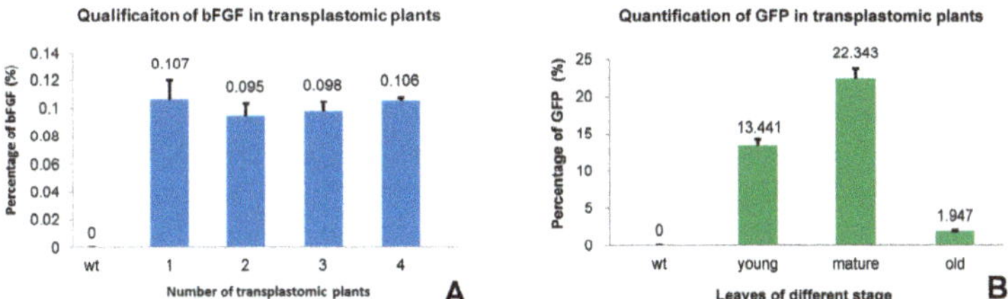

Figure 4. Quantification of bFGF and GFP in transgenic tobacco plants. (**A**) Level of bFGF expression in the leaves of four independent transgenic plants was shown in relation to total soluble protein (mean ± SD). 1–4, independent transplastomic plants; wt, Wild-type plant; (**B**) Level of GFP expression in young, mature and old leaves of transplastomic plants; wt, Wild-type plant. All measurements were made in triplicate.

3. Discussion

Plant-made pharmaceuticals (PMPs), both nuclear and plastid stable transformation, offer several unique advantages over microorganism and mammalian systems, allowing production in large scales but low cost and lack of endotoxins and pyrogens. This had been recognized as a promising potential method for the production of therapeutic proteins [28,29]. Some plants like peas, which have high protein content and excellent storage capacity, are an ideal platform for production of pharmaceuticals [30]. Although no plastid-produced products have been used in clinical trials yet, vaccines produced by nuclear transformation of plants or TMV-based transient expression in plants have been tested in the clinic [23,29].

bFGF has previously been expressed by nuclear transformation system and the yield, 2.3% TSP in soybean seeds and 9.55% TSP in rice seeds [15,16], is rather higher than the production in the present experiment, but considering that: (1) the biomass of tobacco mature leaves with average yield of 2.7 tons per hectare [31]; (2) the tobacco needs approximately 40–50 days from seed to fully-expanded leaves; and (3) it is practicable to harvest five to seven times annually under field condition depending on the growing season, the bFGF productivity of transgenic tobacco might be more competitive compared with soybean and rice. The homoplasmic expression of bFGF was achieved in this study, but accumulation needs to be enhanced greatly (Figure 4A). To do this, some possible measures could be taken including N-terminus and tag-added modification [32]. In addition, chloroplast-expressed proteins are generally soluble [22], making this system more attractive for solving the insolubility of protein expressed by conventional nuclear transformation [16]. This study can also pave the way for bFGF expression in other plants, especially vegetables which is feasible for oral delivery without fermentation, purification, cold storage and transportation [22].

Another puzzling question is the expression of GFP which located in the same cassette with bFGF. GFP expression demonstrated far higher level (Figure 4B) than bFGF. Since all of the possible parameters were considered to maintain the uniformity in Northern hybridization operation except the length of two probe sequences, the similar signal intensity should reflect the amount of mRNA (Figure 3C). The strong disparity in accumulation of two proteins therefore could not be explained by their respective mRNA stability alone. It is reported that protein accumulation in transplastomic plants was very likely controlled by post-translational regulation at the level of protein stability [33–35] and three factors were experimentally recognized to affect their stability, including the penultimate N-terminal amino acid residue, an N-end rule-like protein degradation pathway and additional sequence determinants in the N-terminal region as well [32]. In our study, the different production of bFGF and GFP is somehow caused by protein degradation as there is no detectable difference was visible at least in transcriptional level. We also cannot exclude the possibility of other factors. For example, the initiation of translation might be affected strongly by 3′ UTRs [36] and therefore the 3′UTR in this study, the leader sequences of *g10* and *clpP* genes, might contribute the gap of translation initiation efficiency to a certain extent [37]. In addition, the small-molecule bFGF, unlike well-known stable GFP, is reported to be unstable in the presence of many proteases [38]. To improve the stability of recombinant protein, the small ubiquitin-related modifier fusion could be superior to glutathione *S*-transferase in enhancement of the expression and solubility of difficult-to-express proteins [39]. According to the proposed hypothesis [32], future attempts to express bFGF in tobacco plastids may employ artificially optimized tag(s) to promote the solubility and stability of the products with less energy consumption of the host cells.

4. Materials and Methods

4.1. Plant Materials

Tobacco (*Nicotiana tabacum* cv. Petit Havana) seeds were surface-sterilized and grown on half-strength MS medium supplemented with 8 g/L agar and 30 g/L sucrose with a 16-h light/8-h dark photoperiod. Young leaves were used for plastid DNA extraction and transformation.

4.2. Vector Construction for Chloroplast Transformation

According to the complete tobacco chloroplast genome sequences (accession number: NC-001879), two primers (P1: ACAGAGGATGCAAGCGTTAT and P2: CACTGAGCGATCATTTAGGG) were designed to amplify flanking sequences, representing the 16S-trnI-trnA-23S region of the tobacco plastome, for vector construction. PCR was carried out with 30 cycles of 45 s at 95 °C, 40 s at 50 °C, and 2 min at 72 °C, with a 5 min pre-denaturation before the first cycle at 95 °C and a 10 min final extension at 72 °C. The PCR product was inserted into vector pEASY Blunt Simple (Transgen Biotech Co., Ltd., Beijing, China) and verified by sequencing to generate pEASY-Nt. An expression cassette was obtained, as previously described [25]. The whole cassette was amplified by PCR and the product was inserted into the pEASY-Nt vector to form a new vector named pEASY-synth.

Codon optimization of the *bFGF* encoding target gene was carried out using the software Synthetic Gene Designer (http://www.evolvingcode.net/codon/sgd/index.php). The 474bp optimized *bFGF* gene was inserted into a multiple cloning site (MCS) to yield the final expression vector pXWL-Nt03. (Figure 1). Sequencing revealed that no errors had been introduced.

4.3. Chloroplast Transformation and Homoplastomic Selection

Tobacco chloroplast transformation was carried out according to the protocol previously described [40]. Briefly, young leaves of *in vitro* cultured tobacco plants were bombarded using PDS 1000/He biolistic particle delivery system (Bio-Rad, Hercules, CA, USA) with 50 mg of 0.6 µm gold particles coated with 10 µg of plasmid DNA at 1100 psi. Putative transformants were regenerated in MS medium containing 500 mg/L spectinomycin at the first round of selection. The PCR-confirmed leaves of resistant shoots were cut into pieces and subjected to three additional regeneration cycles in selective medium. T_1 seeds were harvested from T_0 homoplastomic plants proved by molecular identification and sowed the seeds in pots until to T_3 generation.

4.4. Molecular Testing

Total genomic DNA was extracted from fully expanded young leaves of both transplastomic plants and untransformed wild-type plants using the cetyltrimethylammonium bromide (CTAB) protocol [41]. PCR was carried out under the following conditions: 95 °C for 45 s, 50 °C for 40 s, 72 °C for 2 min; 35 cycles with a 5 min pre-denaturation before the first cycle at 95 °C and a 10 min final extension at 72 °C. The sequence of primer pair designed to check expression cassette was Ec1 (GGTCGGAACAAGTTGATAG) and Ec2 (CAGTAGAGTCTTTCAGTGGC), respectively. The expected PCR band will be ~3.4 kb due to insertion of the foreign expression cassette.

Integration of the foreign expression cassette into the tobacco plastid genome was assessed by Southern blot. The total DNA extracted from young leaves was digested with *Bam*H I and *Kpn* I (New England Biolabs, Hitchin, UK) overnight at 37 °C and subjected to 0.8% agarose gel electrophoresis, then transferred onto positive-charged Hybond-N$^+$ nylon membrane (Roche, Basel, Switzerland). Cross-link twice for 30 s each time with 3 min interval. Pre-hybridization and hybridization were carried with the DIG High Prime DNA Labeling and Detection Starter Kit II (Roche) according to the manufacturer's instructions. The DNA-fixed membrane was hybridized at 42 °C for 16 h with the probe in a hybridization oven before washing and visualization.

Total RNA was prepared using RNAiso Reagent (TaKaRa Co., Ltd., Dalian, China) and the manufacturer's instruction from four transgenic lines to test the mRNA transcriptional level by Northern blotting analysis. The *gfp* and *bFGF* genes were amplified as probes. An amount of 5 µg of RNA was loaded in each well and fractionated on a 2.0% denaturing agarose gel, and then transferred onto a nylon membrane for cross-linking. Labeled two probes were hybridized at 65 °C for 16 h in a hybridization oven. Membrane was washed and visualized accordingly.

4.5. Western Blot Assay

Expression of bFGF was measured by Western blot. Briefly, crude protein extracts were extracted from PCR-positive plants, separated by a 15% SDS-PAGE gel, then transferred to a nitrocellulose membrane (0.22 μm, Millipore, Darmstadt, Germany) using a trans-blot semi-dry electrophoretic transfer cell (200 mA, 30 min, Bio-Rad). The nitrocellulose membrane was then saturated with 5% bovine serum albumin (BSA) (*w/v*) in TBS-Tween 20 [25 mM Tris-HCl pH 7.6, 0.15 M NaCl, 0.05% (*w/v*) Tween-20] for 1 h. The membrane was incubated with bFGF-directed antibody (1:100, Santa Cruz Biotechnology Inc., Dallas, TX, USA), prior to incubation with the secondary antibody (1:3000, goat anti-mouse IgG alkaline phosphatase conjugate, Novagen, Madison, WI, USA). Labeled proteins were visualized by adding a substrate BCIP/NBT (Roche).

4.6. ELISA

To measure the expression level of recombinant protein in transplastomic plants, total soluble protein was extracted from fully expanded leaves of both transformed and wild-type plants. The leaves were cut into pieces and grounded by liquid nitrogen, homogenized in PBS buffer (137 mM NaCl, 2.7 mM KCl, 10 mM Na_2HPO_4, 2 mM KH_2PO4). ELISA was performed using commercial anti-bFGF antibodies, and recombinant bFGF as standard (Sigma, St. Louis, MO, USA), as previously described [42]. Similarly, GFP expression was measured in young, mature and old leaves individually, and analyzed the data statistically. Wild-type plant was used as control.

Acknowledgments: We thank Nuo Xu for his kind help in Southern and Northern blot analysis. This research work was founded by Jilin provincial project (C42070501).

Author Contributions: Yun-Peng Wang, Zheng-Yi Wei, Jian Ma and Shao-Chen Xing designed the experiment. Yun-Peng Wang, Xiao-Fang Zhong, Chun-Jing Lin, Yan-Zhi Liu and Yu-Ying Zhang conducted the experiment. Yu-Ying Zhang and Yu-Hong Cai conducted statistical analysis. Yun-Peng Wang, Zheng-Yi Wei and Shao-Chen Xing wrote the paper.

Conflicts of Interest: The authors declare no conflict of interest.

References

1. Yamashita, T.; Yoshioka, M.; Itoh, N. Identification of a novel fibroblast growth factor, FGF-23, preferentially expressed in the ventrolateral thalamic nucleus of the brain. *Biochem. Biophys. Res. Commun.* **2000**, *277*, 494–498. [CrossRef] [PubMed]
2. Gospodarowicz, D.; Bialecki, H.; Greenburg, G. Purification of the fibroblast growth factor activity from bovine brain. *J. Biol. Chem.* **1978**, *253*, 3736–3743. [PubMed]
3. Lavandero, S.; Chappuzeau, A.; Sapag-Hagar, M.; Oka, T. *In vivo* and *in vitro* evidence of basic fibroblast growth factor action in mouse mammary gland development. *FEBS Lett.* **1998**, *439*, 351–356. [CrossRef]
4. Akita, S.; Akino, K.; Imaizumi, T.; Hirano, A. A basic fibroblast growth factor improved the quality of skin grafting in burn patients. *Burns* **2005**, *31*, 855–858. [CrossRef] [PubMed]
5. Zou, H.; Nie, X.H.; Zhang, Y.; Hu, M.; Zhang, Y.A. Effect of basic fibroblast growth factor on the proliferation, migration and phenotypic modulation of airway smooth muscle cells. *Chin. Med. J.* **2008**, *121*, 424–429. [PubMed]
6. Xiong, N.; Yang, H.; Liu, L.; Xiong, J.; Zhang, Z.; Zhang, X.; Jia, M.; Huang, J.; Zhang, Z.; Mohamed, A.A.; Lin, Z.; Wang, T. bFGF promotes the differentiation and effectiveness of human bone marrow mesenchymal stem cells in a rotenone model for parkinson's disease. *Environ. Toxicol. Pharmacol.* **2013**, *36*, 411–422. [CrossRef] [PubMed]
7. Jia, X.; Tian, H.; Tang, L.; Zheng, L.; Zheng, L.; Yang, T.; Yu, B.; Wang, Z.; Lin, P.; Li, X.; *et al.* High-efficiency expression of TAT-bFGF fusion protein in *Escherichia coli* and the effect on hypertrophic scar tissue. *PLoS ONE* **2015**, *10*, e0117448. [CrossRef] [PubMed]
8. Xie, J.X.; Feng, Y.; Yuan, J.M.; You, Z.D.; Lin, H.Y.; Lu, C.L.; Xu, J.J. Positive effects of bFGF modified rat amniotic epithelial cells transplantation on transected rat optic nerve. *PLoS ONE* **2015**, *10*, e0124713. [CrossRef] [PubMed]

9. Gasparian, M.E.; Elistratov, P.A.; Drize, N.I.; Nifontova, I.N.; Dolgikh, D.A.; Kirpichnikov, M.P. Overexpression in Escherichia coli and purification of human fibroblast growth factor (FGF-2). Biochemistry. *Biochemistry (Moscow)* **2009**, *74*, 221–225. [CrossRef] [PubMed]

10. Song, J.A.; Koo, B.K.; Chong, S.H.; Kwak, J.; Ryu, H.B.; Nguyen, M.T.; Vu, T.T.; Jeong, B.; Kim, S.W.; Choe, H. Expression and purification of biologically active human FGF2 containing the b'a' domains of human PDI in *Escherichia coli. Appl. Biochem. Biotechnol.* **2013**, *170*, 67–80. [CrossRef] [PubMed]

11. Kwong, K.W.; Ng, K.L.; Lam, C.C.; Wang, Y.Y.; Wong, W.K. Authentic human basic fibroblast growth factor produced by secretion in *Bacillus subtilis. Appl. Microbiol. Biotechnol.* **2013**, *97*, 6803–6811. [CrossRef] [PubMed]

12. Barr, P.J.; Cousens, L.S.; Lee-Ng, C.T.; Medina-Selby, A.; Masiarz, F.R.; Hallewell, R.A.; Chamberlain, S.H.; Bradley, J.D.; Lee, D.; Steimer, K.S. Expression and processing of biologically active fibroblast growth factors in the yeast *Saccharomyces cerevisiae. J. Biol. Chem.* **1988**, *263*, 16471–16478. [PubMed]

13. Mu, X.; Kong, N.; Chen, W.; Zhang, T.; Shen, M.; Yan, W. High-level expression, purification, and characterization of recombinant human basic fibroblast growth factor in *Pichia pastoris. Protein Express. Purif.* **2008**, *59*, 282–288. [CrossRef] [PubMed]

14. Wu, X.; Kamei, K.; Sato, H.; Sato, S.I.; Takano, R.; Ichida, M.; Mori, H.; Hara, S. High-level expression of human acidic fibroblast growth factor and basic fibroblast growth factor in silkworm (*Bombyx mori* L.) using recombinant baculovirus. *Protein Expr. Purif.* **2001**, *21*, 192–200. [CrossRef] [PubMed]

15. Ding, S.H.; Huang, L.Y.; Wang, Y.D.; Sun, H.C.; Xiang, Z.H. High-level expression of basic fibroblast growth factor in transgenic soybean seeds and characterization of its biological activity. *Biotechnol. Lett.* **2006**, *28*, 869–875. [CrossRef] [PubMed]

16. An, N.; Ou, J.; Jiang, D.; Zhang, L.; Liu, J.; Fu, K.; Dai, Y.; Yang, D. Expression of a functional recombinant human basic fibroblast growth factor from transgenic rice seeds. *Int. J. Mol. Sci.* **2013**, *14*, 3556–3567. [CrossRef] [PubMed]

17. Nausch, H.; Huckauf, J.; Koslowski, R.; Meyer, U.; Broer, I.; Mikschofsky, H. Recombinant production of human interleukin 6 in Escherichia coli. *PLoS ONE* **2013**, *8*, e54933. [CrossRef] [PubMed]

18. Ahmad, N.; Mukhtar, Z. Green factories: Plastids for the production of foreign proteins at high levels. *Gene Ther. Mol. Biol.* **2013**, *15*, 14–29.

19. Bock, R. Genetic engineering of the chloroplast: Novel tools and new applications. *Curr. Opin. Biotechnol.* **2014**, *26*, 7–13. [CrossRef] [PubMed]

20. Sood, P.; Bhattacharya, A.; Sood, A. Problems and possibilities of monocot transformation. *Biol. Plant.* **2011**, *55*, 1–15. [CrossRef]

21. Goldstein, D.A. Tempest in a tea pot: How did the public conversation on genetically modified crops drift so far from the facts? *J. Med. Toxicol.* **2014**, *10*, 194–201. [CrossRef] [PubMed]

22. Daniell, H.; Singh, N.D.; Mason, H.; Streatfield, S.J. Plant-made vaccine antigens and biopharmaceuticals. *Trends Plant Sci.* **2009**, *14*, 669–679. [CrossRef] [PubMed]

23. Lössl, A.G.; Waheed, M.T. Chloroplast-derived vaccines against human diseases: Achievements, challenges and scopes. *Plant Biotechnol. J.* **2011**, *9*, 527–539. [CrossRef] [PubMed]

24. Lu, Y.; Rijzaani, H.; Karcher, D.; Ruf, S.; Bock, R. Efficient metabolic pathway engineering in transgenic tobacco and tomato plastids with synthetic multigene operons. *Proc. Natl. Acad. Sci. USA* **2013**, *110*, 623–632. [CrossRef] [PubMed]

25. Wei, Z.; Liu, Y.; Lin, C.; Wang, Y.; Cai, Q.; Dong, Y.; Xing, S. Transformation of alfalfa chloroplasts and expression of green fluorescent protein in a forage crop. *Biotechnol. Lett.* **2011**, *33*, 2487–2494. [CrossRef] [PubMed]

26. Kittiwongwattana, C.; Lutz, K.; Clark, M.; Maliga, P. Plastid marker gene excision by the phiC31 phage site-specific recombinase. *Plant Mol. Biol.* **2007**, *64*, 137–143. [CrossRef] [PubMed]

27. Magee, A.M.; Horvath, E.M.; Kavanagh, T.A. Pre-screening plastid transgene expression in *Escherichia coli* may be unreliable as a predictor of expression levels in chloroplast-transformed plants. *Plant Sci.* **2004**, *166*, 1605–1611. [CrossRef]

28. He, Y.; Ning, T.; Xie, T.; Qiu, Q.; Zhang, L.; Sun, Y.; Jiang, D.; Fu, K.; Yin, F.; Zhang, W.; *et al.* Large-scale production of functional human serum albumin from transgenic rice seeds. *Proc. Natl. Acad. Sci. USA* **2011**, *108*, 19078–19083. [CrossRef] [PubMed]

29. Sil, B.; Jha, S. Plants: The future pharmaceutical factory. *Am. J. Plant Sci.* **2014**, *5*, 319–327. [CrossRef]

30. Mikschofsky, H.; Broer, I. Feasibility of *Pisum sativum* as an expression system for pharmaceuticals. *Transgenic Res.* **2012**, *21*, 715–24. [CrossRef] [PubMed]

31. Drake, M.P.; Vann, M.C.; Fisher, L.R. Influence of nitrogen application rate on the yield, quality, and chemical components of flue-cured tobacco, part II: Application method. *Tob. Sci.* **2015**, *52*, 26–34. [CrossRef]

32. Apel, W.; Schulze, W.X.; Bock, R. Identification of protein stability determinants in chloroplasts. *Plant J.* **2010**, *63*, 636–650. [CrossRef] [PubMed]

33. Zhou, F.; Badillo-Corona, J.A.; Karcher, D.; Gonzalez-Rabade, N.; Piepenburg, K.; Borchers, A.-M.I.; Maloney, A.P.; Kavanagh, T.A.; Gray, J.C.; Bock, R. High-level expression of HIV antigens from the tobacco and tomato plastid genomes. *Plant Biotechnol. J.* **2008**, *6*, 897–913. [CrossRef] [PubMed]

34. Oey, M.; Lohse, M.; Kreikemeyer, B.; Bock, R. Exhaustion of the chloroplast protein synthesis capacity by massive expression of a highly stable protein antibiotic. *Plant J.* **2009**, *57*, 436–445. [CrossRef] [PubMed]

35. Oey, M.; Lohse, M.; Scharff, L.B.; Kreikemeyer, B.; Bock, R. Plastid production of protein antibiotics against pneumonia via a new strategy for high-level expression of antimicrobial proteins. *Proc. Natl. Acad. Sci. USA* **2009**, *106*, 6579–6584. [CrossRef] [PubMed]

36. Kuersten, S.; Goodwin, E.B. The power of the 3′ UTR: Translational control and development. *Nat. Rev. Genet.* **2003**, *4*, 626–637. [CrossRef] [PubMed]

37. Mertens, N.; Remaut, E.; Fiers, W. Increased stability of phage T7g10 mRNA is mediated by either a 5′- or a 3′-terminal stem-loop structure. *Biol. Chem.* **1996**, *377*, 811–817. [PubMed]

38. Tengood, J.E.; Ridenour, R.; Brodsky, R.; Russell, A.J.; Little, S.R. Sequential delivery of basic fibroblast growth factor and platelet-derived growth factor for angiogenesis. *Tissue Eng. Part A* **2011**, *17*, 1181–1189. [CrossRef] [PubMed]

39. Marblestone, J.G.; Edavettal, S.C.; Lim, T.; Lim, P.; Zuo, X.; Butt, T.R. Comparison of SUMO fusion technology with traditional gene fusion systems: Enhanced expression and solubility with SUMO. *Protein Sci.* **2006**, *15*, 182–189. [CrossRef] [PubMed]

40. Wang, Y.P.; Wei, Z.Y.; Zhang, Y.Y.; Lin, C.J.; Zhong, X.F.; Wang, Y.L.; Ma, J.Y.; Ma, J.; Xing, S.C. Chloroplast-expressed MSI-99 in tobacco improves disease resistance and displays inhibitory effect against rice blast fungus. *Int. J. Mol. Sci.* **2015**, *16*, 4628–4641. [CrossRef] [PubMed]

41. Murray, M.G.; Thompson, W.F. Rapid isolation of high molecular weight plant DNA. *Nucleic Acids Res.* **1980**, *8*, 4321–4325. [CrossRef] [PubMed]

42. Wirth, S.; Calamante, G.; Mentaberry, A.; Bussmann, L.; Lattanzi, M.; Barañao, L.; Bravo-Almonacid, F. Expression of active human epidermal growth factor (hEGF) in tobacco plants by integrative and non-integrative systems. *Mol. Breed.* **2004**, *13*, 23–35. [CrossRef]

 International Journal of
Molecular Sciences

Article

Proteomic Analysis Reveals the Leaf Color Regulation Mechanism in Chimera *Hosta* "Gold Standard" Leaves

Juanjuan Yu [1,2,†], Jinzheng Zhang [3], Qi Zhao [1], Yuelu Liu [3], Sixue Chen [4], Hongliang Guo [5,†], Lei Shi [3] and Shaojun Dai [1,*]

[1]   Development Center of Plant Germplasm Resources, College of Life and Environmental Sciences, Shanghai Normal University, Shanghai 200234, China; yujuan8186@163.com (J.Y.); zhaoqizq@yeah.net (Q.Z.)
[2]   Alkali Soil Natural Environmental Science Center, Northeast Forestry University, Key Laboratory of Saline-alkali Vegetation Ecology Restoration in Oil Field, Ministry of Education, Harbin 150040, China
[3]   Institute of Botany, Chinese Academy of Sciences, Beijing 100093, China; caohua@ibcas.ac.cn (J.Z.); liuyuelu09@sina.com (Y.L.); shilei@ibcas.ac.cn (L.S.)
[4]   Department of Biology, Genetics Institute, Plant Molecular and Cellular Biology Program, Interdisciplinary Center for Biotechnology Research, University of Florida, Gainesville, FL 32610, USA; schen@ufl.edu
[5]   Food Engineering College, Harbin University of Commerce, Harbin 150028, China; guohongliang97@126.com
*   Correspondence: daishaojun@hotmail.com; Tel./Fax: +86-21-6432-4576
†   The authors contributed equally to this work.

Academic Editor: Setsuko Komatsu
Received: 26 December 2015; Accepted: 1 March 2016; Published: 8 March 2016

Abstract: Leaf color change of variegated leaves from chimera species is regulated by fine-tuned molecular mechanisms. *Hosta* "Gold Standard" is a typical chimera *Hosta* species with golden-green variegated leaves, which is an ideal material to investigate the molecular mechanisms of leaf variegation. In this study, the margin and center regions of young and mature leaves from *Hosta* "Gold Standard", as well as the leaves from plants after excess nitrogen fertilization were studied using physiological and comparative proteomic approaches. We identified 31 differentially expressed proteins in various regions and development stages of variegated leaves. Some of them may be related to the leaf color regulation in *Hosta* "Gold Standard". For example, cytosolic glutamine synthetase (GS1), heat shock protein 70 (Hsp70), and chloroplastic elongation factor G (cpEF-G) were involved in pigment-related nitrogen synthesis as well as protein synthesis and processing. By integrating the proteomics data with physiological results, we revealed the metabolic patterns of nitrogen metabolism, photosynthesis, energy supply, as well as chloroplast protein synthesis, import and processing in various leaf regions at different development stages. Additionally, chloroplast-localized proteoforms involved in nitrogen metabolism, photosynthesis and protein processing implied that post-translational modifications were crucial for leaf color regulation. These results provide new clues toward understanding the mechanisms of leaf color regulation in variegated leaves.

Keywords: *Hosta* "Gold Standard"; proteomics; variegated leaves; leaf color; excess nitrogen fertilization

1. Introduction

Leaf color change during plant development and envrionment response is regulated by complicated and fine-tuned molecular mechanisms. Current understandings of leaf color regulatory mechanisms are mainly based on the investigation of leaf color mutants of model plants and crops, such as Arabidopsis [1–3], rice [4], and tomato [5], rather than natural variegated plants. Variegated plants are widely distributed, some of which are cultivated as ornamental plants. Variegated leaves usually consist of green and white/yellow sectors, which are useful resources for investigating chloroplast development and color changes [6]. The chloroplastic or genetic origin of leaf variegation [6], as well as albinism mechanism has been reported [7,8]. Most investigations were focused on the physiological analysis of the white region and green regions [3,9–11], development of *aurea* "golden" leaves, and characteristics of yellow-green variegated leaves [12–18].

Various colors of leaf variegation suggest that different regulatory pathways are responsible in the respective cases [10]. Nitrogen nutrients and light intensities are considered as the main factors for yellow-green variegated leaves and leaves of *aurea* "golden" varieties from some vascular plants. Excess nitrogen fertilization [19], deep shaded [20], and long-term low light illumination [21] lead to green coloration of yellow leaves/regions. The conditional yellow-green variegated leafed plants or *aurea* "golden" varieties have been investigated for chloroplast ultrastructure [19,21–23], construction and function of photosystem I (PSI) and photosystem II (PSII) [21,24,25], chlorophyll synthesis and degradation [21], and plastid regulated expression of nuclear *cab* genes [26–28].

The plant species in genus *Hosta*, belonging to the family Liliaceae, are exceedingly popular perennials in today's gardens. In these species, chimera *Hosta* with variegated leaves has been widely cultivated [29]. *Hosta* "Gold Standard" is a typical chimera *Hosta* species. The margin of its ovate leaves is dark green, and the center of leaves appears light green in the spring and become progressively more golden towards early summer [30]. Excess nitrogen fertilization will turn the golden-green variegated leaves to whole green leaves of *Hosta* "Gold Standard" [19,30]. Therefore, *Hosta* "Gold Standard" is considered as an ideal material to investigate the molecular mechanisms of leaf variegation.

Proteomics approaches provide a high-throughput and systematic analysis of protein networks and molecular regulatory mechanisms of plant development and environmental response [31–35]. Proteomics has been applied to the investigation of leaf coloring mechanisms (albinism) [7,8]. Comparative proteomic studies based on two-dimensional gel electrophoresis (2-DE) have revealed 14 chloroplast proteins changed in the stage albinism line of winter wheat *FA85* as compared with the parent wheat *Aibian 1* [7], and 26 proteins in periodic albinism of White leaf No. 1 (a typical albino tea cultivar grown in China) [8]. These identified proteins are involved in metabolism of nitrogen metabolism, photosynthesis, carbon and energy metabolism, RNA processing, protein processing, signal transduction, and stress and defense response, indicating these physiological processes might play important roles in the albinism [7,8]. The proteomic profiles from various leaf color tissues provide new evidence for leaf color regulation in albinistic leaves, but the molecular mechanisms in golden-green variegated leaves are still to be investigated.

In this study, the margin and center regions of young and mature leaves from *Hosta* "Gold Standard", as well as the leaves after excess nitrogen fertilization were analyzed using physiological and comparative proteomic approaches. Our results indicate that nitrogen metabolism regulation, photosynthesis and energy supply, chloroplast development, and chloroplast protein import/processing play crucial roles in leaf color changes in variegated leaves. The results provide novel insights into understanding the mechanisms of leaf color regulation in variegated leaves.

2. Results

2.1. Variegated Leaf Color Changes in Different Development Stages

The variegated young leaves of *Hosta* "Gold Standard" seedlings were ovate with dark green margins and light green center (Figure 1A). During seedling development, the variegated mature leaves become progressively golden in leaf center, with the leaf margin kept dark green (Figure 1B). After 30 days of excess nitrogen fertilization, the whole leaves turned to green and the phenotype of the variegated leaf disappeared (Figure 1C).

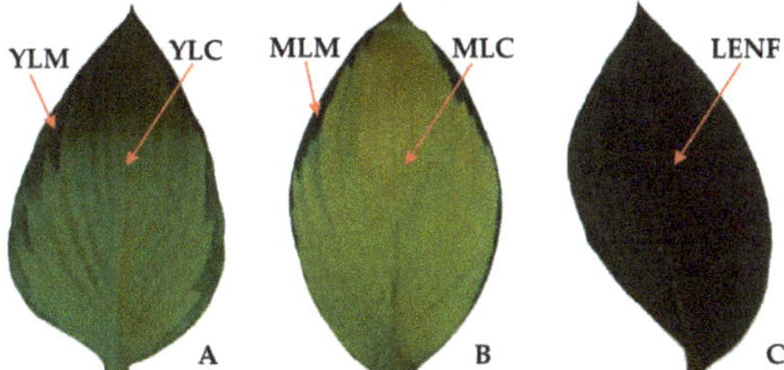

Figure 1. Representative leaves at different development stages of *Hosta* "Gold Standard". (**A**) Young leaf; (**B**) mature leaf; (**C**) leaf under excess nitrogen fertilization. YLM: Young leaf margin; YLC: Young leaf center; MLM: Mature leaf margin; MLC: Mature leaf center; LENF: Leaf under excess nitrogen fertilization.

The color properties of various leaf regions at different development stages of *Hosta* "Gold Standard" were determined. The value of lightness L* reflects the perceived lightness, and values of the color coordinate a* and b* represent the hue from red to green and from yellow to blue, respectively. There were significant differences in lightness L*, color coordinate a*, and color coordinate b* in the marginal regions and central regions either of both young leaves and mature leaves, indicating the variegated leaf phenotype (Figure 2). The lightness L*, color coordinate a*, and color coordinate b* did not have significant changes in mature leaf margin (MLM), when compared with the young leaf margin (YLM) (Figure 2). However, the color properties of the leaf center were changed during leaf development. Lightness L* increased obviously, although color coordinate a* and b* showed no obvious changes in mature leaf center (MLC) compared with young leaf center (YLC). After excess nitrogen fertilization for 30 days, lightness L*, color coordinate a*, and color coordinate b* were decreased significantly in leaves, when compared with MLC (Figure 2).

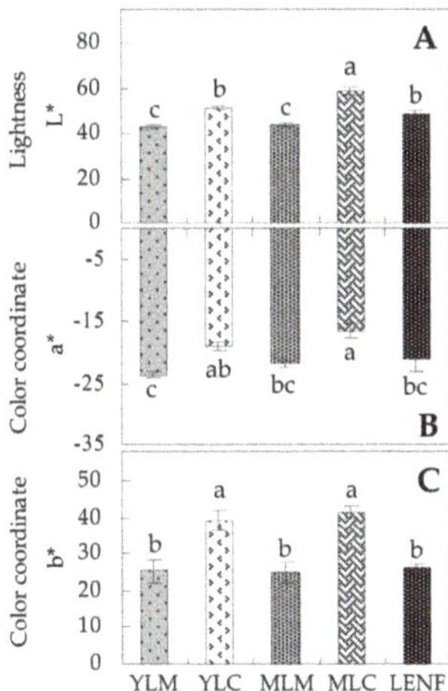

Figure 2. Color properties of various leaf regions at different development stages of *Hosta* "Gold Standard". (**A**) Lightness L*, representing the perceived lightness; (**B**) color coordinate a*, indicating the change in hue from red to green; (**C**) color coordinate b*, indicating the change in hue from yellow to blue. The values were determined in young leaf margin (YLM), young leaf center (YLC), mature leaf margin (MLM), mature leaf center (MLC), and leaf under excess nitrogen fertilization (LENF), which were presented as means $\pm$ SE (n = 3). The different small letters on the columns indicate significant differences among various leaf regions at different development stages ($p < 0.05$).

2.2. Chlorophyll Contents in Differrent Regions of Leaves

To evaluate the leaf color changes, the contents of chlorophylls in various regions of leaves were determined. The contents of total chlorophylls (Chls), chlorophyll *a* (Chl *a*), and chlorophyll *b* (Chl *b*) showed no significant changes in MLM compared with YLM (Figure 3A–C). But the contents of total Chls, Chl *a*, and Chl *b* were lower in the central regions than those in the marginal regions of young leaves and mature leaves, and the decrease of them was significant in mature leaves (Figure 3A–C). Besides, the ratio of Chl *a* to Chl *b* in MLC was the highest among various regions/stages (Figure 3D). After excess nitrogen fertilization for 30 days, the contents of total Chls, Chl *a* and Chl *b* in leaves under excess nitrogen fertilization (LENF) were all increased when compared with MLC, but they still were lower than MLM (Figure 3A–C).

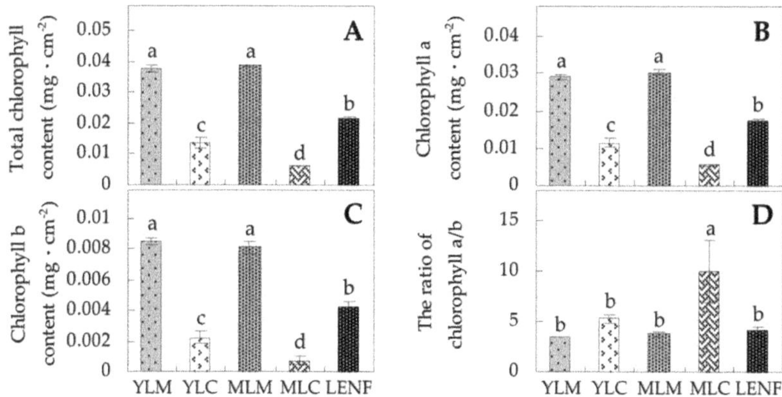

Figure 3. Chlorophyll contents of various leaf regions at different development stages of *Hosta* "Gold Standard". (**A**) Total chlorophyll content; (**B**) chlorophyll *a* content; (**C**) chlorophyll *b* content; (**D**) the ratio of chlorophyll *a*/*b*. The values were determined in young leaf margin (YLM), young leaf center (YLC), mature leaf margin (MLM), mature leaf center (MLC), and leaf under excess nitrogen fertilization (LENF), which were presented as means $\pm$ SE (n = 3). The different small letters on the columns indicate significant differences among various leaf regions at different development stages ($p < 0.05$).

2.3. Nitrogen Contents and Activities of Nitrogen Metabolism-Related Enzymes

To monitor the nitrogen assimilation status in various leaf regions at different development stages from *Hosta* "Gold Standard" or under excess nitrogen fertilization, the nitrogen contents and the activities of nitrogen metabolism related enzymes were determined. During leaf development, the nitrogen contents in the centers of young leaves and mature leaves were all lower than in the margins of these leaves (Figure 4A). Accordingly, the activities of nitrate reductase (NR) and glutamine α-oxoglutarate aminotransferase (GOGAT) in YLC and MLC were reduced when compared with YLM and MLM (Figure 4B,C). The activities of glutamine synthetase (GS) in YLC were lower than YLM, but their activities in MLM and MLC were similar (Figure 4D). Interestingly, the nitrogen levels in LENF were similar with those in leaf centers and obviously lower than in the margins (Figure 4A). The activities of NR, GOGAT and GS in LENF were induced when compared with those in leaf centers, except the GOGAT activity was still lower than in leaf margins (Figure 4B–D).

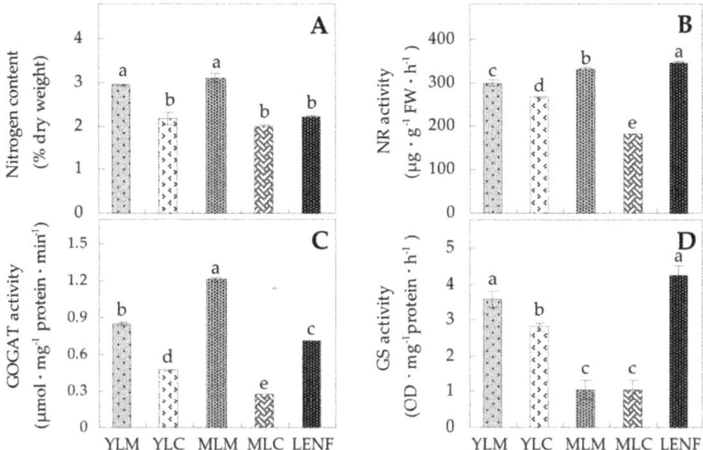

Figure 4. Nitrogen content and activities of nitrogen metabolism-related enzymes in various leaf regions at different development stages of *Hosta* "Gold Standard". (**A**) Nitrogen content; (**B**) nitrate reductase (NR) activity; (**C**) glutamine α-oxoglutarate aminotransferase (GOGAT) activity; (**D**) glutamine synthetase (GS) activity. The values were determined in young leaf margin (YLM), young leaf center (YLC), mature leaf margin (MLM), mature leaf center (MLC), and leaf under excess nitrogen fertilization (LENF), which were presented as means $\pm$ SE (n = 3). The different small letters on the columns indicate significant differences among various leaf regions at different development stages ($p < 0.05$).

2.4. Identification of Differentially Expressed Proteins (DEPs) in Various Leaf Regions

To investigate the DEPs in different leaf regions and nitrogen conditions, the protein profiles of five regions/stages of leaves (YLM, YLC, MLM, MLC, and LENF) were obtained using 2-DE analysis. On Coomassie Brilliant Blue-stained gels (24 cm immobilized pH gradient (IPG) strip, pH 4–7), 908 ± 15, 917 ± 39, 873 ± 30, 856 ± 53, and 839 ± 19 protein spots from YLM, YLC, MLM, MLC, and LENF were detected, respectively (Figure 5, Supplemental Figure S1). After gel image analysis on the basis of the calculated average vol % values of each matched protein spot, 42 reproducibly matched spots showed more than a 1.5-fold change in abundance ($p < 0.05$). The proteins from these spots were digested in gel and identified using mass spectrometry (MALDI TOF-TOF MS). In total, 31 DEPs were identified using Mascot database searching (Figure 5, Table 1, and Supplemental Table S1).

Table 1. Identified differentially expressed proteins of various leaf regions at different development stages of *Hosta* "Gold Standard".

Spot No. [a]	Protein Name [b]	SCL [c]	Plant Species [d]	gi Number [e]	Exp. Mw (Da)/pI [f]	Thr. Mw (Da)/pI [g]	Sco [h]	Cov (%) [i]	QM [j]	V% ± SE [k]
				Nitrogen metabolism (2)						
795	glutamine synthetase isoform GS1c (GS1)	Cyt	*Triticum aestivum*	71361904	39,676/5.58	39,445/5.41	89	15	2	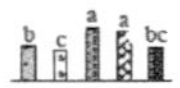
806	glutamine synthetase isoform GS1c (GS1)	Cyt	*T. aestivum*	71361904	39,505/5.75	39,445/5.41	168	15	3	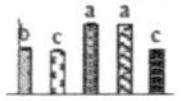
				Photosynthesis (11)						
1120	chlorophyll a/b-binding protein type I, chlorophyll a-b binding protein 6 (CAB6) *	Chl	*Malus x domestica*	33772151	23,327/5.86	15,693/5.05	101	19	3	
881	photosystem II stability/assembly factor HCF136, chloroplastic	Chl	*Vitis vinifera*	225423755	35,844/5.32	44,473/6.92	575	37	12	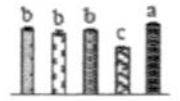
982	O_2 evolving complex 33kD family protein (OEC33)	Chl	*Populus trichocarpa*	224084209	28,767/5.00	35,348/5.89	490	56	12	
740	phosphoglycerate kinase (PGK)	Chl	*Ricinus communis*	255544584	41,980/5.63	50,114/8.74	597	35	11	
338	plastid transketolase (TK)	Chl	*Nicotiana tabacum*	194396261	78,879/5.95	80,451/6.16	141	17	10	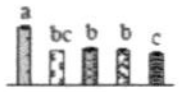
346	transketolase (TK)	Chl	*P. trichocarpa*	224063766	79,837/5.87	81,122/5.97	337	13	9	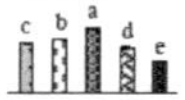
459	hypothetical protein SORBIDRAFT_09g014430, RuBisCO large subunit-binding protein subunit alpha (RBP-α) *	Chl	*Sorghum bicolor*	242090109	66,151/4.84	60,914/5.07	233	31	13	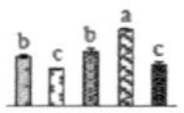
476	RuBisCO large subunit-binding protein subunit beta (RBP-β)	Chl	*V. vinifera*	225442531	63,140/5.18	65,255/5.62	474	17	8	
472	function unknown, RuBisCO large subunit-binding protein subunit beta (RBP-β) *	Chl	*Zea mays*	223948025	64,795/5.24	61,969/5.42	524	30	11	
840	phosphoribulokinase (PRK)	Chl	*Pisum sativum*	1885326	38,189/4.99	39,230/5.41	113	13	3	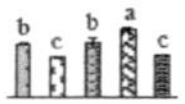

Table 1. *Cont.*

Spot No. [a]	Protein Name [b]	SCL [c]	Plant Species [d]	gi Number [e]	Exp. Mw (Da)/pI [f]	Thr. Mw (Da)/pI [g]	Sco [h]	Cov (%) [i]	QM [j]	V% ± SE [k]
832	function unknown, chloroplastic phosphoribulokinase (PRK) *	Chl	*Glycine max*	255646270	38,858/4.85	45,757/5.89	277	38	11	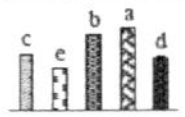
Carbohydrate and energy metabolism (5)										
847	fructose-bisphosphate aldolase (FBA)	Cyt	*Pandanus amaryllifolius*	40716077	37,892/6.55	38,642/6.48	236	26	6	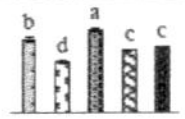
497	glycosyl hydrolase family 1 family protein (GH1)	Chl	*P. trichocarpa*	224135485	61,209/5.61	59,060/9.16	79	24	9	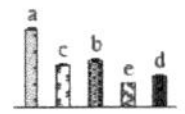
286	NADH-ubiquinone oxidoreductase (complex I)	Mit	*R. communis*	255582280	86,431/5.81	81,687/6.56	630	21	13	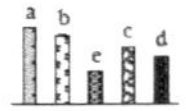
550	chloroplast ATP synthase alpha chain	Chl	*Yucca schidigera*	69214356	55,476/5.31	55,406/5.35	276	44	18	
887	haloalkane dehalogenase (HLD)	Chl	*V. vinifera*	225442892	35,349/5.18	42,368/6.17	99	9	2	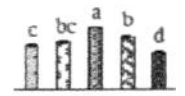
Protein synthesis (2)										
268	function unknown, chloroplastic elongation factor G (cpEF-G) *	Chl	*Arabidopsis thaliana*	14532624	90,396/5.21	86,287/5.48	585	25	15	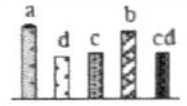
292	mitochondrial glycyl-tRNA synthetase 1 (GRS)	Mit	*V. vinifera*	225456501	86,133/6.16	79,370/5.98	108	5	3	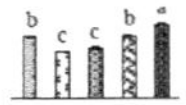
Protein processing and degradation (10)										
355	heat shock cognate 70 kDa protein 1 (Hsc70-1)	Cyt	*V. vinifera*	225449497	77,129/5.15	71,525/5.17	1050	61	26	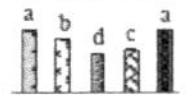
361	heat shock cognate 70 kDa protein 1 (Hsc70-1)	Cyt	*V. vinifera*	225449497	77,663/5.10	71,525/5.17	959	58	25	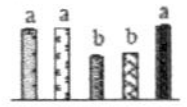
358	heat shock protein, heat shock cognate 70 kDa protein 2 * (Hsc70-2)	Cyt	*R. communis*	255573627	78,743/5.06	71,276/5.14	977	53	23	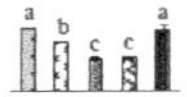
367	heat shock cognate 70 kDa protein 2 (Hsc70-2)	Cyt	*V. vinifera*	225434984	76,864/5.20	71,589/5.17	847	57	25	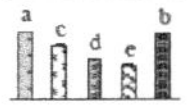

Table 1. *Cont.*

Spot No. [a]	Protein Name [b]	SCL [c]	Plant Species [d]	gi Number [e]	Exp. Mw (Da)/pI [f]	Thr. Mw (Da)/pI [g]	Sco [h]	Cov (%) [i]	QM [j]	V% ± SE [k]
332	hypothetical protein VITISV_006707, stromal 70 kDa heat shock-related protein (cpHsp70) *	Chl	*V. vinifera*	147805297	80,390/4.75	69,198/5.11	722	36	16	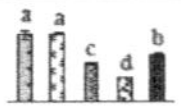
225	hypothetical protein VITISV_009951, chloroplastic chaperone protein ClpB3 *	Chl	*V. vinifera*	147842424	102,884/5.89	89,292/5.28	324	28	16	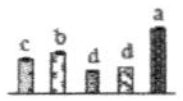
238	ATP-dependent Clp protease ATP-binding subunit clpA homolog CD4B (ClpA)	Chl	*Solanum lycopersicum*	399213	98,201/5.59	102,463/5.86	719	34	27	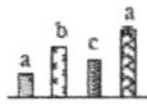
170	zinc metalloprotease (insulinase family), homologous to presequence protease 1 (PreP1) *	Chl/Mit	*A. thaliana*	9294618	113,711/4.99	118,265/5.32	346	13	11	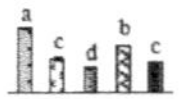
179	zinc metalloprotease (insulinase family), homologous to presequence protease 1 (PreP1) *	Chl/Mit	*A. thaliana*	9294618	113,515/5.07	118,265/5.32	157	11	9	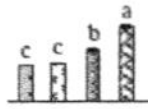
417	AC007592_12 F12K11.22, ATP-dependent zinc metalloprotease FtsH8 *	Chl	*A. thaliana*	6692685	71,862/5.35	71,014/5.81	144	13	5	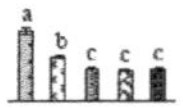
Stress and defence (1)										
1023	dehydroascorbate reductase (DHAR) class glutathione transferase DHAR1	Chl/Cyt	*P. trichocarpa*	283135904	27,322/6.05	24,525/4.93	143	10	2	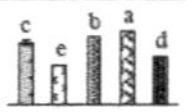

[a]: Assigned spot number as indicated in Figure 5; [b]: The name and functional categories of the proteins identified by MALDI TOF-TOF MS. Protein names marked with an asterisk (*) have been edited based on The Basic Local Alignment Search Tool (BLAST) against National Center for Biotechnology Information non-redundant (NCBInr) protein database. The detailed information of the NCBI BLAST can be found in Table S2; [c]: Protein subcellular localization predicted by software YLoc, LocTree3, Plant-mPLoc, ngLOC, and TargetP. SCL, subcellular location; Chl, chloroplast; Cyt, cytoplasm; Mit, mitochondria; [d]: The plant species that the peptides matched to; [e]: Database accession numbers from NCBInr; [f,g]: Experimental (f) and theoretical (g) molecular weight (Da) and pI of identified proteins. Experimental values were calculated using Image Master 2D Platinum Software. Theoretical values were retrieved from the protein database; [h]: The Mascot score obtained after searching against the NCBInr database; [i]: The amino acid sequence coverage for the identified proteins; [j]: The number of unique peptides identified for each protein; [k]: The mean values of protein spot volumes relative to total volume of all the spots. Five samples (from left to right are YLM, YLC, MLM, MLC, and LENF) of various leaf regions at different development stages of *Hosta* "Gold Standard". YLM, young leaf margin; YLC, young leaf center; MLM, mature leaf margin; MLC, mature leaf center; LENF, leaf under excess nitrogen fertilization. The different small letters on the columns indicate significant differences ($p < 0.05$) among the five samples as determined by one-way Analysis of Variance (ANOVA). Error bars indicate ± standard error (SE).

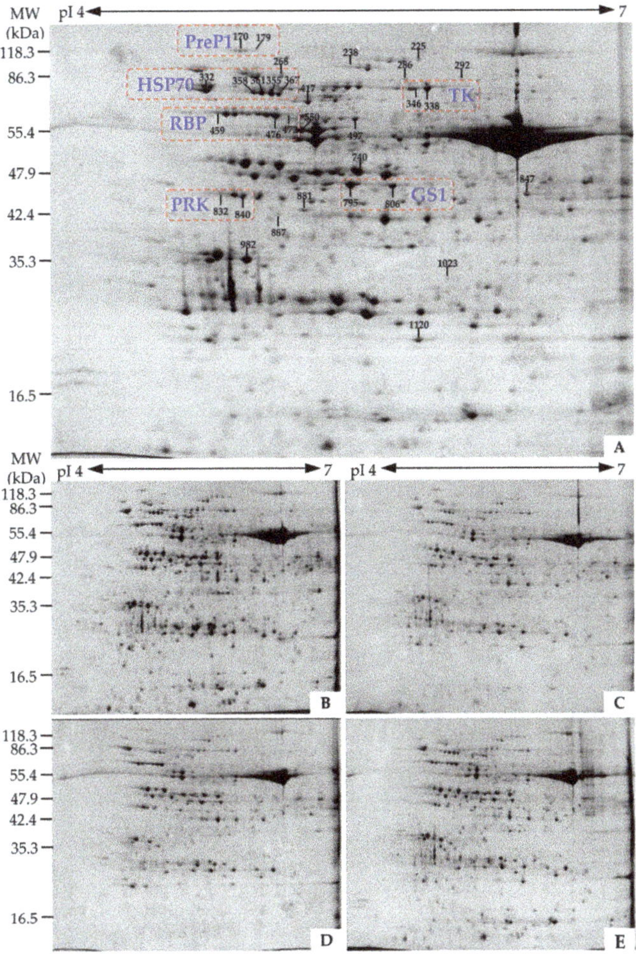

Figure 5. Representative 2-DE gel images of proteins extracted from various leaf regions at different development stages of *Hosta* "Gold Standard". (**A**) Young leaf margin (YLM); (**B**) young leaf center (YLC); (**C**) mature leaf margin (MLM); (**D**) mature leaf center (MLC); (**E**) leaf under excess nitrogen fertilization (LENF). Proteins were separated on 24 cm immobilized pH gradient (IPG) strips (pH 4–7 linear gradient) using isoelectric focusing (IEF) in the first dimension, followed by 12.5% SDS-PAGE gels in the second dimension. The 2-DE gel was stained with Coomassie Brilliant Blue. Molecular weight (*Mw*) in kilodaltons (kDa) and pI of proteins are indicated on the left and top of the gel, respectively. A total of 31 differentially expressed proteins identified by MALDI TOF-TOF MS were marked with numbers on the gel, and the detailed information can be found in Table 1, Supplemental Figure S1, and Supplemental Tables S1 and S2. The protein spots of proteoforms in each group were enclosed in a red rectangle. GS1: Glutamine synthetase isoform GS1c; Hsp70: Heat shock protein 70; PreP1: Presequence protease 1; PRK: Phosphoribulokinase; RBP: RuBisCO large subunit-binding protein subunit; TK, transketolase.

2.5. Annotation and Functional Categorization of DEPs

Among the 31 identified DEPs, seven were originally annotated as unknown, or hypothetical proteins. In this study, they were re-annotated according to The Basic Local Alignment Search

Tool (BLAST) analysis (Table 1, and Supplemental Table S2). Taken together, based on the function information from Gene Ontology, BLAST analysis, and literature, all these proteins were mainly involved in nitrogen metabolism, photosynthesis, carbohydrate and energy metabolism, protein synthesis, protein processing and degradation as well as stress and defense (Table 1). Among them, proteins involved in protein metabolism, including protein synthesis, protein processing and degradation, accounted for the largest group (38% of DEPs) (Figure 6A).

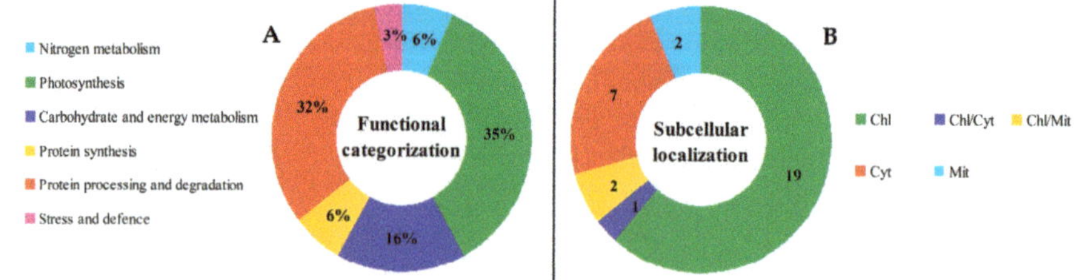

Figure 6. Functional categorization and subcellular localization of the differentially expressed proteins (DEPs) from various leaf regions at different development stages of *Hosta* "Gold Standard". (**A**) A total of 31 DEPs were classified into six functional categories. The percentage of proteins in different functional categories is shown in the pie; (**B**) Subcellular localization categories of the identified proteins. The numbers of proteins with different locations are shown. Chl: Chloroplast; Cyt: Cytoplasm; Mit: Mitochondria.

2.6. Proteoform Analysis of DEPs

The 31 DEPs only represented 21 unique proteins, since six proteins each had multiple proteoforms. These proteins were glutamine synthetase isoform GS1c (GS1) (spots 795 and 806), transketolase (TK) (spots 338 and 346), RuBisCO large subunit-binding protein subunit (RBP) (spots 459, 476, and 472), phosphoribulokinase (PRK) (spots 840 and 832), heat shock protein 70 (Hsp70) (spots 355, 361, 358, 367, and 332), and presequence protease 1 (PreP1) (spots 170 and 179) (Figure 5, Table 1, and Supplemental Table S3).

Interestingly, the proteoforms of each protein were distributed along a horizontal line on the 2DE gel (Figure 5), indicating that they have similar experimental molecular weight, but different isoelectric points (Table 1). Moreover, on the basis of our mass spectrometry (MS) identification, multiple alignments of the amino acid sequences from the six proteins were analyzed. The sequence alignment results indicate that there were high levels of sequence identities among various proteoforms of each protein (e.g., 100% of GS1 and PreP1, over 80% of TK, PRK, and Hsp70, as well as over 70% of RBP) (Supplemental Figure S2). Most importantly, the proteoform expression patterns of each protein were remarkably similar (Table 1, Supplemental Table S3). For example, two proteoforms of GS1 (spots 795 and 806) appeared at higher levels in MLM and MLC than that in YLM, YLC, and LENF. Besides, two proteoforms of PRK (spots 840 and 832) were increased in MLC and decreased in YLC. In addition, five proteoforms of Hsp70 (spots 355, 361, 358, 367, and 332) were all reduced in mature leaves when compared with young leaves, but kept stable level in LENF and YLM (Table 1, Supplemental Table S3).

2.7. Subcellular Localization of DEPs

The analysis of protein subcellular localization is critical for understanding their functions. In this study, the 31 DEPs were submitted to five internet tools (*i.e.*, YLoc, LocTree3, Plant-mPLoc, ngLOC, and TargetP) for subcellular localization (Table 1, Figure 6B, and Supplemental Table S4). Among them, 22 DEPs (71%) were predicted to be localized in chloroplasts, including three DEPs localized in chloroplast and mitochondrion/cytoplasm (Figure 6B). The majority of DEPs were predicted to be localized in chloroplasts, suggesting that the leaf color changes were mainly attributed to the metabolic differences in the chloroplasts. In addition, our results of MS-identified peptides of DEPs provided amino acid sequence information for their subcellular localization. The sequence alignment of four proteins (GS, phosphoglycerate kinase (PGK), fructose-bisphosphate aldolase (FBA), and Hsp70) with MS-identified peptides and their homologs from other plant species (e.g., wheat, rice, and Arabidopsis) supported our prediction of subcellular localization. For example, the MS-identified peptides from two proteoforms of GS1 (spots 795 and 806) were submitted to alignment analysis with their homologs of ten GS family members in wheat [36]. The results indicate that they were well aligned with the cytosolic GS1, but not with the plastidic GS2 (Figure 7, Supplemental Figure S3). For PGK (spot 740) and FBA (spot 847), the MS-identified peptides of PGK were matched with chloroplast-localized homologs from rice (*Oryza sativa* sp. *japonica*) [37] (Supplemental Figure S4), while the MS-identified peptides of FBA were matched with four cytoplasm-localized FBA of eight family members from Arabidopsis [38] (Supplemental Figure S5). For Hsp70, the peptide sequences of five proteoforms were aligned with 14 homologs of Hsp70 in Arabidopsis [39]. The specific MS-identified peptides from four proteoforms (spots 355, 361, 358, and 367) were only matched to the cytoplasm-localized Hsp70s, while one proteoform (spot 332) was well aligned with the Hsp70 localized in the chloroplast (Table 1, Supplemental Figure S6).

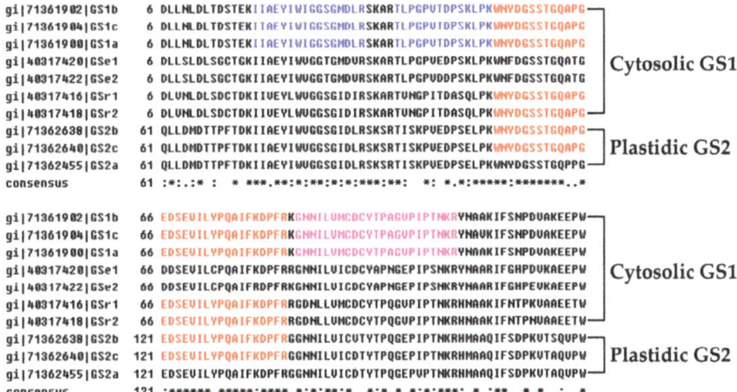

Figure 7. Sequence alignment between the mass spectrometry (MS)-identified peptides from two proteoforms of glutamine synthetase 1 (GS1) (spots 795 and 806) and the amino acid sequences of the ten GS isozymes in wheat (*Triticum aestivum* L.). The amino acid sequences from wheat used for the multiple sequence alignment were from National Center for Biotechnology Information non-redundant (NCBInr) protein database and submitted by Bernard *et al.* [36]. The accession numbers and names of the GS isozymes were listed on the left, and the subfamilies were shown on the right of the amino acid sequences. The asterisk indicates completely conserved residues; colon indicates highly conserved residues; dot indicates a weak conservative region. The MS-identified peptides from spots 795, 806 and both were highlighted in red, blue, and purple fonts, respectively. The results showed that the MS-identified peptides were well aligned with cytosolic GS1, but not plastidic GS2. The detailed information of the identified peptides can be found in Table S1, and the complete alignment can be found in Figure S3.

2.8. Hierarchical Cluster Analysis of DEPs

To better understand the expression patterns of all the coordinately regulated proteins, hierarchical clustering analysis of the 31 DEPs were performed, which revealed two main clusters (Figure 8). Cluster I contained 17 DEPs, the abundances of which in mature leaves were reduced when compared with young leaves (especially in YLM). These proteins were grouped into two distinct subclusters. Subcluster I-1 included the significantly increased proteins in LENF when compared with mature leaves. The proteins in Subcluster I-1 mainly included four proteoforms of heat shock congnate

70 kDa (Hsc70) (spots 335, 358, 367, 361), and a chloroplastic chaperone protein ClpB3 (spot 225), as well as chlorophyll a/b binding protein 6 (CAB6) (spot 1120) and photosystem II stability/assembly factor HCF136 (HCF136) (spot 881). Subcluster I-2 contained the proteins whose abundances were induced in YLM compared with other tissues (e.g., PGK (spot 740) and chloroplastic elongation factor G (cpEF-G) (spot 268)) (Figure 8). Cluster II contained 14 DEPs, which were also grouped into two subclusters. Subcluster II-1 included the proteins that were increased in central regions but decreased in marginal regions of young leaves and mature leaves, as well as in LENF, such as the ClpA subunit (homologous to Hsp93-V/ClpC1 in Arabidopsis) (spot 238), two proteoforms of RBP-β (spots 476 and 472), and NADH-ubiquinone oxidoreductase (spot 286). Subcluster II-2 contained the proteins that were increased in mature leaves but decreased in young leaves and LENF, including two proteoforms of GS1 (spots 795 and 806), two proteoforms of PRK (spots 832 and 840), a proteoform of RBP-α (spot 459), dehydroascorbate reductase (DHAR) (spot 1023), FBA (spot 847), plastid TK (spot 338), and haloalkane dehalogenase (HLD) (spot 887) (Figure 8). These results suggest that the proteins appeared function skewed in different regions of variegated leaves of *Hosta* "Gold Standard".

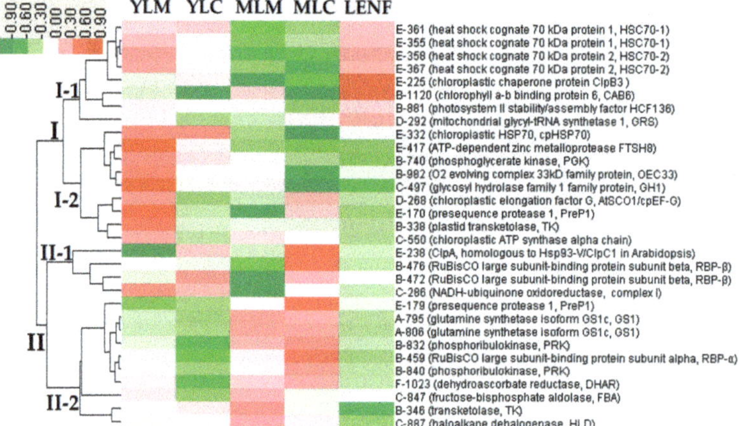

Figure 8. Hierarchical clustering analysis of the expression profiles of the identified 31 proteins. The five columns represent various leaf regions at different development stages of *Hosta* "Gold Standard", including young leaf margin (YLM), young leaf center (YLC), mature leaf margin (MLM), mature leaf center (MLC), leaf under excess nitrogen fertilization (LENF). The rows represent individual proteins. Two main clusters (I and II) and subclusters of I (I-1 and I-2) and subclusters of II (II-1 and II-2) are shown on the left side. Functional categories indicated by capital letters, spot numbers, and protein names are listed on the right side. The scale bar indicates log (base2) transformed protein abundance ratios ranging from −0.9 to 0.9. The ratio was calculated by dividing percent volume (vol %) of each protein spot at various leaf region samples by the average vol % of the five various leaf region samples of the same protein spot. The increased and decreased proteins are represented in red and green, respectively. The color intensity increases with increasing abundant differences. Functional categories: A, nitrogen metabolism; B, photosynthesis; C, carbohydrate and energy metabolism; D, protein synthesis; E, protein processing and degradation; F, stress and defence.

3. Discussion

3.1. Leaf Color Difference of Variegated Leaves Is Determined by the Chlorophyll Contents

It is well known that the leaf color is affected by chlorophyll and carotenoid levels in variegated leaves. In this study, the center color of variegated leaves was gradully turned from light green (in YLC) to golden (in MLC) during normal leaf development, and can be changed to green under excess nitrogen fertilization, but the margin color of variegated leaves has no obvious changes (Figures 1 and 2). *Aurea* yellow leaves were severely deficient in all the pigments and particularly in Chl *a* and Chl *b* [20]. In *Hosta* "Gold Standard" leaves, total Chls, Chl *a*, and Chl *b* in golden regions were obviously lower than in green regions of variegated leaves, and the Chl *b* accumulation was more reduced than Chl *a* (Figure 3A–D). A similar phenomenon has also been observed in *Ligustrum vicaryi* [21] and *Amaranthus tricolor* [18]. *L. vicaryi* is a plant species with chlorophyll-less golden-color leaves on the upper canopy but green leaves under the canopy [21], and *A. tricolor* is a C_4 plant with three-color (green, yellow and red) leaves [18]. After 30 days of excess nitrogen fertilization, the average total Chls, Chl *a* and Chl *b* in LENF were increased obviously when compared with that in central regions, but still lower than that in marginal regions of young leaves and mature leaves (Figure 3A–C). The high abundant Chls in marginal regions ensured higher levels of photosynthesis and other metabolisms, which might compensate for the lower levels in central regions of variegated leaves [17,40].

3.2. Nitrogen Metabolism Level Determined the Chlorophyll Synthesis in Variegated Leaves

Nitrogen nutrition (*i.e.*, nitrate and ammonium) is one of the major determinants of chlorophyll synthesis and chloroplast development in variegated leaves [41]. About 75% of the total nitrogen in plant is required for chlorophyll synthesis, the components of photosynthetic enzymes and thylakoid membranes (reviewed in [41]). Nitrate needs to be reduced to nitrite by NR in roots, and then be reduced to ammonium by nitrite reductase (NiR) in chloroplast. Subsequently, ammonium from various sources (e.g., soil, nitrate reduction, photorespiration, and amino acid decomposition) is fixed into amino acids (*i.e.*, glutamine/glutamate) in the GS/GOGAT pathway [42]. GS catalyzes the ATP-dependent condensation of ammonium to the δ-carboxyl group of glutamate (Glu) to form glutamine (Gln), while GOGAT catalyzes the conversion of Gln and 2-oxoglutarate to two molecules of Glu, which serves as one of the nitrogen donors for the biosynthesis of organic nitrogenous compounds, such as chlorophylls, proteins, and nucleotides [43] (Figure 9). Therefore, NR, GS, and GOGAT are the key enzymes in nitrogen assimilation processes, which play important roles in chlorophyll synthesis and chloroplast development.

In *Hosta* "Gold Standard" leaves, obviously lower levels of nitrogen metabolism in center light green/golden regions than in margin regions of young and mature variegated leaves could be inferred from the results of total nitrogen levels and activities of NR and GOGAT (Figure 4A–C). However, the activities of GS and the levels of two proteoforms of GS1 showed exactly opposite patterns in mature leaves compared to young leaves (Figure 4D, Table 1, Figure 9). In most C_3 plant leaves, GS family includes cytosol-localized GS1 and chloroplast-localized GS2. GS activity in leaves is mainly attributed to GS2 (70%–95%), but not GS1 (5%–30%) [44]. GS1 mainly functions in nitrogen remobilization from protein breakdown in leaves, especially the remobilization of the inorganic form of nitrogen from senescing leaves for grain filling [45]. GS2 mainly functions in the reassimilation of ammonia generated from photorespiration [45] and nitrate reduction [46]. Therefore, although GS1 levels were increased in mature variegated leaves of C3 plant *Hosta*, the decrease of total GS activity in mature leaves was probably due to the decline of GS2 activity (Figure 4D). It has been found that the decreases of GS2 abundance occurred in an albinism line of winter wheat [7] and at the albinistic stage in an albino tea cultivar [8]. This indicates that reassimilation of ammonia released from photorespiration was reduced in mature variegated leaves. This is supported by the decreases of photorespiration and photosynthesis rates due to the decreased abundances of photosynthetic proteins (Table 1). Moreover, the induced abundance of GS1 in mature variegated leaves implied that assimilation of ammonium from the degraded proteins is critical for variegated leave maturation. Interestingly, in leaves during senescence [47], under salt stress [48], or pathogen attack [49], GS activity also decreased, accompanied by the induced GS1 abundance and reduced GS2 abundance.

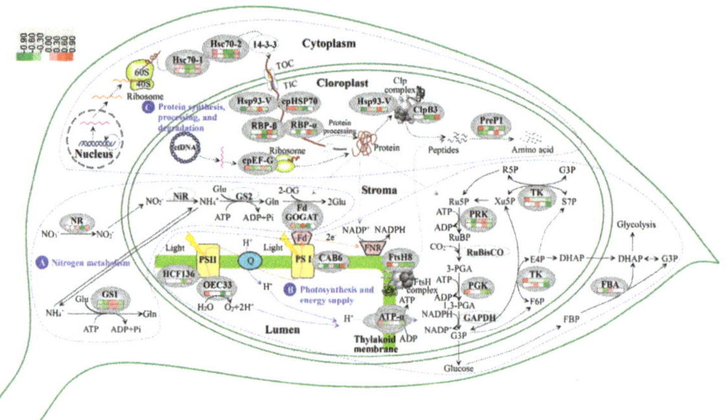

Figure 9. Schematic presentation of the leaf color regulation mechanism in chimera *Hosta* "Gold Standard" leaves. The identified proteins were integrated into subcellular pathways. A, Nitrogen metabolism; B, Photosynthesis and energy supply; C, Protein synthesis, processing and degradation. Protein expression patterns and enzyme activities are marked in the oval shapes with squares and circles in different colors, respectively. The increased and decreased proteins are represented in red and green, respectively. The color intensity increases with increasing abundant differences. The solid line indicates single-step reaction, and the dashed line indicates multistep reaction. Abbreviation: 2-OG, 2-oxoglutarate; 40S, eukaryotic small ribosomal subunit; 60S, eukaryotic large ribosomal subunit; ATP-α, ATP synthase alpha chain; CAB6, chlorophyll a/b binding protein 6; ClpB3, chloroplastic chaperone protein ClpB3; cpEF-G, chloroplastic elongation factor G; cpHsp70, stromal 70 kDa heat shock-related protein; DHAP, dihydroxyacetone phosphate; E4P, erythrose-4-phosphate; F6P, fructose 6-phosphate; FBA, fructose-1,6- bisphosphate aldolase; FBP, fructose 1,6-bisphosphate; Fd, ferredoxin; FNR, ferredoxin-NADP reductase; FtsH8, ATP-dependent zinc metalloprotease FtsH8; G3P, glycerate 3-phosphate; GAPDH, Glyceraldehyde-3-phosphate dehydrogenase; Gln, glutamine; Glu, glutamate; GS1, cytosol-localized glutamine synthetase isoform; GS2, plastid-localized glutamine synthetase isoform; HCF136, photosystem II stability/assembly factor HCF136; Hsc70, heat shock congnate 70 kDa protein; Hsp93-V, heat shock protein 93-V(ClpC1, ClpA subunit); NiR, nitrite reductase; NR, nitrate reductase; OEC33, O_2 evolving complex 33kD family protein; PGA, 3-phosphoglyceric acid; PGK, phosphoglycerate kinase; PreP1, presequence protease 1; PRK, phosphoribulokinase; PSI, photosystem I; PSII, photosystem II; Q, quinine; R5P, ribose-5-phosphate; RBP, RuBisCO large subunit-binding protein subunit; Ru5P, ribulose-5-phosphate; RuBisCO, ribulose-1,5-bisphosphate carboxylase; RuBP, ribulose-1,5-bisphosphate; S7P, sedoheptulose-7-phosphate; TIC, translocon at the inner envelope membrane of chloroplasts; TK, transketolase; TOC, translocon at the outer envelope membrane of chloroplasts; Xu5P, xylulose-5-phosphate.

In addition, nitrogen metabolism was inhibited in central light green/golden regions in variegated leaves, and recovered after excess nitrogen fertilization (Figure 4B–D). Similar nitrogen metabolic processes have been found in chloroplast-ribosome-deficient leaves of *"albostrians"* mutant of barley and *"iojap"* mutant of maize [50]. In these albino mutants, the NR activity was generally 40% lower than wide-type plants [50]. In addition, proteomic analysis has revealed that GS abundance was decreased in albinistic stage, and then increased in the regreening stage in the albino tea cultivar [8]. Among the three enzymes, GS activities in LENF were induced, but GS1 abundances in LENF were lower than those in mature leaves. This indicates that the induced GS activity in LENF was due to the increase of GS2 activity, because it has been shown that excess ammonia can induce GS2 gene transcription in both rice and tobacco leaves [51]. The chloroplast-localized GS2 contributed more Gln for plastid development [52]. Inhibition of GS with the herbicide phosphinothricin (PPT) led to decreases of GS2 mRNA and polypeptides, but increase of GS1, and ultimately rapid chlorosis of leaves [46]. The GS1 subfamily has several members in some plants (e.g., maize [53], Arabidopsis [54], and rice [55]), with different expression patterns and various affinities for NH_4^+ (reviewed in [45]). The alignment analysis indicates that two proteoforms of GS1 in our results were homologous with GS1a/GS1b/ GS1c of wheat and GS1-3/GS1-4 of maize [36]. The two proteoforms of GS1 were consistently expressed in young leaves and LENF, but obviously increased in mature leaves (Table 1, Figure 9). Their homologs in maize seedlings have no obvious changes after external ammonium application [53]. In addition, the two proteoforms of GS1 with similar molecular weight may be generated from post-translational modifications (e.g., phosphorylation), since GS1 activity is regulated by phosphorylation and interaction with 14-3-3 protein during senescence in relation to nitrogen remobilization [47]. All these data indicate that assimilation patterns of ammonium coordinately catalyzed by GS1 and GS2 is critical for many processes of development and stress response (Figure 9).

3.3. Chloroplast Development in Different Stages and Nitrogen Levels

In the leaves of many variegated plant species, chloroplast development is different in various regions, nitrogen levels, light intensities, and other environmental conditions [19,21,22]. Leaves under sufficient nitrogen conditions have large chloroplasts with well-developed grana, stroma lamellae, thylakoid membranes, and photosynthetic enzymes, and starch granules [41,56]. Several PS related proteins (e.g., CAB6, HCF136, and O_2 evolving complex 33 kD family protein (OEC33)) identified in this study are critical for thylakoid membrane structure construction and photosynthesis regulation in variegated leaves. CAB6, HCF136, and OEC33 were decreased in light green/golden regions in young and mature leaves, and induced after excess nitrogen fertilization (Table 1, Figure 9). CAB6 is a constituent of PSI antenna encoded by *LHCa1* gene, and was found to be reduced in the albino leaves compared to the green leaves from bamboo (*Pseudosasa japonica*) [57]. CAB was also increased back to the normal level during leaf recovery from senescence to green leaves [58]. HCF136 is a stroma thylakoid lumenal side-localized PSII assembly factor required for PSII proteins accumulation [59]. The severe mutational defect of HCF136 led to a pale green seedling phenotype [59]. OEC also bounds to the lumen side of PSII, and functions as a water-oxidizing enzyme in the water photooxidation process [60]. It has been reported that some LHCII proteins were deficient in the yellow-leaved mutant of *H. sieboldii* ("Wogan Gold") [22] and maple (*Acer negundo* Hassk. var. *odessanum*) [23], or appeared in low levels in golden leaves from *L. vicaryi* [21]. Importantly, the expression pattern of PS proteins was consistent with our previous observation of chloroplast ultrastructure of *Hosta* "Gold Standard", which proved that thylakoid membrane and grana formation in the golden region was obviously affected, but normal in green region and green leaves after excess nitrogen fertilization [19]. Previous findings in another *Hosta* species (*Hosta sieboldii*) and *L. vicaryi* reported that the grana formation and chloroplast development were blocked in the leaves of yellow mutant, and leaves contain only a few separate thylakoid membranes, and the grana are degraded [22], as well as the chloroplast ultrastructure was different in upper golden leaves, lower green leaves, and regreening upper leaves under different light intensities [21].

3.4. Photosynthesis and Energy Supply Are Different in Various Regions of Variegated Leaves

It is well-known that photosynthesis capacity in the yellow/golden regions is generally lower than in the green regions of variegated leaves of several species, such as willow myrtle (*Agonis flexuosa* (Willd.) Sweet), oleander (*Nerium oleander* L.) [16], agave (*Agave americana*) [17], and fig tree (*Ficus microcarpa* L. f. cv Golden Leaves) [13], but the molecular mechanisms underlying the photosynthetic differences are unknown. In this study, besides the three photosystem proteins (CAB6, HCF136, OEC33), we identified five enzymes related to the Calvin cycle, which were RBP-α, RBP-β, PGK, TK, and PRK (Table 1, Figure 9). The decreases of CAB6, HCF136, and OEC33 in golden regions of variegated leaves would lead to low light reaction activity and shortage of ATP and NADPH for Calvin cycle. Accordingly, PGK and one TK proteoform also appeared to be low in the center golden regions, suggesting the reduced photosynthetic carbon reduction in the golden regions of variegated leaves. In contrast, RBPs and PRKs were induced in the center golden regions of variegated leaves, appearing contradictory to the reduced photosynthesis in the golden regions. Although RBP is normally considered as chloroplast chaperonin 60 (Cpn60), which interacts with RuBisCO for positive regulation of its function in photosynthetic tissues [61], they also can bind to other imported proteins (e.g., stromal plastid division protein and FtsZ polymer), regulating protein folding and dynamics for the formation of normal plastid division apparatus [62,63]. Taken together with the unchanged abundance of RuBisCO in our results, we speculate that the induced RBPs (Cpn60) mainly contribute to other imported protein folding in variegated leaves of *Hosta* "Gold Standard". Additionally, PRK is an essential enzyme in the Calvin cycle for regeneration of the RuBisCO substrate [64], and it was also reported to have negligible control of the Calvin cycle under a range of environmental conditions [65]. The photosynthetic carbon assimilation in golden regions of variegated leaves would be controlled by the rate-limiting enzyme TKs, but not PRKs.

After excess nitrogen fertilization, the levels of these Calvin cycle-related enzymes were decreased, although the photosynthesis system proteins (CAB6, OEC33, and HCF136) were increased in LENF when compared with those in MLC (Table 1, Figure 9). The decreased photosynthesis was also found in *Hosta* "Blue Umbrella" [66] and *Hosta* "Francee" [67] when fertilized with more than 0.5 g/kg urea. All these results indicated that nitrogen level was strongly correlated with photosynthetic capacity [68,69]. Nitrogen assimilation into Glu in leaves is dependent on the consumption of carbon skeletons and supply of ATP and NADPH generated from photosynthesis [70]. In our results, the increased nitrogen assimilation would compete against ATP/NADPH with photosynthetic carbon fixation, leading to the carbon assimilation decrease in LENF [71]. In addition, the difference of photosynthetic CO_2 assimilation would lead to the carbohydrate metabolism changes in various regions of variegated leaves. In the variegated leaves from *A. americana*, the contents of various sugars (e.g., sucrose, glucose, and fructose) were lower in the yellow regions than in the green regions [17]. In this study, we found that the abundance of cytoplasm-located FBA was lower in the central regions than in the marginal regions, indicating glycolysis and gluconeogenesis would be reduced in the golden regions of variegated leaves (Table 1, Figure 9).

3.5. Chloroplast Protein Synthesis, Processing and Degradtaion in Variegated Leaves

Protein synthesis and processing in chloroplasts is crucial for chloroplast biogenesis, development and metabolism [72]. It has been reported that ribosome numbers are much reduced in *aurea* yellow leaves [20]. In this study, we found an induced cpEF-G in golden regions of mature leaves (Table 1, Figure 9), which is encoded by *cpEF-G* in plastid genome. Its homolog in Arabidopsis is snowy cotyledon1 (SCO1). A *sco1-1* mutant exhibited greening inhibition during the early seedling development stage, while the *sco1* null mutant cannot survive [73]. The induced SCO1 functions in the increase of protein translation to compensate for the reduced amount of ribosome in a *clpR4* mutant [74]. Thus, the increased cpEF-G would contribute for the stabilization of protein translation in the golden regions of variegated leaves.

The polypeptide maturation and protein quality control are important for proper plastid differentiation and chloroplast development [75]. It has been found that chaperones and proteases, such as cpHsp70 [76,77], ClpB3 [78], caseinolytic protease, subunit C (ClpC) [79], FtsH8 [80], and PreP1 [81] were involved in leaf variegation. The Arabidopsis mutants of these genes shared some similar characteristics with the variegated leaves of *Hosta* "Gold Standard", such as leaf variegation, decreased pigment content, and reduced chloroplast development [76–81]. In our results, the abundances of ClpB3, Hsp93-V (ClpC1), PreP1, and FtsH8 showed obvious changes in various regions of variegated leaves (Table 1, Figure 9). ClpB3 is a chloroplast-localized chaperone. It functions in the unfolding of aggregated protein cooperating with the DnaK chaperone that is homologous to cpHsp70 [82]. The *ClpB3* mutants (e.g., *albino* or *pale-green*) showed severe defects in chloroplast development [78]. The low abundance of ClpB3 in the variegated leaves of *Hosta* "Gold Standard" indicates that it would contribute to the reduced protein solubility and/or assembly processes in the stroma [79]. Besides, soluble stroma-localized Hsp93 (ClpC) is also thought to be a chaperone component of the Clp protease complexes, which plays a central role in dissolution and degradation of protein aggregates as a stromal housekeeping protease [83,84]. Interestingly, in Arabidopsis, it was also shown that the Hsp93-V (ClpC1) participates in the degradation of chlorophyllide a oxygenase (CAO), which is responsible for the conversion of Chl *a* to Chl *b* [85,86]. Therefore, the increased Hsp93-V (ClpC1) in the golden region of variegated leaves of *Hosta* "Gold Standard" might induce the degradation of CAO, leading to the reduction of conversion of Chl *a* to Chl *b*. This would be the reason of increased ratio of Chl *a*/*b* in the golden regions of the leaves (Figure 3D). In addition, PreP1 is a zinc-dependent metalloendopeptidase. It functions as the organellar peptide degrading protease responsible for degrading free targeting peptides and clearing other unstructured toxic peptides [87]. The protease activity has been shown to be increased in chlorophyll-less leaves, such as albino leaves [8,88] and senescent leaves [89], indicating they may hydrolyze chlorophyll-binding proteins during Chl degradation [58]. The similar trends of ClpB3 and cpHsp70 in this study confirmed the synergistic functions of these chaperones to unfold aggregated proteins followed by protein refolding [90]. Therefore, in our study, the reduced cpHsp70 and ClpB3 might lead to accumulation of unwanted/damaged proteins in the golden regions of variegated leaves especially, and the increased Hsp93-V (ClpC1) and PreP1 might contribute to degrading these aggregates in the golden regions. Additionally, the chloroplast FtsH complex is in charge of the turnover of the Photosystem II reaction center D1 protein, as well as other protein degradation for the development and maintenance of the photosynthetic apparatus [91]. Early chloroplast development requires sufficient FtsH for thylakoid biogenesis [75]. The *ftsh2 ftsh8* double mutants exhibiting an albino-like phenotype emphasize the possibility that both FtsH2 and FtsH8 are necessary for the proper FtsH function [80]. Leaf variegation is suppressed, at least in part, by an increase in FtsH levels [74]. In this study, FtsH8 appeared at lower levels in the variegated mature leaves than in young leaves, suggesting that thylakoid formation was affected in the variegated leaves of *Hosta* "Gold Standard". After excess nitrogen fertilization, the nitrogen assimilation, amino acid metabolism, and protein synthesis are enhanced in the leaves. The increased Hsc70s, cpHsp70 and ClpB3 in LENF would contribute to folding of nascent polypeptide chains, prevention of aggregation, and solubilization/refolding of aggregated proteins [92]. In addition, the decline of Hsp93-V (ClpC1) and PreP1 in LENF indicates that protein degradation was inactive after excess nitrogen fertilization. Similar decreases in protease activities have also been reported in the regreening leaves compared with senescent leaves [58] and albescent leaves [88].

Importantly, protein import from cytosol to developing chloroplasts through translocons at the outer and inner membrane of chloroplasts is necessary for chloroplast differentiation and development [75]. Various cytosolic chaperones (e.g., Hsp70 and Hsp90) are considered to interact with unfolded preproteins for maintaining their import competence and directing them to the translocons [93]. More than 75% of chloroplast transit peptides are predicted to have at least one Hsp70 binding site [94]. cpHsp70, a component of the protein import machinery of the chloroplasts, is crucial for the import of nuclear-encoded polypeptides (e.g., the light harvesting chlorophyll a/b-binding protein (prLHCBP)) [95,96] (Figure 9). When two stromal cpHsc70-1 and cpHsc70-2 were knocked out in Arabidopsis, protein import into the chloroplasts was affected in the early developmental stages of mutants [76]. A T-DNA insertion line of *cpHsc70-1* (Δ*cpHsc70-1*) has variegated cotyledons [77]. In this study, four cytoplasm-located and one chloroplast-located proteoform of Hsp70 appeared at low abundances in variegated leaves of *Hosta* "Gold Standard" (Table 1, Figure 9). Hsp70 proteoforms were also reduced in leaves at albinism stage of winter wheat [7] and the albinistic stage of an albino tea cultivar "White leaf No. 1" [8]. The decrease of Hsp70s in chlorophyll-less leaves may affect chloroplast protein import, leading to non-green leaf color [7,8]. However, we also found Hsp93-V (ClpC1) and RBPs (Cpn60s) have significantly higher abundance in golden leaves, but lower abundance in LENF. RBPs (Cpn60s) contribute to the imported protein to fold into native conformation by interacting with a component of the inner membrane import apparatus Tic110 [97]. The expression patterns of Hsp93-V

(ClpC1) and cpHsp70 in this study proved the previous notion that cpHsp70-1 and Hsp93-V (ClpC1) have redundant functions in chloroplast protein import [76].

4. Materials and Methods

4.1. Materials and Treatment

Hosta "Gold Standard" grew at Beijing Botanical Garden (39°9′N; 116°4′E), Institute of Botany, Chinese Academy of Science. Uniformly-sized field three-year-old seedlings of *Hosta* "Gold Standard" were selected and transplanted to flowerpots (28 cm in diameter, 22 cm in height). Each flowerpot contained one seedling and 5 kg of soils, including garden soil, sand, and humus (2:1:1, *v/v/v*). And its basic contents were as follows: total nitrogen, 0.22%; soluble phosphorus, 0.127%; total kalium, 2.38%; organic matter, 4.36%; and pH 7.403. Owing to reduced biomass accumulation over 50% shadings of the *Hosta* plants [98], 50% of natural light was selected in this study.

Uniformly-sized potted seedlings were then placed in the shade with 50% of natural light by using black shade in mid-April, when YLM and YLC were harvested. About 70 days later, MLM and MLC were harvested from the variegated mature leaves. After this harvest, excess nitrogen fertilizer (2 gram urea per kilogram soil, 46%) was applied for the remaining variegated mature leafed plants according to the methods from Liu *et al.* [19]. After 30 days treatment, LEFN were harvested. All these five samples were used freshly, or immediately frozen in liquid nitrogen and then stored at −80 °C for further experiments.

4.2. Leaf Color Quantification

The leaf color was quantified by using a color meter (NF333 spectrophotometer, Nippon Denshoku Industries Co., Tokyo, Japan). Results were expressed according to CIELAB color coordinates system, L*, a*, and b*. L* represents the perceived lightness, and color coordinate a* and b* indicate the change in hue from red to green and from yellow to blue, respectively [99].

4.3. Chlorophyll Content Determination

The contents of Chl *a*, Chl *b* and total Chls were determined according to the method described by Arnon [100]. Fresh 6 mm diameter leaf discs were homogenized with 80% (*w/v*) cold acetone (Jiangsu Qiangsheng Chemical Co., Ltd., Jiangsu, China), followed by centrifuged at 5000× *g* for 10 min. The chlorophyll contents were calculated from the absorbance of the supernatant at 663 and 646 nm according to the method of Lichtenthaler and Wellburn [101].

4.4. Leaf Nitrogen Content Determination

Fresh leaves were cleaned, dried, smashed, screened through mesh size of 80, and digested with sulfuric acid-hydrogen peroxide. Then the total nitrogen content was determined by using a Kjeltec 2300 Auto analyzer (Foss Tecator AB, Höganas, Sweden).

4.5. Nitrogen Metabolism-Related Enzymes Activity Assay

The activities of NR and GS were determined according to Zhao *et al.* [102]. The activity of GOGAT was determined by the method of Singh and Srivastava [103] with some modification. The extraction buffer contained 1 mM EDTA (Amresco, Solon, OH, USA), 1 mM β-mercaptoethanol (Amresco, Solon, OH, USA), 1 mM $MgCl_2$ (Richjoint Chemical, Shanghai, China), 10 mM Tris (Amresco, Solon, OH, USA)-HCl (pH 8.2). The assay buffer contained 0.4 mL 20 mM L-glutamine (Sigma-Aldrich Co., St. Louis, MO, USA), 0.5 mL 20 mM α-oxoglutarate (Sigma-Aldrich Co., St. Louis, MO, USA), 0.1 mL 10 mM KCI (Richjoint Chemical, Shanghai, China), 0.2 mL 3 mM NADH (F. Hoffmann-La Roche Ltd., Basel, Switzerland) and 0.3 mL of the enzyme preparation. The final volume was completed to 3.0 mL with 25 mM Tris-HCl buffer (pH 7.6) [103]. The reaction was started by the addition of L-glutamine immediately following the enzyme preparation. The absorbance decrease was recorded for 4 min at

340 nm by using a UV-1800 spectrophotometer (Shimadz, Kyoto, Japan). The activity of GOGAT was expressed as the amount of NADH oxidized per minute per milligram protein.

4.6. Protein Sample Preparation, 2-DE, and Image Analysis

The proteins from leaves were extracted using a phenol extraction method described in detail in Wang *et al.* [104]. The experiments were repeated three times using protein samples prepared independently from different batches of plants. Protein concentration was determined using a Quant-kit according to the manufacture's instructions (GE Healthcare Life Science, Uppsala, Sweden). Protein samples were separated using 24 cm, pH 4–7 linear gradient IPG strips through isoelectric focusing (IEF) in the first dimension, followed by 12.5% SDS-PAGE gels in the second dimension [105]. Gel images were acquired by scanning the CBB-stained gel using an ImageScanner III (GE Healthcare Life Science, Uppsala, Sweden) at a resolution of 300 dpi and 16-bit grayscale pixel depth. Then, the images were analyzed using ImageMaster 2D software (version 5.0) (GE Healthcare Life Science, Uppsala, Sweden). For quantitative analysis, the volume of each spot was normalized against the total valid spots. Spots from three biological replicates with more than a 1.5-fold change among the protein samples and a *p* value smaller than 0.05 were considered to be DEPs.

4.7. Protein Identification and Database Searching

The differentially expressed spots were cut from the gels and digested with trypsin according to a previous method [105]. Proteins were identified according to the method described in Sun *et al.* [106]. The MS and MS/MS spectra acquired on a MALDI TOF-TOF mass spectrometer (4800 Proteomics Analyzer, AB SCIEX, Foster City, CA, USA) were searched against the National Center for Biotechnology Information non-redundant (NCBInr) protein databases [107] (10,348,164 sequences entries in NCBInr) using the search engine Mascot (Matrix Sciences, London, UK) [108]. The taxonomic category was Green Plants (730,150 sequences). The searching parameters were set according to Wang *et al.* [104], including mass accuracy of 0.3 Da, one missed cleavage per peptide allowed, carbamidomethylation of cysteine as a fixed modification, and oxidation of methionine as a variable modification. To obtain high confident identification, proteins had to meet the following criteria: the top hits on the database searching report, the MOWSE score greater than 43 ($p < 0.05$), and at least two peptides matched with nearly complete y-ion series and complementary b-ion series present.

4.8. Protein Classification

The identified proteins were searched against the NCBI database [107] and UniProt database [109] to determine if their functions were known. Based on the function information from Gene Ontology, BLAST alignments, and literature, these proteins were grouped into various categories.

4.9. Protein Subcellular Location

The subcellular locations of the identified proteins was predicted using five internet tools according to Zhao *et al.* [110] with some modifications: YLoc [111], confidence score $\geqslant 0.7$; LocTree3 [112], expected accuracy $\geqslant 80\%$; Plant-mPLoc [113], no threshold value in Plant-mPLoc; ngLOC [114], probability $\geqslant 60\%$; and TargetP [115], reliabilityclass $\leqslant 4$. Only the consistent predictions from at least two tools were accepted as a confident result.

4.10. Multiple Sequence Alignment

Multiple sequence alignments of related proteins were created by using ClustalX 1.81, and refined by using BoxShade Server [116].

4.11. Hierarchical Cluster Analysis

Self-organizing tree algorithm hierarchical clustering of the protein expression profiles was performed using Cluster software (version 3.0) as described in the website [117]. Input data was preprocessed by dividing percent volume (vol %) of each protein spot at various leaf region samples by the average vol % of the five various leaf region samples of the same protein spot, and then followed by a log (base 2) transformation.

4.12. Statistical Analysis

All results were presented as means ± standard error (SE) of at least three replicates. Data were analyzed by One-Way Analysis of Variance (ANOVA) using the statistical software SPSS 17.0 (SPSS Inc., Chicago, IL, USA). A *p* value less than 0.05 was considered statistically significant [118].

5. Conclusions

High-throughput proteomics has shown advantages in acquiring large-scale information on individual proteins and their dynamic changes underlying sophisticated leaf color-related cellular processes. We discovered 31 proteins related to the leaf color regulation in *Hosta* "Gold Standard", represented by the induced GS1 and the decreased Hsp70 in mature variegated leaves, as well as the highly abundant cpEF-G and RBP (Cpn60) in the center golden regions of variegated leaves compared to the green regions of variegated leaves. These protein expression patterns implied low levels of nitrogen metabolism, photosynthesis, and energy supply, dissolution and degradation of those unwanted/damaged proteins aggregates in the golden regions, as well as inhibition of protein import from cytosol to developing chloroplasts in the mature variegated leaves. Additionally, diverse proteoforms of several proteins (e.g., GS1, Hsp70, RBP, PRK, and PreP1) implied post-translational modification performs key functions during the development of variegated leaves. All these findings provide useful molecular information for better understanding the complicated leaf color regulation mechanisms. However, further validation and characterization of the proteins identified in this study are needed. Moreover, investigation of low abundant and extremely acidic/basic proteins in the varigated leaves, as well as analysis of the protein post-translational modification and protein-protein interaction using proteomics approaches are also necessary for further understanding of the complex cellular and molecular processes in variegated leaves.

Supplementary Materials: Supplementary materials can be found at http://www.mdpi.com/1422-0067/17/3/346/s1.

Acknowledgments: The project was supported by grants from National Natural Science Foundation of China (No. 31270310) the Program for Professor of Special Appointment (Eastern Scholar) at Shanghai Institutions of Higher Learning (2011), Capacity Construction Project of Local Universities, Shanghai, China (No. 14390502700) to Shaojun Dai; and by the Knowledge Innovation Program of the Chinese Academy of Science (No. KSCX2-EW-B-05 and KSCX2-EW-B-02) and Science and Technology Basic Work of Science and Technology (No. 2014F210300) to Jinzheng Zhang; and by Funding Program for Young Teachers at Universities and Colleges of Shanghai (2014), and General Scientific Research Project of Shanghai Normal University (No. SK201419) to Qi Zhao.

Author Contributions: Shaojun Dai, Lei Shi and Jinzheng Zhang conceived and designed the experiments; Hongliang Guo, Qi Zhao, and Yuelu Liu performed the experiments; Juanjuan Yu, Hongliang Guo, and Qi Zhao analyzed the data; Juanjuan Yu wrote the manuscript with suggestions by Shaojun Dai and Sixue Chen. All authors have read and approved the final manuscript.

Conflicts of Interest: The authors declare no conflict of interest.

References

1. Kumar, A.M.; Söll, D. Antisense *HEMA1* RNA expression inhibits heme and chlorophyll biosynthesis in Arabidopsis. *Plant Physiol.* **2000**, *122*, 49–56. [CrossRef] [PubMed]
2. Chen, M.; Choi, Y.; Voytas, D.F.; Rodermel, S. Mutations in the Arabidopsis *VAR2* locus cause leaf variegation due to the loss of a chloroplast FtsH protease. *Plant J.* **2000**, *22*, 303–313. [CrossRef] [PubMed]

3. Aluru, M.R.; Bae, H.; Wu, D.; Rodermel, S.R. The Arabidopsis immutans mutation affects plastid differentiation and the morphogenesis of white and green sectors in variegated plants. *Plant Physiol.* **2001**, *127*, 67–77. [CrossRef] [PubMed]

4. Wu, Z.; Zhang, X.; He, B.; Diao, L.; Sheng, S.; Wang, J.; Guo, X.; Su, N.; Wang, L.; Jiang, L. A chlorophyll-deficient rice mutant with impaired chlorophyllide esterification in chlorophyll biosynthesis. *Plant Physiol.* **2007**, *145*, 29–40. [CrossRef] [PubMed]

5. Terry, M. Phytochrome chromophore-deficient mutants. *Plant Cell Environ.* **1997**, *20*, 740–745. [CrossRef]

6. Yu, F.; Fu, A.; Aluru, M.; Park, S.; Xu, Y.; Liu, H.; Liu, X.; Foudree, A.; Nambogga, M.; Rodermel, S. Variegation mutants and mechanisms of chloroplast biogenesis. *Plant Cell Environ.* **2007**, *30*, 350–365. [CrossRef] [PubMed]

7. Hou, D.Y.; Xu, H.; Du, G.Y.; Lin, J.T.; Duan, M.; Guo, A.G. Proteome analysis of chloroplast proteins in stage albinism line of winter wheat (*triticum aestivum*) FA85. *BMB Rep.* **2009**, *42*, 450–455. [CrossRef] [PubMed]

8. Li, Q.; Huang, J.; Liu, S.; Li, J.; Yang, X.; Liu, Y.; Liu, Z. Proteomic analysis of young leaves at three developmental stages in an albino tea cultivar. *Proteome Sci.* **2011**, *9*, 44. [CrossRef] [PubMed]

9. Tcherkez, G.; Guérard, F.; Gilard, F.; Lamothe, M.; Mauve, C.; Gout, E.; Bligny, R. Metabolomic characterisation of the functional division of nitrogen metabolism in variegated leaves. *Funct. Plant Biol.* **2012**, *39*, 959–967. [CrossRef]

10. Toshoji, H.; Katsumata, T.; Takusagawa, M.; Yusa, Y.; Sakai, A. Effects of chloroplast dysfunction on mitochondria: White sectors in variegated leaves have higher mitochondrial DNA levels and lower dark respiration rates than green sectors. *Protoplasma* **2012**, *249*, 805–817. [CrossRef] [PubMed]

11. Sakamoto, W.; Uno, Y.; Zhang, Q.; Miura, E.; Kato, Y. Arrested differentiation of proplastids into chloroplasts in variegated leaves characterized by plastid ultrastructure and nucleoid morphology. *Plant Cell Physiol.* **2009**, *50*, 2069–2083. [CrossRef] [PubMed]

12. Yamasaki, H.; Heshiki, R.; Ikehara, N. Leaf-goldenning induced by high light in *Ficus microcarpa* L. f., a tropical fig. *J. Plant Res.* **1995**, *108*, 171–180. [CrossRef]

13. Takahashi, S.; Tamashiro, A.; Sakihama, Y.; Yamamoto, Y.; Kawamitsu, Y.; Yamasaki, H. High-susceptibility of photosynthesis to photoinhibition in the tropical plant *Ficus microcarpa* L. f. cv. Golden Leaves. *BMC Plant Biol.* **2002**, *2*, 2. [CrossRef] [PubMed]

14. Deng, B. Antioxidative response of Golden Agave leaves with different degrees of variegation under high light exposure. *Acta Physiol. Plant.* **2012**, *34*, 1925–1933. [CrossRef]

15. Hung, C.; Xie, J. A comparison of plants regenerated from a variegated *Epipremnum aureum*. *Biol. Plant.* **2009**, *53*, 610–616. [CrossRef]

16. Downton, W.; Grant, W. Photosynthetic and growth responses of variegated ornamental species to elevated CO_2. *Funct. Plant Biol.* **1994**, *21*, 273–279. [CrossRef]

17. Raveh, E.; Wang, N.; Nobel, P.S. Gas exchange and metabolite fluctuations in green and yellow bands of variegated leaves of the monocotyledonous CAM species *Agave americana*. *Physiol. Plant.* **1998**, *103*, 99–106. [CrossRef]

18. Wang, Y.; Meng, Y.-L.; Ishikawa, H.; Hibino, T.; Tanaka, Y.; Nii, N.; Takabe, T. Photosynthetic adaptation to salt stress in three-color leaves of a C_4 plant *Amaranthus tricolor*. *Plant Cell Physiol.* **1999**, *40*, 668–674. [CrossRef]

19. Liu, Y.; Zhang, J.; Li, X.; Liu, H.; Sun, G. Effects of nitrogen fertilizer rate on leaf color of chimera *Hosta* "Gold standard". *Acta Prataculturae Sin.* **2011**, *20*, 93–100.

20. Fulgosi, H.; Ljubešić, N.; Wrischer, M. Regreening of yellow leaves. In *Plastid Development in Leaves during Growth and Senescence*; Springer: Berlin, Germany, 2013; pp. 589–599.

21. Yuan, M.; Xu, M.Y.; Yuan, S.; Chen, Y.E.; Du, J.B.; Xu, F.; Zhang, Z.W.; Guo, Z.C.; Zhao, Z.Y.; Lin, H.H. Light regulation to chlorophyll synthesis and plastid development of the chlorophyll-less golden-leaf privet. *J. Integr. Plant Biol.* **2010**, *52*, 809–816. [CrossRef] [PubMed]

22. Vaughn, K.C.; Wilson, K.G.; Stewart, K.D. Light-harvesting pigment-protein complex deficiency in *Hosta* (Liliaceae). *Planta* **1978**, *143*, 275–278. [CrossRef] [PubMed]

23. Fulgosi, H.; Jurić, S.; Lepeduš, H.; Hazler-Pilepić, K.; Prebeg, T.; Ljubešić, N. Thylakoid system disassembly during bleaching of aurea mutants of maple *Acer negundo* Hassk. var. *Odessanum*. *Croat. Chem. Acta* **2008**, *81*, 89–95.

24. Okabe, K.; Schmid, G.H.; Straub, J. Genetic characterization and high efficiency photosynthesis of an aurea mutant of tobacco. *Plant Physiol.* **1977**, *60*, 150–156. [CrossRef] [PubMed]

25. Kawata, E.E.; Cheung, A.Y. Molecular analysis of an aurea photosynthetic mutant (*Su/Su*) in tobacco: LHCP depletion leads to pleiotropic mutant phenotypes. *EMBO J.* **1990**, *9*, 4197–4203. [PubMed]

26. Oelmüller, R.; Kendrick, R. Blue light is required for survival of the tomato phytochrome-deficient aurea mutant and the expression of four nuclear genes coding for plastidic proteins. *Plant Mol. Biol.* **1991**, *16*, 293–299. [CrossRef] [PubMed]

27. Oelmüller, R.; Kendrick, R.; Briggs, W. Blue-light mediated accumulation of nuclear-encoded transcripts coding for proteins of the thylakoid membrane is absent in the phytochrome-deficient aurea mutant of tomato. *Plant Mol. Biol.* **1989**, *13*, 223–232. [CrossRef] [PubMed]

28. Palomares, R.; Herrmann, R.; Oelmüller, R. Different blue-light requirement for the accumulation of transcripts from nuclear genes for thylakoid proteins in *Nicotiana tabacum* and *Lycopersicon esculentum*. *J. Photochem. Photobiol. B Biol.* **1991**, *11*, 151–162. [CrossRef]

29. American Hosta Society. Registrations. *Hosta J.* **2009**, *40*, 3–70.

30. Zhang, J.; Liu, Y.; Li, X.; Liu, H.; Sun, G.; He, Q. Effects of excessive application of nitrogen fertilizer on leaf color, key enzymes activities of nitrogen metabolism and chloroplast ultrastructure of a chimera *Hosta* "Gold Standard". *Acta Prataculturae Sin.* **2011**, *20*, 93–101.

31. Chen, S.; Harmon, A.C. Advances in plant proteomics. *Proteomics* **2006**, *6*, 5504–5516. [CrossRef] [PubMed]

32. Yu, J.; Chen, S.; Zhao, Q.; Wang, T.; Yang, C.; Diaz, C.; Sun, G.; Dai, S. Physiological and proteomic analysis of salinity tolerance in *Puccinellia tenuiflora*. *J. Proteome Res.* **2011**, *10*, 3852–3870. [CrossRef] [PubMed]

33. Suo, J.; Zhao, Q.; Zhang, Z.; Chen, S.; Cao, J.; Liu, G.; Wei, X.; Wang, T.; Yang, C.; Dai, S. Cytological and proteomic analyses of *Osmunda cinnamomea* germinating spores reveal characteristics of fern spore germination and rhizoid tip-growth. *Mol. Cell. Proteom.* **2015**, *14*, 2510–2534. [CrossRef] [PubMed]

34. Zhao, Q.; Zhang, H.; Wang, T.; Chen, S.; Dai, S. Proteomics-based investigation of salt-responsive mechanisms in plant roots. *J. Proteom.* **2013**, *82*, 230–253. [CrossRef] [PubMed]

35. Zhang, H.; Han, B.; Wang, T.; Chen, S.; Li, H.; Zhang, Y.; Dai, S. Mechanisms of plant salt response: Insights from proteomics. *J. Proteome Res.* **2011**, *11*, 49–67. [CrossRef] [PubMed]

36. Bernard, S.M.; Møller, A.L.B.; Dionisio, G.; Kichey, T.; Jahn, T.P.; Dubois, F.; Baudo, M.; Lopes, M.S.; Tercé-Laforgue, T.; Foyer, C.H. Gene expression, cellular localisation and function of glutamine synthetase isozymes in wheat (*Triticum aestivum* L.). *Plant Mol. Biol.* **2008**, *67*, 89–105. [CrossRef] [PubMed]

37. Joshi, R.; Karan, R.; Singla-Pareek, S.L.; Pareek, A. Ectopic expression of Pokkali phosphoglycerate kinase-2 (OsPGK2-P) improves yield in tobacco plants under salinity stress. *Plant Cell Rep.* **2016**, *35*, 37–41. [CrossRef] [PubMed]

38. Lu, W.; Tang, X.; Huo, Y.; Xu, R.; Qi, S.; Huang, J.; Zheng, C.; Wu, C.-A. Identification and characterization of fructose 1,6-bisphosphate aldolase genes in *Arabidopsis* reveal a gene family with diverse responses to abiotic stresses. *Gene* **2012**, *503*, 65–74. [CrossRef] [PubMed]

39. Sung, D.Y.; Vierling, E.; Guy, C.L. Comprehensive expression profile analysis of the Arabidopsis Hsp70 gene family. *Plant Physiol.* **2001**, *126*, 789–800. [CrossRef] [PubMed]

40. Weisberg, L.A.; Wimmers, L.E.; Turgeon, R. Photoassimilate-transport characteristics of nonchlorophyllous and green tissue in variegated leaves of *Coleus blumei* Benth. *Planta* **1988**, *175*, 1–8. [CrossRef] [PubMed]

41. Joshi, P.; Nayak, L.; Misra, A.N.; Biswal, B. Response of mature, developing and senescing chloroplasts to environmental stress. In *Plastid Development in Leaves during Growth and Senescence*; Springer: Berlin, Germany, 2013; pp. 641–668.

42. Tischner, R. Nitrate uptake and reduction in higher and lower plants. *Plant Cell Environ.* **2000**, *23*, 1005–1024. [CrossRef]

43. Bernard, S.M.; Habash, D.Z. The importance of cytosolic glutamine synthetase in nitrogen assimilation and recycling. *New Phytol.* **2009**, *182*, 608–620. [CrossRef] [PubMed]

44. McNally, S.F.; Hirel, B.; Gadal, P.; Mann, A.F.; Stewart, G.R. Glutamine synthetases of higher plants evidence for a specific isoform content related to their possible physiological role and their compartmentation within the leaf. *Plant Physiol.* **1983**, *72*, 22–25. [CrossRef] [PubMed]

45. Thomsen, H.C.; Eriksson, D.; Møller, I.S.; Schjoerring, J.K. Cytosolic glutamine synthetase: A target for improvement of crop nitrogen use efficiency? *Trends Plant Sci.* **2014**, *19*, 656–663. [CrossRef] [PubMed]

46. Pérez-García, A.; de Vicente, A.; Cantón, F.R.; Cazorla, F.M.; Codina, J.C.; García-Gutiérrez, Á.; Cánovas, F.M. Light-dependent changes of tomato glutamine synthetase in response to *Pseudomonas syringae* infection or phosphinothricin treatment. *Physiol. Plant.* **1998**, *102*, 377–384. [CrossRef]

47. Finnemann, J.; Schjoerring, J.K. Post-translational regulation of cytosolic glutamine synthetase by reversible phosphorylation and 14-3-3 protein interaction. *Plant J.* **2000**, *24*, 171–181. [CrossRef] [PubMed]

48. Santos, C.; Pereira, A.; Pereira, S.; Teixeira, J. Regulation of glutamine synthetase expression in sunflower cells exposed to salt and osmotic stress. *Sci. Hortic. Amst.* **2004**, *103*, 101–111. [CrossRef]

49. Tavernier, V.; Cadiou, S.; Pageau, K.; Laugé, R.; Reisdorf-Cren, M.; Langin, T.; Masclaux-Daubresse, C. The plant nitrogen mobilization promoted by *Colletotrichum lindemuthianum* in *Phaseolus* leaves depends on fungus pathogenicity. *J. Exp. Bot.* **2007**, *58*, 3351–3360. [CrossRef] [PubMed]

50. Borner, T.; Mendel, R.; Schiemann, J. Nitrate reductase is not accumulated in chloroplast-ribosome-deficient mutants of higher plants. *Planta* **1986**, *169*, 202–207. [CrossRef] [PubMed]

51. Kozaki, A.; Sakamoto, A.; Takeba, G. The promoter of the gene for plastidic glutamine synthetase (GS2) from rice is developmentally regulated and exhibits substrate-induced expression in transgenic tobacco plants. *Plant Cell Physiol.* **1992**, *33*, 233–238.

52. Cren, M.; Hirel, B. Glutamine synthetase in higher plants regulation of gene and protein expression from the organ to the cell. *Plant Cell Physiol.* **1999**, *40*, 1187–1193. [CrossRef]

53. Sukanya, R.; Li, M.-G.; Snustad, D.P. Root-and shoot-specific responses of individual glutamine synthetase genes of maize to nitrate and ammonium. *Plant Mol. Biol.* **1994**, *26*, 1935–1946. [CrossRef] [PubMed]

54. Ishiyama, K.; Inoue, E.; Watanabe-Takahashi, A.; Obara, M.; Yamaya, T.; Takahashi, H. Kinetic properties and ammonium-dependent regulation of cytosolic isoenzymes of glutamine synthetase in Arabidopsis. *J. Biol. Chem.* **2004**, *279*, 16598–16605. [CrossRef] [PubMed]

55. Ishiyama, K.; Inoue, E.; Tabuchi, M.; Yamaya, T.; Takahashi, H. Biochemical background and compartmentalized functions of cytosolic glutamine synthetase for active ammonium assimilation in rice roots. *Plant Cell Physiol.* **2004**, *45*, 1640–1647. [CrossRef] [PubMed]

56. Bondada, B.R.; Syvertsen, J.P. Leaf chlorophyll, net gas exchange and chloroplast ultrastructure in citrus leaves of different nitrogen status. *Tree Physiol.* **2003**, *23*, 553–559. [CrossRef] [PubMed]

57. Yang, H.; Xia, X.; Fang, W.; Fu, Y.; An, M.; Zhou, M. Identification of genes involved in spontaneous leaf color variation in *Pseudosasa japonica*. *Genet. Mol. Res.* **2015**, *14*, 11827–11840. [CrossRef] [PubMed]

58. Zavaleta-Mancera, H.; Franklin, K.; Ougham, H.; Thomas, H.; Scott, I. Regreening of senescent *Nicotiana* leaves I. Reappearance of NADPH-protochlorophyllide oxidoreductase and light-harvesting chlorophyll *a/b*-binding protein. *J. Exp. Bot.* **1999**, *50*, 1677–1682.

59. Plücken, H.; Müller, B.; Grohmann, D.; Westhoff, P.; Eichacker, L.A. The HCF136 protein is essential for assembly of the photosystem II reaction center in *Arabidopsis thaliana*. *FEBS Lett.* **2002**, *532*, 85–90. [CrossRef]

60. Yamamoto, Y. Quality control of photosystem II. *Plant Cell Physiol.* **2001**, *42*, 121–128. [CrossRef] [PubMed]

61. Hemmingsen, S.M. The plastid chaperonin. *Semin. Cell Biol.* **1990**, *1*, 47–54. [PubMed]

62. Gutteridge, S.; Gatenby, A.A. Rubisco synthesis, assembly, mechanism, and regulation. *Plant Cell* **1995**, *7*, 809–819. [CrossRef] [PubMed]

63. Suzuki, K.; Nakanishi, H.; Bower, J.; Yoder, D.W.; Osteryoung, K.W.; Miyagishima, S.-Y. Plastid chaperonin proteins Cpn60α and Cpn60β are required for plastid division in *Arabidopsis thaliana*. *BMC Plant Biol.* **2009**, *9*, 38. [CrossRef] [PubMed]

64. Raines, C.A.; Lloyd, J.C.; Dyer, T.A. Molecular biology of the C$_3$ photosynthetic carbon reduction cycle. *Photosynth. Res.* **1991**, *27*, 1–14. [CrossRef] [PubMed]

65. Raines, C.A. The Calvin cycle revisited. *Photosynth. Res.* **2003**, *75*, 1–10. [CrossRef] [PubMed]

66. Zhang, J.; Zhou, M.; Li, X.; Yu, X.; Jiang, C.; Sun, G. The single and interactive effects of nitrogen application rate and light condition on *Hosta* "Blue Umbrella" growth and photosynthetic characteristics. *Acta Hortic. Sin.* **2007**, *34*, 1497.

67. Zhou, M.; Zhang, J.; Li, M.; Wang, Z. The single and interactive effects of nitrogen application rate and light condition on *Hosta* "Francee" growth and photosynthetic characteristics. *J. Hebei For. Sci. Technol.* **2007**, 1–4.

68. Hikosaka, K. Interspecific difference in the photosynthesis-nitrogen relationship: patterns, physiological causes, and ecological importance. *J. Plant Res.* **2004**, *117*, 481–494. [CrossRef] [PubMed]

69. Luo, J.; Zhou, J.; Li, H.; Shi, W.; Polle, A.; Lu, M.; Sun, X.; Luo, Z.-B. Global poplar root and leaf transcriptomes reveal links between growth and stress responses under nitrogen starvation and excess. *Tree Physiol.* **2015**. [CrossRef] [PubMed]

70. Mokhele, B.; Zhan, X.; Yang, G.; Zhang, X. Review: Nitrogen assimilation in crop plants and its affecting factors. *Can. J. Plant Sci.* **2012**, *92*, 399–405. [CrossRef]

71. Guo, S.; Zhou, Y.; Gao, Y.; Li, Y.; Shen, Q. New insights into the nitrogen form effect on photosynthesis and photorespiration. *Pedosphere* **2007**, *17*, 601–610. [CrossRef]

72. Pogson, B.J.; Albrecht, V. Genetic dissection of chloroplast biogenesis and development: An overview. *Plant Physiol.* **2011**, *155*, 1545–1551. [CrossRef] [PubMed]

73. Albrecht, V.; Ingenfeld, A.; Apel, K. Characterization of the *snowy cotyledon 1* mutant of *Arabidopsis thaliana*: the impact of chloroplast elongation factor G on chloroplast development and plant vitality. *Plant Mol. Biol.* **2006**, *60*, 507–518. [CrossRef] [PubMed]

74. Wu, W.; Zhu, Y.; Ma, Z.; Sun, Y.; Quan, Q.; Li, P.; Hu, P.; Shi, T.; Lo, C.; Chu, I.K. Proteomic evidence for genetic epistasis: ClpR4 mutations switch leaf variegation to virescence in Arabidopsis. *Plant J.* **2013**, *76*, 943–956. [CrossRef] [PubMed]

75. Kato, Y.; Sakamoto, W. Plastid protein degradation during leaf development and senescence: Role of proteases and chaperones. In *Plastid Development in Leaves during Growth and Senescence*; Springer: Berlin, Germany, 2013; pp. 453–477.

76. Su, P.-H.; Li, H.-M. Stromal Hsp70 is important for protein translocation into pea and Arabidopsis chloroplasts. *Plant Cell* **2010**, *22*, 1516–1531. [CrossRef] [PubMed]

77. Latijnhouwers, M.; Xu, X.-M.; Møller, S.G. Arabidopsis stromal 70-kDa heat shock proteins are essential for chloroplast development. *Planta* **2010**, *232*, 567–578. [CrossRef] [PubMed]

78. Lee, U.; Rioflorido, I.; Hong, S.W.; Larkindale, J.; Waters, E.R.; Vierling, E. The Arabidopsis ClpB/Hsp100 family of proteins: chaperones for stress and chloroplast development. *Plant J.* **2007**, *49*, 115–127. [CrossRef] [PubMed]

79. Sjögren, L.L.; MacDonald, T.M.; Sutinen, S.; Clarke, A.K. Inactivation of the *clpC1* gene encoding a chloroplast Hsp100 molecular chaperone causes growth retardation, leaf chlorosis, lower photosynthetic activity, and a specific reduction in photosystem content. *Plant Physiol.* **2004**, *136*, 4114–4126. [CrossRef] [PubMed]

80. Zaltsman, A.; Ori, N.; Adam, Z. Two types of FtsH protease subunits are required for chloroplast biogenesis and photosystem II repair in Arabidopsis. *Plant Cell* **2005**, *17*, 2782–2790. [CrossRef] [PubMed]

81. Cederholm, S.N.; Bäckman, H.G.; Pesaresi, P.; Leister, D.; Glaser, E. Deletion of an organellar peptidasome PreP affects early development in *Arabidopsis thaliana*. *Plant Mol. Biol.* **2009**, *71*, 497–508. [CrossRef] [PubMed]

82. Goloubinoff, P.; Mogk, A.; Zvi, A.P.B.; Tomoyasu, T.; Bukau, B. Sequential mechanism of solubilization and refolding of stable protein aggregates by a bichaperone network. *PNAS* **1999**, *96*, 13732–13737. [CrossRef] [PubMed]

83. Clarke, A.K.; MacDonald, T.M.; Sjögren, L.L. The ATP-dependent Clp protease in chloroplasts of higher plants. *Physiol. Plant.* **2005**, *123*, 406–412. [CrossRef]

84. Guo, F.; Maurizi, M.R.; Esser, L.; Xia, D. Crystal structure of ClpA, an Hsp100 chaperone and regulator of ClpAP protease. *J. Biol. Chem.* **2002**, *277*, 46743–46752. [CrossRef] [PubMed]

85. Nakagawara, E.; Sakuraba, Y.; Yamasato, A.; Tanaka, R.; Tanaka, A. Clp protease controls chlorophyll *b* synthesis by regulating the level of chlorophyllide *a* oxygenase. *Plant J.* **2007**, *49*, 800–809. [CrossRef] [PubMed]

86. Yamasato, A.; Nagata, N.; Tanaka, R.; Tanaka, A. The N-terminal domain of chlorophyllide *a* oxygenase confers protein instability in response to chlorophyll *b* accumulation in Arabidopsis. *Plant Cell* **2005**, *17*, 1585–1597. [CrossRef] [PubMed]

87. Kmiec, B.; Glaser, E. A novel mitochondrial and chloroplast peptidasome, PreP. *Physiol. Plant.* **2012**, *145*, 180–186. [CrossRef] [PubMed]

88. Hao, L.S.M.F. Studies on the stage albescent phenomenon in tea the changes of rubpcase and proteinase. *Sci. Agric. Sin.* **1999**, *32*, 33–38.

89. Buchanan-Wollaston, V.; Earl, S.; Harrison, E.; Mathas, E.; Navabpour, S.; Page, T.; Pink, D. The molecular analysis of leaf senescence—A genomics approach. *Plant Biotechnol. J.* **2003**, *1*, 3–22. [CrossRef] [PubMed]

90. Olinares, P.D.B.; Kim, J.; van Wijk, K.J. The Clp protease system: A central component of the chloroplast protease network. *BBA Bioenerg.* **2011**, *1807*, 999–1011. [CrossRef] [PubMed]

91. Liu, X.; Yu, F.; Rodermel, S. Arabidopsis chloroplast FtsH, *var2* and suppressors of *var2* leaf variegation: A review. *J. Integr. Plant Biol.* **2010**, *52*, 750–761. [CrossRef] [PubMed]

92. Mayer, M.; Bukau, B. Hsp70 chaperones: cellular functions and molecular mechanism. *Cell. Mol. Life Sci.* **2005**, *62*, 670–684. [CrossRef] [PubMed]

93. Ling, Q.; Trösch, R.; Jarvis, P. The ins and outs of chloroplast protein transport. In *Plastid Development in Leaves during Growth and Senescence*; Springer: Berlin, Germany, 2013; pp. 239–280.

94. Rial, D.V.; Arakaki, A.K.; Ceccarelli, E.A. Interaction of the targeting sequence of chloroplast precursors with Hsp70 molecular chaperones. *Eur. J. Biochem.* **2000**, *267*, 6239–6248. [CrossRef] [PubMed]

95. Waegemann, K.; Paulsen, H.; Soll, J. Translocation of proteins into isolated chloroplasts requires cytosolic factors to obtain import competence. *FEBS Lett.* **1990**, *261*, 89–92. [CrossRef]

96. Kourtz, L.; Ko, K. The early stage of chloroplast protein import involves Com70. *J. Biol. Chem.* **1997**, *272*, 2808–2813. [CrossRef] [PubMed]

97. Kessler, F.; Blobel, G. Interaction of the protein import and folding machineries of the chloroplast. *PNAS* **1996**, *93*, 7684–7689. [CrossRef] [PubMed]

98. Zhang, J.; Shi, L.; Shi, A.; Zhang, Q. Photosynthetic responses of four *Hosta* cultivars to shade treatments. *Photosynthetica* **2004**, *42*, 213–218. [CrossRef]

99. Willis, O.O.; Mouti, M.E.; Sila, D.N.; Mwasaru, M.; Thiongo, G.; Murage, H.; Ojijo, N.O. Physico-chemical properties and antioxidant potential of syrup prepared from "Madhura"sweet sorghum (*Sorghum bicolor* L. Moench) cultivar grown at different locations in Kenya. *Sugar Tech.* **2013**, *15*, 263–270. [CrossRef]

100. Arnon, D.I. Copper enzymes in isolated chloroplasts. Polyphenoloxidase in *Beta vulgaris*. *Plant Physiol.* **1949**, *24*, 1–15. [CrossRef] [PubMed]

101. Lichtenthaler, H.; Wellburn, A. Determinations of total carotenoids and chlorophylls a and b of leaf extracts in different solvent. *Biochem. Soc. Trans.* **1983**, *603*, 591–593. [CrossRef]

102. Zhao, S.; Shi, G.; Dong, X. *Experiment Instruction of Plant Physiology*; China Agriculture Technology Press: Beijing, China, 2002.

103. Singh, R.P.; Srivastava, H.S. Increase in glutamate synthase (NADH) activity in maize seedlings in response to nitrate and ammonium nitrogen. *Physiol. Plant.* **1986**, *66*, 413–416. [CrossRef]

104. Wang, X.; Chen, S.; Zhang, H.; Shi, L.; Cao, F.; Guo, L.; Xie, Y.; Wang, T.; Yan, X.; Dai, S. Desiccation tolerance mechanism in resurrection fern-ally *Selaginella tamariscina* revealed by physiological and proteomic analysis. *J. Proteome Res.* **2010**, *9*, 6561–6577. [CrossRef] [PubMed]

105. Dai, S.; Li, L.; Chen, T.; Chong, K.; Xue, Y.; Wang, T. Proteomic analyses of *Oryza sativa* mature pollen reveal novel proteins associated with pollen germination and tube growth. *Proteomics* **2006**, *6*, 2504–2529. [CrossRef] [PubMed]

106. Sun, Y.; Wang, Q.; Li, Z.; Hou, L.; Dai, S.; Liu, W. Comparative proteomics of peanut gynophore development under dark and mechanical stimulation. *J. Proteome Res.* **2013**, *12*, 5502–5511. [CrossRef] [PubMed]

107. National Center for Biotechnology Information non-redundant (NCBInr) protein databases. Available online: http://www.ncbi.nlm.nih.gov/protein/ (accessed on 20 January 2010).

108. Matrix Science. Available online: http://www.matrixscience.com (accessed on 20 January 2010).

109. UniProt. Available online: http://www.uniprot.org/ (accessed on 15 August 2015).

110. Zhao, Q.; Gao, J.; Suo, J.; Chen, S.; Wang, T.; Dai, S. Cytological and proteomic analyses of horsetail (*Equisetum arvense* L.) spore germination. *Front. Plant Sci.* **2015**, *6*, 441. [CrossRef] [PubMed]

111. YLoc: Interpretable Subcellular Localization Prediction. Available online: http://abi.inf.uni-tuebingen.de/Services/YLoc/webloc.cgi (accessed on 27 August 2015).

112. LocTree3: Protein Subcellular Localization Prediction System. Available online: https://rostlab.org/services/loctree3/ (accessed on 27 August 2015).

113. Plant-mPLoc: Predicting Subcellular Localization of Plant Proteins Including Those with Multiple Sites. Available online: http://www.csbio.sjtu.edu.cn/bioinf/plant-multi/# (accessed on 27 August 2015).

114. ngLOC: A Bayesian Method for Predicting Protein Subcellular Localization. Available online: http://genome.unmc.edu/ngLOC/index.html (accessed on 27 August 2015).

115. TargetP 1.1 Server. Available online: http://www.cbs.dtu.dk/services/TargetP/ (accessed on 27 August 2015).

116. BoxShade: Pretty Printing and Shading of Multiple-Alignment Files. Available online: http://www.ch.embnet.org/software/BOX_form.html (accessed on 12 October 2015).

117. Cluster 3.0 Software. Available online: http://bonsai.hgc.jp/~mdehoon/software/cluster/software.htm (accessed on 20 October 2015).
118. Yu, J.; Chen, S.; Wang, T.; Sun, G.; Dai, S. Comparative proteomic analysis of *Puccinellia tenuiflora* leaves under Na$_2$CO$_3$ stress. *Int. J. Mol. Sci.* **2013**, *14*, 1740–1762. [CrossRef] [PubMed]

International Journal of
Molecular Sciences

Communication

Changes in the *Arabidopsis thaliana* Proteome Implicate cAMP in Biotic and Abiotic Stress Responses and Changes in Energy Metabolism

May Alqurashi [1,2], Chris Gehring [1] and Claudius Marondedze [2,*]

[1] Biological and Environmental Sciences and Engineering Division,
    King Abdullah University of Science and Technology, Thuwal 23955-6900, Saudi Arabia;
    may.qurashi@kaust.edu.sa (M.A.); christoph.gehring@kaust.edu.sa (C.G.)
[2] Cambridge Centre for Proteomics, Cambridge System Biology, Department of Biochemistry,
    University of Cambridge, Tennis Court Road, Cambridge CB2 1QR, UK
*   Correspondence: cm833@cam.ac.uk or cmarondedze@gmail.com; Tel.: +44-783-141-9678

Academic Editor: Setsuko Komatsu
Received: 20 April 2016; Accepted: 24 May 2016; Published: 1 June 2016

Abstract: The second messenger 3′,5′-cyclic adenosine monophosphate (cAMP) is increasingly recognized as having many different roles in plant responses to environmental stimuli. To gain further insights into these roles, *Arabidopsis thaliana* cell suspension culture was treated with 100 nM of cell permeant 8-bromo-cAMP for 5 or 10 min. Here, applying mass spectrometry and comparative proteomics, 20 proteins were identified as differentially expressed and we noted a specific bias in proteins with a role in abiotic stress, particularly cold and salinity, biotic stress as well as proteins with a role in glycolysis. These findings suggest that cAMP is sufficient to elicit specific stress responses that may in turn induce complex changes to cellular energy homeostasis.

Keywords: cAMP-dependent proteome; biotic stress; abiotic stress; glycolysis; TCA cycle

1. Introduction

Environmental factors such as biotic and abiotic stresses can cause constraints on the growth, development and productivity of plants. These stresses also disturb cellular homeostasis, and consequently, a rapid response is initiated to alleviate the impact of stress. Signaling molecules including second messengers such as cyclic nucleotides play an important role in this early phase. Cyclic nucleotides and 3′,5′-cyclic adenosine monophosphate (cAMP) in particular have long been established as important messengers in prokaryotes as well as in lower and higher eukaryotes [1]. In plants, cAMP has been reported to have direct and/or indirect roles in many developmental processes including pollen growth [2] and response to biotic stress [3–5].

The levels of cAMP have been shown to increase in response to biotic stress and subsequently influence calcium (Ca^{2+}) influx by targeting membrane cyclic nucleotide-gated channels (CNGCs) [6], thereby increasing cytosolic free Ca^{2+}. Plant CNGCs are ligand and voltage-gated channels functioning in sensory signal transduction and have been proposed to regulate Ca^{2+} influx into the cytosol [7,8]. The CNGCs also have a role in plant development and plant responses to biotic and abiotic stress [9]. CNGCs open upon binding of either cAMP or 3′,5′-cyclic guanosine monophosphate (cGMP) which act as potential activating ligands [8] and close upon binding of Ca^{2+}/calmodulin (Figure 1). The cyclic nucleotides and calmodulin competitively bind to overlapping binding sites at the C-terminus in the cytosolic part of the channel [10,11]. In turn, Ca^{2+} amplifies the signal as part of the cellular response [12]. In *Arabidopsis*, there are 20 annotated CNGCs [13,14] and CNGC2 in particular is an inward-rectifying potassium (K^{2+}) channel that is blocked by Ca^{2+} [8]. Mutation of CNGC2 results in an impaired hypersensitive response to avirulent pathogens and cAMP-/cGMP-dependent cytosolic Ca^{2+} elevation [15].

At the structural level, voltage-independent channels (VICs) in the plasma membrane of *Arabidopsis* root cells have been reported to be highly sensitive to cAMP and these channels are also implicated in salt tolerance in *Arabidopsis* seedlings by reducing sodium (Na^+) influx [16]. Furthermore, an increasing number of plant mononucleotide cyclases, including adenylate cyclases (enzymes that synthesis cAMP from ATP), have been predicted [17,18] and experimentally confirmed [2,19].

Mass spectrometry-based approaches have been employed previously to study cAMP in plants, firstly to determine the concentration of cAMP [20], secondly to look at the interactome network of cAMP [21], and finally to look at the cAMP-dependent responses [22]. In the latter study, systems-level analyses showed that upon treatment of *Arabidopsis* plants with 1 or 10 µM of cAMP, proteins involved in responses to temperature, light and photosynthesis were changing in abundance at 1 and 3 h after treatment. However, previous studies have used high concentrations of cAMP, which may not be physiologically relevant. Therefore, we have undertaken studying cAMP-dependent proteome changes at low cAMP concentration and monitoring early cellular responses with a view to gain further insight into the possible physiological implications at cellular levels.

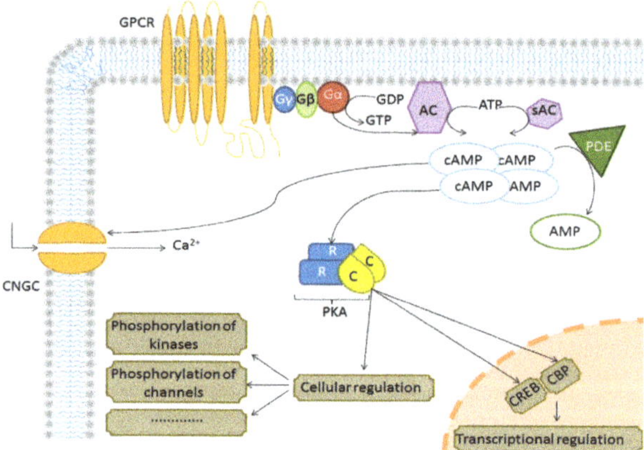

Figure 1. An illustration of some signal transduction pathways mediated by the G-protein coupled receptor based on the animal system. When adenylyl cyclase (AC) is activated by the Gα-subunit of the G-protein coupled receptor (GPCR), it catalyzes the formation of cAMP. Cyclic AMP then activates many substrates and kinases such as protein kinase A (PKA) which will regulate many biological processes. Cyclic nucleotide-gated channel (CNGC); soluble adenylyl cyclase (sAC); phosphodiesterase (PDE); cAMP response element-binding (CREB); CREB-binding protein (CBP); catalytic subunit of PKA (C); regulatory subunit of PKA (R).

2. Results and Discussion

In order to obtain a cAMP-dependent proteome, Arabidopsis ecotype Columbia-0 (Col-0) cell suspension cultures were treated with 100 nM of cell permeant 8-Br-cAMP and samples were collected at 5 and 10 min after treatment. Extracted proteins from three biological replicates were digested with trypsin and the peptides were labeled with tandem mass tag (TMT) six-plex for quantitative analysis. A total of 1023 quantifiable proteins were identified (false discovery rate: 0.8%). Proteins present in at least two biological replicates from each treatment time point were compared to their corresponding controls. Twenty proteins were detected as differentially expressed, with at least a ±1.5-fold change (±0.6 in $\log_2$ transformation) and a statistically significant p-value of ≤0.05 (Table 1). After 5 min of cAMP treatment, three proteins increased in abundance and four proteins decreased in abundance, while after 10 min of cAMP treatment, 10 proteins increased in abundance and four proteins decreased in abundance. In this study, only a few of the proteins fulfilled the set thresholds and it is of particular note that in addition to each protein being detected in at least two out of three biological replicates, proteins from cAMP-treated cells were compared with mock-treated cells collected at the same time point.

Table 1. 3′,5′-Cyclic adenosine monophosphate (cAMP)-responsive proteins after 5 and 10 min treatment.

Accession Number	Protein Name	Fold Change (Log$_2$)	p-Value	GO Term
Proteins identified changing after 5 min of cAMP treatment				
At3g16460	Jacalin-related lectin 34	4.495	0.01358	A
At2g01140	Fructose-bisphosphate aldolase 3	3.532	0.01473	B, D, E
At1g20450	Early response to dehydration 10 (ERD10)	0.971	0.04245	A, B, E
At2g37220	RNA-binding protein	−0.607	0.00180	A, C
At1g08110	Glyoxalase I	−0.651	0.00912	B, D, E
At1g14980	Chaperonin 10	−0.733	0.02203	
At2g27710	60S acidic ribosomal protein family	−1.476	0.03411	A
Proteins identified changing after 10 min of cAMP treatment				
At1g23100	GroES-like family protein	39.566	0.02049	
At1g24360	3-Oxoacyl-[acyl-carrier-protein] reductase	9.238	0.01110	A
At1g28200	FH interacting protein 1	2.979	0.04999	
At1g14980	Chaperonin 10	1.771	0.04747	
At1g48920	Nucleolin like 1	1.434	0.03283	
At1g53240	Mitochondrial malate dehydrogenase	0.846	0.03931	A, B, C, D, E
At2g38540	Lipid transfer protein 1	0.844	0.04096	
At2g41430	Early response to dehydration 15 (ERD15)	0.756	0.03081	B, C, D
At2g47730	Glutathione *S*-transferase Φ 8	0.726	0.04412	A, B, C
At3g16450	Jacalin-related lectin 33	0.628	0.02041	A
At1g11580	Methylesterase PCR A	−0.606	0.03673	C
At4g21860	Methionine sulfoxide reductase B2	−0.697	0.04911	
At4g38740	Rotamase cyclophilin 1 (ROC1)	−0.706	0.00733	D
At5g47200	RAB GTPase homolog 1A	−0.856	0.03223	

GO, gene ontology; A, response to cold (GO:0009409); B, response to salt stress (GO:0009651); C, response to bacterium (GO:0009617); D, response to cadmium ion (GO:0046686); E, glycolytic process (GO:0006096).

Gene ontology (GO) analysis using FatiGO$^+$ allowed classification of identified proteins in each group based on their enrichments as compared to the normal distribution [23,24]. Eight of the 20 differentially expressed proteins are in the category "response to cold", six proteins are in the category "response to salt stress", five proteins are in both "response to bacterium" and "response to cadmium ion" and four proteins are enriched in the category "glycolytic process" (Table 1). Although 24 proteins that are involved in the glycolytic process were identified (Table S1), only four were significantly changing in abundance at either 5 or 10 min after cAMP treatment. The four proteins include fructose-bisphosphate aldolase 3 (FBA3; At2g01140), mitochondrial malate dehydrogenase (MDH; At1g53240) and early response to dehydration 10 (ERD10; At1g20450) which increased in abundance while glyoxalase I (At1g08110) decreased (Figure 2).

Fructose-bisphosphate aldolase catalyzes two reversible reactions. The first involves the enzymatic conversion of fructose 1,6-bisphosphate into the triose phosphates, dihydroxyacetone phosphate and glyceraldehyde 3-phosphate in gluconeogenesis. The second reaction involves the condensation reaction of fructose-1,6-biphosphate and sedoheptulose-1,7-biphosphate in the Calvin cycle [25]. The mRNA levels of FBA3 have been shown to increase in response to abscisic acid (ABA) and salicylic acid (SA) and to various abiotic stresses such as salinity, drought, cold and heat, but the expression levels were noted to decrease in response to cadmium (Cd^{2+}) ions [26]. Increased abundance of FBA3 was also observed at 5 min after cAMP treatment (Table 1) and this implicates cAMP in abiotic stress responses.

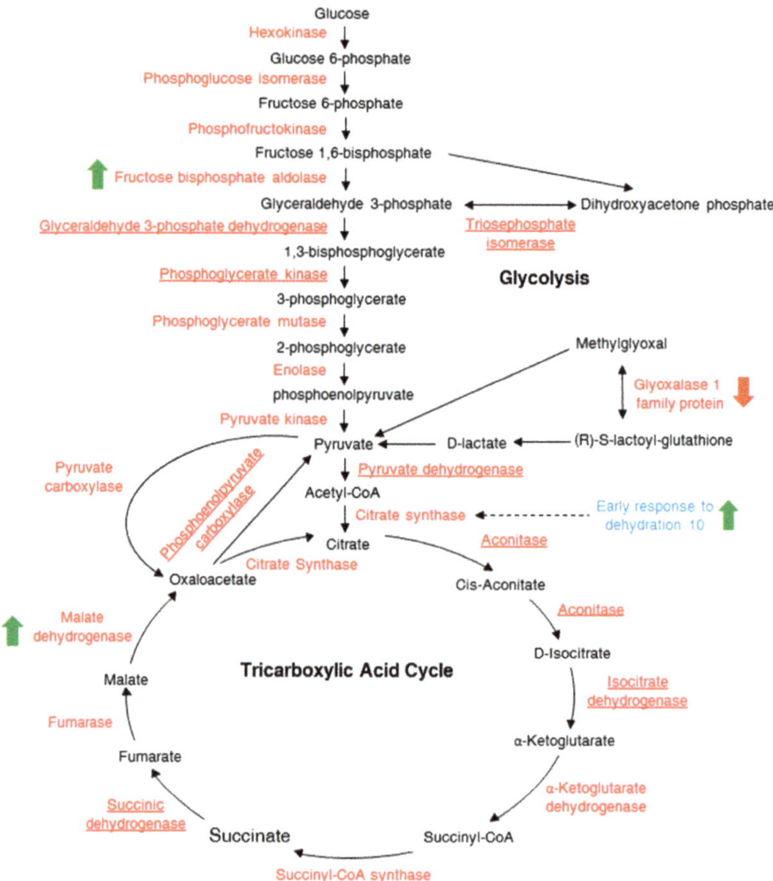

Figure 2. A schematic diagram of the glycolysis and tricarboxylic acid (TCA) cycle pathways showing proteins differentially expressed post cAMP treatment. The glycolysis and TCA metabolic products are shown in black, the enzymes are in red and a protein associated with the TCA cycle is shown in blue. The green arrows indicate proteins that accumulate in response to cAMP while the red arrow indicates a protein that is reduced in quantity. All the other proteins underlined are some of the proteins identified in the current study that are not changing in abundance after cAMP treatment.

Malate dehydrogenase reversibly catalyzes the oxidation of malate to oxaloacetate using the reduction of NAD^+ to NADH in the tricarboxylic acid (TCA) cycle [27]. The control of the TCA cycle activity is dependent on several enzymes, including aconitase, fumarase, succinate dehydrogenase, 2-oxoglutarate dehydrogenase and MDH, which all have different flux control coefficients [28]. A flux control coefficient is a quantifiable parameter measuring the contribution or effect of enzymes on the overall steady-state flux of a metabolic pathway. Mitochondrial MDH has the highest flux control coefficient (1.76) and is considered one of the rate-limiting steps [28]. Plant mitochondrial MDH is also important in oxidizing NADH in the TCA cycle and in the malate/aspartate shuttle in photorespiration [29]. Malate dehydrogenase has been shown to increase in abundance in response to oxidative stress [30] and flooding [31]. In tomato (*Solanum lycopersicum*), mutations in mitochondrial MDH affect growth and fruit yield [32], while in *Arabidopsis,* single and double knockouts of the two mitochondrial *MDH* genes appear to have little effect on growth and development [33]. In addition to MDH increasing in abundance in response to cAMP, ERD10 also increased in abundance in response to cAMP treatment (Table 1). An increase in abundance of ERD10 has previously been shown to have a protective role by slowing the heat-induced aggregation and/or inactivation rate of various substrates such as alcohol dehydrogenase and citrate synthase, the enzyme that catalyses the first step of the TCA cycle [34]. Given that increases in both MDH and ERD10 abundance have been observed to increase the respiration rate [29,34], an increase in MDH and ERD10 in response to cAMP implicates cAMP directly or indirectly in the modulation of the TCA cycle and, with it, energy metabolism.

In contrast, glyoxalase I, a protein that detoxifies methylglyoxal, a cytotoxic by-product of glycolysis [35], decreases in abundance in response to cAMP. In *Arabidopsis*, glyoxalase 1 (At1g08110) has been shown to decrease in abundance under abiotic stress conditions including salinity, drought, osmotic and cold [36]. Given that proteins ERD10, mitochondrial MDH and FBA3 increase and glyoxalase I decreases, we hypothesize that cAMP has a regulatory function in abiotic stress responses in general and the modulation of the TCA cycle in particular.

Of the 20 differentially expressed proteins identified in this study, five proteins, 3-oxoacyl-[acyl-carrier-protein] reductase (At1g24360), rotamase cyclophilin 1 (ROC1; At4g38740), RNA-binding protein (At2g37220), early response to dehydration 15 (ERD15; At2g41430) and mitochondrial MDH, have been associated with the response to ABA. Abscisic acid controls several aspects of development and adaptation to stress [37–39]. The 3-oxoacyl-[acyl-carrier-protein] reductase that catalyzes the first reduction step in fatty acid biosynthesis [40] is repressed by ABA in guard cells of *Arabidopsis* [41], but it increases in abundance in response to cAMP treatment in *Arabidopsis* cell suspension culture (Table 1). On the other hand, ROC1, a signaling protein component controlling plant responses to light, is an important link between phytochrome signaling and brassinosteroid sensitivity [42]. Rotamase cyclophilin 1 has been shown to decrease in abundance in response to abiotic stress and after ABA treatment [43], and this is consistent with the response to cAMP (Table 1).

In *Arabidopsis*, RNA-binding proteins have also been shown to play a role in ABA signaling during germination and drought tolerance [44], and ABA affects tyrosine dephosphorylation of a chloroplast-localized RNA-binding protein (At2g37220) in particular, and may modify its RNA-binding activity and thereby regulate gene expression [45]. The RNA-binding protein is related to the maize glycine-rich RNA-binding protein A, and the encoding gene has been shown to be induced by ABA and this is also a target for phosphorylation [46,47]. The protein has also been observed to increase in abundance at least two-fold in response to cold treatment [48]. However, in response to cAMP, the RNA-binding protein decreased in abundance (Table 1). It is therefore important to follow up on the phosphorylation status of this RNA-binding protein to see whether cAMP affects tyrosine dephosphorylation much like in the response to ABA.

Another protein that is involved in ABA responses is ERD15. In addition to being a negative regulator of ABA responses, ERD15 has a role in the defense against pathogens [38]. Besides the central role of ABA in controlling responses to abiotic stress stimuli, ABA also influences biotic stress responses and may interfere with signaling that is regulated by other hormones including SA [49]. ABA treatment prior to infection can increase the susceptibility of *Arabidopsis* to *Pseudomonas syringae* pv. *tomato*, while decreased ABA levels can improve SA-dependent defenses, suggesting that ABA modulates SA-dependent defense responses [50]. Expression of *ERD15* was observed to increase at least 50 times more in *Arabidopsis* inoculated with *Paenibacillus polymyxa* than in untreated plants [51]. Given that ERD15 is induced by biotic stress, rapidly but transiently induced in response to ABA [52] and increased in abundance in response to cAMP treatment, this again is consistent with the hypothesis that cAMP is part of the biotic stress response.

Glutathione *S*-transferase Φ8 (GSTF8, At2g47730), a marker for early stress and defense responses, also increases in abundance in response to cAMP (Table 1). The expression of GSTF8 can be induced by a range of biotic and abiotic stresses, hydrogen peroxide and in response to SA [53–57]. Overall this increase in abundance of GSTF8 upon induction by various stresses is also an indirect support for a role of cAMP in both biotic and abiotic stress responses.

The response to cAMP also leads to enrichment in the GO categories "response to cold" and "response to salt stress" (Table 1) and the proteins in these categories include the mitochondrial MDH, GSTF8, RNA-binding protein, 3-oxoacyl-(acyl-carrier-protein) reductase and ERD10. Interestingly, *ERD10* has been reported to be a general anti-stress protein that is up-regulated in response to a broad range of abiotic stresses [58] and was previously shown to increase in abundance after treatment of *Arabidopsis* leaves with 1 or 10 µM cAMP [22]. Furthermore, it is not uncommon for salinity stress-responsive proteins to also be responsive to Cd^{2+} ion stress and for both responses to confer a degree of cross-protection [59]. It is therefore not surprising to see that five out of the six proteins enriched in the category "response to salt stress" are also enriched in the category "response to cadmium ion" (Table 1).

Treatment of *Arabidopsis* roots with Cd^{2+} ions (10 µM) has been shown to alter the root proteome, e.g., an increase in the accumulation of proteins involved in the synthesis of glutathione-derived metal-binding proteins such as ATP sulfurylase, glutathione *S*-transferase, latex allergen-like proteins [60]. Interestingly, treatment of *Arabidopsis* seedlings with Cd^{2+} ions (50 µM) has been observed to increase the cellular cAMP concentration [61], and this in turn can induce the expression of enzymes involved in the phenylpropanoid pathway [62]. The phenylpropanoid pathway synthesizes precursors and metabolites protecting against abiotic stress including Cd^{2+} [63,64]. Taken together, the changing abundance of at least some proteins in response to Cd^{2+} may indeed require a direct or indirect interaction with cAMP.

What is the link between cAMP, Cd^{2+} and salinity? In *Arabidopsis*, an increase in cAMP levels significantly reduced Na^+ influx in roots and it was shown that VIC-mediated Na^+ currents are down-regulated by cAMP [16]. Further, treatment of *Arabidopsis* roots with 100 mM NaCl causes a decrease in cellular levels of cAMP, leading to a deactivation of protein kinase A. Thus, during salinity stress, cAMP suppresses expression of the Na^+ efflux pump [16]. In addition, five of the differentially expressed proteins enriched in the category "response to salt stress" (Table 1) have been previously observed to undergo differential expression in response to low Cd^{2+} stress [65]. Most of the proteins enriched in "response to cadmium ion" are also enriched in the category "response to salt stress", suggesting that cAMP has a role in both salinity and Cd^{2+} stress responses.

Twelve of the differentially expressed proteins identified (Table 1) have previously been linked to cabbage leaf curl virus infection [66]. Of these, seven show a similar differential accumulation pattern to cAMP (Table 2) consistent with a generalized role in plant defense against pathogens.

Table 2. Comparison between proteins responsive to cAMP and proteins involved in pathogen response.

Accession Number	Protein Name	cAMP Treatment		Pathogen Response *	
		Fold Change (Log₂)	*p*-Value	Fold Change (Log₂)	*p*-Value
AT3G16460	Jacalin-related lectin 34	4.495	0.01358	0.274	4.90×10^{-5}
AT2G01140	Fructose-bisphosphate aldolase 3	3.532	0.01473	0.372	1.48×10^{-5}
AT1G20450	Early response to dehydration 10 (ERD10)	0.971	0.04245	−0.756	7.00×10^{-6}
AT2G37220	RNA-binding protein	−0.607	0.00180	−0.539	2.35×10^{-5}
AT2G27710	60S acidic ribosomal protein family	−1.476	0.03411	−0.692	2.25×10^{-7}
AT1G24360	3-Oxoacyl-[acyl-carrier-protein] reductase	9.238	0.01110	−0.265	3.97×10^{-4}
AT1G53240	Mitochondrial malate dehydrogenase	0.846	0.03931	−0.599	1.11×10^{-5}
AT2G38540	Lipid transfer protein 1	0.844	0.04096	−2.701	5.04×10^{-25}
AT1G11580	Methylesterase PCR A	−0.606	0.03673	−0.194	1.86×10^{-3}
AT4G21860	Methionine sulfoxide reductase B2	−0.697	0.04911	−0.766	2.55×10^{-8}
AT4G38740	Rotamase cyclophilin 1 (ROC1)	−0.706	0.00733	−0.827	1.40×10^{-7}
AT5G47200	RAB GTPase homolog 1A	−0.856	0.03223	0.498	2.11×10^{-7}

* Proteins reported in [66].

Finally, a previous study using 1 or 10 μM cAMP treatments of *Arabidopsis* leaves for 1 or 3 h implicated cAMP in light and temperature responses [22]. Despite the much higher concentration and later time points used in the earlier study, three proteins, methylesterase PCR A (At1g11580), 60S acidic ribosomal protein family (At2g27710) and RNA-binding protein (At2g37220), were common and decreased in abundance in response to cAMP, adding to the growing evidence for a key role of cAMP in the transduction and/or modulation of environmental stimuli.

3. Materials and Methods

3.1. Plant Material and Growth Conditions

Arabidopsis thaliana ecotype Columbia-0 (Col-0) cell suspension culture was grown in 250 mL Erlenmeyer flasks containing 100 mL of Gamborg's B5 medium with vitamins (Sigma-Aldrich, St. Louis, MO, USA) [67] supplemented with 3% (w/v) sucrose, 0.05 μg·mL^{-1} (v/v) kinetin, 1 mg·mL^{-1} 2,4-dichlorophenoxyacetic acid and 0.05% (w/v) MES. Cells were grown in a growth chamber (Innova® 43, New Brunswick Scientific Co., Edison, NJ, USA) under photosynthetic light with 12 h light/12 h dark cycles at 23 °C and orbital agitation at 120 rpm and sub-cultured every 1-days.

3.2. cAMP Treatment and Protein Extraction

At 10 days post-subculturing, three biological replicate flasks were treated with 100 nM of 8-bromo-cAMP and cells of each mock (equal volume of water) or cAMP treated were collected at 0, 5 and 10 min post-treatment. Media were drained off using Stericup® filter unit (Millipore, Billerica, MA, USA), and the cells were immediately snap frozen in liquid nitrogen and stored at −80 °C until use. Approximately 1 g of cells was ground to a fine powder with mortar and pestle in liquid nitrogen and proteins were precipitated in trichloroacetic acid in acetone, vortexed and incubated overnight. Precipitated proteins were pelleted, washed and re-suspended in urea lysis buffer (7 M urea, 2 M thiourea, 4% (w/v) 3-[(3-Cholamidopropyl)dimethylammonio]-1-propanesulfonate). Approximately 100 μg of total soluble protein extract was reduced, alkylated, digested with trypsin and purified using Sep-Pak Vac tC18 100 mg cartridge (Waters, Milford, MA, USA), as described previously [68], prior to drying in a Speed Vac concentrator (Thermo Scientific, Bremen, Germany).

3.3. Peptide Labeling Using Tandem Mass Tag and Peptide Fractionation by OFFGEL Fractionator

Purified and dried tryptic peptides were labeled with tandem mass tag (TMT™) six-plex (Thermo Scientific, Bremen, Germany) according to manufacturer's instructions. Each biological replicate was labeled separately, pooled together and then fractionated using a 3100 OFFGEL fractionator (Agilent Technologies, Santa Clara, CA, USA) using 24-well high resolution immobilized pH gradient strips, pH 3–10, as described previously [68].

3.4. Protein Identification by LTQ Orbitrap and Quantification of Differentially Expressed Proteins

Dried peptide fractions were re-suspended in a solution containing 5% (v/v) acetonitrile and 0.1% (v/v) formic acid and analyzed by an LTQ-Orbitrap Velos (Thermo Scientific, Bremen, Germany) coupled with a nanoelectrospray ion source (Proxeon Biosystems, Odense, Denmark) for nano-LC–MS/MS analyses. The MS scan range was 350 to 1600 m/z with the normalized collision-induced dissociation at 35.0 V. The top 10 precursor ions were selected in the MS scan with resolution R = 60,000 for fragmentation in the linear ion trap. All spectra were submitted for protein identification to MASCOT search engine (Matrix Science, London, UK) as described previously [22], except that TMT labeling was added as a fixed modification. Identified proteins were evaluated and quantitated using Scaffold Q+ software, version 4.0.4 (Proteome Software, Portland, OR, USA). Proteins were considered as positive identifications if they were identified with a minimum of two unique peptides, a MASCOT ion score $\geqslant$26, a peptide probability of 95% and a protein threshold of 99%. Abundance levels of positively identified proteins from cAMP-treated cells that were present in at least two technical replicate were compared with mock-treated cells collected at the same time point. Differential expression of a protein was considered significant if the fold change was at least ± 1.5-fold change (± 0.6 in $\log_2$ transformation) and statistical significance p-value of $\leqslant 0.05$.

3.5. Computational Analysis of Functional Enrichment

The gene ontology (GO) and functional categorization analyses of the differentially expressed proteins were performed using FatiGO$^+$ tool in Babelomics version 5 suite [69,70].

4. Conclusions

cAMP treatment causes changes in the proteome that are diagnostic for biotic and abiotic responsive proteins. Furthermore, cAMP also affects abundance levels of enzymes in the glycolytic pathway and the TCA cycle and this is likely to have direct implications for the energy-transducing pathways and ATP generation. Finally, cAMP may conceivably link biotic and abiotic stress responses in stress-dependent changes of energy metabolism.

Supplementary Materials: Supplementary materials can be found at http://www.mdpi.com/1422-0067/17/6/852/s1.

Acknowledgments: We thank Ludivine Thomas for her intellectual input during the project.

Author Contributions: Chris Gehring, Claudius Marondedze and May Alqurashi designed the research. May Alqurashi performed the experiments and data analysis. All authors participated in writing and revising the manuscript and approved the final version.

Conflicts of Interest: The authors declare no conflict of interest.

References

1. Gancedo, J.M. Biological roles of cAMP: Variations on a theme in the different kingdoms of life. *Biol. Rev.* **2013**, *88*, 645–668. [CrossRef] [PubMed]

2. Moutinho, A.; Hussey, P.; Trewavas, A.; Malhó, R. Cyclic AMP acts as a second messenger in pollen tube growth and reorientation. *Proc. Natl. Acad. Sci. USA* **2001**, *98*, 10481–10486. [CrossRef] [PubMed]

3. Cooke, C.J.; Smith, C.J.; Walton, T.J.; Newton, R.P. Evidence that cyclic AMP is involved in the hypersensitive response of *Medicago sativa* to a fungal elicitor. *Phytochemistry* **1994**, *35*, 889–895. [CrossRef]

4. Kim, H.-S.; Park, S.-Y.; Lee, S.; Adams, E.L.; Czymmek, K.; Kang, S. Loss of cAMP-dependent protein kinase A affects multiple traits important for root pathogenesis by *Fusarium oxysporum*. *Mol. Plant Microbe Interact.* **2011**, *24*, 719–732. [CrossRef] [PubMed]

5. Lemtiri-Chlieh, F.; Thomas, L.; Marondedze, C.; Irving, H.; Gehring, C. Cyclic nucleotides and nucleotide cyclases in Plant stress responses. In *Abiotic Stress Response in Plants—Physiological, Biochemical and Genetic Perspectives*; Shanker, A., Venkateswarlu, B., Eds.; InTech: Vienna, Austria, 2011; p. 346p.

6. Ma, W.; Qi, Z.; Smigel, A.; Walker, R.; Verma, R.; Berkowitz, G. Ca^{2+}, cAMP, and transduction of non-self perception during plant immune responses. *Proc. Natl. Acad. Sci. USA* **2009**, *106*, 20995–21000. [CrossRef] [PubMed]

7. Lemtiri-Chlieh, F.; Berkowitz, G.A. Cyclic adenosine monophosphate regulates calcium channels in the plasma membrane of *Arabidopsis* leaf guard and mesophyll cells. *J. Biol. Chem.* **2004**, *279*, 35306–35312. [CrossRef] [PubMed]

8. Qi, Z.; Verma, R.; Gehring, C.; Yamaguchi, Y.; Zhao, Y.; Ryan, C.; Berkowitz, G. Ca^{2+} signaling by plant *Arabidopsis thaliana* Pep peptides depends on AtPepR1, a receptor with guanylyl cyclase activity, and cGMP-activated Ca^{2+} channels. *Proc. Natl. Acad. Sci. USA* **2010**, *107*, 21193–21198. [CrossRef] [PubMed]

9. Kugler, A.; Köhler, B.; Palme, K.; Wolff, P.; Dietrich, P. Salt-dependent regulation of a CNG channel subfamily in *Arabidopsis*. *BMC Plant Biol.* **2009**, *9*, 1–11. [CrossRef] [PubMed]

10. Kaplan, B.; Sherman, T.; Fromm, H. Cyclic nucleotide-gated channels in plants. *FEBS Lett.* **2007**, *581*, 2237–2246. [CrossRef] [PubMed]

11. Ladwig, F.; Dahlke, R.I.; Stührwohldt, N.; Hartmann, J.; Harter, K.; Sauter, M. Phytosulfokine regulates growth in *Arabidopsis* through a response module at the plasma membrane that includes CYCLIC NUCLEOTIDE-GATED CHANNEL17, H^+-ATPase, and BAK1. *Plant Cell* **2015**, *27*, 1718–1729. [CrossRef] [PubMed]
12. DeWald, D.B.; Torabinejad, J.; Jones, C.A.; Shope, J.C.; Cangelosi, A.R.; Thompson, J.E.; Prestwich, G.D.; Hama, H. Rapid Accumulation of Phosphatidylinositol 4,5-Bisphosphate and Inositol 1,4,5-Trisphosphate Correlates with Calcium Mobilization in Salt-Stressed *Arabidopsis*. *Plant Physiol.* **2001**, *126*, 759–769. [CrossRef] [PubMed]
13. Frietsch, S.; Wang, Y.-F.; Sladek, C.; Poulsen, L.R.; Romanowsky, S.M.; Schroeder, J.I.; Harper, J.F. A cyclic nucleotide-gated channel is essential for polarized tip growth of pollen. *Proc. Natl. Acad. Sci. USA* **2007**, *104*, 14531–14536. [CrossRef] [PubMed]
14. Konrad, K.R.; Hedrich, R. The use of voltage-sensitive dyes to monitor signal-induced changes in membrane potential–ABA triggered membrane depolarization in guard cells. *Plant J.* **2008**, *55*, 161–173. [CrossRef] [PubMed]
15. Ma, W.; Berkowitz, G.A. Ca^{2+} conduction by plant cyclic nucleotide gated channels and associated signaling components in pathogen defense signal transduction cascades. *New Phytol.* **2011**, *190*, 566–572. [CrossRef] [PubMed]
16. Maathuis, F.J.M.; Sanders, D. Sodium uptake in *Arabidopsis* roots is regulated by cyclic nucleotides. *Plant Physiol.* **2001**, *127*, 1617–1625. [CrossRef] [PubMed]
17. Gehring, C. Adenyl cyclases and cAMP in plant signaling—Past and present. *Cell Commun. Signal.* **2010**, *8*, 15–19. [CrossRef] [PubMed]
18. Marondedze, C.; Wong, A.; Thomas, L.; Irving, H.; Gehring, C. Cyclic Nucleotide Monophosphates in Plants and Plant Signaling. In *Handbook of Experimental Pharmacology*; Barrett, J.E., Flockerzi, V., Frohman, M.A., Geppetti, P., Hofmann, F.B., Michel, M.C., Page, C.P., Thorburn, A.M., Wang, K., Eds.; Springer: Berlin, Germany; Heidelberg, Germany, 2016; pp. 1–17.
19. Al-Younis, I.; Wong, A.; Gehring, C. The *Arabidopsis thaliana* K^+-uptake permease 7 (*AtKUP7*) contains a functional cytosolic adenylate cyclase catalytic centre. *FEBS Lett.* **2015**, *589*, 3848–3852. [CrossRef] [PubMed]
20. Newton, R.; Gibbs, N.; Moyse, C.; Wiebers, J.; Brown, E. Mass spectrometric identification of adenosine 3′:5′-cyclic monophosphate isolated from a higher plant tissue. *Phytochemistry* **1980**, *19*, 1909–1911. [CrossRef]
21. Donaldson, L.; Meier, S.; Gehring, C. The *Arabidopsis* cyclic nucleotide interactome. *Cell Commun. Signal.* **2016**, *14*, 10–27. [CrossRef] [PubMed]
22. Thomas, L.; Marondedze, C.; Ederli, L.; Pasqualini, S.; Gehring, C. Proteomic signatures implicate cAMP in light and temperature responses in *Arabidopsis thaliana*. *J. Proteom.* **2013**, *83*, 47–59. [CrossRef] [PubMed]
23. Meier, S.; Gehring, C. A guide to the integrated application of on-line data mining tools for the inference of gene functions at the systems level. *Biotechnol. J.* **2008**, *3*, 1375–1387. [CrossRef] [PubMed]
24. Alqurashi, M.; Meier, S. Inferring biological functions of guanylyl cyclases with computational methods. In *Cyclic Nucleotide Signaling in Plants*; Gehring, C., Ed.; Humana Press: Totowa, NJ, USA, 2013; Volume 1016, pp. 225–234.
25. Flechner, A.; Gross, W.; Martin, W.F.; Schnarrenberger, C. Chloroplast class I and class II aldolases are bifunctional for fructose-1,6-biphosphate and sedoheptulose-1,7-biphosphate cleavage in the Calvin cycle. *FEBS Lett.* **1999**, *447*, 200–202. [CrossRef]
26. Lu, W.; Tang, X.; Huo, Y.; Xu, R.; Qi, S.; Huang, J.; Zheng, C.; Wu, C. Identification and characterization of fructose 1,6-bisphosphate aldolase genes in *Arabidopsis* reveal a gene family with diverse responses to abiotic stresses. *Gene* **2012**, *503*, 65–74. [CrossRef] [PubMed]
27. Nunes-Nesi, A.; Carrari, F.; Lytovchenko, A.; Smith, A.M.O.; Ehlers Loureiro, M.; Ratcliffe, R.G.; Sweetlove, L.J.; Fernie, A.R. Enhanced photosynthetic performance and growth as a consequence of decreasing mitochondrial malate dehydrogenase activity in transgenic tomato plants. *Plant Physiol.* **2005**, *137*, 611–622. [CrossRef] [PubMed]
28. Araújo, W.L.; Nunes-Nesi, A.; Nikoloski, Z.; Sweetlove, L.J.; Fernie, A.R. Metabolic control and regulation of the tricarboxylic acid cycle in photosynthetic and heterotrophic plant tissues. *Plant Cell Environ.* **2012**, *35*, 1–21. [CrossRef] [PubMed]
29. Scheibe, R. Malate valves to balance cellular energy supply. *Physiol. Plant.* **2004**, *120*, 21–26. [CrossRef] [PubMed]

30. Taylor, N.L.; Heazlewood, J.L.; Day, D.A.; Millar, A.H. Differential impact of environmental stresses on the pea mitochondrial proteome. *Mol. Cell. Proteom.* **2005**, *4*, 1122–1133. [CrossRef] [PubMed]

31. Komatsu, S.; Yamamoto, A.; Nakamura, T.; Nouri, M.-Z.; Nanjo, Y.; Nishizawa, K.; Furukawa, K. Comprehensive analysis of mitochondria in roots and hypocotyls of soybean under flooding stress using proteomics and metabolomics techniques. *J. Proteome Res.* **2011**, *10*, 3993–4004. [CrossRef] [PubMed]

32. Nunes-Nesi, A.; Araújo, W.L.; Fernie, A.R. Targeting mitochondrial metabolism and machinery as a means to enhance photosynthesis. *Plant Physiol.* **2011**, *155*, 101–107. [CrossRef] [PubMed]

33. Tronconi, M.A.; Fahnenstich, H.; Gerrard Weehler, M.C.; Andreo, C.S.; Flügge, U.-I.; Drincovich, M.F.; Maurino, V.G. *Arabidopsis* NAD-malic enzyme functions as a homodimer and heterodimer and has a major impact on nocturnal metabolism. *Plant Physiol.* **2008**, *146*, 1540–1552. [CrossRef] [PubMed]

34. Kovacs, D.; Kalmar, E.; Torok, Z.; Tompa, P. Chaperone activity of ERD10 and ERD14, two disordered stress-related plant proteins. *Plant Physiol.* **2008**, *147*, 381–390. [CrossRef] [PubMed]

35. Thornalley, P.J. Glyoxalase I—Structure, function and a critical role in the enzymatic defence against glycation. *Biochem. Soc. Trans.* **2003**, *31*, 1343–1348. [CrossRef] [PubMed]

36. Mustafiz, A.; Singh, A.; Pareek, A.; Sopory, S.; Singla-Pareek, S. Genome-wide analysis of rice and *Arabidopsis* identifies two glyoxalase genes that are highly expressed in abiotic stresses. *Funct. Integr. Genom.* **2011**, *11*, 293–305. [CrossRef] [PubMed]

37. Finkelstein, R.R.; Rock, C.D. Abscisic acid biosynthesis and response. *Arab. Book* **2002**, *1*, e0058. [CrossRef] [PubMed]

38. Cutler, S.R.; Rodriguez, P.L.; Finkelstein, R.R.; Abrams, S.R. Abscisic acid: Emergence of a core signaling network. *Annu. Rev. Plant Biol.* **2010**, *61*, 651–679. [CrossRef] [PubMed]

39. Zhu, J.-K. Salt and drought stress signal transduction in plants. *Annu. Rev. Plant Biol.* **2002**, *53*, 247–273. [CrossRef] [PubMed]

40. Harwood, J.L. Fatty acid metabolism. *Annu. Rev. Plant Physiol. Plant Mol. Biol.* **1988**, *39*, 101–138. [CrossRef]

41. Leonhardt, N.; Kwak, J.M.; Robert, N.; Waner, D.; Leonhardt, G.; Schroeder, J.I. Microarray expression analyses of *Arabidopsis* guard cells and isolation of a recessive abscisic acid hypersensitive protein phosphatase 2C mutant. *Plant Cell* **2004**, *16*, 596–615. [CrossRef] [PubMed]

42. Trupkin, S.A.; Mora-García, S.; Casal, J.J. The cyclophilin ROC1 links phytochrome and cryptochrome to brassinosteroid sensitivity. *Plant J.* **2012**, *71*, 712–723. [CrossRef] [PubMed]

43. Matsui, A.; Ishida, J.; Morosawa, T.; Mochizuki, Y.; Kaminuma, E.; Endo, T.A.; Okamoto, M.; Nambara, E.; Nakajima, M.; Kawashima, M.; *et al. Arabidopsis* transcriptome analysis under drought, cold, high-salinity and ABA treatment conditions using a tiling array. *Plant Cell Physiol.* **2008**, *49*, 1135–1149. [CrossRef] [PubMed]

44. Kuhn, J.M.; Schroeder, J.I. Impacts of altered RNA metabolism on abscisic acid signaling. *Curr. Opin. Plant Biol.* **2003**, *6*, 463–469. [CrossRef]

45. Ghelis, T.; Bolbach, G.; Clodic, G.; Habricot, Y.; Miginiac, E.; Sotta, B.; Jeannette, E. Protein tyrosine kinases and protein tyrosine phosphatases are involved in abscisic acid-dependent processes in *Arabidopsis* seeds and suspension cells. *Plant Physiol.* **2008**, *148*, 1668–1680. [CrossRef] [PubMed]

46. Gomez, J.; Sanchez-Martinez, D.; Stiefel, V.; Rigau, J.; Puigdomenech, P.; Pages, M. A gene induced by the plant hormone abscisic acid in response to water stress encodes a glycine-rich protein. *Nature* **1988**, *334*, 262–264. [CrossRef] [PubMed]

47. Testi, M.G.; Croce, R.; Laureto, P.P.-D.; Bassi, R. A CK2 site is reversibly phosphorylated in the photosystem II subunit CP29. *FEBS Lett.* **1996**, *399*, 245–250. [CrossRef]

48. Amme, S.; Matros, A.; Schlesier, B.; Mock, H.-P. Proteome analysis of cold stress response in *Arabidopsis thaliana* using DIGE-technology. *J. Exp. Bot.* **2006**, *57*, 1537–1546. [CrossRef] [PubMed]

49. Mauch-Mani, B.; Mauch, F. The role of abscisic acid in plant–pathogen interactions. *Curr. Opin. Plant Biol.* **2005**, *8*, 409–414. [CrossRef] [PubMed]

50. Mohr, P.G.; Cahill, D.M. Abscisic acid influences the susceptibility of *Arabidopsis thaliana* to *Pseudomonas syringae* pv. *tomato* and *Peronospora parasitica. Funct. Plant Biol.* **2003**, *30*, 461–469. [CrossRef]

51. Timmusk, S.; Wagner, E.G.H. The plant-growth-promoting rhizobacterium *Paenibacillus polymyxa* induces changes in *Arabidopsis thaliana* gene expression: A possible connection between biotic and abiotic stress responses. *Mol. Plant-Microbe Interact.* **1999**, *12*, 951–959. [CrossRef] [PubMed]

52. Kariola, T.; Brader, G.; Helenius, E.; Li, J.; Heino, P.; Palva, E.T. Early Response to Dehydration 15. A negative Regulator of ABA-responses in *Arabidopsis*. *Plant Physiol.* **2006**, *142*, 1559–1573. [CrossRef] [PubMed]

53. Chen, W.; Chao, G.; Singh, K.B. The promoter of a H_2O_2-inducible, *Arabidopsis* glutathione *S*-transferase gene contains closely linked OBF- and OBP1-binding sites. *Plant J.* **1996**, *10*, 955–966. [CrossRef] [PubMed]

54. Chen, W.; Singh, K.B. The auxin, hydrogen peroxide and salicylic acid induced expression of the *Arabidopsis GST6* promoter is mediated in part by an ocs element. *Plant J.* **1999**, *19*, 667–677. [CrossRef] [PubMed]

55. Wagner, U.; Edwards, R.; Dixon, D.; Mauch, F. Probing the diversity of the *Arabidopsis* glutathione *S*-transferase gene family. *Plant Mol. Biol.* **2002**, *49*, 515–532. [CrossRef] [PubMed]

56. Uquillas, C.; Letelier, I.; Blanco, F.; Jordana, X.; Holuigue, L. NPR1-Independent Activation of Immediate Early Salicylic Acid-Responsive Genes in *Arabidopsis*. *Mol. Plant Microbe Interact.* **2004**, *17*, 34–42. [CrossRef] [PubMed]

57. Perl-Treves, R.; Foley, R.C.; Chen, W.; Singh, K.B. Early Induction of the *Arabidopsis GSTF8* Promoter by Specific Strains of the Fungal Pathogen *Rhizoctonia solani*. *Mol. Plant Microbe Interact.* **2004**, *17*, 70–80. [CrossRef] [PubMed]

58. Kimura, M.; Yamamoto, Y.Y.; Seki, M.; Sakurai, T.; Sato, M.; Abe, T.; Yoshida, S.; Manabe, K.; Shinozaki, K.; Matsui, M. Identification of *Arabidopsis* genes regulated by high light-stress using cDNA microarray. *Photochem. Photobiol.* **2003**, *77*, 226–233. [CrossRef]

59. Smýkalová, I.; Zámečníková, B. The relationship between salinity and cadmium stress in barley. *Biol. Plant.* **2003**, *46*, 269–273. [CrossRef]

60. Roth, U.; von Roepenack-Lahaye, E.; Clemens, S. Proteome changes in *Arabidopsis thaliana* roots upon exposure to Cd^{2+}. *J. Exp. Bot.* **2006**, *57*, 4003–4013. [CrossRef] [PubMed]

61. Pietrowska-Borek, M.; Nuc, K.; Chadzinikolau, T.; Lechowska, K. Cadmium stress causes accumulation of cAMP and this signal molecule induces phenylpropanoid pathway in *Arabidopsis thaliana* seedlings. In Proceedings of the First Polish-German Biochemical Societies Joint Meeting, Poznon, Poland, 11–14 September 2012. poster 4.30.

62. Pietrowska-Borek, M.; Nuc, K. Both cyclic-AMP and cyclic-GMP can act as regulators of the phenylpropanoid pathway in *Arabidopsis thaliana* seedlings. *Plant Physiol. Biochem.* **2013**, *70*, 142–149. [CrossRef] [PubMed]

63. Dixon, R.A.; Paiva, N.L. Stress-induced phenylpropanoid metabolism. *Plant Cell* **1995**, *7*, 1085–1097. [CrossRef] [PubMed]

64. Ferrer, J.-L.; Austin, M.B.; Stewart, C., Jr.; Noel, J.P. Structure and function of enzymes involved in the biosynthesis of phenylpropanoids. *Plant Physiol. Biochem.* **2008**, *46*, 356–370. [CrossRef] [PubMed]

65. Skórzyńska-Polit, E.; Drażkiewicz, M.; Krupa, Z. The Activity of the Antioxidative System in Cadmium-Treated *Arabidopsis thaliana*. *Biol. Plant.* **2003**, *47*, 71–78. [CrossRef]

66. Ascencio-Ibáñez, J.T.; Sozzani, R.; Lee, T.J.; Chu, T.M.; Wolfinger, R.D.; Cella, R.; Hanley-Bowdoin, L. Global analysis of *Arabidopsis* gene expression uncovers a complex array of changes impacting pathogen response and cell cycle during geminivirus infection. *Plant Physiol.* **2008**, *148*, 436–454. [CrossRef] [PubMed]

67. Gamborg, O.L.; Miller, R.A.; Ojima, K. Nutrient requirements of suspension cultures of soybean root cells. *Exp. Cell Res.* **1968**, *50*, 151–158. [CrossRef]

68. Marondedze, C.; Wong, A.; Groen, A.; Serrano, N.; Jankovic, B.; Lilley, K.; Gehring, C.; Thomas, L. Exploring the *Arabidopsis* proteome: Influence of protein solubilization buffers on proteome coverage. *Int. J. Mol. Sci.* **2015**, *16*, 857–870. [CrossRef] [PubMed]

69. Al-Shahrour, F.; Díaz-Uriarte, R.; Dopazo, J. FatiGO: A web tool for finding significant associations of Gene Ontology terms with groups of genes. *Bioinformatics* **2004**, *20*, 578–580. [CrossRef] [PubMed]

70. BabelomicsV5. Available online: http://babelomics.bioinfo.cipf.es/ (accessed on 1 September 2015).

International Journal of
Molecular Sciences

Article

Aluminum Toxicity-Induced Alterations of Leaf Proteome in Two Citrus Species Differing in Aluminum Tolerance

Huan Li [1,2], Lin-Tong Yang [1,2], Yi-Ping Qi [3], Peng Guo [1,2], Yi-Bin Lu [1,2] and Li-Song Chen [1,2,4,*]

[1] Institute of Plant Nutritional Physiology and Molecular Biology, College of Resources and Environment, Fujian Agriculture and Forestry University, Fuzhou 350002, China; lihuanenglish123@126.com (H.L.); talstoy@163.com (L.-T.Y.); 6253730@163.com (P.G.); yibin.07@163.com (Y.-B.L.)

[2] Fujian Provincial Key Laboratory of Soil Environmental Health and Regulation, College of Resources and Environment, Fujian Agriculture and Forestry University, Fuzhou 350002, China

[3] Institute of Materia Medica, Fujian Academy of Medical Sciences, Fuzhou 350001, China; qiyiping2008@hotmail.com

[4] The Higher Educational Key Laboratory of Fujian Province for Soil Ecosystem Health and Regulation, Fujian Agriculture and Forestry University, Fuzhou 350002, China

[*] Correspondence: lisongchen2002@hotmail.com; Tel.: +86-591-8385-4966; Fax: +86-591-8377-6849

Academic Editor: Setsuko Komatsu
Received: 13 May 2016; Accepted: 13 July 2016; Published: 21 July 2016

Abstract: Seedlings of aluminum-tolerant 'Xuegan' (*Citrus sinensis*) and Al-intolerant 'sour pummelo' (*Citrus grandis*) were fertigated for 18 weeks with nutrient solution containing 0 and 1.2 mM $AlCl_3 \cdot 6H_2O$. Al toxicity-induced inhibition of photosynthesis and the decrease of total soluble protein only occurred in *C. grandis* leaves, demonstrating that *C. sinensis* had higher Al tolerance than *C. grandis*. Using isobaric tags for relative and absolute quantification (iTRAQ), we obtained more Al toxicity-responsive proteins from *C. sinensis* than from *C. grandis* leaves, which might be responsible for the higher Al tolerance of *C. sinensis*. The following aspects might contribute to the Al tolerance of *C. sinensis*: (a) better maintenance of photosynthesis and energy balance via inducing photosynthesis and energy-related proteins; (b) less increased requirement for the detoxification of reactive oxygen species and other toxic compounds, such as aldehydes, and great improvement of the total ability of detoxification; and (c) upregulation of low-phosphorus-responsive proteins. Al toxicity-responsive proteins related to RNA regulation, protein metabolism, cellular transport and signal transduction might also play key roles in the higher Al tolerance of *C. sinensis*. We present the global picture of Al toxicity-induced alterations of protein profiles in citrus leaves, and identify some new Al toxicity-responsive proteins related to various biological processes. Our results provide some novel clues about plant Al tolerance.

Keywords: aluminum toxicity; *Citrus grandis*; *Citrus sinensis*; iTRAQ; leaves; proteome

---

1. Introduction

Aluminum is the most abundant metallic element in the Earth's crust [1]. In neutral or slightly acidic soils, Al is mainly in the form of insoluble deposits and is biologically inactive. In acidic solutions (pH < 5.0), Al exists in the forms of Al^{3+} and $Al(OH)^{2+}$, which are soluble and available to plants [2]. Because micromolar concentration of Al^{3+} can rapidly inhibit root growth, Al toxicity is a major factor limiting crop productivity in many acidic soils through the tropics and subtropics. Over 50% of the world's potential arable lands are acidic [3,4]. Moreover, the acidity of the soils is gradually increasing due to environmental problems, including acid deposition, improper application of chemical fertilizers, intensive agriculture and monoculture [5].

Higher plants have evolved two main mechanisms of Al detoxification (i.e., external and internal detoxification mechanisms) that enable them to tolerate high levels of Al. However, the molecular mechanisms for Al tolerance in higher plants are not fully understood [4,6–9]. The alteration of proteins is an important process to cope with elevated Al in higher plants [7,10–13]. Proteome analysis is becoming a powerful tool for the investigation of protein roles in higher plants. There have been several reports investigating Al toxicity-responsive proteins using isobaric tags for relative and absolute quantification (iTRAQ) or two-dimensional gel electrophoresis (2-DE). Wang et al. used iTRAQ to isolate 106 differentially-abundant proteins from Al-treated rice roots [11], indicating that Al toxicity-induced activation of the glycolysis/gluconeogenesis shunt might be a rapid and effective way to balance the available energy levels to prevent the Al toxicity-induced shortage of intracellular energy. Using iTRAQ, Jiang et al. isolated more Al toxicity-responsive proteins from Al-tolerant *Citrus sinensis* roots than from Al-intolerant *Citrus grandis* roots [10], suggesting that the higher metabolic flexibility might contribute to the higher Al tolerance of *C. sinensis*. Zhen et al. used 2-DE to investigate the Al toxicity-induced alterations of root proteome in an Al-resistant soybean cultivar and concluded that proteins related to stress/defense, signal transduction, transport, protein folding, gene regulation and primary metabolism were critical for plant survival under Al toxicity [14]. Yang et al. used 2-DE to identify 12 upregulated and five downregulated proteins from Al toxicity-exposed rice roots. Further analysis showed that cysteine synthase (CS) played a key role in rice Al tolerance [12]. Duressa et al. observed that Al toxicity-induced alterations of root protein profiles differed between Al-tolerant and -sensitive soybean genotypes, concluding that enzymes involved in organic acid biosynthesis and detoxification systems played crucial roles in soybean Al tolerance [15]. Dai et al. isolated 35 proteins related to Al tolerance from Al-tolerant wild barley roots [16]. In tomato roots, proteins involved in detoxification were induced by Al toxicity [17]. Oh et al. showed that Al toxicity increased the abundance of 19 proteins, including *S*-adenosylmethionine synthetase, oxalate oxidase (OXO), malate dehydrogenase (MDH), citrate synthase and ascorbate peroxidase (APX), and decreased the abundance of 28 proteins, including heat shock protein (HSP) 70, *O*-methyltransferase 4 and enolase, in wheat roots [18]. All of these studies, however, have focused on root proteomics, because the inhibition of root growth is one of the earliest and most easily-observed symptoms of Al toxic damage [1]. To our knowledge, data available on the effects of Al toxicity on leaf proteomics are rare [7]. In a study, Rahman et al. used 2-DE to identify eight upregulated proteins, including glutamine synthetase and peroxiredoxin, and nine downregulated ones, including ribulose-1,5-bisphosphate carboxylase/oxygenase (RuBisCO) in high Al-treated leaves [19]. In another study, Yang et al. identified 92 proteins in Al or NO-treated rice leaves. Further analyses confirmed that NO alleviated Al toxicity-induced reactive oxygen species (ROS) and reactive nitrogen species (RNS) toxicities by increasing the activity and abundance of antioxidant enzymes [13].

Citrus is mainly grown in acidic and strong acidic soils. Low pH and high Al are the factors causing the poor growth and decreased lifespan of citrus trees [20]. In 2011, we investigated the pH of 319 soil samples from Pinghe pummelo (*Citrus grandis*) orchards, located in Zhangzhou, China. The pH ranged from 3.26 to 6.22 with an average value of 4.34. Up to 90.0% of the soils displayed a pH of less than 5.0 [21]. During 1998–1999, Huang et al. assayed the pH of 200 soil samples from Pinghe pummelo orchards. The pH ranged from 3.57 to 7.25 with an average value of 4.63. Eighty-five point five percent of the soils had a lower pH than 5.0 [22]. Soil acidification has been occurring rapidly in pummelo orchards in the last decade. Therefore, understanding the mechanisms of Al toxicity and Al tolerance in citrus plants is very important for citrus production. Previously, we investigated the Al toxicity-responsive proteins in the roots of two citrus species differing in Al tolerance [10]. On this basis, we further examined the Al toxicity-induced alterations of gas exchange, Al and total soluble protein concentrations and protein profiles revealed by iTRAQ in Al-tolerant *C. sinensis* and Al-intolerant *C. grandis* leaves [23]. The objectives were (a) to understand the molecular mechanisms of citrus in dealing with Al toxicity at the protein level and (b) to identify proteins conferring Al tolerance in citrus.

2. Results

2.1. Leaf Gas Exchange, Al and Total Soluble Protein Concentrations

We found that +Al *C. grandis* leaves had decreased CO_2 assimilation and stomatal conductance, but increased intercellular CO_2 concentration (Figure 1A–C), implying that Al toxicity-induced inhibition of photosynthesis in *C. grandis* was mainly limited by non-stomatal factors. However, Al toxicity had no significant influence on *C. sinensis* leaf gas exchange. Leaf gas exchange parameters did not significantly differ between the two citrus species at each given Al level, except that both CO_2 assimilation and stomatal conductance were lower in +Al *C. grandis* leaves than in +Al *C. sinensis* leaves (Figure 1A–C). In *C. grandis*, the Al level was higher in +Al leaves than in controls. In *C. sinensis*, the Al level showed an increased trend in +Al leaves, but did not significantly differ between +Al and control leaves. No significant difference was observed in leaf Al concentration between the two citrus species at each given Al treatment (Figure 1D). Besides, Al toxicity only decreased the total soluble protein level in *C. grandis* leaves (Figure 1E). Therefore, *C. sinensis* was more tolerant to Al toxicity than *C. grandis*. A similar result has been obtained by Yang et al. [23] and Jiang et al. [10].

2.2. Leaf Al Toxicity-Responsive Proteins

We produced a total of 642,359 and 656,455 spectra from *C. sinensis* and *C. grandis* leaves respectively using iTRAQ (Figures S1A and S2A). Unlike total spectra, the numbers of peptide spectrum matches, peptides identified, unique peptides, proteins identified and protein group(s) identified were higher in *C. sinensis* leaves than in *C. grandis* ones. The protein number in the two citrus species decreased with increased number of peptides that matched to proteins. The number of proteins with the same number of peptides was lower in *C. grandis* leaves than in *C. sinensis* ones (Figures S1B and S2B).

Protein mass distribution, the distribution of sequence coverage and the distribution of peptide length were similar between *C. sinensis* and *C. grandis* leaves. Proteins with 20–30 kDa were the most abundant (Figures S1C and S2C). The protein number in the two citrus species decreased with increased sequence coverage (Figures S2D and S3D). The number of peptides increased as amino acid residues increased from six to eight, was less changed as amino acid residues increased from eight to ten, then decreased with increased amino acid residues (Figures S1E and S2E).

We isolated 176 proteins with increased abundance and 134 proteins with decreased abundance from +Al *C. sinensis* leaves. These proteins were related to energy and carbohydrate metabolism, sulfur metabolism, stress response, low-phosphorus (P) response, nucleic acid metabolism, protein and amino acid metabolism, cell wall and cytoskeleton metabolism, cellular transport, lipid metabolism, signal transduction and other processes. By contrast, we only obtained six up- and 11 downregulated proteins from +Al *C. grandis* leaves. These proteins were associated with energy and carbohydrate metabolism, S metabolism, stress response, low-P response, nucleic acid metabolism, protein and amino acid metabolism, cellular transport, signal transduction and other processes (Figure 2A,B and Tables S1 and S2).

As shown in Figure 2C and Tables S1 and S2, 324 differentially-abundant proteins were found in +Al *C. sinensis* and *C. grandis* leaves. Among these proteins, 307 differentially-abundant proteins only presented in *C. sinensis* and 14 only presented in *C. grandis*, and only three differentially-abundant proteins with the same accession number (i.e., Ciclev10026096 m, Ciclev10015568 m and Ciclev10009194 m) were shared by the two species.

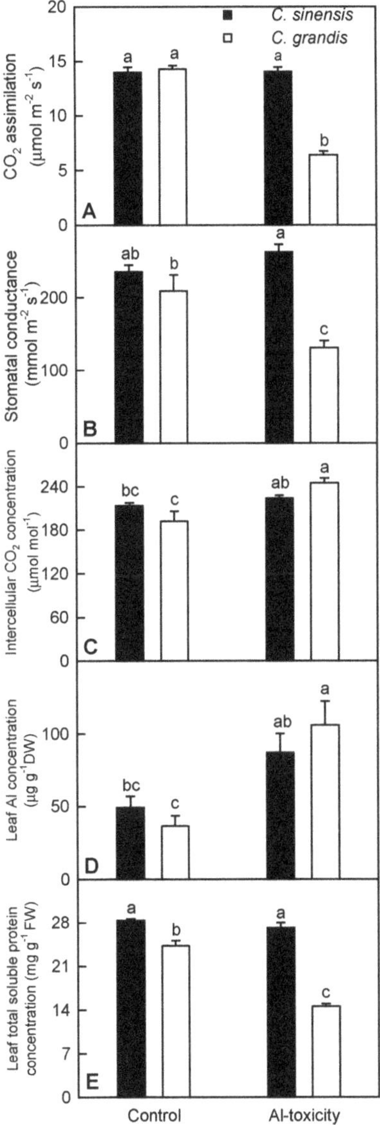

Figure 1. Effects of Al toxicity on leaf CO_2 assimilation (**A**); stomatal conductance (**B**); intercellular CO_2 concentration (**C**); Al (**D**) and total soluble protein (**E**) concentrations. Bars represent the means ± standard error SE (n = 5). DW: dry weight; FW: fresh weight. Differences among four treatment combinations (two species × two Al) were analyzed by two-way analysis of variance. Means were separated by Duncan's new multiple range test. Different letters above the bars indicate a significant difference at $p < 0.05$.

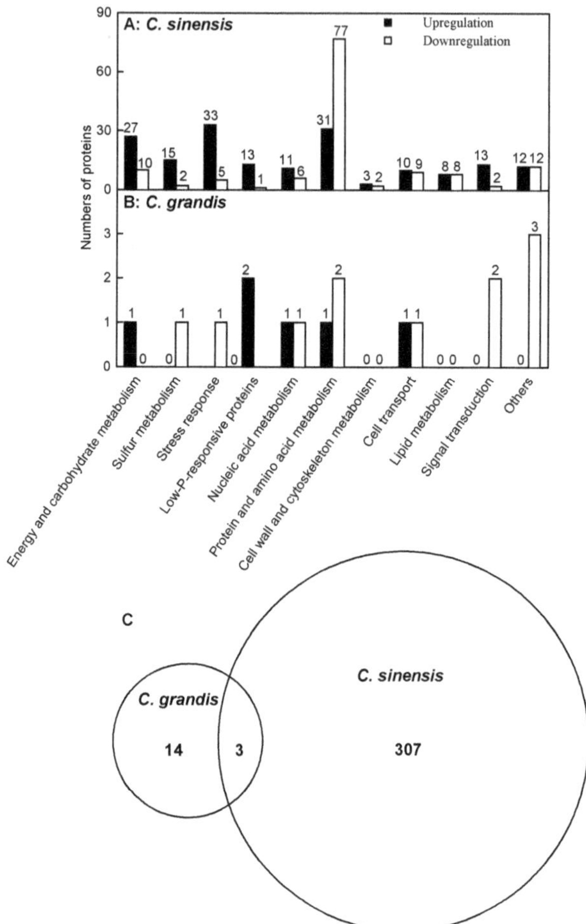

Figure 2. Classification of Al toxicity-responsive proteins in *C. sinensis* (**A**) and *C. grandis* (**B**) leaves; and Venn diagram analysis of Al toxicity-responsive proteins (**C**).

2.3. Transcriptional Analysis of Genes for Some Differentially-Abundant Proteins

qRT-PCR was applied to analyze the mRNA levels of genes for 20 differentially-abundant proteins, which were selected in a randomized manner from +Al *C. grandis* and *C. sinensis* leaves. Except for three genes (i.e., Ciclev10008649m, Ciclev10000951m and Ciclev10022212m), the expression profiles of all of these genes matched with our iTRAQ data (Figure S3; Tables S1 and S2), indicating that 85% of these differentially-abundant proteins were regulated at the transcriptional level.

2.4. Analysis of Five Al Toxicity-Responsive Enzymes in C. sinensis Leaves

In this study, we measured the activities of five enzymes related to ROS scavenging (i.e., superoxide dismutase (SOD), APX, catalase (CAT) and monodehydroascorbate reductase (MDAR)) and jasmonic acid (JA) biosynthesis (i.e., lipoxygenase (LOX)) in *C. sinensis* leaves in order to validate the differentially-abundant enzymes. As shown in Figure 3 and Table S1, the activities of all five enzymes matched well with our iTRAQ data.

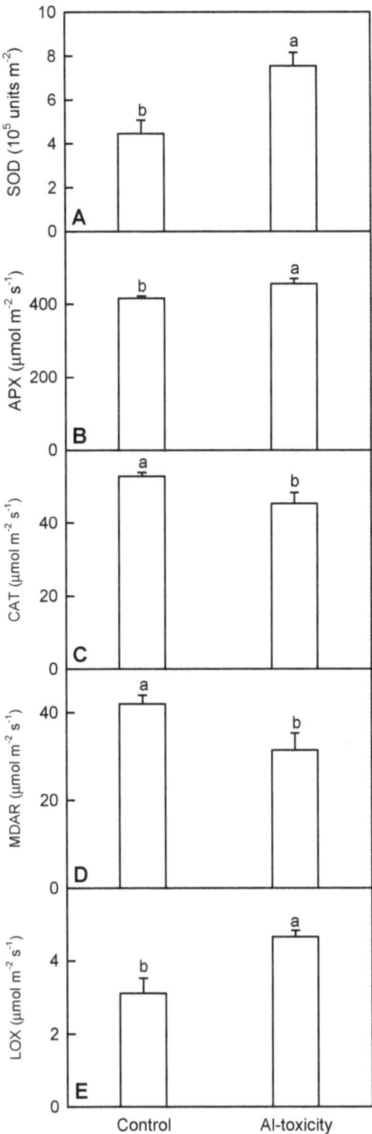

Figure 3. Effects of Al toxicity on the activities of superoxide dismutase (SOD) (**A**); ascorbate peroxidase (APX) (**B**); catalase (CAT) (**C**); monodehydroascorbate reductase (MDAR) (**D**) and lipoxygenase (LOX) (**E**) in *C. sinensis* leaves. Bars represent the means $\pm$ SE (n = 4). Significance tests for two means (control and Al toxicity) were carried out by the unpaired t-test at the $p < 0.05$ level. Different letters above the bars indicate a significant difference at $p < 0.05$.

3. Discussion

3.1. C. sinensis Displayed Higher Metabolic Flexibility than C. grandis

Alterations of proteome compositions are required for plants to deal with Al [7,10–13]. As shown in Tables S1 and S2, the number of Al toxicity-responsive proteins was higher in *C. sinensis* leaves than in *C. grandis* leaves, meaning that *C. sinensis* had higher metabolic flexibility than *C. grandis*. This agrees with the report that more Al toxicity-responsive proteins were isolated from *C. sinensis* roots than from *C. grandis* ones [10]. Thus, we concluded that the metabolic flexibility contributed to the higher Al tolerance of *C. sinensis*.

3.2. Al Toxicity-Induced Alterations of Energy and Carbohydrate Metabolism-Related Proteins Contribute to the Higher Al-Tolerance of C. sinensis

We isolated 27 up- and 10 downregulated proteins related to energy and carbohydrate metabolism from +Al *C. sinensis* leaves, but only one upregulated proteins from +Al *C. grandis* ones (Tables S1 and S2). A similar result has been obtained on +Al *C. sinensis* and *C. grandis* roots [10]. Energy deprivation is a general symptom of photosynthetic plants under most types of stress. A close relationship exists between energy availability and stress tolerance [24]. Here, we identified eight upregulated proteins in photosystem (PS) II and Calvin cycle from +Al *C. sinensis* leaves, which agrees with the report that Al toxicity increased the activities of Calvin cycle enzymes in citrus leaves [25]. The upregulation of these proteins might be responsible for the higher photosynthesis and for the maintenance of energy homeostasis. This is also supported by our data that CO_2 assimilation was not reduced in +Al *C. sinensis* leaves (Figure 1A).

Seven increased (i.e., AT5G11720.1, AT3G26720.1, AT4G39010.1, AT4G34480.1, AT3G57240.1, AT3G57270.1 and AT5G24090.1) and two decreased (i.e., AT5G49360.1 and AT3G48950.1) proteins in abundances related to cell wall polysaccharide degradation were isolated from +Al *C. sinensis* leaves. Furthermore, four upregulated proteins in cell wall polysaccharide (i.e., AT1G22360.1 and AT1G22380.1) and starch (i.e., AT2G36390.1 and AT1G32900.1) biosynthesis, one downregulated protein (i.e., AT1G29050.1) in cellulose synthesis and one downregulated protein (i.e., AT5G19220.1) in starch biosynthesis were identified from +Al *C. sinensis* leaves (Table S1). Thus, the remodeling of the cell wall structure might occur in these leaves by selective degradation and/or biosynthesis of some polysaccharide components, thus contributing to their Al tolerance [26]. Besides, the degradation products of the reserve polysaccharides might be utilized as a source of carbon or energy for other key metabolic processes [27].

We found that all four differentially-abundant proteins (i.e., AT3G04120.1, AT5G56350.1, AT3G08590.1 and AT5G15140.1) in glycolysis were upregulated in +Al *C. sinensis* leaves (Table S1). Wang et al. observed that most of the Al toxicity-responsive proteins in glycolysis were upregulated in Al-tolerant rice roots, concluding that Al-tolerant roots could maintain their basic respiration and provide more glycolytically-produced ATP under Al stress [11]. Thus, the Al toxicity-induced upregulation of glycolysis might play a role in the higher Al tolerance of *C. sinensis*.

3.3. Al Toxicity-Induced Upregulation of Antioxidant Systems and Other Stress-Related Proteins Played a Role in the Al Tolerance of C. sinensis

Aluminum toxicity can lead to the excessive generation of ROS, thus causing lipid peroxidation in plants [28,29]. Plants have evolved diverse non-enzymatic and enzymatic defense mechanisms to minimize cellular damage caused by ROS. S metabolism is a core pathway for the synthesis of S-containing compounds [7,12]. Through producing different S-containing compounds, such as reduced glutathione (GSH), cysteine, cysteine-rich metal-chelating proteins, glutaredoxins (GRXs) and thioredoxins (TRXs), ATP sulfurylase (ATPS) and other S-metabolism-related enzymes play key roles in plant tolerance to abiotic stresses, including Al toxicity [30,31]. Here, we obtained 15 up- and two downregulated proteins associated with S metabolism from +Al *C. sinensis* leaves, indicating that S metabolism was enhanced in these leaves (Table S1; Figure S4). However, we only isolated one downregulated S metabolism-related protein from +Al *C. grandis* leaves (Table S2). This agrees with our report that S metabolism was upregulated in +Al *C. sinensis* and *C. grandis* roots, particularly in the former [10].

Besides S metabolism-related proteins, the abundance of four other antioxidant enzymes was higher in +Al *C. sinensis* leaves (Table S1), which agrees with our report that the activities of antioxidant enzymes were enhanced in +Al citrus leaves [32]. However, the abundances of CAT, ATP1a/ATP1b and probable MDAR were decreased in +Al *C. sinensis* leaves. To validate the reliability of iTRAQ data, we assayed the activities of four differentially-abundant antioxidant enzymes (i.e., SOD, APX, CAT and MDAR) in *C. sinensis* leaves. The activities of SOD and APX were higher in +Al leaves than in controls, while the reverse was the case for the activities of CAT and MDAR (Figure 3A–D). This fully agrees with the data obtained by iTRAQ (Table S1). Furthermore, we isolated 11 other upregulated proteins (i.e., AT4G10720.1, AT5G54620.1, AT2G42590.3, AT3G23400.1, AT4G22240.1, AT3G47860.1, AT1G09560.1, AT4G03240.1 and three AT1G17100.1), which can protect plants against oxidative stress, from +Al *C. sinensis* leaves.

Al toxicity increases the production of aldehydes [29], which can result in a rapid and excessive accumulation of ROS in plant cells [33]. Yin et al. demonstrated the involvement of aldehydes in +Al tobacco roots [29]. Transgenic plants and yeasts overexpressing aldehyde scavenging enzyme genes, such as aldo-keto reductase [34], alcohol dehydrogenase [35] and aldehyde dehydrogenase [33], displayed enhanced tolerance to oxidative stress resulting from various environmental stresses through aldehyde detoxification. We observed that the abundances of one probable aldo-keto reductase 2 and two alcohol dehydrogenases were increased in +Al *C. sinensis* leaves, indicating that aldo-keto reductase and alcohol dehydrogenase might play a role in the adaptation of *C. sinensis* to Al toxicity. However, the abundance of aldehyde dehydrogenase 22A1 was reduced in +Al *C. sinensis* leaves (Table S1).

To sum up, the total ability of detoxification was greatly enhanced in +Al *C. sinensis* leaves. However, only one downregulated S metabolism-related protein was identified from +Al *C. grandis* leaves (Table S2). Therefore, the Al toxicity-induced upregulation of detoxification systems might play a role in the higher Al tolerance of *C. sinensis*.

Chaperones/HSPs play key roles in protecting plants against various stresses, including Al toxicity [7]. We obtained eight upregulated chaperones/HSPs from +Al *C. sinensis* leaves (Table S1), which agrees with the reports that there are one low molecular weight (LMW)-HSP and three DnaJ-like proteins in Al-tolerant soybean roots [14], and two dnaK-type molecular chaperone hsc70.1 (At5g02500) in Al-tolerant *Arabidopsis* ecotype (Col-0) roots increased in response to Al toxicity [36]. Therefore, chaperones/HSPs might play a role in the higher Al tolerance of *C. sinensis* via re-establishing normal protein conformation and maintaining cellular homeostasis. In addition, Al toxicity also induced another seven stress-related proteins (i.e., AT1G01470.1, AT5G54110.1, AT1G17020.1, AT3G04720.1, AT2G21620.1, AT3G53990.1 and AT1G24020.2) in *C. sinensis* leaves.

3.4. Low P-Responsive Proteins Were Induced by Al Toxicity, Particularly in C. sinensis Leaves

We isolated one down- and 14 upregulated low P-responsive proteins from +Al *C. sinensis* leaves, but only two upregulated ones from +Al *C. grandis* leaves (Tables S1 and S2). This agrees with our reports that Al toxicity reduced the P level in citrus roots, stem and leaves [23,37]. Inorganic pyrophosphatases, which catalyze the hydrolysis of pyrophosphate (PPi) to phosphate (Pi), may play a role in plant adaptation to Pi- limitation [38]. Ribonucleases have a role in the remobilization of Pi during Pi limitation [39]. Al toxicity leads to plant P deficiency, thus inducing purple acid phosphatases in order to enhance Pi acquisition and utilization in plants [11]. The glycerophosphodiester phosphodiesterase (GDPD)-mediated lipid metabolic pathway might function in the release of Pi from phospholipids during Pi deprivation [40]. Gregory et al. reported that in vivo phosphorylation activation of phosphoenolpyruvate carboxylase contributed to the metabolic adaptation of Pi-starved *Arabidopsis* [41]. Caparrós-Martín et al. found that the expression of *AtSgpp* (At2g38740) encoding haloacid dehalogenase (HAD)-like hydrolase was affected by (a)biotic stresses, being the greatest under Pi starvation, concluding that *AtSgpp* might function in maintaining the homeostatic balance of Pi in the cell [42]. Thus, the Al toxicity-induced upregulation of low P-responsive proteins might play a role in the maintenance of cellular P homeostasis through the conversion of organic P and/or PPi into available Pi.

3.5. RNA Regulations Might Play a Role in the Higher Al Tolerance of C. sinensis

Gene expression is regulated at the transcriptional and the post-transcriptional levels. We identified five upregulated transcription factors (TFs) (i.e., probable WRKY TF 50, basic-leucine zipper (bZIP) TF family protein, nuclear factor Y (NFY), subunit C11, C2H2-like zinc finger protein and HAP3-like protein), one downregulated TF (Myb domain protein 15), two upregulated (i.e., methyl-CPG-binding domain 11 and GLNB1 homolog) and two downregulated (i.e., zinc knuckle (CCHC-type) family protein and mitochondrial transcription termination factor family protein) proteins related to transcription regulation in +Al *C. sinensis* leaves (Table S1). Regulatory genes have crucial roles in plant tolerance to abiotic stresses, including Al toxicity [6,7,43,44]. Three zinc finger TFs, *AtSTOP1* (sensitive to proton rhizotoxicity), *OsART1* (Al resistance TF) and *TaSTOP1*, have been cloned from *Arabidopsis*, rice and bread wheat, respectively. They play a key role in Al detoxification via regulating multiple genes responsible for Al tolerance [44,45]. The expression level of bZIP94 TF was higher in Al-tolerant soybean roots than in Al-sensitive ones 48 h after Al treatment [46]. Plant NFY confers maize, rice and *Arabidopsis* stress tolerance [47,48]. The post-transcriptional regulations, which include RNA capping, RNA polyadenylation, RNA splicing, RNA transport and RNA stability, play a role in plant adaptation to abiotic stress, including Al toxicity [49,50]. We got two upregulated RNA-binding (RRM/RBD/RNP motifs) family proteins for mRNA stability, one upregulated pentatricopeptide repeat (PPR) superfamily protein mainly involved in regulating post-transcriptional processes [51], one upregulated putative pre-mRNA splicing factor and one downregulated spliceosome-associated protein 130 for RNA splicing from +Al *C. sinensis* leaves. An *Arabidopsis* chloroplast PPR protein SVR7 was reported to be important for normal photosynthesis and oxidative tolerance [52]. SOAR1, a cytosolic-nuclear PPR protein, had key roles in plant tolerance [51]. Lee et al. showed that STA1, a pre-mRNA splicing factor, was essential for stress tolerance in *Arabidopsis* [53]. To conclude, most of these differentially-abundant proteins related to RNA regulations were induced in +Al *C. sinensis* leaves. However, we only isolated one upregulated methyl-CpG-binding domain 9 and one downregulated DEK domain-containing chromatin-associated protein related to RNA regulations from +Al *C. grandis* leaves (Table S2). Thus, RNA regulation might play a role in the higher Al tolerance of *C. sinensis*.

3.6. Protein Metabolism Was More Adaptive to Al Toxicity in C. sinensis than in C. grandis

We isolated 66 down- and nine upregulated proteins in protein synthesis from +Al *C. sinensis* leaves (Table S1), implying that protein synthesis was impaired in these leaves. However, the total soluble protein concentration did not significantly differ between +Al and control *C. sinensis* leaves (Figure 1E). It seemed that other causes were involved in regulating the total soluble protein level in +Al leaves. Cao et al. showed that Al decreased the ATP level in *Pinus massoniana* needles [54]. Thus, the ATP level might be decreased in +Al *C. sinensis* leaves. Protein synthesis, a major consumer of ATP, is subject to strict regulation under conditions where ATP production becomes limited [55]. Thus, the Al toxicity-induced downregulation of protein synthesis-related proteins might be an adaptive response to Al toxicity by saving ATP due to reasonable regulation of protein translation in these leaves. Similar results have been reported on +Al *C. sinensis*, rice and *Arabidopsis* roots [10,11,56]. Furthermore, the Al toxicity-induced inhibition of many ribosomal proteins in *C. sinensis* leaves might imply a redistribution of the resources to meet the increased requirement for amino acids in non-ribosomal peptide (i.e., GSH and phytochelation for Al complexation) synthesis. This is also supported by the above inference that S metabolism was induced in +Al *C. sinensis* leaves and by the report that +Al *Citrus reshni* leaves had higher levels of GSH and oxidized glutathione (GSSG) [32]. However, we only identified two downregulated protein synthesis-related proteins in +Al *C. grandis* leaves (Table S2).

Plant proteases are required for the strict control of protein quality and the selective degradation of specific proteins in response to biotic and abiotic stresses [57]. Futile and inactive (i.e., incorrectly folded) proteins are targeted by ubiquitin for degradation [58]. As expected, we identified 15 upregulated proteinase-related proteins and two upregulated (i.e., AT4G10790.1 and AT4G17510.1) ubiquitination-related proteins in +Al *C. sinensis* leaves (Table S1). In addition to degrading mature proteins into free amino acids, the proteolytic cleavage of proteins by proteinases and ubiquitination also plays a role in the modification and maturation of proteins. Because no difference was observed in the total soluble protein level between +Al and control *C. sinensis* leaves (Figure 1E), the Al toxicity-induced upregulation of proteases and ubiquitination-related proteins might mainly function in the modification and maturation of proteins, which might provide an adaptive response to Al toxicity by maintaining the stability of protein complexes and/or the recycling of nitrogen. By contrast, we only obtained one upregulated ubiquitin-protein ligase 1 from +Al *C. grandis* leaves (Table S2).

3.7. Cell Wall and Cytoskeleton Metabolism-Related Proteins

Expansins, which enable the growing cell wall to extend by weakening noncovalent bonding between the matrix and cellulose microfibrils, are believed to be the key regulators of wall extension during growth [59]. The upregulation of expansin-like B1 in +Al *C. sinensis* leaves (Table S1) might be helpful to plant growth by loosening the cell wall. This is supported by our data that the abundance of seven proteins involved in cell wall polysaccharide degradation were enhanced in +Al *C. sinensis* leaves. Al toxicity inhibits cytoskeletal dynamics and Al interacts with the microtubules and actin filaments [60]. As expected, we identified four cytoskeleton-related Al toxicity-responsive proteins from *C. sinensis* leaves. The higher abundance of actin depolymerizing factor 1 in +Al *C. sinensis* leaves means that depolymerization of actin filaments might be enhanced in these leaves. The polymerization and depolymerization of actin filaments may provide cells with the ability to rapidly remodel the cytoskeleton in response to endogenous cues or external signals [61]. Thus, the induction of actin depolymerizing factor 1 in +Al *C. sinensis* leaves might be an adaptive response to Al toxicity.

3.8. Cellular Transport-Related Proteins

We identified ten up- and nine downregulated proteins related to cellular transport in +Al *C. sinensis* leaves (Table S1), indicating that cellular transport was altered under Al-stress. Hamilton et al. showed that V-ATPase was induced by Al toxicity in an Al-resistant wheat cultivar [62]. Further study suggested that V-ATPase activity played a role in wheat Al resistance [63]. Ferritins not only are important for iron homeostasis, but also play key roles in preventing oxidative damage by sequestering highly reactive intracellular Fe and inhibiting the production of hydroxyl radicals [64]. Besides their role in the transport of O_2, plant hemoglobins are plausible targets for enhancing stress tolerance. Over-expression of hemoglobin 1 (Hb1) gene conferred stress tolerance to plants by maintaining the cellular energy status and growth, as well as improving the survival of plants under stress conditions [65–68]. CDGSH Fe-S domain-containing protein NEET plays crucial roles in plant development, senescence, Fe homeostasis/metabolism and ROS homeostasis [69]. Wang et al. reported that clathrin light chains (CLCs) played a key role in regulating clathrin-mediated trafficking, auxin signaling and development in *Arabidopsis* [70]. Lam et al. demonstrated that the secretory carrier-associated membrane proteins (SCAMPs) highlighted the developing cell plate during cytokinesis in tobacco BY-2 cells [71]. Voltage-dependent anion channels (VDACs) mediate the exchange of metabolites, such as ATP, NADH and ions between mitochondria and cytoplasm. AtVDAC1 is essential for the maintenance of mitochondrial functions associated with energy transaction in *Arabidopsis* [72]. Heavy metal-associated domain-containing protein plays a role in heavy metal transport and/or detoxification [73,74]. Therefore, the Al toxicity-induced upregulation of these proteins might be an adaptive strategy. However, we only isolated one upregulated transport-related protein (AT1G80310.1) from + Al *C. grandis* leaves (Table S2).

3.9. Lipid Metabolism-Related Proteins

JA one of the most important signaling molecules, is an oxylipin. Oxylipins and JA derivatives have active roles in plant tolerance to (a)biotic stresses [75]. We found that the levels of three enzymes involved in JA biosynthesis (i.e., lipoxygenase 2 (LOX2), allene oxide cyclase 3 (AOC3) and acyl-activating enzyme 7) were increased in +Al *C. sinensis* leaves, but unaffected in +Al *C. grandis* leaves (Tables S1 and S2). Furthermore, the activity of LOX was increased in +Al *C. sinensis* leaves (Figure 3D). This agrees with the reports that both JA biosynthesis and level might be enhanced in +Al *C. sinensis* roots [10], that both the shoot LOX activity and Al toxicity-induced increase in shoot LOX activity were higher in the Al-tolerant sorghum cultivar than in the Al-sensitive one [76] and that NO enhanced the Al toxicity-induced increase in AOC3 in rice leaves [13]. Therefore, JA signaling might be activated in +Al *C. sinensis* leaves, thus contributing to the higher Al tolerance of *C. sinensis*. Similarly, the levels of two acyl-CoA thioesterases, which presumably participate in the release of JA from JA-CoA [77], were increased in +Al *C. sinensis* leaves. However, the level of long chain acyl-CoA synthetase 4, which activates free fatty acids to acyl-CoA thioesters, was decreased in +Al *C. sinensis* leaves (Table S1). GDSL esterases/lipases perform crucial roles in plant abiotic responses [78]. Gujjar et al. reported that GDSL20 were downregulated more strongly in a drought-sensitive tomato line than in a drought-tolerant one [79]. Thus, the higher level of GDSL esterase/lipase 5 in +Al *C. sinensis* leaves (Table S1) might contribute to the higher Al tolerance of *C. sinensis*.

3.10. Signal Transduction-Related Proteins

Signal transduction is altered by Al toxicity [6,7,13]. As expected, we isolated five upregulated and one downregulated protein related to protein phosphorylation/dephosphorylation, four upregulated Ca signal-related proteins and another five proteins (i.e., four upregulated and one downregulated) involved in signal transduction from +Al *C. sinensis* leaves (Table S1). Clay and Nelson found that the loss of *VH1* caused premature leaf senescence and defective vascular transport in *Arabidopsis* [80]. The upregulation of the protein phosphatase 2C (PP2C) family protein in +Al *C. sinensis* leaves agrees with the reports that PP2A was elevated only in Al-tolerant soybean roots by Al toxicity [15] and that the expression of *PP2C* was upregulated in Al-tolerant soybean roots relative to Al-sensitive ones [46]. Studies in transgenic plants demonstrated that serine/threonine-protein phosphatases 2A and 5 positively regulated the responses of plants to abiotic stresses [81,82]. The increased abundance of four Ca signal-related proteins in +Al *C. sinensis* leaves agrees with the report that three Ca-binding proteins were enhanced in +Al rice leaves [13], implying the involvement of Ca in *C. sinensis* Al tolerance. The upregulation of the auxin-responsive family protein in +Al *C. sinensis* leaves indicates that the auxin signal pathway might be involved in plant Al tolerance. This agrees with the report that NO improved rice Al tolerance possibly through its interaction with auxin and GA signals [13]. In addition, we obtained two proteins (i.e., one upregulated farnesylcysteine lyase and one downregulated farnesyltransferase A) involved in abscisic acid (ABA) signaling from +Al *C. sinensis* leaves. To sum up, the responses of *C. sinensis* to Al toxicity were regulated in multiple signal pathways, thus contributing to the higher Al tolerance, while only two downregulated proteins related to signal transduction were isolated from +Al *C. grandis* ones (Table S2).

4. Materials and Methods

4.1. Plant Materials and Al Treatments

'Sour pummelo' (*Citrus grandis* (L.) Osbeck) seeds were collected from Fujian Academy of Forestry Sciences, Fuzhou, China. 'Xuegan' (*Citrus sinensis* (L.) Osbeck) seeds were collected from Minan village, Tingjiang town, Mawei district, Fuzhou, China. This study was carried out at Fujian Agriculture and Forestry University, Fuzhou, China (26°5′ N, 119°14′ E). Plant culture and Al treatments were performed according to Jiang et al. [10]. In late May (five weeks after seed germination), uniform seedlings of *C. grandis* and *C. sinensis* were transported to 6 L pots (two plants to a pot) containing clean river sand and grown in a greenhouse under a natural photoperiod. Each pot was supplied with 500 mL of nutrient solution every two days. The nutrient solution contained the following macronutrients (in mM): KNO_3, 1; $Ca(NO_3)_2$, 1; KH_2PO_4, 0.1; $MgSO_4$, 0.5; and micronutrients (in μM): H_3BO_3, 20; $MnCl_2$, 2; $ZnSO_4$, 2; $CuSO_4$, 0.5; $(NH_4)_6Mo_7O_{24}$, 0.065; and Fe-ethylenediaminetetraacetic acid (EDTA), 20. Six weeks after transplanting, each pot was supplied daily with a nutrient solution containing 0 (control) or 1.2 mM $AlCl_3 \cdot 6H_2O$ (+Al) until the sand was saturated. The pH of the nutrient solutions was adjusted to 4.1–4.2 using HCl or NaOH. Eighteen weeks after the beginning of Al treatments, fully-expanded (about seven-week-old, midribs and petioles removed) leaves were collected at noon under full sun from different replicates and treatments and immediately frozen in liquid N_2. Samples were stored at $-80\,^\circ\text{C}$ until they were used for protein extraction, qRT-PCR analysis, total soluble protein concentration and the enzyme activity assay. The remaining seedlings that were not sampled were used to measure leaf Al concentration and gas exchange.

4.2. Measurements of Leaf Gas Exchange, Total Soluble Protein and Al Concentrations

Leaf gas exchange was measured with a CIARS-2 portable photosynthesis system (PP systems, Herts, U.K.) at ambient CO_2 concentration under a controlled light intensity of ca. $1000\ \text{m}^{-2} \cdot \text{s}^{-1}$ between 9:00 and 11:00 on a clear day. During all of the measurements, leaf temperature and relative humidity were $31.9 \pm 0.2\,^\circ\text{C}$ and $71.4\% \pm 0.6\%$, respectively. There were five replicates per treatment.

Leaf total soluble protein concentration was assayed according to Bradford [83] after being extracted with 50 mM Na_2HPO_4-KH_2PO_4 (pH 7.0) and 5% (w/v) insoluble polyvinylpyrrolidone. There were five replicates per treatment.

About seven-week-old leaves (midribs and petioles removed) were collected and dried at $70\,^\circ\text{C}$ for 48 h. Leaf Al concentration was determined colorimetrically by the aluminon (the triammonium salt of aurintricarboxylic acid) after being digested in a mixture of HNO_3:$HClO_4$ [84]. There were five replicates per treatment.

4.3. Protein Extraction

Proteins were extracted from frozen leaves using a phenol extraction procedure according to Yang et al. [85]. Briefly, equal amounts of frozen leaves from six plants (one plant per pot) were mixed as a biological replicate. There was one biological replicate for each treatment. About 1 g frozen mixed samples was well ground in liquid N_2 with a mortar and pestle. Four milliliters of ice-cooled buffer containing 100 mM Tris-HCl pH 7.8, 100 mM KCl, 50 mM L-ascorbic acid, 1% (v/v) Triton X-100, 1% (v/v) β-mercaptoethanol and 1 mM phenylmethanesulfonyl fluoride (PMSF) were added to the powder and gently pulverized. The mixture was allowed to thaw slowly on ice. The resulting suspension was transferred to a 10-mL tube, then an equal volume of Tris-phenol (pH 8.0) was added. The mixture was thoroughly vortexed before centrifuging at $13,000 \times g$ for 15 min at $4\,^\circ\text{C}$. The upper phenolic phase was transferred to a 50-mL tube, then five volumes of 100 mM ammonium acetate/methanol were added. After mixing carefully, the mixture was stored at $-20\,^\circ\text{C}$ overnight. The supernatant was removed carefully after centrifugation at $13,000 \times g$ for 15 min at $4\,^\circ\text{C}$, then the protein pellets were suspended in 25 mL of ice-cooled methanol for 2 h at $-20\,^\circ\text{C}$. Protein pellets were collected by centrifugation at $13,000 \times g$ for 15 min at $4\,^\circ\text{C}$ and then were resuspended in 25 mL of ice-cooled acetone containing 0.1% β-mercaptoethanol and kept at $-20\,^\circ\text{C}$ for 2 h. After centrifugation at $13,000 \times g$ for 15 min at $4\,^\circ\text{C}$, the pellets were washed twice with 25 mL of ice-cooled acetone, then dried by lyophilization and finally stored at $-80\,^\circ\text{C}$ until use. Lyophilized proteins were dissolved in buffer containing 8 M urea, 50 mM triethylammonium bicarbonate (TEAB), pH 8.5, 0.1% sodium dodecyl sulfate (SDS) and protease inhibitor cocktail (Roche, Indianapolis, IN, USA) for 1 h at $4\,^\circ\text{C}$ under constant mixing. After being centrifuged at $16,000 \times g$ for 20 min at $4\,^\circ\text{C}$, the supernatant was collected and quantified using the BCA assay kit (Pierce, Rockford, IL, USA).

4.4. iTRAQ Analysis

iTRAQ analysis was performed according to the manufacturer's instructions (AB Sciex Inc., Framingham, MA, USA) at CapitalBio Technology, Beijing. Six volumes of pre-cooled acetone ($-20\,^\circ\text{C}$) were added to each sample tube (100 μg total protein). After being inverted thrice, the tube was incubated at $-20\,^\circ\text{C}$ until a flocculent was formed (4–16 h). Thereafter, the acetone was decanted after centrifugation at $10,000 \times g$ for 15 min. Protein from +Al and control leaves was dissolved in a mixture (100 μL) containing 35 μL 8 M urea and 9 μL 500 mM TEAB. Protein reduction, cysteine block, trypsin digestion and iTRAQ™ labels were performed according to manufacturer's protocol for iTRAQ® Reagents–8plex (AB Sciex Inc., MA, USA; Sciex iChemistry® Product Number 4390812). Al-treated and control samples for *C. sinensis* were labeled with 121 and 119 tags; samples of *C. grandis* were labeled with 117 and 114 tags, respectively. Each label contained a reporter group, a peptide reactive group (PRG) and a balance group. When the labeled peptide was fragmented along the peptide backbone by MS/MS fragmentation, the iTRAQ™ reporter groups broke off and yielded distinct ions at m/z 113, 114, 115, 116, 117, 118, 119 and 121. The relative intensities of the reporter ions were shown to be directly proportional to the relative levels of each peptide in the samples. The peptides labeled with the isobaric tags were incubated at room temperature for 2 h. The labeled peptide mixtures were then pooled and stored at -80 °C until use.

For strong cationic exchange (SCX) chromatography using an Agilent 1260 Infinity high-performance liquid chromatography (HPLC) (Agilent Technologies, Palo Alto, CA, USA), the labeled peptides were first lyophilized and reconstituted in solvent A (2% acetonitrile (ACN), pH 10), then the samples were loaded onto the XBridge C18, 5 μm, 250 × 4.6 mm column (Waters, Milford, MA, USA) and eluted using a gradient of 5%–45% Solvent B (90% ACN, pH 10) for 40 min. A total of 40 fractions was collected, which were then concatenated to 20 fractions, vacuum dried and stored at -80 °C until further LC-MS analysis.

The LC-MS/MS analysis was carried out in CapitalBio Technology using a Q Exactive mass spectrometer (Thermo Scientific, San Jose, CA, USA). The peptide mixture was separated by reversed phase chromatography on a DIONEX nano-UPLC system using an Acclaim C18 PepMap100 nano-Trap column (75 μm × 2 cm) connected to an Acclaim PepMap RSLC C18 analytical column (75 μm × 25 cm, 2 μm particle size) (Thermo Scientific). Before loading, the sample was dissolved in Mobile Phase A, containing 2% ACN and 0.1% formic acid. A linear gradient of Mobile Phase B (0.1% formic acid in 99.9% ACN) from 2%–35% in 45 min was followed by a steep increase to 80% Mobile Phase B in 1 min at a flow rate of 300 nL· min^{-1}. The nano-LC was coupled online with the Q Exactive mass spectrometer using a stainless steel Emitter coupled to a nanospray ion source.

Mass spectrometry analysis was made in a data-dependent manner with full scans (350–1600 m/z) acquired using an Orbitrap mass analyzer (Thermo Fisher Scientific, Carlsbad, CA, USA) at a mass resolution of 70,000 FWHM at 400 m/z in Q Exactive. The twenty most intense precursor ions from a survey scan were selected for MS/MS from each duty cycle and detected at a mass resolution of 17,500 FWHM at m/z of 400 in the Orbitrap analyzer. All of the tandem mass spectra were produced by the higher energy collision dissociation (HCD) method. Dynamic exclusion was set for 20 s.

Proteome discoverer (1.4) software (Thermo Scientific) was used to perform database searching against the *Citrus clementina* database (https://www.citrusgenomedb.org/species/clementina/genome1.0) using the Sequest algorithms. The following settings were applied: precursor mass tolerance of 20 ppm, fragment mass tolerance of 0.02 Da. Trypsin was specified as the digesting enzyme, and 2 missed cleavages were allowed. Cysteine carbamidomethylation and iTRAQ modifications (N-terminus and lysine residues) were defined as fixed modifications, and methionine oxidation was the variable modification. The results were filtered using the following settings: only high confident peptides with a global false discovery rate (FDR) <1% based on a target-decoy approach were included in the results. In the iTRAQ quantitation workflow, the most confident centroid method was used with an integration window of 20 ppm. For protein quantitation, only proteins that contained at least two unique peptides were used to quantify proteins. The quantitative protein ratios were weighted and normalized by the median ratio in Mascot [10,86]. In this study, a protein was considered differentially abundant when it had a fold change of >2 and a p value of <0.05.

Bioinformatic analysis of proteins was performed according to Yang et al. [85] and Gan et al. [87].

4.5. qRT-PCR Analysis of Gene Expression

About 300 mg of frozen leaves collected equally from six pots (one plant per pot, one leaf per plant) were mixed as a biological replicate. There were three biological replicates for each treatment. Total RNAs were independently extracted thrice from +Al and control frozen leaves using the Recalcitrant Plant Total RNA Extraction Kit (Centrifugal column type, Bioteke Corporation, Beijing, China) according to the manufacturer's instructions. Gene-specific primers were designed using Primer Software Version 5.0 (PREMIER Biosoft International, Palo Alto, CA, USA) according to the corresponding sequences of selected proteins in the citrus genome (http://www.phytozome.net/cgi-bin/gbrowse/citrus/). The sequences of the F and R primers used are given in Table S3. qRT-PCR was performed according to Zhou et al. [88]. Each sample was run in two technical replicates. For the normalization of gene expression, citrus *actin* (GU911361.1) was used as an internal standard, and the leaves from control plants were used as the reference sample, which was set to 1.

4.6. Analysis of SOD, APX, CAT, MDAR and LOX Activities in C. sinensis Leaves

Leaf SOD, APX, CAT and MDAR were assayed according to Li et al. [89]. Leaf LOX was assayed by the formation of conjugated dienes from linoleic acid according to Axelrod et al. [90].

4.7. Experimental Design and Statistical Analysis

There were 40 seedlings (20 pots) in a completely randomized design. Experiments were performed with 3–5 replicates except for iTRAQ analysis. For each treatment, only one biological sample was used to perform iTRAQ analysis. The replicates represented material from individual plants, except for iTRAQ and qRT-PCR analysis, in which each biological replicate was created by pooling equal samples from six different plants (one plant per pot). Differences among four treatment combinations (two species × two Al) were analyzed by two-way analysis of variance. Four means were separated by Duncan's new multiple range test at $p < 0.05$. Significant tests between two means (control and Al toxicity) were carried out by the unpaired t-test at the $p < 0.05$ level.

4.8. Data Deposit

The mass spectrometry proteomics data have been deposited to the ProteomeXchange Consortium via the PRIDE Proteomics Identification (PRIDE) partner repository with the dataset identifier PXD002916

5. Conclusions

Al toxicity only lowered *C. grandis* leaf CO_2 assimilation and total soluble protein concentration, demonstrating that *C. sinensis* had higher Al tolerance than *C. grandis*. Here, we used iTRAQ to investigate comparatively Al toxicity-responsive proteins in Al-tolerant *C. sinensis* and Al-intolerant *C. grandis* leaves and obtained more differentially-abundant proteins from +Al *C. sinensis* leaves than from +Al *C. grandis* leaves. The majority of the differentially-abundant proteins only presented in *C. sinensis* or *C. grandis*; only three Al toxicity-responsive proteins were shared by both. *C. sinensis* displayed higher metabolic flexibility than *C. grandis*, possibly contributing to the higher Al tolerance of *C. sinensis*. The higher Al tolerance of *C. sinensis* might include several aspects: (a) photosynthesis and energy-related proteins were more adaptive to Al toxicity in *C. sinensis* than in *C. grandis*, which might account for the better maintenance of photosynthesis and energy balance in +Al *C. sinensis* leaves; (b) less increased requirement for detoxification of ROS and other toxic compounds, such as aldehydes, due to Al toxicity-induced inhibition of photosynthesis, because CO_2 assimilation was not significantly altered, and great improvement of the total ability of detoxification via inducing proteins related to detoxification of ROS (i.e., S metabolism related proteins, SOD, peroxidase, plastid-lipid-associated proteins, germin-like proteins and frataxin) and aldehydes (i.e., aldo-keto reductase and alcohol dehydrogenase) in +Al *C. sinensis* leaves; (c) induction of low P-responsive proteins in +Al *C. sinensis* leaves. In addition, Al toxicity-responsive proteins related to RNA regulations, protein metabolism, cellular transport and signal transduction might contribute to the higher Al tolerance of *C. sinensis*. Here, we presented the global picture of Al toxicity-induced protein alterations in Al-tolerant *C. sinensis* and Al-intolerant *C. grandis* leaves and identified some new Al toxicity-responsive proteins involved in carbohydrate and energy metabolism (i.e., ferredoxin 3 and aldose 1-epimerase), detoxification (i.e., ATPS, lipocalin, aldo-keto reductase, ankyrin repeat-containing protein and frataxin), low P-response (i.e., ribonuclease and purple acid phosphatase), nucleic acid metabolism (i.e., bZIP TF and HAP3-like protein), protein and amino acid metabolism (i.e., cystatin, arginase and cysteine proteinases), cellular transport (i.e., ferritin and CDGSH iron-sulfur domain-containing protein NEET) and signal transduction (i.e., VH1-interacting kinase and farnesylcysteine lyase) from plant leaves. Thus, our findings will increase our understanding of the molecular mechanisms on citrus Al toxicity and Al tolerance at the protein level.

Supplementary Materials: Supplementary materials can be found at http://www.mdpi.com/1422-0067/17/7/1180/s1.

Acknowledgments: This work was financially supported by the National Natural Science Foundation of China (No. 31301740), the Specialized Research Fund for the Doctoral Program of Higher Education of China (No. 20123515110017), the Natural Science Foundation of Fujian Province of China (No. 2014J05033) and the earmarked fund for the China Agriculture Research System (No. CARS-27).

Author Contributions: Li-Song Chen, Lin-Tong Yang and Yi-Ping Qi conceived of and designed the experiments. Huan Li performed the experiments. Huan Li, Peng Guo and Yi-Bin Lu analyzed the data. Huan Li and Li-Song Chen wrote the paper. All authors have given approval of the final version of the manuscript.

Conflicts of Interest: The authors declare no conflict of interest.

References

1. Foy, C.D.; Chaney, R.L.; White, W.C. The physiology of metal toxicity in plants. *Annu. Rev. Plant Physiol.* **1978**, *29*, 511–566. [CrossRef]
2. Kinraide, T.B. Identity of the rhizotoxic aluminium species. *Plant Soil* **1991**, *134*, 167–178.
3. Kochian, L.V. Cellular mechanisms of aluminum toxicity and resistance in plants. *Annu. Rev. Plant Physiol. Plant Mol. Biol.* **1995**, *46*, 237–260. [CrossRef]
4. Yang, L.T.; Qi, Y.P.; Jiang, H.X.; Chen, L.S. Roles of organic acid anion secretion in aluminium tolerance of higher plants. *BioMed Res. Int.* **2013**, *2013*, 173682. [CrossRef] [PubMed]
5. Wu, D.M.; Fu, Y.Q.; Yu, Z.W.; Shen, H. Status of red soil acidification and aluminum toxicity in south China and prevention. *Soils* **2013**, *45*, 577–584.
6. Wang, L.Q.; Yang, L.T.; Guo, P.; Zhou, X.X.; Ye, X.; Chen, E.J.; Chen, L.S. Leaf cDNA-AFLP analysis reveals novel mechanisms for boron-induced alleviation of aluminum toxicity in *Citrus grandis* seedlings. *Ecotox. Environ. Saf.* **2015**, *120*, 349–359. [CrossRef] [PubMed]
7. Zheng, L.; Lan, P.; Shen, R.F.; Li, W.F. Proteomics of aluminum tolerance in plants. *Proteomics* **2014**, *14*, 566–578. [CrossRef] [PubMed]
8. Zhou, X.X.; Yang, L.T.; Qi, Y.P.; Guo, P.; Chen, L.S. Mechanisms on boron-induced alleviation of aluminum toxicity in *Citrus grandis* seedlings at a transcriptional level revealed by cDNA-AFLP analysis. *PLoS ONE* **2015**, *10*, e0115485. [CrossRef] [PubMed]
9. Zhu, H.; Wang, H.; Zhu, Y.; Zou, J.; Zhao, F.J.; Huang, C.F. Genome-wide transcriptomic and phylogenetic analyses reveal distinct aluminum tolerance mechanisms in the aluminum-accumulating species buckwheat (*Fagopyrum tataricum*). *BMC Plant Biol.* **2015**, *15*, 16. [CrossRef] [PubMed]
10. Jiang, H.X.; Yang, L.T.; Qi, Y.P.; Lu, Y.B.; Huang, Z.R.; Chen, L.S. Root iTRAQ protein profile analysis of two citrus species differing in aluminum tolerance in response to long-term aluminum toxicity. *BMC Genom.* **2015**, *16*, 949. [CrossRef] [PubMed]
11. Wang, Z.Q.; Xu, X.Y.; Gong, Q.Q.; Xie, C.; Fan, W.; Yang, J.L.; Lin, Q.S.; Zheng, S.J. Root proteome of rice studied by iTRAQ provides integrated insight into aluminum stress tolerance mechanisms in plants. *J. Proteom.* **2014**, *98*, 189–205. [CrossRef] [PubMed]
12. Yang, Q.; Wang, Y.; Zhang, J.; Shi, W.; Qian, C.; Peng, X. Identification of aluminum-responsive proteins in rice roots by a proteomic approach: Cysteine synthase as a key player in Al response. *Proteomics* **2013**, *7*, 737–749. [CrossRef] [PubMed]
13. Yang, L.; Tian, D.; Todd, C.D.; Luo, Y.; Hu, X. Comparative proteome analyses reveal that nitric oxide is an important signal molecule in the response of rice to aluminum toxicity. *J. Proteome Res.* **2013**, *12*, 1316–1330. [CrossRef] [PubMed]

14. Zhen, Y.; Qi, J.L.; Wang, S.S.; Su, J.; Xu, G.H.; Zhang, M.S.; Miao, L.; Peng, X.X.; Tian, D.; Yang, Y.H. Comparative proteome analysis of differentially expressed proteins induced by Al toxicity in soybean. *Physiol. Plant.* **2007**, *131*, 542–554. [CrossRef] [PubMed]
15. Duressa, D.; Soliman, K.; Taylor, R.; Senwo, Z. Proteomic analysis of soybean roots under aluminum stress. *J. Plant Genom.* **2011**, *2011*, 282531. [CrossRef] [PubMed]
16. Dai, H.; Cao, F.; Chen, X.; Zhang, M.; Ahmed, I.M.; Chen, Z.H.; Li, C.; Zhang, G.; Wu, F. Comparative proteomic analysis of aluminum tolerance in Tibetan wild and cultivated barleys. *PLoS ONE* **2013**, *8*, e63428. [CrossRef] [PubMed]
17. Zhou, S.; Sauvé, R.; Thannhauser, T.W. Proteome changes induced by aluminium stress in tomato roots. *J. Exp. Bot.* **2009**, *60*, 1849–1857. [CrossRef] [PubMed]
18. Oh, M.W.; Roy, S.K.; Kamal, A.H.; Cho, K.; Cho, S.W.; Park, C.S.; Choi, J.S.; Komatsu, S.; Woo, S.H. Proteome analysis of roots of wheat seedlings under aluminum stress. *Mol. Biol. Rep.* **2014**, *41*, 671–681. [CrossRef] [PubMed]
19. Rahman, M.A.; Kim, Y.G.; Lee, B.H. Proteomic response of alfalfa subjected to aluminum (Al) stress at low pH soil. *J. Korean Grassl. Forage Sci.* **2014**, *34*, 262–268. [CrossRef]
20. Lin, Z.; Myhre, D.L. Citrus root growth as affected by soil aluminum level under field conditions. *Soil Sci. Soc. Am. J.* **1990**, *54*, 1340–1344. [CrossRef]
21. Li, Y.; Han, M.Q.; Lin, F.; Ten, Y.; Lin, J.; Zhu, D.H.; Guo, P.; Weng, Y.B.; Chen, L.S. Soil chemical properties, 'Guanximiyou' pummelo leaf mineral nutrient status and fruit quality in the southern region of Fujian province, China. *J. Soil Sci. Plant Nutr.* **2015**, *15*, 615–628. [CrossRef]
22. Huang, Y.Z.; Li, J.; Wu, S.H.; Pang, D.M. Nutrition condition of the orchards in the main production areas of Guanxi honeypomelo trees (Pinhe county). *J. Fujian Agric. Univ.* **2001**, *30*, 40–43.
23. Yang, L.T.; Jiang, H.X.; Tang, N.; Chen, L.S. Mechanisms of aluminum tolerance in two species of citrus: Secretion of organic acid anions and immobilization of aluminum by phosphorus in roots. *Plant Sci.* **2011**, *180*, 521–530. [CrossRef] [PubMed]
24. Baena-González, E.; Sheen, J. Convergent energy and stress signaling. *Trends Plant Sci.* **2008**, *13*, 474–482. [CrossRef] [PubMed]
25. Chen, L.S.; Qi, Y.P.; Smith, B.R.; Liu, X.H. Aluminum-induced decrease in CO_2 assimilation in citrus seedlings is unaccompanied by decreased activities of key enzymes involved in CO_2 assimilation. *Tree Physiol.* **2005**, *25*, 317–324. [CrossRef] [PubMed]
26. Minic, Z.; Jouanin, L. Plant glycosyl hydrolases involved in cell wall polysaccharide degradation. *Plant Physiol. Biochem.* **2006**, *44*, 435–449. [CrossRef] [PubMed]
27. Minic, Z. Physiological roles of plant glycoside hydrolases. *Planta* **2008**, *227*, 723–740. [CrossRef] [PubMed]
28. Shen, X.; Xiao, X.; Dong, Z.; Chen, Y. Silicon effects on antioxidative enzymes and lipid peroxidation in leaves and roots of peanut under aluminum stress. *Acta Physiol. Plant.* **2014**, *36*, 3063–3069. [CrossRef]
29. Yin, L.; Mano, J.; Wang, S.; Tsuji, W.; Tanaka, K. The involvement of lipid peroxide-derived aldehydes in aluminum toxicity of tobacco roots. *Plant Physiol.* **2010**, *152*, 1406–1417. [CrossRef] [PubMed]
30. Anjum, N.A.; Gill, R.; Kaushik, M.; Hasanuzzaman, M.; Pereira, E.; Ahmad, I.; Tuteja, N.; Gill, S.S. ATP-sulfurylase, sulfur-compounds, and plant stress tolerance. *Front. Plant Sci.* **2015**, *6*, 210. [CrossRef] [PubMed]
31. Ezaki, B.; Gardner, R.C.; Ezaki, Y.; Matsumoto, H. Expression of aluminum induced genes in transgenic Arabidopsis plants can ameliorate aluminum stress and/or oxidative stress. *Plant Physiol.* **2000**, *122*, 657–665. [CrossRef] [PubMed]
32. Chen, L.S.; Qi, Y.P.; Liu, X.H. Effects of aluminum on light energy utilization and photoprotective systems in citrus leaves. *Ann. Bot.* **2005**, *96*, 35–41. [CrossRef] [PubMed]
33. Sunkar, R.; Bartels, D.; Kirch, H.H. Overexpression of a stress-inducible aldehyde dehydrogenase gene from *Arabidopsis thaliana* in transgenic plants improves stress tolerance. *Plant J.* **2003**, *35*, 452–464. [CrossRef] [PubMed]
34. Turóczy, Z.; Kis, P.; Török, K.; Cserháti, M.; Lendvai, A.; Dudits, D.; Horváth, G.V. Overproduction of a rice aldo-keto reductase increases oxidative and heat stress tolerance by malondialdehyde and methylglyoxal detoxification. *Plant Mol. Biol.* **2011**, *75*, 399–412. [CrossRef] [PubMed]
35. Zeng, T.; Liu, S.; Luo, R.; Gong, P.; Zhao, D.; Fang, X. Cloning and expression of an alcohol dehydrogenase from *Lotus japonicus* and characterization of LjADH1. *Legume Genom. Genet.* **2011**, *2*, 6–13.

36. Karuppanapandian, T.; Rhee, S.J.; Kim, E.J.; Han, B.K.; Hoekenga, O.A.; Lewe, G.P. Proteomic analysis of differentially expressed proteins in the roots of Columbia-0 and Landsberg erecta ecotypes of *Arabidopsis thaliana* in response to aluminum toxicity. *Can. J. Plant Sci.* **2012**, *92*, 1267–1282. [CrossRef]

37. Jiang, H.X.; Tang, N.; Zheng, J.G.; Li, Y.; Chen, L.S. Phosphorus alleviates aluminum-induced inhibition of growth and photosynthesis in *Citrus grandis* seedlings. *Physiol. Plant.* **2009**, *137*, 298–311. [CrossRef] [PubMed]

38. Hernández-Domíguez, E.E.; Valencia-Turcotte, L.G.; Rodríguez-Sotres, R. Changes in expression of soluble inorganic pyrophosphatases of *Phaseolus vulgaris* under phosphate starvation. *Plant Sci.* **2012**, *187*, 39–48. [CrossRef] [PubMed]

39. Bariola, P.A.; Howard, C.J.; Taylor, C.B.; Verburg, M.T.; Jaglan, V.D.; Green, P.J. The *Arabidopsis* ribonuclease gene RNS1 is tightly controlled in response to phosphate limitation. *Plant J.* **1994**, *6*, 673–685. [CrossRef] [PubMed]

40. Cheng, Y.; Zhou, W.; El Sheery, N.I.; Peters, C.; Li, M.; Wang, X.; Huang, J. Characterization of the *Arabidopsis* glycerophosphodiester phosphodiesterase (GDPD) family reveals a role of the plastid-localized AtGDPD1 in maintaining cellular phosphate homeostasis under phosphate starvation. *Plant J.* **2011**, *66*, 781–795. [CrossRef] [PubMed]

41. Gregory, A.L.; Hurley, B.A.; Tran, H.T.; Valentine, A.J.; She, Y.M.; Knowles, V.L.; Plaxton, W.C. In vivo regulatory phosphorylation of the phosphoenolpyruvate carboxylase AtPPC1 in phosphate-starved *Arabidopsis thaliana*. *Biochem. J.* **2009**, *420*, 57–65. [CrossRef] [PubMed]

42. Caparrós-Martín, J.A.; McCarthy-Suárez, I.; Culiáñez-Macià, F.A. HAD hydrolase function unveiled by substrate screening: Enzymatic characterization of *Arabidopsis thaliana* subclass I phosphosugar phosphatase AtSgpp. *Planta* **2013**, *237*, 943–954. [CrossRef] [PubMed]

43. Delhaize, E.; Ma, J.F.; Ryan, P.R. Transcriptional regulation of aluminium tolerance genes. *Trends Plant Sci.* **2012**, *17*, 341–348. [CrossRef] [PubMed]

44. García-Oliveira, A.L.; Benito, C.; Prieto, P.; de Andrade Menezes, R.; Rodrigues-Pousada, C.; Guedes-Pinto, H.; Martins-Lopes, P. Molecular characterization of TaSTOP1 homoeologues and their response to aluminium and proton (H$^+$) toxicity in bread wheat (*Triticum aestivum* L.). *BMC Plant Biol.* **2013**, *13*, 134. [CrossRef] [PubMed]

45. Iuchi, S.; Koyama, H.; Iuchi, A.; Kobayashi, A.; Kitabayashi, S.; Kobayashi, Y.; Ikka, T.; Hirayama, T.; Shinozaki, K.; Kobayashi, M. Zinc finger protein STOP1 is critical for proton tolerance in *Arabidopsis* and coregulates a key gene in aluminum tolerance. *Proc. Natl. Acad. Sci. USA* **2007**, *104*, 9900–9905. [CrossRef] [PubMed]

46. Duressa, D.; Soliman, K.M.; Taylor, R.W.; Chen, D. Gene expression profiling in soybean under aluminum stress: Genes differentially expressed between Al-tolerant and Al-sensitive genotypes. *Am. J. Mol. Biol.* **2011**, *1*, 156–173. [CrossRef]

47. Chen, M.; Zhao, Y.; Zhuo, C.; Lu, S.; Guo, Z. Overexpression of a NF-YC transcription factor from bermudagrass confers tolerance to drought and salinity in transgenic rice. *Plant Biotechnol. J.* **2015**, *13*, 482–491. [CrossRef] [PubMed]

48. Nelson, D.E.; Repetti, P.P.; Adams, T.R.; Creelman, R.A.; Wu, J.; Warner, D.C.; Anstrom, D.C.; Bensen, R.J.; Castiglioni, P.P.; Donnarummo, M.G.; et al. Plant nuclear factor Y (NF-Y) B subunits confer drought tolerance and lead to improved corn yields on water-limited acres. *Proc. Natl. Acad. Sci. USA* **2007**, *104*, 16450–16455. [CrossRef] [PubMed]

49. Floris, M.; Mahgoub, H.; Lanet, E.; Robaglia, C.; Menand, B. Post-transcriptional regulation of gene expression in plants during abiotic stress. *Int. J. Mol. Sci.* **2009**, *10*, 3168–3185. [CrossRef] [PubMed]

50. Liu, J.; Piñeros, M.A.; Kochian, L.V. The role of aluminum sensing and signaling in plant aluminum resistance. *J. Integr. Plant Biol.* **2014**, *56*, 221–230. [CrossRef] [PubMed]

51. Jiang, S.C.; Mei, C.; Liang, S.; Yu, Y.T.; Lu, K.; Wu, Z.; Wang, X.F.; Zhang, D.P. Crucial roles of the pentatricopeptide repeat protein SOAR1 in *Arabidopsis* response to drought, salt and cold stresses. *Plant Mol. Biol.* **2015**, *88*, 369–385. [CrossRef] [PubMed]

52. Lv, H.X.; Huang, C.; Guo, G.Q.; Yang, Z.N. Roles of the nuclear encoded chloroplast SMR domain-containing PPR protein SVR7 in photosynthesis and oxidative stress tolerance in *Arabidopsis*. *J. Plant Biol.* **2014**, *57*, 291–301. [CrossRef]

53. Lee, B.H.; Kapoor, A.; Zhu, J.; Zhu, J.K. STABILIZED1, a stress-upregulated nuclear protein, is required for pre-mRNA splicing, mRNA turnover, and stress tolerance in *Arabidopsis*. *Plant Cell* **2006**, *18*, 1736–1749. [CrossRef] [PubMed]

54. Cao, H.F.; Gao, J.X.; Shu, J.M. Study on the response of *Pinus massoniana* seedling to aluminum. *Acta Ecol. Sin.* **1992**, *12*, 239–246.

55. Hait, W.N.; Versele, M.; Yang, J.M. Surviving metabolic stress: Of mice (squirrels) and men. *Cancer Discov.* **2014**, *4*, 646–649. [CrossRef] [PubMed]

56. Kumari, M.; Taylor, G.J.; Deyholos, M.K. Transcriptomic responses to aluminum stress in roots of *Arabidopsis thaliana*. *Mol. Genet. Genom.* **2008**, *279*, 339–357. [CrossRef] [PubMed]

57. García-Lorenzo, M.; Sjödin, A.; Jansson, S.; Funk, C. Protease gene families in *Populus* and *Arabidopsis*. *BMC Plant Biol.* **2006**, *6*, 30. [CrossRef] [PubMed]

58. Lyzenga, W.J.; Stone, S.L. Abiotic stress tolerance mediated by protein ubiquitination. *J. Exp. Bot.* **2012**, *63*, 599–616. [CrossRef] [PubMed]

59. Link, B.M.; Cosgrove, D.J. Acid-growth response and α-expansins in suspension cultures of bright yellow 2 tobacco. *Plant Physiol.* **1988**, *118*, 907–916. [CrossRef]

60. Zhang, J.; He, Z.; Tian, H.; Zhu, G.; Peng, X. Identification of aluminium-responsive genes in rice cultivars with different aluminium sensitivities. *J. Exp. Bot.* **2007**, *58*, 2269–2278. [CrossRef] [PubMed]

61. Ramachandran, S.; Christensen, H.E.M.; Ishimaru, Y.; Dong, C.H.; Wen, C.M.; Cleary, A.L.; Chua, N.H. Profilin plays a role in cell elongation, cell shape maintenance, and flowering in *Arabidopsis*. *Plant Physiol.* **2000**, *124*, 1637–1647. [CrossRef] [PubMed]

62. Hamilton, C.A.; Good, A.G.; Taylor, J.T. Induction of vacuolar ATPase and mitochondrial ATP synthase by aluminum in an aluminum-resistant cultivar of wheat. *Plant Physiol.* **2001**, *125*, 2068–2077. [CrossRef] [PubMed]

63. Hamilton, C.A.; Good, A.G.; Taylor, J.T. Vacuolar H^+-ATPase, but not mitochondrial F_1F_0-ATPase, is required for aluminum resistance in *Saccharomyces cerevisiae*. *FEMS Microbiol. Lett.* **2001**, *205*, 231–236. [CrossRef] [PubMed]

64. Briat, J.F.; Ravet, K.; Arnaud, N.; Duc, C.; Boucherez, J.; Touraine, B. New insights into ferritin synthesis and function highlight a link between iron homeostasis and oxidative stress in plants. *Ann. Bot.* **2010**, *105*, 811–822. [CrossRef] [PubMed]

65. Lira-Ruan, V.; Ross, E.J.H.; Sarath, G.; Klucas, R.V.; Arredondo-Peter, R. Mapping and analysis of a hemoglobin gene family from *Oryza sativa*. *Plant Physiol. Biochem.* **2002**, *40*, 199–202. [CrossRef]

66. Vigeolas, H.; Hühn, D.; Geigenberger, P. Nonsymbiotic hemoglobin-2 leads to an elevated energy state and to a combined increase in polyunsaturated fatty acids and total oil content when overexpressed in developing seeds of transgenic *Arabidopsis* plants. *Plant Physiol.* **2011**, *155*, 1435–1444. [CrossRef] [PubMed]

67. Nie, X.; Hill, R.D. Mitochondrial respiration and hemoglobin gene expression in barley aleurone tissue. *Plant Physiol.* **1997**, *114*, 835–840. [PubMed]

68. Hunt, P.W.; Klok, E.J.; Trevaskis, B.; Watts, R.A.; Ellis, M.H.; Peacock, W.J.; Dennis, E.S. Increased level of hemoglobin 1 enhances survival of hypoxic stress and promotes early growth in *Arabidopsis thaliana*. *Proc. Natl. Acad. Sci. USA* **2002**, *99*, 17197–17202. [CrossRef] [PubMed]

69. Nechushtai, R.; Conlan, A.R.; Harir, Y.; Song, L.; Yogev, O.; Eisenberg-Domovich, Y.; Livnah, O.; Michaeli, D.; Rosen, R.; Ma, V.; et al. Characterization of *Arabidopsis* NEET reveals an ancient role for NEET proteins in iron metabolism. *Plant Cell* **2007**, *24*, 2139–2154. [CrossRef] [PubMed]

70. Wang, C.; Yan, X.; Chen, Q.; Jiang, N.; Fu, W.; Ma, B.; Liu, J.; Li, C.; Bednarek, S.Y.; Pan, J. Clathrin light chains regulate clathrin-mediated trafficking, auxin signaling, and development in *Arabidopsis*. *Plant Cell* **2013**, *25*, 499–516. [CrossRef] [PubMed]

71. Lam, S.K.; Cai, Y.; Hillmer, S.; Robinson, D.G.; Jiang, L.W. SCAMPs highlight the developing cell plate during cytokinesis in tobacco BY-2 cells. *Plant Physiol.* **2008**, *147*, 1637–1645. [CrossRef] [PubMed]

72. Pan, X.; Chen, Z.; Yang, X.; Liu, G. *Arabidopsis* voltage-dependent anion channel 1 (AtVDAC1) is required for female development and maintenance of mitochondrial functions related to energy-transaction. *PLoS ONE* **2014**, *9*, e106941. [CrossRef] [PubMed]

73. Sun, X.H.; Yu, G.; Li, J.T.; Jia, P.; Zhang, J.C.; Jia, C.G.; Zhang, Y.H.; Pan, H.Y. A heavy metal-associated protein (AcHMA1) from the halophyte, *Atriplex canescens* (Pursh) Nutt.; confers tolerance to iron and other abiotic stresses when expressed in *Saccharomyces cerevisiae*. *Int. J. Mol. Sci.* **2014**, *15*, 14891–14906. [CrossRef] [PubMed]

74. Tian, L.; Peel, G.J.; Lei, Z.; Aziz, N.; Dai, X.; He, J.; Watson, B.; Zhao, P.X.; Sumner, L.W.; Dixon, R.A. Transcript and proteomic analysis of developing white lupin (*Lupinus albus* L.) roots. *BMC Plant Biol.* **2009**, *9*. [CrossRef] [PubMed]

75. Santino, A.; Taurino, M.; De Domenico, S.; Bonsegna, S.; Poltronieri, P.; Pastor, V.; Flors, V. Jasmonate signaling in plant development and defense response to multiple (a)biotic stresses. *Plant Cell Rep.* **2013**, *32*, 1085–1098. [CrossRef] [PubMed]

76. Peixoto, P.H.P.; Cambraia, J.; Sant'Anna, R.; Mosquim, P.R.; Moreira, M.A. Aluminum effects on lipid peroxidation and on the activities of enzymes of oxidative metabolism in sorghum. *R. Bras. Fisiol. Veg.* **1999**, *11*, 137–143.

77. Li, C.; Schilmiller, A.L.; Liu, G.; Lee, G.I.; Jayanty, S.; Sageman, C.; Vrebalov, J.; Giovannoni, J.J.; Yagi, K.; Kobayashi, Y.; et al. Role of β-oxidation in jasmonate biosynthesis and systemic wound signaling in tomato. *Plant Cell* **2005**, *17*, 971–986. [CrossRef] [PubMed]

78. Naranjo, M.A.; Forment, J.; Roldan, M.; Serrano, R.; Vicente, O. Overexpression of *Arabidopsis thaliana* LTL1, a salt-induced gene encoding a GDSL-motif lipase, increases salt tolerance in yeast and transgenic plants. *Plant Cell Environ.* **2006**, *29*, 1890–1900. [CrossRef] [PubMed]

79. Gujjar, R.S.; Akhtar, M.; Rai, A.; Singh, M. Expression analysis of drought induced genes in wild tomato line (*Solanum habrochaites*). *Curr. Sci.* **2014**, *107*, 496–502.

80. Clay, N.K.; Nelson, T. VH1, a provascular cell-specific receptor kinase that influences leaf cell patterns in *Arabidopsis*. *Plant Cell* **2002**, *14*, 2707–2722. [CrossRef] [PubMed]

81. País, S.M.; Téllez-Iñón, M.; Capiati, D.A. Serine/threonine protein phosphatases type 2A and their roles in stress signaling. *Plant Signal. Behav.* **2009**, *4*, 1013–1015. [CrossRef] [PubMed]

82. Park, J.H.; Lee, S.Y.; Kim, W.Y.; Jung, Y.J.; Chae, H.B.; Jung, H.S.; Kang, C.H.; Shin, M.R.; Kim, S.Y.; Suudi, M.; et al. Heat-induced chaperone activity of serine/threonine protein phosphatase 5 enhances thermotolerance in *Arabidopsis thaliana*. *New Phytol.* **2011**, *191*, 692–705. [CrossRef] [PubMed]

83. Bradford, M.M. A rapid and sensitive method for quantitation of microgram quantities of protein utilizing the principle of protein-dye binding. *Anal. Biochem.* **1976**, *72*, 248–254. [CrossRef]

84. Hsu, P.H. Effect of initial pH, phosphate, and silicate on the determination of aluminum with aluminon. *Soil Sci.* **1963**, *96*, 230–238. [CrossRef]

85. Yang, L.T.; Qi, Y.P.; Lu, Y.B.; Guo, P.; Sang, W.; Feng, H.; Zhang, H.X.; Chen, L.S. iTRAQ protein profile analysis of *Citrus sinensis* roots in response to long-term boron-deficiency. *J. Proteom.* **2013**, *93*, 179–206. [CrossRef] [PubMed]

86. Yang, L.T.; Lu, Y.B.; Zhang, Y.; Guo, P.; Chen, L.S. Proteomic profile of *Citrus grandis* roots under long-term boron-deficiency revealed by iTRAQ. *Trees Struct. Funct.* **2016**, *30*, 1057–1071. [CrossRef]

87. Gan, C.S.; Chong, P.K.; Pham, T.K.; Wright, P.C. Technical, experimental, and biological variations in isobaric tags for relative and absolute quantitation (iTRAQ). *J. Proteome Res.* **2007**, *6*, 821–827. [CrossRef] [PubMed]

88. Zhou, C.P.; Qi, Y.P.; You, X.; Yang, L.T.; Guo, P.; Ye, X.; Zhou, X.X.; Ke, F.J.; Chen, L.S. Leaf cDNA-AFLP analysis of two citrus species differing in manganese tolerance in response to long-term manganese toxicity. *BMC Genom.* **2013**, *14*, 621. [CrossRef] [PubMed]

89. Li, Q.; Chen, L.S.; Jiang, H.X.; Tang, N.; Yang, L.T.; Lin, Z.H.; Li, Y.; Yang, G.H. Effects of manganese-excess on CO_2 assimilation, ribulose-1,5-bisphosphate carboxylase/oxygenase, carbohydrates and photosynthetic electron transport of leaves, and antioxidant systems of leaves and roots in *Citrus grandis* seedlings. *BMC Plant Biol.* **2010**, *10*, 42. [CrossRef] [PubMed]

90. Axelrod, B.; Cheesbrough, T.M.; Laakso, S. Lipoxygenase from soybeans. *Methods Enzymol.* **1981**, *71*, 441–451.

International Journal of
Molecular Sciences

Article

Isobaric Tags for Relative and Absolute Quantitation (iTRAQ)-Based Comparative Proteome Analysis of the Response of Ramie under Drought Stress

Xia An [1,2], Jingyu Zhang [1], Lunjin Dai [1], Gang Deng [3], Yiwen Liao [1], Lijun Liu [1], Bo Wang [1,*] and Dingxiang Peng [1,*]

[1] Key Laboratory of Crop Ecophysiology and Farming Systems in the Middle Reaches of the Yangtze River, Ministry of Agriculture, College of Plant Science and Technology, Huazhong Agricultural University, Wuhan 430070, China; anxia@webmail.hzau.edu.cn (X.A.); zhangjingyu0408@gmail.com (J.Z.); dailunjin@gmail.com (L.D.); lywen0902@gmail.com (Y.L.); liulijun@mail.hzau.edu.cn (L.L.)
[2] Flower Research and Development Centre, Zhejiang Academy of Agricultural Sciences, Hangzhou 311202, China
[3] School of Agricultural Science, Yunnan University, Kunming 650091, China; denggang1986@ynu.edu.cn
* Correspondence: wangbo@mail.hzau.edu.cn (B.W.); pdxiang@mail.hzau.edu.cn (D.P.); Tel./Fax: +86-27-8728-7136 (B.W. & D.P.)

Academic Editor: Setsuko Komatsu
Received: 18 May 2016; Accepted: 15 September 2016; Published: 27 September 2016

Abstract: In this study, we conducted the first isobaric tags for relative and absolute quantitation (*isobaric tags for relative and absolute quantitation* (iTRAQ))-based comparative proteomic analysis of ramie plantlets after 0 (minor drought stress), 24 (moderate drought stress), and 72 h (severe drought stress) of treatment with 15% (*w/v*) poly (ethylene glycol)6000 (PEG6000) to simulate drought stress. In our study, the association analysis of proteins and transcript expression revealed 1244 and 968 associated proteins identified in leaves and roots, respectively. L1, L2, and L3 are leaf samples which were harvested at 0, 24, and 72 h after being treated with 15% PEG6000, respectively. Among those treatment groups, a total of 118, 216, and 433 unique proteins were identified as differentially expressed during L1 vs. L2, L2 vs. L3, and L1 vs. L3, respectively. R1, R2, and R3 are root samples which were harvested at 0, 24, and 72 h after being treated with 15% PEG6000, respectively. Among those treatment groups, a total of 124, 27, and 240 unique proteins were identified as differentially expressed during R1 vs. R2, R2 vs. R3, and R1 vs. R3, respectively. Bioinformatics analysis indicated that glycolysis/gluconeogenesis was significantly upregulated in roots in response to drought stress. This enhancement may result in more glycolytically generated adenosine triphosphate (ATP) in roots to adapt to adverse environmental conditions. To obtain complementary information related to iTRAQ data, the mRNA levels of 12 proteins related to glycolysis/gluconeogenesis in leaves and 7 in roots were further analyzed by qPCR. Most of their expression levels were higher in R3 than R1 and R2, suggesting that these compounds may promote drought tolerance by modulating the production of available energy.

Keywords: comparative proteome analysis; drought stress; ramie; isobaric tags for relative and absolute quantitation (iTRAQ)

---

1. Introduction

Among several factors controlling plant growth, water plays a vital role [1]. A global water shortage is a very serious environmental problem. A looming water crisis, which is lead by poor water management, increased competition of limited water resources, and the uncertain consequences of global warming, is threatening agricultural productivity world widely [2]. With increasingly limited water resources of agriculture, developing the tolerance of crops to water shortages might be the most economical way to improve agricultural productivity [3]. Therefore, an urgent need exists to enhance crop tolerance to drought stress.

Plants can respond to adverse environments with several physiological and biochemical strategies that were derived from a long-term domestication process [2]. Plant response to drought stress is a complex course, and several mechanisms include drought resistance, which include drought escape via a developmental plasticity, drought avoidance through reducing water loss and enhancing water uptake, and drought tolerance by means of antioxidant capacity and osmotic adjustment [4]. In order to defend against drought stress, plants undergo a process of stress acclimation. This process may require changes in a large number of stress-related gene expressions [5–7] and synthesis of diverse functional proteins [8–10]. Recently, expanding transcriptome data sets has uncovered that many genes were induced or repressed upon drought stress in *Arabidopsis* [11], maize [12], rice [13], soybean [14], and ramie [15]. This study was lead to an understanding of the drought stress regulatory mechanism. However, several transcripts will experience transcriptional, translational, and post-translational modifications, revealing that the potential drought stress molecular mechanisms via differentially expressed gene identification are not comprehensive enough [16].

Ramie (*Boehmeria nivea* L.) is an important natural fiber crop. Ramie is grown on about 80,000 ha with an annual fiber production of 150,000 t in 2012 (FAOSTAT, http://faostat3.fao.org). In China, ramie is the second most important fiber crop, behind cotton in crop acreage and fiber production [15]. Ramie grows vigorously in well-watered cultivation environments, resulting in a high yield of vegetative fiber extracted from stem bast. Good irrigation is essential for this crop, as fiber yield is reduced under drought stress [17]. To avoid competing land with food crops, ramie can be transferred to arid or semiarid hilly mountainous areas, where it will face a more serious drought threat in the future. High fiber yields were obtained in drought-tolerant cultivars of ramie with root systems, leaf responses, cellular responses, and biochemical activities that allowed high levels of photosynthesis and carbon deposition under stress [17]. Twelve transcription factors involved in the drought response were found by Illumina tag-sequencing and qRT-PCR in ramie [15]. However, the levels of mRNA and proteins did not always correlate well [10]. Protein expression changes in response to drought stress have been studied in some other plants, and drought stress-induced proteins involved in photosynthesis [18], signaling pathways [19], oxidative stress detoxification [20], and transport [21] have been identified. However, the specific proteins induced in ramie under drought conditions remain unknown.

Here, we provide an iTRAQ-based comparative analysis of the drought-resistant response of ramie. This is the first use of iTRAQ to research on the molecular mechanisms of ramie related to drought stress. Identifying and quantifying multiple sample proteins simultaneously is the advantage of this approach [22]. Proteins that are too large or small, too acidic or basic, too hydrophobic, or in low abundance are difficult to observe via 2D gel electrophoresis, but can be identified by iTRAQ [22,23]. We imposed drought stress by PEG to evaluate plant drought-tolerance preliminarily [24]. In this study, the leaves and roots of "Huazhu No. 5" were harvested 0 (L1 and R1), 24 (L2 and R2), and 72 (L3 and R3) hours after being treated with 15% (*w/v*) PEG6000. Our results provide new insights into the ramie response to drought stress.

2. Results and Discussion

2.1. Analytical Strategy for Proteome Identification under Drought Stress

The leaves only slightly yellowed and curled after 24 h of drought treatment (Figure 1c) but showed severe chlorosis and stopped growing after 72 h (Figure 1d). The relative water content (RWC) of leaves seemed to decline consistently [25].

Many abiotic stresses trigger the production of reactive oxygen species (ROS), which disrupts normal metabolism by causing oxidative damage to membrane lipids, proteins, and nucleic acids [26].

Peroxidase activity (POD) can eliminate these harmful molecules [27]. The POD activity first increased but later decreased [25]. Based on the critical time-points for RWC and the POD activity [25], three critical time-points (0, 24, and 72 h after drought stress) were screened for morphological (Figure 1) and physiological results [25]. An iTRAQ-based quantitative proteome analysis was performed for a global view of the proteome responses to different durations of drought treatments (Figure S1). Protein mass distribution in ramie is shown in Figure S2.

Figure 1. (**a**) After two-week propagation from stem cuttings, the plantlets of "Huazhu No. 5" were transplanted into half-strength Hoagland's solution for 20 days. Leaf and root samples of the same sizes were harvested at 0 h (**b**), 24 h(**c**), and 72 h (**d**) after materials had been treated with 15% (*w/v*) PEG6000 to induce drought stress. They were cultured under cool white fluorescent light in 16/8 h (light/dark) with a relative humidity of 50%–70% and temperatures about 25 ± 2 °C in the daytime and 20 ± 2 °C at night. Scale bar = 2 cm.

2.2. Correlation Coefficients of Biological Replicates

To determine differentially expressed proteins in leaves (L1, L2, and L3) and roots (R1, R2, and R3), the correlation coefficients between pairs of biological replicates were first evaluated (Figure 2). The two L1 and R1 controls were used as denominators, respectively. We used the proteins that were quantified with iTRAQ ratios to calculate correlation coefficients. The ratios (L2L3 vs. L1 and R2R3 vs. R1) were then log-transformed and plotted against each other. As illustrated in Figure 2, all correlation coefficients of the biological replicates were equal to or greater than 0.8 [28], indicating the excellent biological reproducibility of drought-regulated protein expression.

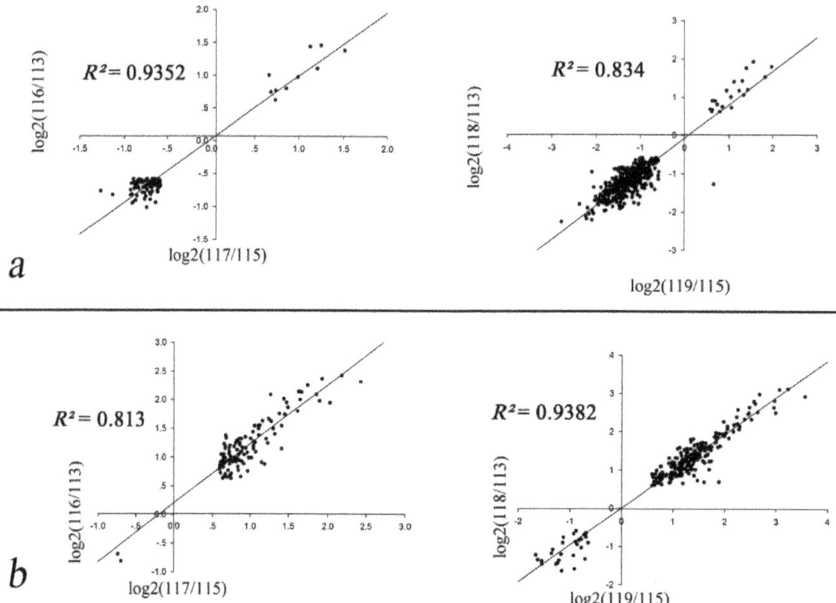

Figure 2. The correlation coefficients were calculated between two biological replicates. The ratios of quantified proteins were log-transformed and plotted. (**a**) Leaves; (**b**) roots. Tags 113 and 115 in Figure 2a represent repeat (1) and repeat (2) of 0 h of leaf treatment with 15% (*w/v*) PEG6000, respectively. Tags 116 and 117 in Figure 2a represent repeat (1) and repeat (2) of 24 h of leaf treatment with 15% (*w/v*) PEG6000, respectively. Tags 118 and 119 in Figure 2a represent repeat (1) and repeat (2) of 72 h of leaf treatment with 15% (*w/v*) PEG6000, respectively. Tags 113 and 115 in Figure 2b represent repeat (1) and repeat (2) of 0 h of root treatment with 15% (*w/v*) PEG6000, respectively. Tags 116 and 117 in Figure 2b represent repeat (1) and repeat (2) of 24 h of root treatment with 15% (*w/v*) PEG6000, respectively. Tags 118 and 119 in Figure 2b represent repeat (1) and repeat (2) of 72 h of root treatment with 15% (*w/v*) PEG6000, respectively. The ratios (L2L3 vs. L1 and R2R3 vs. R1) were then log-transformed and plotted against each other.

2.3. Functional Classification and Annotation

We conducted gene ontology (GO) functional annotation analysis for all identified proteins. The results cover a wide range of biological processes, cellular components, and molecular functions, including 44 important functional groups (Figure 3; Table S1). The largest subcategory in the biological process category was "metabolic processes" and the second was "cellular processes". In the cellular component category, "cell", "cell part", and "organelle" were the main categories. "Binding" and "catalytic activity" were the main categories of molecular function.

Identified proteins were classified according to their biological functions using the NCBI COG (Cluster of Orthologous Groups of proteins) database (http://www.ncbi.nlm.nih.gov/COG/). All leaf proteins were classified into 22 COG subcategories (Figure 4a; Table S2) including general function prediction only (13.92%); posttranslational modification, protein turnover, and chaperones (13.57%); carbohydrate transport and metabolism (12.30%); energy production and conversion (10.79%); and translation, ribosomal structure, and biogenesis (10.10%). All root proteins were classified into 21 subcategories (Figure 4b; Table S2) including general function prediction only (12.66%); carbohydrate transport and metabolism (12.34%); posttranslational modification, protein turnover, and chaperones (12.34%); energy production and conversion (11.87%); translation, ribosomal structure, and biogenesis (10.00%), and amino acid transport and metabolism (9.38%). In this study, the identified proteins were mainly involved in general function prediction only (R); carbohydrate transport and metabolism (G); energy production and conversion (C), posttranslational modification, protein turnover, and chaperones (O); translation, ribosomal structure, and biogenesis (J), and amino acid transport and metabolism (E) (Figure 4). Proteins related to energy metabolism, stress resistance, "cell growth, differentiation and structure", and metabolism-related proteins have also been reported in ramie under N, P, and K deficiency [29]. The other identified proteins with COG categories are shown in Figure 4.

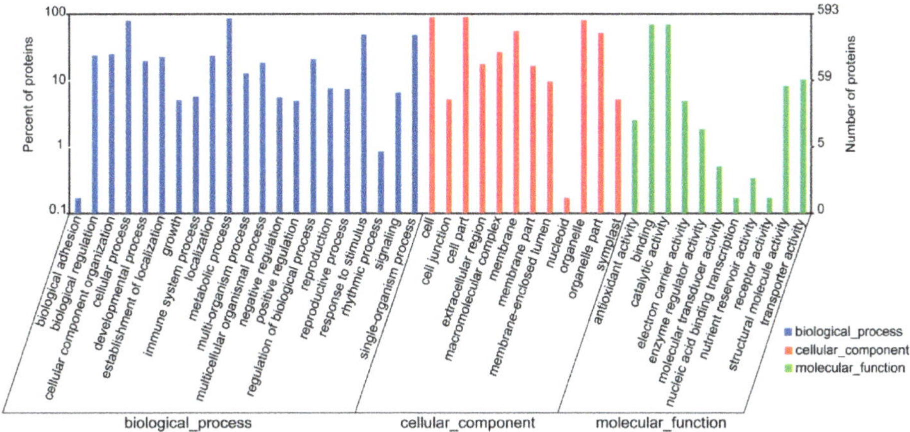

Figure 3. The categorization of proteins is based on gene ontology (GO) annotation. The category number is displayed with biological process, cellular components, and molecular functions. *y*-axis (**left**) represents percentages of proteins identified, *y*-axis (**right**) represents the protein number.

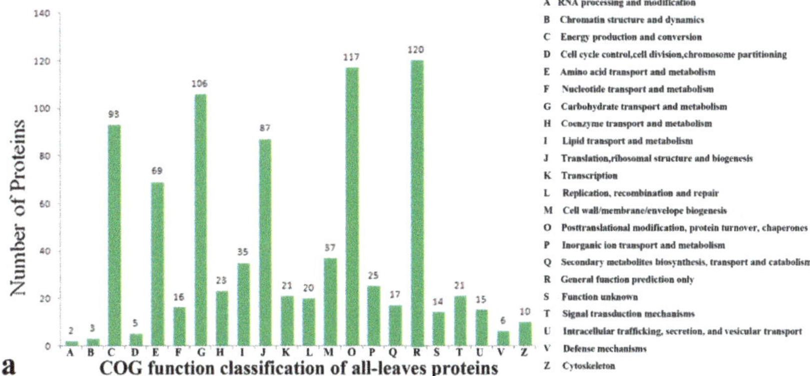

Figure 4. *Cont.*

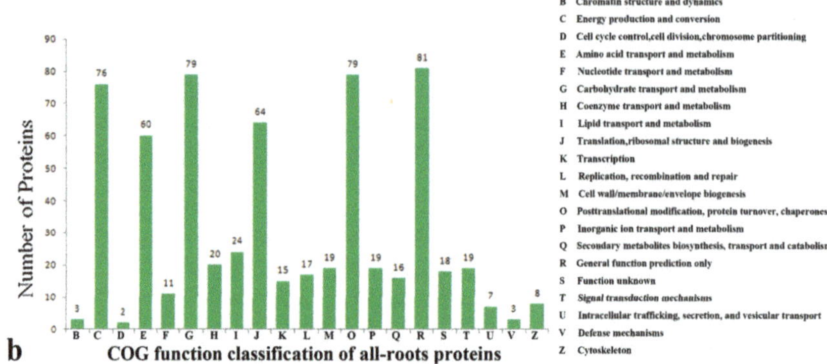

Figure 4. COG function classification of all leaf (**a**) and root (**b**) proteins.

2.4. Effects of Drought Stress on Expression Changes of the Ramie Leaf and Root Proteomes

The results showed that most proteins in leaves were downregulated under drought stress. Compared to L1 (control), 10 proteins in L2 were upregulated, and 108 proteins were downregulated, while 20 proteins in L3 were upregulated, and 413 were downregulated (Table 1). The upregulated proteins in L3 included all 10 that were upregulated in L2, and the downregulated proteins in L3 included 107 that were also downregulated in L2. Proteins that were upregulated or downregulated in L3 were selected for further analysis.

Table 1. Numbers of differently expressed proteins during drought stress. The "upregulated" row indicates the number of upregulated proteins in posterior samples compared with anterior samples. The "downregulated" row indicates the number of downregulated proteins in posterior samples compared with anterior samples.

Samples	Upregulated	Downregulated
L1–L2	10	108
L2–L3	0	216
L1–L3	20	413
R1–R2	122	2
R2–R3	20	7
R1–R3	211	29

In contrast, most root proteins were upregulated under drought stress. Compared to R1 (control), 122 proteins in R2 were upregulated, and 2 were downregulated; 211 proteins in R3 were upregulated, and 29 were downregulated (Table 1). The upregulated proteins in R3 included all 122 proteins that were upregulated in R2, and the downregulated proteins in R3 contained 2 proteins that were downregulated in R2. Proteins that were upregulated or downregulated in R3 were selected for further analysis.

The differentially regulated proteins from leaves and roots were clustered according to similarities in change profiles across all conditions. A dendrogram and colored image were produced as a cluster analysis of different samples using Cluster 3.0 (Michael Eisen, Stanford, CA, USA). Dark boxes indicate no change in expression pattern compared to the control. In Figure 5 (Table S3) and Figure 6 (Table S4), each row represents a single protein and each column represents a treatment (Figure 5: L2 on left, L3 on right; Figure 6: R2 on left, R3 on right). Cluster analysis revealed that the differentially expressed proteins in leaves could be generally divided into two groups: continuously upregulated (4.6%; Cluster I) and downregulated (95.4%; Cluster II) in response to drought stress (Figure 5). The differentially expressed proteins in roots also formed two upregulated (87.9%; Cluster I) and downregulated (12.1%; Cluster II) groups in response to drought stress (Figure 6). For Cluster II, proteins in ramie leaves were mainly involved in energy metabolism and photosynthesis. Other protein functions included secondary metabolism, starch and sucrose metabolism, disease/defense, signal transduction, cell structure, and protein synthesis. Plants need a considerable amount of ATP for sufficient energy for growth, development, and stress responses [30]. Under drought stress, the ATP synthesis process in ramie was influenced significantly. Large quantities of ATP-related proteins were downregulated, including ATP synthase alpha subunit, ATP synthase beta subunit, and ATP synthase CF1 alpha subunit. When plants are under abiotic stress, the initial response is to lower energy metabolism by reducing ATP synthesis in cells. In accordance with these results, protein abundance of ATP synthase was decreased under drought stress in spring wheat varieties Ningchun 4 [31]. Photosynthesis is sensitive to drought and other types of stress (e.g., nutrient stress) [29]. For Cluster I, proteins in ramie roots were mainly involved in energy metabolism. Other protein functions included secondary metabolism, amino acid metabolism, disease/defense, sucrose metabolism, and protein synthesis. Large quantities of ATP-related proteins were upregulated, including ATP synthase alpha subunit, ATP synthase beta subunit, and V-type proton ATPase catalytic subunit A. ATP synthase beta subunit was upregulated in drought-tolerant Tibetan wild barley genotype XZ5 but downregulated in drought-sensitive XZ54 [32]. In addition, roots and leaves of ramie also showed differential responses in accumulation of ATP synthesis-related proteins. It has been reported that these enzymes play an important role in the removal of abnormal or damaged proteins and in the fine control of some key cellular components, combining a peptidase and a chaperone activity [33]. Energy deprivation is a general symptom of photosynthetic plants under stress and ultimately arrests growth and causes cell death. Many proteins were upregulated in roots under drought stress to induce alternative glycolysis pathways to maintain energy levels. Energy deficit often enhances inherent pathways of carbohydrate metabolisms [34].

2.5. Association and Differential Expression Analysis of Proteome and Transcriptome Data

To investigate association and differential expression analysis of proteome data produced in this research and transcriptome data produced in our earlier published Illumina Paired-End sequencing project, hierarchical cluster analyses were conducted in this study. All differently expressed proteins were represented including data from L3 and R3. Approximately, 138,000 transcripts were detected from ramie under PEG6000 simulated drought stress in our earlier published Illumina Paired-End sequencing project. The expression profiles of all differentially expressed proteins and corresponding transcripts under drought stress are shown in Figure 7. The correlations of gene expression at the transcript and protein levels for the leaves and roots were 65 and 12, respectively (Table S5). The results indicated that more differentially expressed proteins were observed in ramie leaves than in ramie roots. Differentially expressed proteins in ramie leaves were mainly involved in photosynthesis and energy metabolism, disease/defense, cell structure, and protein synthesis. It is well known that the photosynthetic system and its maintenance will badly affect plant survival under abiotic stress environment [35,36]. Among the differentially expressed proteins, it was found that photosystem II (PSII) proteins were downregulated in transcription and protein levels. Drought stress has been reported to damage the photosynthetic system by causing severe disruption of the PSII complex [37] and the chloroplast envelope [38]. Similar studies have been reported in rice under drought stress [37]. Water, as the reducing agent, involves absorbed photons to provide fundamental energy for photosynthesis in green plants, so drought might be the most intense of all abiotic stresses affecting the photosynthesis process. In the transcriptional level, ATP synthase CF1 α subunit and ATP synthase γ chain chloroplastic-like isoform 1 were found to be upregulated, but downregulated in the protein level. Fructose 1,6-bisphosphate aldolase was found to be upregulated at the transcriptional level but at the protein level be downregulated. We found that most proteins differentially expressed under drought stress showed contrary trends with their corresponding transcripts (Figure 7) in ramie leaves. Plants may respond to drought stress by changing post-transcriptional regulation.

Post-transcriptional regulation is a potential target mechanism that can be deeply studied in order to elucidate the drought response in plants [39]. We found that most proteins differentially expressed under drought stress showed similar trends with their corresponding transcripts (Figure 7) in ramie roots. Chitinase was found to be upregulated at the transcriptional and protein levels in ramie roots. Previous research has indicated that the expression level of the drought-induced protein 3 (DIP3) protein obtained in roots of upland rice was up in a short time under drought tolerance. Thus, our results reveal that the class III chitinases member DIP3 may be a stress-induced protein when plants respond to stress conditions [40].

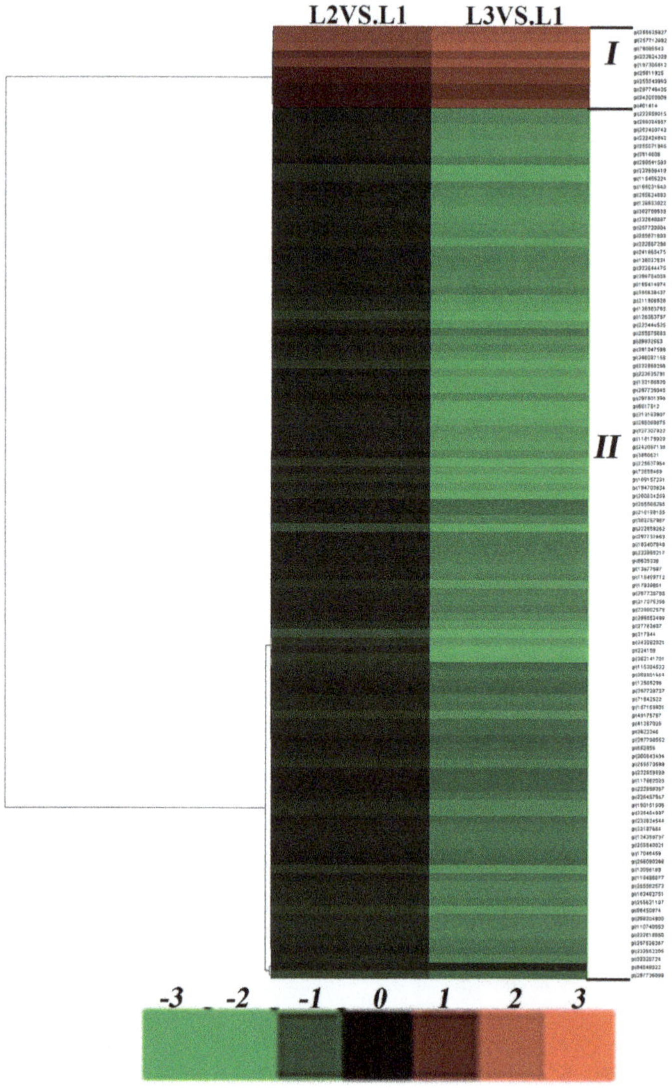

Figure 5. Hierarchical display of data from differentially expressed protein of leaves under drought stress. Upregulated proteins are in red; downregulated proteins are in green (for interpretation of the color references in the figure legend).

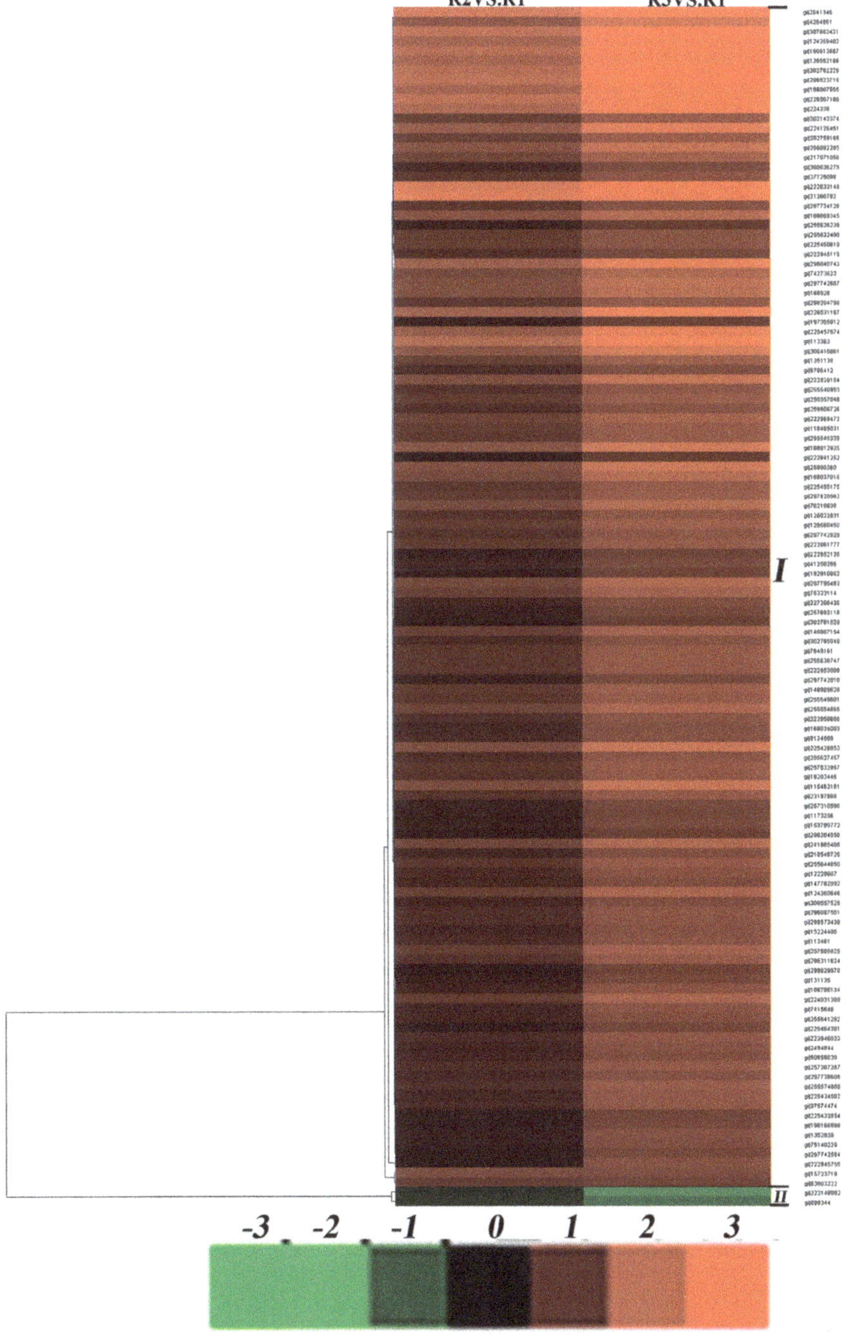

Figure 6. Hierarchical display of data from differentially expressed protein of roots under drought stresses. Upregulated proteins are in red; downregulated proteins are in green (for interpretation of the color references in the figure legend).

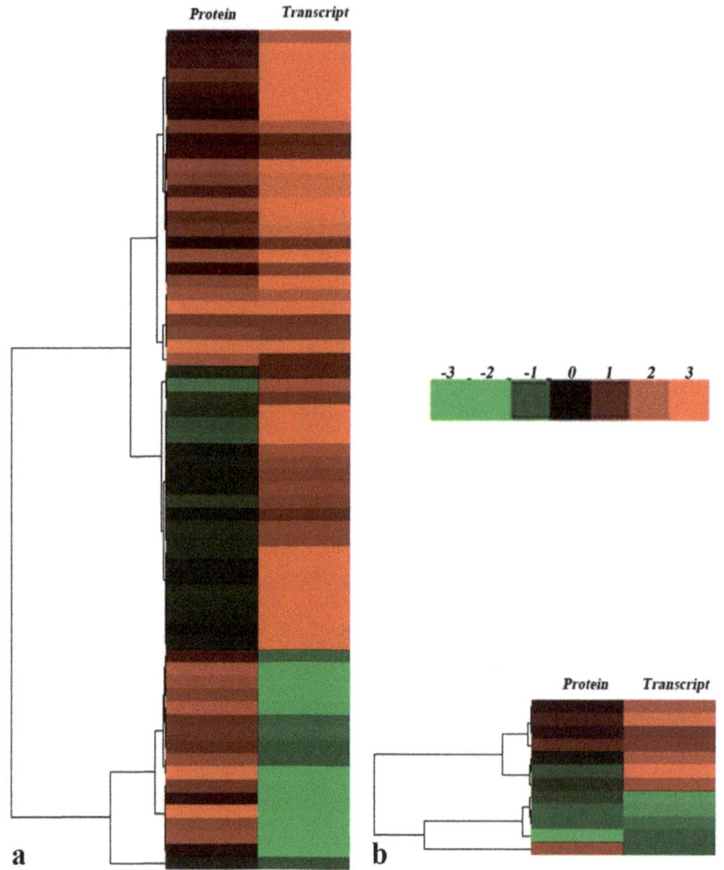

Figure 7. Association clustering analysis of differentially expressed proteins and its corresponding transcripts. Protein (**left**) represents the expression levels of differentially expressed proteins, Transcript (**right**) represents the expression profile of the corresponding genes encoding differentially expressed proteins. (**a**) leaf of ramie; (**b**) root of ramie. Upregulated proteins are in red; downregulated proteins are in green (for interpretation of the color references in the figure legend).

2.6. Identification Biochemical Reactions Significantly Upregulated in Roots by Drought Stress

ABA plays an important role in the expression of genes related to drought stress in nearly all cells [41–43]. Drought-stressed ramie roots upregulated heat shock protein (gi | 4204861 and gi | 255582806), ribosomal protein (gi | 241865406, gi | 18203445, gi | 148807154, gi | 50659630, and gi | 133793), ubiquitin-conjugating enzyme family protein (gi | 192910862), and chaperonin (gi | 108706134), all of which involved ABA signaling to enhance the cellular dehydration tolerance. In addition, many ribosome proteins have been linked to cell structure, protein translation, protein biosynthesis, and plant development in wheat [44]. Heat shock proteins play broad roles in many cellular processes in *Arabidopsis* subjected to heat stress [45].

Excessive accumulation of ROS under drought stress can disrupt normal metabolism by oxidative damage of membrane lipids, proteins, and nucleicacids [25]. Catalase (CAT) and POD were all upregulated at the protein level under drought stress. CAT and POD (among other so-called scavengers) are able to eliminate these harmful molecules. Therefore, the mechanisms of the

ROS-reducing system and the antioxidant enzyme-increasing system can play important roles in enhancing tolerance to drought stress.

Kyoto Encyclopedia of Genes and Genomes (KEGG) Orthology-Based Annotation System (http://kobas.cbi.pku.edu.cn) [46], was used to identify significant pathways involved in the response to drought stress in ramie. The drought-responsive proteins in roots represented a wide range of pathways, including metabolic pathways, glycolysis/gluconeogenesis, biosynthesis of secondary metabolites, ribosomes, oxidative phosphorylation, pyruvate metabolism, glyoxylate and dicarboxylate metabolism, phenylalanine metabolism, phenylpropanoid biosynthesis, and the citrate cycle (TCA cycle). Energy deprivation is a general symptom of stressed photosynthetic plants [47]. Photosynthesis, respiration rates, or both are dramatically reduced under stress [47–49], causing energy deprivation and growth arrest [34,49]. The energy deprivation often enhances inherent pathways of carbohydrate metabolism and induces alternative pathways of glycolysis to maintain energy [34]. Glycolysis/gluconeogenesis is an alternative bioenergetic pathway in stressed organisms (Table S6). To obtain information complementary to the iTRAQ data, the mRNA levels of 12 proteins in leaves and 7 proteins in roots related to glycolysis/gluconeogenesis were further analyzed by qPCR. Details of the glycolytic/gluconeogenetic pathway are shown in Figure 8. Eleven proteins in leaves, gi|168035690 (EC:5.4.2.2), gi|302142655 (EC:2.7.1.1), gi|297735045 (EC:4.1.2.13), gi|302774424 (EC:4.1.2.13), gi|82941449 (EC:4.1.2.13), gi|298541583 (EC:2.7.2.3), gi|129915 (EC:2.7.2.3), gi|222868326 (EC:5.4.2.1), gi|297746511 (EC:2.7.1.40), gi|118489203 (EC:1.8.1.4), and gi|298552499 (EC:1.2.1.3), were involved in glycolysis/gluconeogenesis (green in Figure 8). Six proteins in roots, gi|118481158 (EC:2.7.2.3), gi|296523718 (EC:4.1.1.1), gi|298552499 (EC:1.2.1.3), gi|2641346 (EC:1.1.1.1), gi|222845119 (EC:1.2.4.1), and gi|225450619 (EC:1.8.1.4), were also involved (red in Figure 8), while one protein, gi|297735045 (EC:4.1.2.13) in roots, was upregulated and colored by green. The upregulated expressions of these proteins may play a vital role in initiating the glycolytic/gluconeogenetic pathway under unfavorable conditions. This hypothesis is consistent with a previous observation that several glycolysis/gluconeogenesis-related genes were induced under aluminum stress in wheat [50]. Glycolysis could also be used to generate ATP to meet the energy requirement [51]. Under drought treatments, 30% of ramie proteins were related to metabolism and energy conversion. Roots of ramie appear to be able to perceive and convert stress signaling into energy status and induce alternative metabolic pathways to adjust their growth and development in response to drought stress [34]. Thus, enhancement of the glycolytic/gluconeogenetic pathway could result in more glycolytically generated ATP in roots to adapt to adverse environmental conditions [52].

2.7. Verification of Isobaric Tags for Relative and Absolute Quantitation (iTRAQ) Data on Selected Candidates by qPCR

To obtain information complementary to the iTRAQ data and KOBAS results, we also examined expression levels of genes involved in glycolysis/gluconeogenesis. RNA of the leaves and roots was extracted after 0 h, 24 h, and 72 h of PEG6000 treatments and subjected to qPCR analysis. qPCR data values for Figures 9 and 10 are shown in Table S7, using *GAPDH* as an internal control. The expression patterns of the twelve genes (gi|302774424, gi|82941449, gi|298541583, gi|129915, gi|222868326, gi|297746511, gi|118489203, gi|298552499, gi|297735045, gi|222845119, gi|168035690, and gi|302142655) were summarized in Figure 8. The qPCR results show that, upon drought stress, most genes encoding key enzymes related to glycolysis/gluconeogenesis were significantly downregulated in leaves of ramie (Figure 9), in agreement with the iTRAQ data. However, gi|82941449 and gi|297735045 were downregulated after 24 h of treatment but upregulated after 72 h. These data indicate that the transcript and protein levels of differentially expressed genes are not always consistent. The expression patterns of the seven genes in roots (gi|298552499, gi|297735045, gi|118481158, gi|222845119, gi|296523718, gi|225450619, and gi|2641346) are summarized in Figure 10. Upon drought stress, most genes encoding key enzymes related to glycolysis/gluconeogenesis were significantly upregulated in roots (Figure 10), in line with the iTRAQ data. Such remarkable activation of glycolytic/gluconeogenetic pathway suggests a strong promotion of biosynthesis of available energy [34].

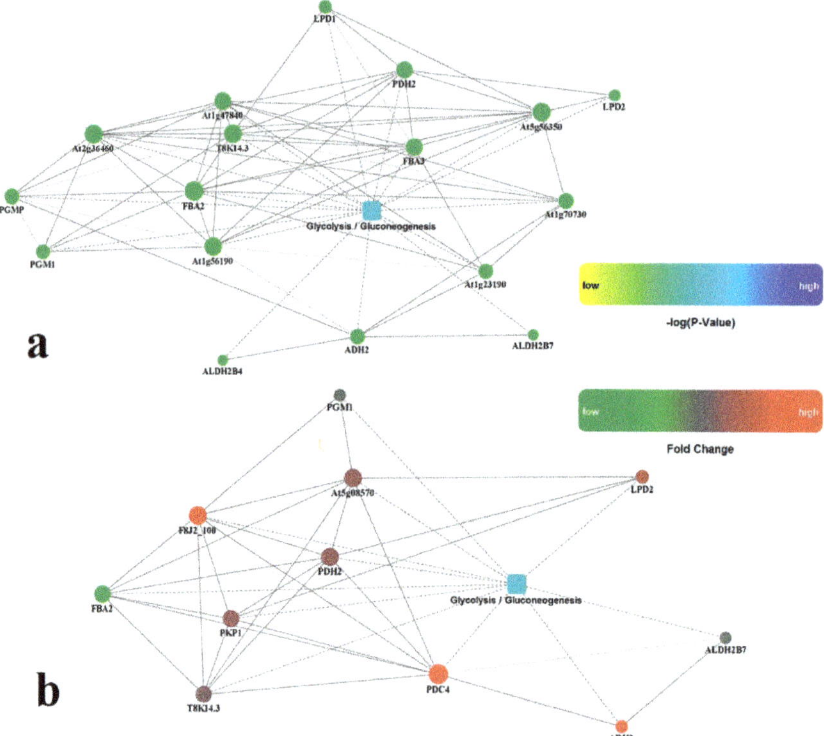

Figure 8. Regulatory changes in the pathway of glycolysis/gluconeogenesis. Colored circle nodes correspond with the ramie genes detected in the isobaric tags for relative and absolute quantitation (iTRAQ) data. (**a**) upregulated or downregulated genes in leaves; (**b**) upregulated or downregulated genes in roots. The above network model is generated with a cytoscape web application, based on information gained from up to four levels of functional analysis: fold change of gene/protein, protein-protein interaction, Kyoto Encyclopedia of Genes and Genomes (KEGG) pathway enrichment, and biological process enrichment. Circle nodes: genes/proteins; rectangle nodes: KEGG pathway or biological process. Pathways are colored in a gradient color from yellow to blue; yellow indicates a lower *p*-value, and blue indicates a higher *p*-value. Biological processes are colored in red. In the case of fold change analysis, genes/proteins are colored in red (upregulation) and green (downregulation). A default confidence cutoff of 400 was used: interactions with a higher confident score are shown as solid lines between genes/proteins; dashed lines indicate otherwise.

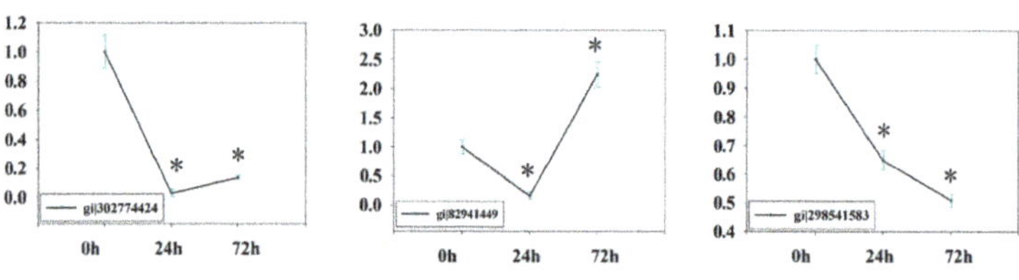

Figure 9. *Cont.*

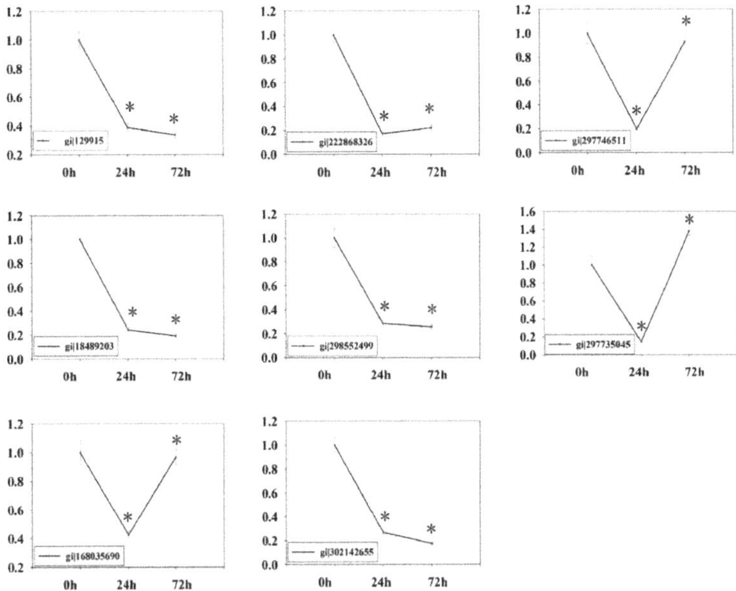

Figure 9. qPCR data for the mRNA expression levels of genes for drought-responsive proteins mapped in glycolytic/gluconeogenetic pathway in ramie leaves. The values represent relative mRNA levels against control groups (0 h samples), values of which were all set to 1 unit. Statistically significant differences in gene expression are indicated with asterisks: * $p < 0.05$.

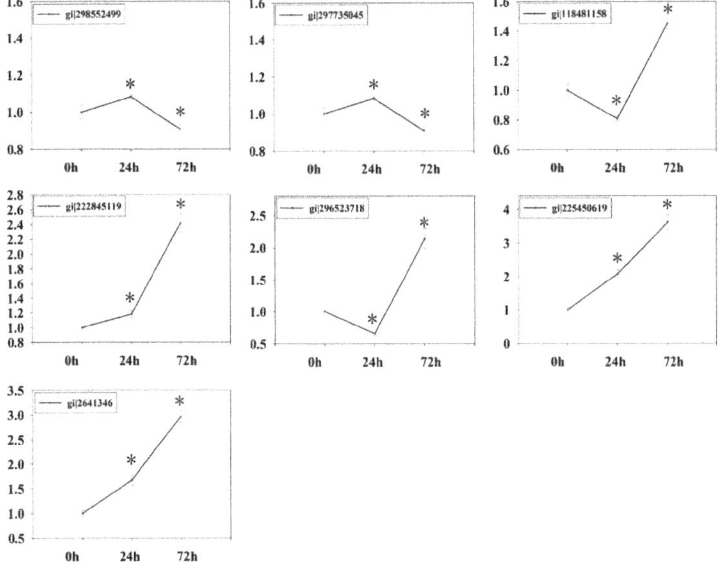

Figure 10. qPCR data for the mRNA expression levels of genes for drought-responsive proteins mapped in glycolytic/gluconeogenetic pathway in roots of ramie. The values represent relative mRNA levels against control groups (0 h samples), values of which were all set to 1 unit. Statistically significant differences in gene expression are indicated with asterisks: * $p < 0.05$.

3. Materials and Methods

3.1. Plant Materials and Stress Treatments

"Huazhu No. 5" is an elite ramie variety [53] with characteristics of high yield, good fiber quality, and high drought resistance levels. Two weeks after planting, the "Huazhu No. 5" plantlets were propagated from stem cuttings, which were transplanted into a half-strength Hoagland's solution for 20 days. The seedling stage was considered to last until the plants reached about 10 cm in height (Figure 1a). Stem cuttings were prepared for ramie plantlets. Plantlets were used as sources of the leaves and roots. They were cultured under cool white fluorescent light in 16/8 h (light/dark) with a relative humidity of 50%–70% and temperatures about 25 ± 2 °C in the daytime and 20 ± 2 °C at night.

Leaf and root samples of the same size were harvested and photographed at 0 (L1 and R1; Figure 1b), 24 (L2 and R2; Figure 1c), and 72 (L3 and R3; Figure 1d) hours after the plantlets had been treated with 15% (w/v) PEG6000 to induce drought stress with three replicates. Half-strength Hoagland's solution was replaced each day with freshly prepared solutions. Leaves and roots were collected and frozen in liquid nitrogen and stored at -80 °C prior to analysis.

3.2. Protein Extraction

Proteins from two biological replicates in each treatment were extracted and prepared as previously described with some modifications [29]. Dry protein powder was treated with a 0.5-mL lysis buffer (8 M of urea, 4% CHAPS, 40 mM of Tris-HCL, 5 mM of EDTA (Sigma, St. Louis, MO, USA), 1 mM of PMSF (Sigma, St. Louis, MO, USA), and 10 mM of Dithiothreitol (DTT) (Sigma, St. Louis, MO, USA), pH 8.0). The samples were sonicated thrice for 5 min on ice using a high intensity ultrasonic processor. The remaining debris was removed by centrifugation at $30,000 \times g$ at 4 °C for 15 min. The supernatant was transferred to a new tube, reduced with 10 mM of DTT for 1 h, and alkylated with 55 m of Miodoacetamide for 45 min at room temperature in darkness. Then, 4 volumes of prechilled acetone were added to the protein, which was precipitated for 30 min at -20 °C. After centrifugation, the pellet was then dissolved in 0.5 M of TEAB (Sigma, St. Louis, MO, USA) and sonicated for 5 min. The centrifugation step was repeated, and the supernatant was collected. Protein content was determined with a 2-D Quant kit (GE Healthcare, Piscataway, NJ, USA) according to the manufacturer's instructions.

3.3. Digestion and iTRAQ Labeling

Approximately 100 µg of protein for each sample was digested with trypsin (Promega, Madison, WI, USA) overnight at 37 °C in a 1:20 trypsin-to-protein mass ratio. After trypsin digestion, peptides were dried by vacuum centrifugation, reconstituted in 0.5 M of TEAB, and processed with an 8-plex iTRAQ kit (Applied Biosystems, Foster City, CA, USA). Briefly, one unit of iTRAQ reagent (defined as the amount of reagent required to label 100 µg of peptides) was thawed and reconstituted in 70 µL of isopropanol. Peptides from treatment and control subgroups were labeled with different iTRAQ tags by incubation at room temperature for 2 h. The iTRAQ-labeled peptide mixtures were then pooled, dried by vacuum centrifugation (Speed-Vac, Savant) and fractionated by strong cationic exchange (SCX) chromatography (Phenomenex, Guangzhou, China).

3.4. Fractionation by Strong Cationic Exchange (SCX) Chromatography

For SCX chromatography using a Shimadzu LC-20AB HPLC Pump system (Kyoto, Japan), the iTRAQ-labeled peptide mixture was reconstituted with 4 mL of buffer A (25 mM of NaH_2PO_4 in 25% ACN, pH 3.0) and loaded onto a 4.6×250-mm Ultremex SCX column containing 5-µm particles (Phenomenex, Torrance, CA, USA). The peptides were eluted at a flow rate of 1 mL/min with a gradient of buffer A for 10 min, 5%–35% buffer B (25 mM of NaH_2PO_4, 1 M of KCl in 25% ACN, pH 3.0) for 11 min, and 35%–80% buffer B for 1 min. The system was then maintained in 80% buffer B for 3 min before equilibrating with buffer A for 10 min prior to the next injection. Elution was monitored by measuring absorbance at 214 nm, and fractions were collected every 1 min. The eluted peptides were pooled as 10 fractions, desalted with a Strata X C18 column (Phenomenex), and vacuum-dried.

3.5. Liquid Chromatography-Electrospray Ionization-Tandem Mass Spectrometry (LC-ESI-MS/MS) Analysis

Each fraction was resuspended in 10 μL of buffer A (5% ACN, 0.1% FA) and centrifuged at 20,000 g for 10 min. In each fraction, the final concentration of peptide was about 0.5 μg/uL on average using a NanoDrop (Thermo Fisher Scientific, Wilmington, DE, USA) in conjunction with the Scopes method [54]. Supernatant (8 μL) was loaded on a Dionex Ultimate 3000 UPLC system by the autosampler onto a C18 trap column at 8 μL/min for 4 min, and the peptides were eluted onto an analytical C18 column (inner diameter 75 μm) packed in-house. The 40-min gradient was run at 300 nL/min starting from 2% to 35% B (95% ACN, 0.1% FA), followed by a 5-min linear gradient to 80%, maintenance at 80% B for 4 min, and a return to 5% in 1 min.

The peptides were subjected to nanoelectrospray ionization followed by tandem mass spectrometry (MS/MS) in a nano-ESI LTQ-Velos Pro Orbitrap-Elite mass spectrometer (ThermoFisher Scientific, San Jose, CA, USA) coupled online to the ultra performance liquid chromatography (UPLC). Intact peptides were detected in the Orbitrap at a resolution of 60,000. MS scans ranged from 350 to 2000 m/z. MS/MS was performed using a high energy collision dissociation (HCD) operating mode with a normalized collision energy setting of 45%. MS/MS spectra of up to 15 of the most intense ions were acquired. Isolation width was set as 2 m/z and dynamic exclusion was set as 30 s. The electrospray voltage applied was 1.8 kV. The mass spectrometry proteomics data have been deposited to the ProteomeXchange Consortium [55] via the PRIDE partner repository with the dataset identifier PXD001940 and 10.6019/PXD001940.

3.6. Association Analysis of Proteomics and Transcriptomics

In the ramie transcriptome sequencing, the de novo assembly often carried out under the circumstances, neither genomic information nor directly linked transcriptome information were available. In this study, we used previously published Illumina Paired-End sequencing project data, because the drought stress treatment on the material is the same as the material treatment of ramie transcriptome analysis we published previously [25]. Cluster analysis was used to identify groups of similarly differentially expressed proteins and transcripts at three drought stress stages, and the results output through the software Java Treeview in graphic form. Cluster analysis of association expression in differentially expressed proteins with corresponding transcript levels data was analyzed with Cluster 3.0.

3.7. Quantitative Realtime PCR (qRT-PCR)

Each sample comprised material from three plants that were mixed. Leaf and roots were collected and frozen in liquid nitrogen and stored at −80 °C prior to analysis for subsequent RNA extraction. Total RNA was separately isolated from the six samples (L1, L2, L3, R1, R2, and R3) with the RNAprep Pure Plant Kit (Tiangen Biotech, Beijing, China), following the manufacturer's instructions. RNA quality was confirmed by gel electrophoresis and with the NanoDrop 2000 spectrophotometer (Thermo Fisher Scientific, Wilmington, DE, USA). The ramie *GAPDH* gene was selected as an internal control in each reaction. The primers used in this study were listed in Table 1. PCR amplification was performed as previously described [25]. The reactions were performed in triplicate, and the results were averaged.

3.8. Data Analysis

The raw mass data were converted to anmgf.file with Proteome Discover 1.3 (Thermo Fisher Scientific) with in-house MASCOT software 2.3.02 (Matrix Science, London, UK). In the database search, full tryptic specificity was required with tolerance set at one missed cleavage. The FDR (false positive rate) was cutoff with 1%. In this study, hierarchical cluster analyses were conducted according to a previous report [28]. Carbamido methylation of cysteine was set as a fixed modification. Gln->pyro-Glu of the N terminus, deamination of the N terminus, and oxidation of methionine were set as variable modifications. The initial precursor mass tolerance was set to 15 ppm, and the fragment ion level was set to 0.02 Da. iTRAQ 8-plex was set as quantitation. In this study, the database was obtained from transcription sequence [25]. The transcription sequence data have been deposited at the NCBI in the Short Read Archive database under accession SRP041143. After stringent quality checking and data cleaning, approximately 33,976,322,460 bp (30 G) of high-quality data (94.02% of the raw data) were generated under the Q20 standard (Q30 = 87.19%). Assembly of the high-quality sequencing reads yielded 138,381 unigenes, with an average length of 730.6 bp and a range from 201 to 20,860 bp. The N50 scaffold size was 972 bp. The unigene sequences were compared to the non-redundant (nr) protein database with a cutoff E-value of 1×10^{-5}. As a result, 47,565 unigenes (34%) were annotated. The cutoff value for downregulated proteins was 0.67-fold and for the upregulated proteins was 1.5-fold [56]. The p value threshold was <0.05. The differentially expressed proteins were used for GO terms/KEGG pathway enrichment analyses using the hyper geometric test to measure significantly enriched terms:

$$P = 1 - \sum_{i=0}^{m-1} \frac{\binom{M}{i} \binom{N-M}{n-i}}{\binom{N}{n}}$$

where N is the number of proteins with GO/KEGG annotations, and n represents the number of differentially expressed proteins in N. The variables, M and m, represent the total number of proteins and the number of differentially expressed proteins, respectively, in each GO/KEGG term. The threshold for significant enrichment of protein sets was corrected to a p value of ≤ 0.05. Every experiment was carried out with three biological replicates, except iTRAQ experiments, carried out with two biological replicates.

4. Conclusions

The iTRAQ-based comparative proteome analysis and qPCR data presented here will help in further understanding the responses of ramie to drought stress and improving the drought tolerance of this fiber crop. KOBAS analysis indicated that glycolysis/gluconeogenesis was the main metabolic pathway upregulated in ramie roots in response to drought stress. The activation of glycolysis/gluconeogenesis in ramie roots appeared to be a rapid and effective way to balance the levels of available energy to prevent intracellular energy shortages. In conclusion, this study provides a valuable source for proteomic studies in ramie plants, especially in those under drought stress. Our work will assist in breeding drought-resistant ramie varieties.

Supplementary Materials: The mass spectrometry proteomics data have been deposited to the Proteome Xchange Consortium [55] via the PRIDE partner repository with the dataset identifier PXD001940 and 10.6019/PXD001940. Supplementary materials can be found at www.mdpi.com/1422-0067/17/10/1607/s1.

Acknowledgments: This study was supported by the National Natural Science Funds (31171594) and China Agriculture Research System (CARS-19-E12).

Author Contributions: Xia An, Jingyu Zhang, Lunjin Dai, and Yiwei Liao carried out the sample preparing, results output, and drafting of the manuscript. Xia An, Bo Wang, and Gang Deng conceived and designed the experiments. Lijun Liu and Dingxiang Peng contributed reagents and materials. All authors read and approved the final manuscript.

Conflicts of Interest: The authors declare no conflict of interest.

Abbreviations

iTRAQ	isobaric tag for relative and absolute quantification
SCX	strong cation exchange
HPLC	high pressure liquid chromatography
KEGG	Kyoto Encyclopedia of Genes and Genomes
KOBAS	KEGG Orthology-Based Annotation System
PEG	polyethylene glycol
RWC	relative water content
POD	peroxidaseactivity
qRT-PCR	quantitative realtime PCR

References

1. Mohammadi, P.P.; Moieni, A.; Komatsu, S. Comparative proteome analysis of drought-sensitive and drought-tolerant rapeseed roots and their hybrid F1 line under drought stress. *Amino Acids* **2012**, *43*, 2137–2152. [CrossRef] [PubMed]
2. Hu, H.; Dai, M.; Yao, J.; Xiao, B.; Li, X.; Zhang, Q.; Xiong, L. Overexpressing a NAM, ATAF, and CUC (NAC) transcription factor enhances drought resistance and salt tolerance in rice. *Proc. Natl. Acad. Sci. USA* **2006**, *103*, 12987–12992. [CrossRef] [PubMed]
3. Alam, I.; Lee, D.G.; Kim, K.H.; Park, C.H.; Sharmin, S.A.; Lee, H.; Oh, K.W.; Yun, B.W.; Lee, B.H. Proteome analysis of soybean roots under waterlogging stress at an early vegetative stage. *J. Biosci.* **2010**, *35*, 49–62. [CrossRef] [PubMed]
4. Zhang, Q. Strategies for developing green super rice. *Proc. Natl. Acad. Sci. USA* **2007**, *104*, 16402–16409. [CrossRef] [PubMed]
5. Zhu, J.K. Salt and drought stress signal transduction in plants. *Annu. Rev. Plant Biol.* **2002**, *53*, 247–273. [CrossRef] [PubMed]
6. Shinozaki, K.; Yamaguchi-Shinozaki, K.; Seki, M. Regulatory network of gene expression in the drought and cold stress responses. *Curr. Opin. Plant Biol.* **2003**, *6*, 410–417. [CrossRef]
7. Seki, M.; Kamei, A.; Yamaguchi-Shinozaki, K.; Shinozaki, K. Molecular responses to drought, salinity and frost: Common and different paths for plant protection. *Curr. Opin. Biotechnol.* **2003**, *14*, 194–199. [CrossRef]
8. Plomion, C.; Lalanne, C.; Claverol, S.; Meddour, H.; Kohler, A.; Bogeat-Triboulot, M.B.; Barre, A.; Le Provost, G.; Dumazet, H.; Jacob, D.; et al. Mapping the proteome of poplar and application to the discovery of drought-stress responsive proteins. *Proteomics* **2006**, *6*, 6509–6527. [CrossRef] [PubMed]
9. Aranjuelo, I.; Molero, G.; Erice, G.; Avice, J.C.; Nogues, S. Plant physiology and proteomics reveals the leaf response to drought in alfalfa (*Medicago sativa* L.). *J. Exp. Bot.* **2011**, *62*, 111–123. [CrossRef] [PubMed]
10. Ashoub, A.; Beckhaus, T.; Berberich, T.; Karas, M.; Bruggemann, W. Comparative analysis of barley leaf proteome as affected by drought stress. *Planta* **2013**, *237*, 771–781. [CrossRef] [PubMed]
11. Matsui, A.; Ishida, J.; Morosawa, T.; Mochizuki, Y.; Kaminuma, E.; Endo, T.A.; Okamoto, M.; Nambara, E.; Nakajima, M.; Kawashima, M.; et al. *Arabidopsis* transcriptome analysis under drought, cold, high-salinity and aba treatment conditions using a tiling array. *Plant Cell Physiol.* **2008**, *49*, 1135–1149. [CrossRef] [PubMed]
12. Zheng, J.; Fu, J.; Gou, M.; Huai, J.; Liu, Y.; Jian, M.; Huang, Q.; Guo, X.; Dong, Z.; Wang, H.; et al. Genome-wide transcriptome analysis of two maize inbred lines under drought stress. *Plant Mol. Biol.* **2010**, *72*, 407–421. [CrossRef] [PubMed]
13. Cal, A.J.; Liu, D.; Mauleon, R.; Hsing, Y.I.; Serraj, R. Transcriptome profiling of leaf elongation zone under drought in contrasting rice cultivars. *PLoS ONE* **2013**, *8*, e54537. [CrossRef] [PubMed]
14. Le, D.T.; Nishiyama, R.; Watanabe, Y.; Tanaka, M.; Seki, M.; Ham le, H.; Yamaguchi-Shinozaki, K.; Shinozaki, K.; Tran, L.S. Differential gene expression in soybean leaf tissues at late developmental stages under drought stress revealed by genome-wide transcriptome analysis. *PLoS ONE* **2012**, *7*, e49522. [CrossRef] [PubMed]
15. Liu, T.; Zhu, S.; Tang, Q.; Yu, Y.; Tang, S. Identification of drought stress-responsive transcription factors in ramie (*Boehmeria nivea* L. Gaud.). *BMC Plant Biol.* **2013**, *13*, 130. [CrossRef] [PubMed]

16. Budak, H.; Akpinar, B.A.; Unver, T.; Turktas, M. Proteome changes in wild and modern wheat leaves upon drought stress by two-dimensional electrophoresis and nanolc-esi-ms/ms. *Plant Mol. Biol.* **2013**, *83*, 89–103. [CrossRef] [PubMed]

17. Liu, F.; Liu, Q.; Liang, X.; Huang, H.; Zhang, S. Morphological, anatomical, and physiological assessment of ramie (*Boehmeria nivea* (L.) Gaud.) tolerance to soil drought. *Genet. Resour. Crop Evol.* **2005**, *52*, 497–506. [CrossRef]

18. Bonhomme, L.; Monclus, R.; Vincent, D.; Carpin, S.; Lomenech, A.M.; Plomion, C.; Brignolas, F.; Morabito, D. Leaf proteome analysis of eight populus xeuramericana genotypes: Genetic variation in drought response and in water-use efficiency involves photosynthesis-related proteins. *Proteomics* **2009**, *9*, 4121–4142. [CrossRef] [PubMed]

19. Ali, G.M.; Komatsu, S. Proteomic analysis of rice leaf sheath during drought stress. *J. Proteome Res.* **2006**, *5*, 396–403. [CrossRef] [PubMed]

20. Zang, X.; Komatsu, S. A proteomics approach for identifying osmotic-stress-related proteins in rice. *Phytochemistry* **2007**, *68*, 426–437. [CrossRef] [PubMed]

21. Nouri, M.Z.; Komatsu, S. Comparative analysis of soybean plasma membrane proteins under osmotic stress using gel-based and LC MS/MS-based proteomics approaches. *Proteomics* **2010**, *10*, 1930–1945. [CrossRef] [PubMed]

22. Yang, L.T.; Qi, Y.P.; Lu, Y.B.; Guo, P.; Sang, W.; Feng, H.; Zhang, H.X.; Chen, L.S. iTRAQ protein profile analysis of citrus sinensis roots in response to long-term boron-deficiency. *J. Proteom.* **2013**, *93*, 179–206. [CrossRef] [PubMed]

23. Zieske, L.R. A perspective on the use of itraq reagent technology for protein complex and profiling studies. *J. Exp. Bot.* **2006**, *57*, 1501–1508. [CrossRef] [PubMed]

24. Verslues, P.E.; Agarwal, M.; Katiyar-Agarwal, S.; Zhu, J.; Zhu, J.K. Methods and concepts in quantifying resistance to drought, salt and freezing, abiotic stresses that affect plant water status. *Plant J.* **2006**, *45*, 523–539. [CrossRef] [PubMed]

25. An, X.; Chen, J.; Zhang, J.; Liao, Y.; Dai, L.; Wang, B.; Liu, L.; Peng, D. Transcriptome profiling and identification of transcription factors in ramie (*Boehmeria nivea* L. Gaud) in response to peg treatment, using illumina paired-end sequencing technology. *Int. J. Mol. Sci.* **2015**, *16*, 3493–3511. [CrossRef] [PubMed]

26. Ashraf, M. Biotechnological approach of improving plant salt tolerance using antioxidants as markers. *Biotechnol. Adv.* **2009**, *27*, 84–93. [CrossRef] [PubMed]

27. Chen, S.; Cui, X.; Chen, Y.; Gu, C.; Miao, H.; Gao, H.; Chen, F.; Liu, Z.; Guan, Z.; Fang, W. Cgdreba transgenic chrysanthemum confers drought and salinity tolerance. *Environ. Exp. Bot.* **2011**, *74*, 255–260. [CrossRef]

28. Zi, J.; Zhang, J.; Wang, Q.; Zhou, B.; Zhong, J.; Zhang, C.; Qiu, X.; Wen, B.; Zhang, S.; Fu, X.; et al. Stress responsive proteins are actively regulated during rice (*Oryza sativa*) embryogenesis as indicated by quantitative proteomics analysis. *PLoS ONE* **2013**, *8*, e74229. [CrossRef] [PubMed]

29. Deng, G.; Liu, L.J.; Zhong, X.Y.; Lao, C.Y.; Wang, H.Y.; Wang, B.; Zhu, C.; Shah, F.; Peng, D.X. Comparative proteome analysis of the response of ramie under N, P and K deficiency. *Planta* **2014**, *239*, 1175–1186. [CrossRef] [PubMed]

30. Jiang, Y.; Yang, B.; Harris, N.S.; Deyholos, M.K. Comparative proteomic analysis of nacl stress-responsive proteins in *Arabidopsis* roots. *J. Exp. Bot.* **2007**, *58*, 3591–3607. [CrossRef] [PubMed]

31. Ge, P.; Ma, C.; Wang, S.; Gao, L.; Li, X.; Guo, G.; Ma, W.; Yan, Y. Comparative proteomic analysis of grain development in two spring wheat varieties under drought stress. *Anal. Bioanal. Chem.* **2012**, *402*, 1297–1313. [CrossRef] [PubMed]

32. Wang, N.; Zhao, J.; He, X.; Sun, H.; Zhang, G.; Wu, F. Comparative proteomic analysis of drought tolerance in the two contrasting tibetan wild genotypes and cultivated genotype. *BMC Genom.* **2015**, *16*, 432. [CrossRef] [PubMed]

33. Peng, Z.; Wang, M.; Li, F.; Lv, H.; Li, C.; Xia, G. A proteomic study of the response to salinity and drought stress in an introgression strain of bread wheat. *Mol. Cell. Proteom.* **2009**, *8*, 2676–2686. [CrossRef] [PubMed]

34. Wang, Z.Q.; Xu, X.Y.; Gong, Q.Q.; Xie, C.; Fan, W.; Yang, J.L.; Lin, Q.S.; Zheng, S.J. Root proteome of rice studied by iTRAQ provides integrated insight into aluminum stress tolerance mechanisms in plants. *J. Proteom.* **2014**, *98*, 189–205. [CrossRef] [PubMed]

35. Nogues, S.; Baker, N.R. Effects of drought on photosynthesis in mediterranean plants grown under enhanced uv-b radiation. *J. Exp. Bot.* **2000**, *51*, 1309–1317. [CrossRef] [PubMed]

36. Ramachandra Reddy, A.; Chaitanya, K.V.; Vivekanandan, M. Drought-induced responses of photosynthesis and antioxidant metabolism in higher plants. *J. Plant Physiol.* **2004**, *161*, 1189–1202. [CrossRef] [PubMed]
37. Maksup, S.; Roytrakul, S.; Supaibulwatana, K. Physiological and comparative proteomic analyses of thai jasmine rice and two check cultivars in response to drought stress. *J. Plant Int.* **2012**, *9*, 43–55. [CrossRef]
38. Vassileva, V.; Demirevska, K.; Simova-Stoilova, L.; Petrova, T.; Tsenov, N.; Feller, U. Long-term field drought affects leaf protein pattern and chloroplast ultrastructure of winter wheat in a cultivar-specific manner. *J. Agron. Crop Sci.* **2012**, *198*, 104–117. [CrossRef]
39. Caruso, G.; Cavaliere, C.; Foglia, P.; Gubbiotti, R.; Samperi, R.; Laganà, A. Analysis of drought responsive proteins in wheat (*Triticum durum*) by 2D-page and maldi-tof mass spectrometry. *Plant Sci.* **2009**, *177*, 570–576. [CrossRef]
40. Guo, X.L.; Bai, L.R.; Su, C.Q.; Shi, L.R.; Wang, D.W. Molecular cloning and expression of drought-induced protein 3 (DIP3) encoding a class III chitinase in upland rice. *Genet. Mol. Res.* **2013**, *12*, 6860–6870. [CrossRef] [PubMed]
41. Shinozaki, K.; Yamaguchi-Shinozaki, K. Molecular responses to dehydration and low temperature: Differences and cross-talk between two stress signaling pathways. *Curr. Opin. Plant Biol.* **2000**, *3*, 217–223. [CrossRef]
42. Finkelstein, R.R.; Gampala, S.S.; Rock, C.D. Abscisic acid signaling in seeds and seedlings. *Plant Cell* **2002**, *14*, S15–S45. [PubMed]
43. Xiong, L.; Schumaker, K.S.; Zhu, J.K. Cell signaling during cold, drought, and salt stress. *Plant Cell* **2002**, *14*, S165–S183. [PubMed]
44. Yao, Y.; Ni, Z.; Du, J.; Wang, X.; Wu, H.; Sun, Q. Isolation and characterization of 15 genes encoding ribosomal proteins in wheat (*Triticum aestivum* L.). *Plant Sci.* **2006**, *170*, 579–586. [CrossRef]
45. Swindell, W.R.; Huebner, M.; Weber, A.P. Transcriptional profiling of *Arabidopsis* heat shock proteins and transcription factors reveals extensive overlap between heat and non-heat stress response pathways. *BMC Genom.* **2007**, *8*, 125. [CrossRef] [PubMed]
46. Xie, C.; Mao, X.; Huang, J.; Ding, Y.; Wu, J.; Dong, S.; Kong, L.; Gao, G.; Li, C.Y.; Wei, L. Kobas 2.0: A web server for annotation and identification of enriched pathways and diseases. *Nucleic Acids Res.* **2011**, *39*, W316–W322. [CrossRef] [PubMed]
47. Baena-Gonzalez, E.; Rolland, F.; Thevelein, J.M.; Sheen, J. A central integrator of transcription networks in plant stress and energy signalling. *Nature* **2007**, *448*, 938–942. [CrossRef] [PubMed]
48. Smith, A.M.; Stitt, M. Coordination of carbon supply and plant growth. *Plant Cell Environ.* **2007**, *30*, 1126–1149. [CrossRef] [PubMed]
49. Baena-Gonzalez, E.; Sheen, J. Convergent energy and stress signaling. *Trends Plant Sci.* **2008**, *13*, 474–482. [CrossRef] [PubMed]
50. Hamilton, C.A.; Good, A.G.; Taylor, G.J. Induction of vacuolar atpase and mitochondrial ATP synthase by aluminum in an aluminum-resistant cultivar of wheat. *Plant Physiol.* **2001**, *125*, 2068–2077. [CrossRef] [PubMed]
51. Shi, D.Y.; Xie, F.Z.; Zhai, C.; Stern, J.S.; Liu, Y.; Liu, S.L. The role of cellular oxidative stress in regulating glycolysis energy metabolism in hepatoma cells. *Mol. Cancer* **2009**, *8*, 32. [CrossRef] [PubMed]
52. Ouyang, S.Q.; Liu, Y.F.; Liu, P.; Lei, G.; He, S.J.; Ma, B.; Zhang, W.K.; Zhang, J.S.; Chen, S.Y. Receptor-like kinase ossik1 improves drought and salt stress tolerance in rice (*Oryza sativa*) plants. *Plant J.* **2010**, *62*, 316–329. [CrossRef] [PubMed]
53. An, X.; Wang, B.; Liu, L.; Jiang, H.; Chen, J.; Ye, S.; Chen, L.; Guo, P.; Huang, X.; Peng, D. Agrobacterium-mediated genetic transformation and regeneration of transgenic plants using leaf midribs as explants in ramie (*Boehmeria nivea* (L.) Gaud). *Mol. Biol. Rep.* **2014**, *41*, 3257–3269. [CrossRef] [PubMed]
54. Villeneuve, L.M.; Stauch, K.L.; Fox, H.S. Proteomic analysis of the mitochondria from embryonic and postnatal rat brains reveals response to developmental changes in energy demands. *J. Proteom.* **2014**, *109*, 228–239. [CrossRef] [PubMed]
55. Vizcaino, J.A.; Deutsch, E.W.; Wang, R.; Csordas, A.; Reisinger, F.; Rios, D.; Dianes, J.A.; Sun, Z.; Farrah, T.; Bandeira, N.; et al. Proteomexchange provides globally coordinated proteomics data submission and dissemination. *Nat. Biotechnol.* **2014**, *32*, 223–226. [CrossRef] [PubMed]

56. Fan, N.J.; Gao, C.F.; Wang, C.S.; Zhao, G.; Lv, J.J.; Wang, X.L.; Chu, G.H.; Yin, J.; Li, D.H.; Chen, X.; et al. Identification of the up-regulation of TP-α, collagen α-1(VI) chain, and S100A9 in esophageal squamous cell carcinoma by a proteomic method. *J. Proteom.* **2012**, *75*, 3977–3986. [CrossRef] [PubMed]

International Journal of
Molecular Sciences

MDPI

Article

Drought-Induced Leaf Proteome Changes in Switchgrass Seedlings

Zhujia Ye [1], Sasikiran Sangireddy [1], Ikenna Okekeogbu [1], Suping Zhou [1,*], Chih-Li Yu [2],
Dafeng Hui [2], Kevin J. Howe [3], Tara Fish [3] and Theodore W. Thannhauser [3,*]

[1] Department of Agricultural Sciences, Tennessee State University, 3500 John Merritt Blvd, Nashville,
    TN 37209, USA; zye@my.tnstate.edu (Z.Y.); sangisasi@gmail.com (S.S.); iyk_oc@yahoo.com (I.O.)
[2] Department of Biological Sciences, Tennessee State University, 3500 John Merritt Blvd, Nashville, TN 37209,
    USA; cyu@my.tnstate.edu (C.-L.Y.); dhui@tnstate.edu (D.H.)
[3] Functional & Comparative Proteomics Center, USDA-ARS, Cornell University, Ithaca, NY 14853, USA;
    kjh46@cornell.edu (K.J.H.); tlf26@cornell.edu (T.F.)
*   Correspondence: zsuping@tnstate.edu (S.Z.); tt34@cornell.edu (T.W.T.);
    Tel.: +1-615-963-2465 (S.Z.); +1-607-255-8808 (T.W.T.)

Academic Editor: Setsuko Komatsu
Received: 21 June 2016; Accepted: 27 July 2016; Published: 2 August 2016

Abstract: Switchgrass (*Panicum virgatum*) is a perennial crop producing deep roots and thus highly tolerant to soil water deficit conditions. However, seedling establishment in the field is very susceptible to prolonged and periodic drought stress. In this study, a "sandwich" system simulating a gradual water deletion process was developed. Switchgrass seedlings were subjected to a 20-day gradual drought treatment process when soil water tension was increased to 0.05 MPa (moderate drought stress) and leaf physiological properties had expressed significant alteration. Drought-induced changes in leaf proteomes were identified using the isobaric tags for relative and absolute quantitation (iTRAQ) labeling method followed by nano-scale liquid chromatography mass spectrometry (nano-LC-MS/MS) analysis. Additionally, total leaf proteins were processed using a combinatorial library of peptide ligands to enrich for lower abundance proteins. Both total proteins and those enriched samples were analyzed to increase the coverage of the quantitative proteomics analysis. A total of 7006 leaf proteins were identified, and 257 (4% of the leaf proteome) expressed a significant difference ($p < 0.05$, fold change <0.6 or >1.7) from the non-treated control to drought-treated conditions. These proteins are involved in the regulation of transcription and translation, cell division, cell wall modification, phyto-hormone metabolism and signaling transduction pathways, and metabolic pathways of carbohydrates, amino acids, and fatty acids. A scheme of abscisic acid (ABA)-biosynthesis and ABA responsive signal transduction pathway was reconstructed using these drought-induced significant proteins, showing systemic regulation at protein level to deploy the respective mechanism. Results from this study, in addition to revealing molecular responses to drought stress, provide a large number of proteins (candidate genes) that can be employed to improve switchgrass seedling growth and establishment under soil drought conditions (Data are available via ProteomeXchange with identifier PXD004675).

Keywords: physiological properties; isobaric tags for relative and absolute quantitation (iTRAQ); ProteoMiner; functional pathways; abscisic acid (ABA) signaling; "Sandwich" plant growth system

---

1. Introduction

Switchgrass (*Panicum virgatum*), has been selected as a model herbaceous bioenergy species in the USA due to its high biomass yield, strong tolerance to drought and flooding conditions, relatively low herbicide and fertilizer input requirements, and widespread adaptability to temperate climate [1–3].

Recently, a shortage of fresh water and increasingly severe drought have become a significant challenge to crop production [4]. Based on data from the National Weather Service Centers for Environmental Prediction [5], soil moisture contents in the topsoil layer have declined over the past decade (2005–2015) in many regions of the USA, especially in central states.

Drought tolerance is one of the most striking physiological properties of switchgrass. Mature plants have a very deep root system and a highly efficient C_4 metabolic pathway [6]. However, switchgrass plants are slow to establish in the field, often requiring two to three growing seasons to develop deep root systems. During the early stages of growth when seedlings have a relatively shallow root distribution (0–15 cm) in the top soil, these plants are very susceptible to both periodic and long-term drought conditions [7]. A field trial shows that drought significantly affected seedling growth of switchgrass in the first year. Furthermore, biomass yield declined greatly after three consecutive years of drought [8]. Thus, developing switchgrass plants with strong drought tolerance during the early stages of growth is an effective strategy to ensure high biomass yields during subsequent years in the field.

Plant growth depends on cell division, cell enlargement, and differentiation [9]. Under drought conditions, cell elongation and division are both suppressed by the reduced photosynthesis driven by diminished CO_2 influx and limitation of carboxylation by abscisic acid (ABA)-dependent stomatal closure [9–12]. On the other hand, stomatal closing has been viewed as a drought tolerance mechanism to avoid excess water loss via transpiration. A set of physiological parameters related to drought tolerance has been identified including leaf relative water content (RWC), electrolyte leakage (EL), photosynthetic rate (Pn), stomatal conductance (g_s), transpiration rate (Tr), intercellular CO_2 concentration (Ci), and water use efficiency (WUE) [13]. Thus, whether plants can sustain active growth or just survive the water-deficient conditions depends on how efficiently they regulate these complex processes.

Proteins are the primary molecules that carry out various biological functions in cells and in an entire organism [14]. Alterations in proteome composition provide the basis for a plant to perform different biological functions, including adapting to changing and/or suboptimal environmental conditions [15–21]. With the rapid development of proteomic technologies, two-dimensional liquid chromatography, in combination with multiplexed quantitative techniques such as isobaric tags for relative and absolute quantitation (iTRAQ), provides the ability to perform relative or absolute quantification of proteomes [22–26]. Quantitative proteomics using the shot-gun bottom-up approach has been used to evaluate drought-responsive proteins in important crop species, such as rice, maize, wheat, cotton, amaranth, alfalfa, sugar beets, and tomatoes [18,20,27–37]. Conclusively, these proteomics studies have significantly increased our understanding of molecular regulation at the translational and post-translational levels in plants.

The separation and detection of all proteins contained in any given proteome remains a challenge because the analysis of low-abundance proteins is difficult in the presence of the highly abundant proteins. Characterization of the photosynthetically active leaf proteome is a very difficult task as the ribulose-1,5-biphosphate carboxylase/oxygenase (Rubisco) proteins would account for approximately 40% of total protein content [38]. An earlier study using immunoaffinity subtraction of Rubisco was able to increase the resolution of more protein species in leaf protein samples [39]. However, those antibodies are very expensive, which limits their usage in large quantitative proteomics experiments (unpublished data, Zhou, Tennessee State University, Nashville, TN, USA, 2016).

The ProteoMiner protein depletion/enrichment technology, which employs a large, highly diverse bead-based library of combinatorial peptide ligands, has proven to be a powerful tool for uncovering low-abundance proteins. Using this approach, Fasoli et al. detected 79% more proteins from spinach leaves than could be detected without the depletion/enrichment process [40]. More importantly, the ProteoMiner protein enrichment method produces highly stable and reproducible results, which is extremely important in quantitative proteomics where two or more samples are analyzed in each treatment condition [41,42].

This study was carried out with a goal to understand the changes in leaf proteome in switchgrass under drought stress and to develop the association between the expression of these proteins and the physiological properties that give rise to drought tolerance. As described above, removal of highly abundant Rubisco protein is an effective strategy for increasing the overall number of identified proteins, thus the ProteoMiner depletion/enrichment procedure was performed to reduce the scale of dynamic range in protein abundance. By enabling the identification of low-abundance proteins and increasing the number of proteins quantified, this study provides an in-depth understanding of systemic changes in the drought-induced proteomes in switchgrass seedlings.

2. Results

2.1. Drought-Induced Physiological Properties of Switchgrass

Sixteen days after the initiation of water withholding, young leaves on drought-treated plants started to show signs of wilting as the soil water tension of treated groups reached 0.05 MPa. Twenty days after water withholding, soil water tension increased to 0.08 ± 0.02 MPa (Table 1). At this time, the relative growth of drought-treated plants was reduced significantly (a 20% decrease), as well as the stomatal conductance and transpiration rate ($p < 0.01$), compared to the untreated control plants. The water use efficiency, which is defined as the ratio of the photosynthetic rate to the transpiration rate [43], showed an 7.1% increase in the drought-treated group, which was significantly higher than the untreated control plants ($p < 0.01$). Changes in these physiological properties showed that leaves and plants as a whole experienced a progressive drought-stress during the 20 days of withholding water. At this time-point (20 days after withholding water), the drought treatments were terminated and tissues were harvested for further analysis (Figure 1).

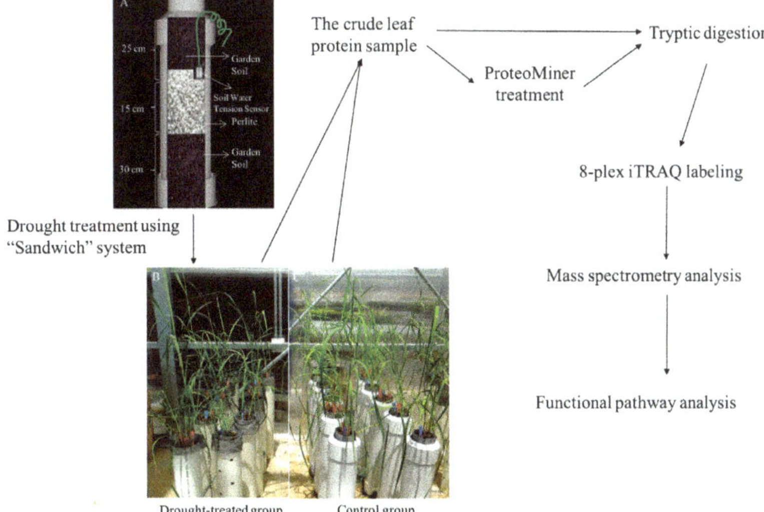

Figure 1. Flow chart of the drought treatments and quantitative proteomics procedure. Plants were grown in a "sandwich" system (**A**); During the 20th day of the water withholding period, physiological data were recorded on both drought-treated (**B**); and well-watered control plants (**C**). Leaf protein samples were extracted followed by the ProteoMiner enrichment. Quantitative proteomics analysis was performed using the crude leaf protein extracts and the ProteoMiner-enriched samples. Functional pathways were developed using information on the drought-induced changes in the leaf proteomes, and the association between protein expression and physiological properties was developed focusing on drought stress tolerance.

Table 1. Effects of drought treatments on physiological properties of switchgrass.

Treatment		Control	Drought
	Soil Water Tension (MPa)	0.00 ± 0.00 [A,†]	0.08 ± 0.02 [B,†]
	Leaf Relative Water Content	77.35 ± 0.01 [A]	71.08 ± 0.02 [B]
Plant Height (cm)	0 Day drought treatment	18.31 ± 6.18 [A]	19.08 ± 4.97 [A]
	20 days drought treatment	43.26 ± 9.11 [A]	39.75 ± 8.49 [B]
	Relative plant height	24.96 ± 6.21 [A]	20.67 ± 6.22 [B]
Photosynthesis	Leaf photosynthetic rate (μmol CO_2/m^2/s)	22.96 ± 3.22 [A]	21.69 ± 7.17 [A]
	Stomatal conductance (mol H_2O/m^2/s)	0.138 ± 0.03 [A]	0.125 ± 0.05 [B]
	Transpiration rate (mmol H_2O/m^{-2}/s)	6.88 ± 1.11 [A]	6.09 ± 2.15 [B]
	Water use efficiency (μmol CO_2/mmol H_2O)	3.35 ± 0.20 [A]	3.59 ± 0.25 [B]

Data for all the measurements except plant height were collected after 20 days of the water withholding treatments. Data are presented as means $\pm$ standard deviations (SD) of four independent replicates. Within columns, means followed by the same letter are not significantly different ($p < 0.01$). Leaf relative water content (W_r) was calculated using the following equation: $W_r = (W_f - W_d)/W_f$, where fresh weight (W_f) was taken immediately after harvest, and dry weight was measured after drying tissues at 70 °C for three days until a constant dry weight (W_d). Plant height was measured from the bottom of the tiller (start point) to the top of the latest node (end point). [†] Means within columns followed by the same letter are not different at the 1% level.

2.2. Effects of the ProteoMiner Enrichment Process on the Identification of Leaf Proteome

The ProteoMiner enrichment method is used to increase the relative concentration of low-abundance proteins by depleting those high-abundance proteins in a protein sample, and thus to increase the depth of coverage of the leaf proteomes to be identified in a proteomics analysis. In plant leaf proteomes, more than 40% of the total leaf protein content consists of Rubisco [38]. One-dimensional gel electrophoresis showed that in the ProteoMiner-treated (PMT) samples, the band intensity of high abundance proteins (i.e., Rubisco protein) was reduced, whereas the intensity of several weaker protein bands was increased, compared to the counterparts of the Crude Leaf Extracts (CLE) protein samples (Figure S1). Analysis of the peptide numbers for the proteins detected in the PMT sample indicates that 1101 proteins were identified by a greater number of peptides after ProteoMiner enrichment—for instance, the number of peptides in an adenylate kinase protein (Pavir.Fa02159.1) was increased from 29 in CLE to 189 in PMT samples. On the other hand, 1876 proteins were identified by fewer peptides. For example, the number of peptides in Rubisco subunits including Pavir.Cb01593.1, Pavir.Cb01387.1, and Pavir.J32704.1 was decreased (Figure S2, Table S1-1,2). These results demonstrate that ProteoMiner did deplete the concentration of the highly abundant proteins while simultaneously enriching low-abundance proteins.

2.3. Identification of Quantified Proteins

In this study, 7006 proteins were identified in the switchgrass leaf proteome with the assistance of the ProteoMiner enrichment method (Table 2, Tables S1-1–3 and S2). A total of 5493 proteins were identified in the CLE samples and 4839 unique proteins were identified in the PMT samples. Between the CLE and PMT samples, 3326 proteins overlapped. The use of a ProteoMiner enrichment step resulted in the identification of 1513 proteins that were not found in the CLE samples. It appears that the ProteoMiner enrichment is complementary to the analysis of the crude leaf protein extracts, and a combination of both approaches was shown to quantify more proteins than either individually.

Table 2. The number of proteins identified in the proteomes identified using the crude leaf protein extracts and ProteoMiner-enriched samples.

	Protein Classification	CLE [a]	PMT [b]	The Number of Proteins from CLE and PMT
Proteins identified with one or more peptides	The total number of proteins	5493	4839	7006
	The number of proteins overlapped in CLE and PMT	3326		
	The number of proteins identified in CLE	2167	-	
	The number of protein identified in PMT	-	1513	
Quantified proteins with two or more peptides	The total number of proteins	4746	4134	5680
	The number of proteins overlapped in CLE and PMT	3200		
	The number of protein in CLE	1546	-	
	The number of proteins in PMT	-	934	
Differentially expressed proteins (FDR < 0.01, fold change < 0.06 or > 1.7)	The total number of proteins	205	107	257
	The number of proteins in CLE and PMT	55		
	The number of proteins in CLE	150	-	
	The number of proteins in PMT	-	52	

[a] The number of proteins identified in the crude leaf protein extracts; [b] The number of proteins identified in the ProteoMiner enriched samples; CLE: Crude Leaf Extracts; PMT: ProteoMiner-treated; FDR: false discovery rate.

Among the total identified proteomes, 81.1% of them (5680/7006) contained at least two unique peptides (Table 2). Quantitative analysis revealed that 257 proteins, which was approximately 4% of the total quantified proteomes (257/7006), passed the threshold value of $\pm 2\sigma$ (standard deviation), $p < 0.05$ (t-test and false discovery rate (FDR) corrections), and fold change <0.6 or >1.7. These proteins were considered significantly changed under the drought treatment conditions. Among the 257 drought-induced significant proteins, 55 proteins showed consistent changes in both the CLE and PMT protein samples, 150 proteins were found only in CLE samples, and 52 proteins were only identified in PMT samples (Tables S1-5 and S3). In addition, the false negative rate (β) was calculated as 0.02 by summing the probabilities that each of the proteins judged to be unchanged was in fact differentially expressed. This suggests that the power of the experiment was very high ($p = 1 - \beta = 0.98$).

MapMan is a bioinformatics tool for developing the associations between gene (protein) expression and cellular processes, but this offline program only performs analysis of genomes contained in the MapMan Store. As the annotated switchgrass genome database is not listed, the program will not recognize the protein accession identity and therefore cannot map the protein expression data to biological functions. Instead, in this study, the *Arabidopsis thaliana* accessions annotated for those drought-induced switchgrass proteins were used when developing the functional pathways (Figure S3). Results showed that each functional group contained upregulated and downregulated proteins. A large number of the significantly changed proteins are associated with RNA transcription/processing, protein synthesis, and protein degradation pathways (Table 3).

Table 3. The number of proteins identified in the crude leaf protein extracts and ProteoMiner-treated samples.

	Classification	CLE [a]	PMT [b]	CLE and PMT [c]
Molecular Function	Abiotic/biotic stress	72	25	116
	Cell division/cell cycle	11	7	42
	Cell organization	26	11	47
	Cell vesicle transport	21	6	31
	Development	41	16	46
	DNA repair	4	2	7
	DNA synthesis	20	15	28
	Functional enzyme	62	64	180
	Metal binding	4	1	11
	Phyto-hormone metabolism	21	11	36
	Protein and amino acids activation	15	13	35
	Protein degradation	88	39	172
	Protein post-translation	27	12	41
	Protein synthesis	61	54	209
	Protein targeting	21	22	81
	Redox balance	31	18	98
	RNA transcription/processing	113	74	212
	Signaling regulation	82	39	98
	Transport	24	35	65
Cellular Metabolism	Amino acid metabolism	39	38	91
	C1-metabolism	4	5	16
	Cell wall synthesis/modification	13	13	20
	Fermentation	3	3	6
	Glycolysis	9	12	41
	Glyoxylate cycle	1	0	10
	Lipid metabolism	22	30	51
	Major CHO metabolism	11	10	35
	Minor CHO metabolism	7	0	26
	Mitochondrial electron transport/ATP synthesis	9	9	57
	N-metabolism	2	2	7
	Nucleotide metabolism	24	14	53
	Oxidative pentose phosphate (OPP) pathway	7	3	12
	Photosystem. Calvin cycle	4	6	36
	Photosystem. Light reaction	15	10	82
	Photorespiration	3	1	14
	S-assimilation	2	2	5
	Secondary metabolism	18	29	67
	TCA cycle	8	10	52
	Tetrapyrrole synthesis	13	9	20
	Others and not assigned proteins	588	264	944
	Total	1546	934	3200

[a] The number of proteins identified in the crude leaf protein extracts (CLE); [b] The number of proteins identified in the ProteoMiner-treated samples (PMT); [c] The number of proteins combining the proteomes identified in CLE and PMT.

2.4. Proteins in Regulation of Transcription and Translation

For proteins involved in gene transcription, several members of the G2-like, myeloblastosis (MYB) and bZIP transcription factors (TFs) were identified. A MYB-related transcription factor TRY (Triptychon) (Pavir.Eb02165.1) and a G2-like transcription factor APL (altered phloem development) (Pavir.Fa01260.1) were significantly reduced under drought stress. The former TF did not pass the threshold as a significant protein in CLE, and the latter TF was identified only in the PMT samples. The GBF (G-box binding factor) (Pavir.Ea03718.1), a member of the bZIP TFs family involved in ABA and stress signaling [44], was significantly increased (>2-fold) (Table S1-4), and it was identified in both CLE and PMT samples.

Proteins involved in protein synthesis and degradation were altered. The chloroplast-targeted FtsH protease (Pavir.J13145.1) was up-regulated at a higher than four-fold level. Moreover, the relative abundance of a senescence-specific Cys-protease protein (SAG) (Pavir.J08126.1) markedly declined in response to drought stress. Regarding to changes in protein synthesis, drought stress induced a plastid-specific 50S ribosomal protein (PSRP) (Pavir.Ea00033.1), which is an important member of the translation machinery in chloroplasts. However, the drought-induced significant change was found only in PMT samples, not in CLE samples (Table S1-4).

2.5. Cell Division and Cell Wall Modification

Two proteins involved in the cell cycle and cell division were identified. Pavir.Gb00127.1, a regulator of chromosome condensation (RCC), was significantly decreased, whereas prohibitin (PHB) (Pavir.Aa01476.1) was significantly increased (Table S1-4). UDP-glucose 4-epimerase (UGE) (Pavir.J14539.1), with a proven function in cell wall carbohydrate metabolism [45], was upregulated more than six-fold. A cell-wall-modifying xyloglucan endotransglycosylase/hydrolase (XET) (Pavir.Fa01211.1) [46] was increased 3.39-fold. These two cell-wall-related proteins were not identified in PMT (Table S1-4).

2.6. Phyto-Hormone Metabolism and Signaling Transduction Pathways

Four proteins involved in the metabolism of auxin and ethylene were all induced by drought stress (Pavir.Ga00273.1, Pavir.J01120.1, Pavir.J01160.1, and Pavir.Ia03739.1). Of the significantly changed proteins in the ABA-metabolic pathway, the upregulated proteins were classified as GRAM domain-containing proteins (Pavir.Ca02189.1 and Pavir.Cb00761.1), ABA-responsive elements-binding factor (ABF) (Pavir.J00256.1), and 9-cis-epoxycarotenoid dioxygenases (NCED) (Pavir.Ba03791.1) (Table S1-4).

Six calcium-binding proteins that interact with the second messenger "Ca^{2+}" to transduce stress signals into plant cells were identified, five of them markedly upregulated (Pavir.Ea00612.1, Pavir.Ca00053.1, Pavir.Eb03832.1, Pavir.Ib02894.1, and Pavir.J09383.1). These proteins were identified in both CLE and PMT, or in CLE but not in PMT. The one reduced (Pavir.Da01126.1) protein was identified in PMT but not in CLE (Table S1-4).

2.7. Stress-Responsive Proteins

The drought treatments induced 31 abiotic/biotic stress responsive proteins. These stress proteins include six biotic stress responsive proteins (Pavir.Bb00478.1, Pavir.Fb02059.1, Pavir.Ga02124.1, Pavir.Ha00419.1, Pavir.J09667.1, and Pavir.J00406.1), five dehydrins (DHNs) (Pavir.Bb03589.1, Pavir.Ca01575.1, Pavir.Aa00887.1, Pavir.J04551.1, and Pavir.J13075.1), 13 heat shock proteins (HSP) (Pavir.Ea00289.1, Pavir.J35929.1, Pavir.J33423.1, Pavir.J24160.1, Pavir.Ia03665.1, Pavir.J40704.1, Pavir.Ib01136.1, Pavir.Ab00778.1, Pavir.J19824.1, Pavir.Fa01476.1, Pavir.Aa00282.1, Pavir.Hb01472.1, and Pavir.J21349.1), one cold stress-related protein (Pavir.J31919.1), and six other stress responsive proteins. Of them, the HSP20-like protein (Pavir.J21349.1) increased more than 9.73-fold (Table S1-4).

2.8. Carbohydrate Metabolism

The relative abundance level of proteins in carbohydrate metabolic pathways, such as gluconeogenesis, starch metabolism, and the biosynthesis of raffinose family oligosaccharides (RFO), were altered in response to the drought treatments. The induced proteins include malate synthase (Pavir.Gb01372.1), β amylase protein (Pavir.J18576.1), and two galactinol synthase proteins (Pavir.J07018.1 and Pavir.J40731.1), but a starch synthase (Pavir.J06822.1) was repressed under the drought-treated conditions (Table S1-4).

2.9. Nitric Acid Metabolism

Under moderate drought stress, three proteins involved in the biosynthesis of free amino acids were markedly upregulated, Δ1-pyrroline-5-carboxylate synthetase protein (P5CS) (Pavir.J02344.1), methionine-γ-lyase protein (MGL) (Pavir.Ib03758.1), and L-asparagine amidohydrolase (Pavir.Gb00328.1). An enzyme-catalyzing β-oxidation of fatty acids, 3-ketoacyl-CoA thiolase-2 (KAT2/PED1/PKT3) (Pavir.J16366.1) which is involved in ABA signaling, was also significant increased (Table S1-4).

3. Discussion

Among all the drought tolerance mechanisms, an increased ABA content in leaves has been shown to play a key role in activating signaling pathways that control stomatal closure, thus reducing transpirational water loss [47,48].

During the 20 days of drought treatment period, the switchgrass leaves showed a gradual decline in stomatal conductance and transpiration rates, which are indications of a reduced stomatal aperture. This prediction of stomatal behavior is supported by the upregulation of PHB (Pavir.Aa01476.1), which regulates the level of nitric oxide accumulation that induces stomatal closure and thus enhances the adaptive plant responses against drought stress [49,50].

Changes in protein expression support the elevated biosynthesis of ABA and the induction of ABA-mediated signal transduction pathways during the drought treatment period (Figure 2). In the ABA biosynthesis pathway, 9-*cis*-epoxycarotenoid dioxygenase (NCED) catalyzes the step to convert 9-*cis*-xanthophylls to xanthoxin, which is the direct precursor of ABA [51]. The regulatory role of NCED in ABA biosynthesis in leaves under stress conditions has been clearly demonstrated in many studies, showing that the abundance of NCED proteins is directly correlated with ABA content [51–55]. The same metabolic changes may have occurred in switchgrass leaves where the significant increase (4.5-fold) of an NCED protein (Pavir.Ba03791.1) may result in an elevated ABA content in the drought-treated leaves (Table S1-4).

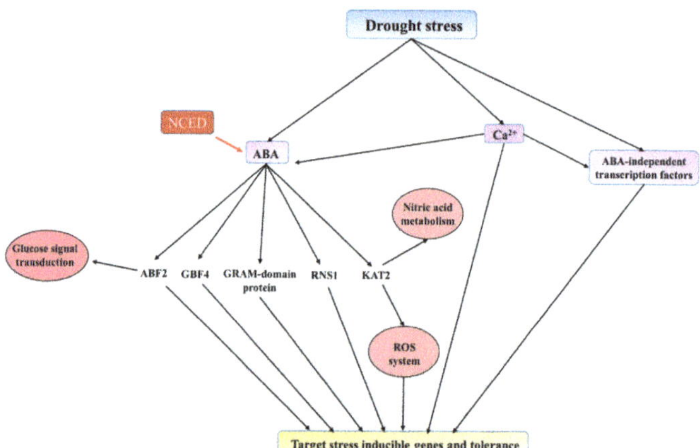

Figure 2. Schematic of the drought-induced signaling pathway based on proteome changes in switchgrass leaves. The biosynthesis of abscisic acid (ABA) was increased due to the elevated level of 9-*cis*-epoxycarotenoid dioxygenases (NCED) protein in drought-treated leaves. The elevated ABA level concurs with the induction of several ABA-responsive transcription factors, such as ABF2 (ABA-responsive elements-binding factor 2), GBF4 (G-box binding factor 4), GRAM, and ABA-responsive proteins including RNS (secreted ribonuclease) and KAT2 (3-ketoacyl-CoA thiolase-2). The ABA-independent signal transduction pathway appears to also play a role in drought-induced molecular regulation in switchgrass leaves. Several signal transduction processes may involve a second messenger (Ca^{2+}).

In the ABA-dependent signaling pathway, bZIP transcription factors is one of the major families that have been described to be associated with plant responses to stress conditions [44]. In this study, two members of bZIP proteins, GBF (G-box binding factor) (Pavir.Ea03718.1) and ABF (ABA-responsive elements-binding factor) (Pavir.J00256.1), were significantly increased under drought stress. The overexpression of ABF can alter ABA sensitivity, dehydration tolerance, and the expression levels of ABA/stress-regulated genes [56]. Furthermore, the GBF and ABF protein, and an ABA-responsive GRAM domain-containing protein (Pavir.Cb00761.1) were identified as drought-induced proteins in both CLE and PMT samples, which validates the high confidence of these significant proteins. Taken together, we have shown for the first time that the ABA-dependent pathway are regulated at protein level, which in turn may have a significant role in activating the transcription of drought tolerance genes in switchgrass.

Ribonuclease S1 (RNS1) (Pavir.Fa00890.1) plays a very important part in both wound- and ABA-responsive signaling pathways, and RNS1 itself is a target for post-transcriptional regulation by ABA [57]. The upregulated enzyme 3-ketoacyl-CoA thiolase-2 (KAT2/PED1/PKT3) (Pavir.J16366.1) has an important role in regulating reactive oxygen species (ROS) production in response to ABA [58]. In addition, three proteins annotated to the regulatory components of ABA receptor 3 (Pavir.Ab01039.1, Pavir.Ca00496.1, and Pavir.Cb01723.1) showed varied changes (0.78–1.48-fold), but none of them passed the threshold criteria for significantly changed proteins in this study. These results indicate a very dynamic adjustment system regulating the expression of proteins in ABA biosynthesis and signaling pathways, which in turn modulates the activation of drought tolerance mechanism in switchgrass leaves (Figure 2).

Environmental stimuli usually require a second messenger, such as Ca^{2+}, to transduce the signals into a plant cell. Under stress conditions, calcium-binding proteins (e.g., calmodulin or calmodulin-related protein) are induced in response to elevated levels of free Ca^{2+} in cells, and then they, in turn, activate signal transduction pathways with an impact on the activity of a variety of target enzymes [59–68]. The dynamic changes in the isoforms of these calcium-binding proteins quantified in this study represent the complex network of drought stress-induced signal transduction in switchgrass (Figure 2).

The drought-induced metabolic rearrangement is one of the major components for plants to acquire tolerance to stress conditions. Soluble sugars can accumulate to function as osmolytes to maintain cell turgor and have the ability to protect membranes and proteins from stress damage [69–71]. In the drought-treated switchgrass leaves, the induced proteins include malate synthase (Pavir.Gb01372.1), which is a key enzyme in the glyoxylate cycle for the regeneration of glucose from organic acids (Table S1-4). Maruyama et al. detected an increased level of malate synthase transcripts in rice plants subjected to drought stress, and their data implied that regulation of the glyoxylate cycle may be involved in glucose accumulation in response to dehydration in rice [69]. In the starch metabolic pathway, two proteins showed a significant alteration under drought treatment condition: the downregulated starch synthase protein (Pavir.J06822.1), which participates in starch biosynthesis, and the upregulated β amylase protein (Pavir.J18576.1) involved in the hydrolysis of starch into sugars (Table S1-4). Starch is the main form of carbohydrate storage in most plants and can be rapidly mobilized into soluble sugars. Drought and salt stress generally lead to an active conversion of starch into soluble sugars in leaves [71–73].

Plants experiencing environmental stress like cold, heat, drought, or salinity accumulate raffinose family oligosaccharides (RFO) in leaves [71,73–79]. These sugars have been implicated in membrane protection and radical scavenging [80,81]. In this study, two galactinol synthase proteins (Pavir.J07018.1 and Pavir.J40731.1) were induced in drought-treated leaves (Table S1-4), and these enzymes catalyze formation of galactinol from myo-inositol and UDP-galactose in the biosynthesis of RFO [82]. In summary, the drought-induced proteome changes seem to favor accumulation of soluble sugars, which might serve a role in protecting against cellular dehydration under drought treatment conditions.

Additionally, proteins associated with the biosynthesis of free amino acids were markedly upregulated in drought-treated leaves, which include Δ1-pyrroline-5-carboxylate synthetase protein (P5CS) (Pavir.J02344.1), the rate-limiting enzyme in proline biosynthesis, and methionine-γ-lyase protein (MGL) (Pavir.Ib03758.1), which is a precursor in isoleucine (Ile) biosynthesis (Table S1-4). Accumulation of proline (Pro) and branched-chain amino acids is commonly observed in plants subjected to osmotic stress [83,84]. Proline can serve as a free radical scavenger to overcome the oxidative stress by abiotic stress, and the accumulation of this amino acid enhances the ability of plants to grow in water-restricted or saline environments [85]. The accumulation of free isoleucine was induced in response to drought stress in *A. thaliana* [86]. The activation of these biosynthesis pathways leading to proline and isoleucine accumulation may also serve a critical role in amino acid homeostasis in drought-treated switchgrass leaves.

In the drought-treated leaves, the downregulated expression of a regulator of chromosome condensation (RCC; Pavir.Gb00127.1) may have affected the cell division, since it can bind to chromatin and generate a Ras-related nuclear protein (RAN)-guanosine triphosphate (GTP)/RAN-guanosine diphosphate (GDP) (Ran-GTP/Ran-GDP) gradient across the nuclear envelope that is required both to drive nucleocytoplasmic transport and to regulate processes associated with progression of the cell cycle and mitosis [87,88]. This might be a mechanism underlying the smaller leaf areas on drought-treated plants. Additionally, the elevated level of xyloglucan endotransglycosylase (XET) (Pavir.Fa01211.1) may assist in the process of cell wall remodeling with an impact on strengthening the wall layers and protecting mesophyll cells against physiological dehydration stress [89].

4. Materials and Methods

As described in Figure 1, the experiment is comprised of four major steps: drought treatments, protein sample preparation, proteomics analysis, and functional pathway classification of the drought-induced leaf proteomes.

4.1. Construction of a "Sandwich" Drought Treatment System

The "sandwich" treatment system was structured to simulate the process of a gradual decline in water content in the surface soil during drought under field conditions. It is comprised of double PVC pipes (an outer pipe and an inner pipe), a PVC sewer and drain coupling, and a PVC sewer and drain cap (Steinhouse Supply Company, Nashville, TN, USA). A fiberglass screen (New York Wire®, Grand Island, NY, USA) was placed inside the inner pipe, which assisted when pulling out the plants for checking root length.

The "sandwich" treatment system is divided into three layers: garden soil (25 m), perlite (15 cm), and garden soil (30 cm). The garden soil and perlite were products of Scotts Miracle-Gro Company (Marysville, OH, USA). After withholding water, the top layer of soil becomes drier gradually as the middle perlite layer drains the water quickly and cuts off moisture movement upward, and the moist bottom soil layer serves to induce root growth downward. The water depletion process in the top layer would induce gradual drought stress on plants. A 200SS WATERMARK Soil Moisture Sensor (IRROMETER Company Inc., Riverside, CA, USA) was placed at the bottom of the top soil layer in each growth tube.

4.2. Preparation of Seedling Plants

Switchgrass "Alamo" seeds were surface disinfected in 50% household bleach followed by three rinses in deionized water. Seeds were germinated in Magenta boxes partially filled with water and placed on an incubator shaker (50 rpm) at 25 °C for three days. Germinating seeds bearing 1 cm long radicals were transferred into seed cubes (Smithers-Oasis Company, Kent, OH, USA). These seedlings were watered every three days until they had grown to the three-leaf stage. At that stage, they were transplanted into the "sandwich" system and maintained in an open-roofed greenhouse at ambient temperature. The moisture content of the growing medium was maintained (soil water tension <0.01 MPa) until seedling roots reached the perlite layer. Each biological replicate contained 10 tubes each growing two plants, and four biological replicates were set up for drought-treated and non-treated control groups. A randomized block design was used in this study.

4.3. Drought Treatment and Physiological Measurements

Two weeks after transplanting, the root length was evaluated every three days. Three samples from each replicate group were selected, randomly, in each inspection. Once the longest roots reached the perlite layer, drought treatment was initiated by withholding water to these test plants. The control groups received normal watering at the rate of 4 L of water every three days for each "sandwich" system. The drought treatment was initiated on 25 April and ended on 15 May 2013.

Leaf photosynthetic rate, stomatal conductance, and transpiration rate were measured using a LI-COR 6400 Portable Photosynthesis System (Li-Cor Inc., Lincoln, NE, USA). Two fully expanded young leaves randomly selected from each plant were measured between 10:00 am and 3:00 pm. Light in the leaf chamber was set at 2000 μmol photons/m^2/s. Water use efficiency (WUE) was calculated by WUE = leaf photosynthetic rate (Pn)/transpiration (Tr) [43]. Soil water tension was recorded daily. Plant height was collected before (H_b) and after (H_a) the drought treatment and was measured from the bottom of the tiller (start point) to the bottom of the latest node (end point). Relative plant height ($H_r = H_b - H_a$) was used to compare the difference of plant relative growth rate during the drought treatment. Fresh weight (W_f) of leaves were measured at harvest, and they were dried at 70 °C for three days until a constant dry weight (W_d). Relative water content was calculated by $W_r = (W_f - W_d)/W_f$. Data analysis was performed using PROC GLM procedure of SAS software (Version: 9.3. SAS Inc., Cary, NC, USA). The effect of drought treatments was analyzed using a randomized block design analysis of variance (ANOVA). When a significant effect of drought treatment was detected, least significant difference (LSD) was used for multiple comparisons.

4.4. Tissue Harvest and Preparation of Protein Samples

Twenty days after water withholding, when the top-layer soil moisture declined to below 0.05 MPa and stomatal conductance, respiration, and water use efficiency of leaves showed a significant difference between drought-treated and non-treated control groups, plants were considered to have activated the drought-induced physiological process. The top three fully expanded leaves were cut into approximately 2 cm long pieces, wrapped with aluminum foil, frozen in liquid nitrogen, and stored at −80 °C until protein extraction.

Frozen samples were ground into a fine powder under liquid nitrogen using a Retsch Mixer Mill MM 400 (Retsch GmbH, Haan, Germany). Protein extraction followed a previously described protocol [18]. Briefly, leaf tissue powder was washed sequentially in 10% trichloroacetic acid (TCA) in acetone, 80% methanol in 0.1 M ammonium acetate, and 80% acetone with centrifugation to pellet the powder after each step. Protein was then extracted in a phenol (pH 8.0) and dense sodium dodecyl sulfate (SDS) buffer (30% sucrose, 2% SDS, 5% β-mercaptoethanol (v/w) in 0.1 M Tris-HCl, pH 8.0). After incubation at 4 °C for 2 h, the mixture was centrifuged at 16,000× g at 4 °C for 20 min. Protein in the upper phenol phase was precipitated in 0.1 M ammonium acetate in methanol after incubation overnight at −20 °C. After washes in methanol and then acetone, the air-dried protein pellets were wetted in a buffer containing 500 mM triethylammonium bicarbonate (TEAB), 2 M urea, 0.1% sodium dodecyl sulfate (SDS), and a protease inhibitor cocktail for plant cell and tissue extracts (100× dilution in the extraction buffer) (Part #9599; Sigma, St. Louis, MO, USA).

For enrichment of low-abundance proteins, the individual protein extracts were processed using a ProteoMiner Protein Enrichment kit (Bio-Rad, Hercules, CA, USA). One milliliter of each protein sample was added to the ProteoMiner columns. Proteins were bound to beads after shaking in the columns using a Mini LabRoller overnight at room temperature. Columns were then washed three times with a wash buffer (150 mM NaCl, 10 mM NaH$_2$PO$_4$, pH 7.4). Then, the columns were incubated at room temperature for 15 min in rehydrated elution reagent (8 M urea, 2% 3-((3-cholamidopropyl) dimethylammonium)-1-propanesulfonate (CHAPS) and 5% acetic acid) before eluting the proteins. Proteins were concentrated using 5 KDa Corning Spin-X UF centrifugal concentrator (Sigma, St. Louis, MO, USA).

Protein concentration was determined using a Bradford Assay Kit (Bio-Rad). Protein quality was examined by separating 15 µg of proteins on 10%–20% precast Criterio TGX polyacrylamide gels (Bio-Rad).

4.5. Isobaric Tags for Relative and Absolute Quantification (iTRAQ) Labeling and Mass Spectrometry Analysis

For iTRAQ labeling, protein samples containing 100 µg protein each were diluted using a buffer containing 500 mM TEAB, 0.1% SDS, and the same protease inhibitor as described above at the same concentration to reduce urea concentration to below 1 M. Then the protein sample was processed following the instructions of the 8-plex iTRAQ labeling kit [21]. Protein tryptic digestion was conducted using sequence grade modified trypsin (Promega, Madison, WI, USA) after incubation at 37 °C for 16 h. The control samples were labeled with tags 113, 115, 117, and 118 and the treated samples with 114, 116, 119, and 121. After combining all the labeled samples, unbound tags and SDS were removed through cation exchange cartridge (AB SCIEX). Salts and other impurities were removed using reverse-phase (RP) solid-phase extraction procedure involving 1-cm^3, 50-mg Sep-Pak C18 cartridges following the manufacturer's instructions (Waters; Milford, MA, USA). Peptides were eluted in 500 µL 50% (v/v) acetonitrile with 0.1% trifluoroacetic acid (TFA). Samples were dried at reduced pressure using a CentiVac Concentrator (labConco, Kansas City, MO, USA).

The peptide samples were subjected to a first dimension of high-pH Ultra Performance Liquid. Chromatography (UPLC) separation using an Acquity UPLC System (Waters) coupled with a robotic fraction collector (Probot; Dionex, Sunnyvale, CA, USA) [21]. One hundred micrograms of the multiplexed sample were injected and fractionated into 48 fractions in a 96-well plate. The 48 fractions were concatenated to yield 22 samples as follows: samples 1–4 and 45–48 were combined to yield two 2nd dimension fractions (samples 1–4 not analyzed in 2nd dimension); then for the remaining samples (5–44), every 20th fraction was combined. For the low-pH second dimension, low-pH RP chromatography was employed. Dried samples were reconstituted with 15 µL of 2% acetonitrile with 0.5% formic acid. Nano-LC separations of tryptic peptides were performed as described previously. The eluent from the analytical column was delivered to the LTQ-Orbitrap Elite (Thermo-Fisher Scientific, Waltham, MA, USA) via a "Plug and Play" nano ion source (CorSolutions LLC, Ithaca, NY, USA). The mass spectrometer was externally calibrated across the m/z range from 375–1800 with Ultramark 1621 for the Fourier transform (FT) mass analyzer, and individual runs were internally calibrated with the background polysiloxane ion at m/z 445.1200025 as a lock mass [24,90,91].

The Orbitrap Elite was operated in the positive ion mode with nanosource voltage set at 1.7 kV and capillary temperature at 250 °C. A parallel data-dependent acquisition (DDA) mode was used to obtain one MS survey scan with the FT mass analyzer, followed by isolation and fragmentation of the 15 most abundant, multiply-charged precursor ions with a threshold ion count higher than 50,000 in both the LTQ mass analyzer and the high energy collisionally induced dissociation (HCD)-based FT mass analyzer at a resolution of 15,000 full width at half maximum (FWHM) and m/z 400. MS survey scans were acquired with resolution set at 60,000 across the survey scan range (m/z 375–1800). Dynamic exclusion was utilized with repeat count set to 1 with a 40 s repeat duration; exclusion list size was set to 500, 20 s exclusion duration, and low and high exclusion mass widths set to 1.5. Fragmentation parameters were set with isolation width at 1.5 m/z, normalized collision energy at 37%, and activation Q at 0.25. Activation time for HCD analysis was 0.1 min. All data were acquired using XCalibur 2.1 (Thermo-Fisher Scientific) [18,24]. Proteins were identified using the MS data to query the switchgrass annotated database (http://www.phytozome.net/) via Mascot v2.3.02 (Matrix Sciences, Boston, MA, USA). The mass spectrometry proteomics data have been deposited to the ProteomeXchange Consortium via the PRIDE [92] partner repository with the dataset identifier PXD004675 and 10.6019/PXD004675.

4.6. Protein Identification and Quantification, and Statistics Analysis

For a protein to be included in the quantitative analysis, it was required that at least two unique peptides have to be identified in all eight biological samples. The intensities of reporter ions of constituent peptides were log2-transformed. Then, log2 fold values from all constituent peptides were subjected to *t*-test (general linear model procedure) followed by false discovery rate (FDR) corrections to test the statistical significance of the difference in normalized abundance of each protein between the drought-treated and control sample groups [21]. The log2 transformed abundance ratios were then fit to a normal distribution ($p < 0.01$) [93]. Two standard deviations (i.e., a 95% confidence level) of the log2 fold transformed protein abundance ratio (treated/control) were used as the cutoff for significantly changed proteins. The antilog conversion was used to represent the fold change of proteins. Statistical analyses were performed using SAS (version 9.3; SAS Institute, Cary, NC, USA) [18].

4.7. Functional Pathway Analysis of Drought-Induced Proteins

In the annotated switchgrass database (*Panicum virgatum v1.1*, Phytozome v11.0), each accession is associated with a unigene accession in *Arabidopsis thaliana*. The switchgrass annotated genome is not included in the database of the MapMan pathway tools. Therefore, in this study, the *A. thaliana* database in MapMan (MapMan, version 3.5.1R2, Max Planck Institute of Molecular Plant Physiology, Potsdam-Golm, Germanry) was used to develop the functional pathways [94]. Additional literature and database searches were conducted to develop the association between drought-induced proteins and drought tolerance, and highlight new discoveries using proteomics analysis.

4.8. Statistical Analysis

All independent experiments were repeated four times. Experimental data were presented as means and standard deviations (SD). The SAS version 9.0 software (SAS Inc., Cary, NC, USA) was used to perform the analysis of variance (ANOVA) and least significant difference (LSD) tests for the physiological data, and *t*-tests and FDR tests in the analysis of quantitative proteomics data.

5. Conclusions

This study has identified drought-induced changes in leaf proteomes that occurred when plants have shown significant physiological changes from drought-treated to non-treated control conditions. The identified proteins are involved in both ABA-dependent and ABA-independent signaling pathways, and diverting metabolic pathways toward increasing cellular concentrations of soluble sugars and stress-related amino acids (proline and isoleucine). The accumulation of a diverse species of stress proteins can be considered as the hallmark for switchgrass plants to acquire drought tolerance. Information provided in this paper advanced our understanding of molecular mechanisms underlying drought tolerance in C_4 plants.

Supplementary Materials: Supplementary materials can be found at http://www.mdpi.com/1422-0067/17/8/1251/s1.

Acknowledgments: The authors wish to thank Sheng Zhang of the Proteomics and Mass Spectrometry Facility of the Cornell University Institute of Biotechnology for expert technical assistance and helpful discussion; Roger Sauve, Jason de Koff, Fur-Chi Chen, and George Smith at Tennessee State University for discussions in experimental design; and Mrs. Sarabjit Bhatti and Long Zhang for assisting in carrying out the study. This work was supported by the National Institute of Food and Agriculture, U.S. Department of Agriculture, Grant No. 2012-02466, project TENX-1507-SE, and the U.S. Department of Agriculture Agricultural Research Service, Grant No. 1907-21000-036/037-00D. ARS disclaimer: "Mention of trade names or commercial products in this publication is solely for the purpose of providing specific information and does not imply recommendation or endorsement by the U.S. Department of Agriculture." Seeds of switchgrass "Alamo" were kindly provided by Jason de Koff, Tennessee State University, Nashville, TN, USA.

Author Contributions: Zhujia Ye designed and performed the experiments, and prepared the manuscript as part of her Ph.D. thesis. Kevin J. Howe and Tara Fish conducted mass-spectrometry analysis. Chih-Li Yu conducted photosynthesis measurements. Dafeng Hui designed the program for statistical analysis. Sasikiran Sangireddy and Ikenna Okekeogbu contributed equally and both conducted the drought treatment experiments. Suping Zhou and Theodore W. Thannhauser developed the experimental system and revised the paper. All authors read and agreed with the final manuscript. Suping Zhou and Theodore W. Thannhauser are the corresponding authors and are responsible for all contacts and correspondence.

Conflicts of Interest: The authors declare no conflict of interest.

References

1. Wright, L.I.; Cushman, J.H.; Ehrenshaft, A.R.; McLaughlin, S.B.; McNabb, W.A.; Martin, S.A.; Ranney, J.W.; Tuskan, A.G.; Turhollow, A.F. *Biofuels Feedstock Development Program Annual Progress Report for 1992*; ORNL-6781; Environmental Sciences Division Publication: Washington, DC, USA, 1993.
2. Parrisha, D.J.; Fike, J.H. The biology and agronomy of switchgrass for biofuels. *CRC. Crit. Rev. Plant Sci.* **2005**, *24*, 423–459. [CrossRef]
3. Wright, L.L. *Historical Perspective on How and Why Switchgrass Was Selected as a "Model" High-Potential Energy Crop*; ORNL/TM-2007/109; Oak Ridge National Laboratory: Oak Ridge, TN, USA, 2007.
4. Nezhadahmadi, A.; Prodhan, Z.H.; Faruq, G. Drought tolerance in wheat. *Sci. World J.* **2013**, *2013*, 610721. [CrossRef] [PubMed]
5. National Weather Service Centers for Environmental Prediction. Available online: http://www.cpc.ncep. noaa.gov/products/monitoring_and_data/topsoil.shtml (accessed on 1 January 2016).
6. Keyser, P.; Harper, C.; Bates, G.; Waller, J.; Doxon, E. Native warm-season grasses for mid-south forage production. *UT Ext.* **2011**, SP731-A.
7. Sanderson, M.A.; Reed, R.L. Switchgrass growth and development: Water, nitrogen, and plant density effects. *J. Range Manag.* **2000**, *53*, 221–227. [CrossRef]
8. Vogel, K.P. Improved plant & production practices for grasslands & biomass crops in the mid-continental USA. In Proceedings of the DOE/USDA Biomass Feedstock Gate Review Meeting, Washington, DC, USA, 14–16 March 2005.
9. Farooq, M.; Wahid, A.; Kobayashi, N.; Fujita, D.; Basra, S.M.A. Plant drought stress: Effects, mechanisms and management. *Agron. Sustain. Dev.* **2009**, *29*, 185–212. [CrossRef]
10. Nonami, H. Plant water relations and control of cell elongation at low water potentials. *J. Plant Res.* **1998**, *111*, 373–382. [CrossRef]
11. Kaya, M.D.; Okçub, G.; Ataka, M.; Çıkılıc, Y.; Kolsarıcıa, Ö. Seed treatments to overcome salt and drought stress during germination in sunflower (*Helianthus annuus* L.). *Eur. J. Agron.* **2006**, *24*, 291–295. [CrossRef]
12. Hussain, M.; Malik, M.A.; Farooq, M.; Ashraf, M.Y.; Cheema, M.A. Improving drought tolerance by exogenous application of glycinebetaine and salicylic acid in sunflower. *J. Agron. Crop Sci.* **2008**, *194*, 193–199. [CrossRef]
13. Liu, Y.; Zhang, X.; Tran, H.; Shan, L.; Kim, J.; Childs, K.; Ervin, E.H.; Frazier, T.; Zhao, B. Assessment of drought tolerance of 49 switchgrass (*Panicum virgatum*) genotypes using physiological and morphological parameters. *Biotechnol. Biofuels* **2015**, *8*. [CrossRef] [PubMed]
14. Alberts, B.; Johnson, A.; Lewis, J. *Molecular Biology of the Cell*, 4th ed.; Garland Science: New York, NY, USA, 2002.
15. Barkla, B.J.; Castellanos-Cervantes, T.; de León, J.L.D.; Matros, A.; Mock, H.P.; Perez-Alfocea, F.; Salekdeh, G.H.; Witzel, K.; Zörb, C. Elucidation of salt stress defense and tolerance mechanisms of crop plants using proteomics—current achievements and perspectives. *Proteomics* **2013**, *13*, 1885–1900. [CrossRef] [PubMed]
16. Ghosh, D.; Xu, J. Abiotic stress responses in plant roots: A proteomics perspective. *Front. Plant Sci.* **2014**, *5*, 1–13. [CrossRef] [PubMed]
17. Ngara, R.; Ndimba, B.K. Understanding the complex nature of salinity and drought-stress response in cereals using proteomics technologies. *Proteomics* **2014**, *14*, 611–621. [CrossRef] [PubMed]
18. Okekeogbu, I.; Ye, Z.; Sangireddy, S.; Li, H.; Bhatti, S.; Hui, D.; Zhou, S.; Howe, K.; Fish, T.; Yang, Y.; et al. Effect of aluminum treatment on proteomes of radicles of seeds derived from AL-treated tomato plants. *Proteomes* **2014**, *2*, 169–190. [CrossRef]
19. Zhou, S.; Sauve, R.; Fish, T.; Thannhauser, T.W. Salt-induced and Salt-suppressed Proteins in Tomato Leaves. *J. Am. Soc. Hortic. Sci.* **2009**, *134*, 289–294.

20. Zhou, S.; Sauvé, R.; Liu, Z.; Reddy, S.; Bhatti, S. Heat-induced proteome changes in tomato leaves. *J. Am. Soc. Hortic. Sci.* **2011**, *136*, 219–226.
21. Zhou, S.; Palmer, M.; Zhou, J.; Bhatti, S.; Howe, K.; Fish, T.; Thannhauser, T.W. Differential root proteome expression in tomato genotypes with contrasting drought tolerance exposed to dehydration. *J. Am. Soc. Hort. Sci.* **2013**, *138*, 131–141.
22. Pottiez, G.; Wiederin, J.; Fox, H.S.; Ciborowski, P. Comparison of 4-plex to 8-plex iTRAQ quantitative measurements of proteins in human plasma samples. *J. Proteome Res.* **2012**, *11*, 3774–3781. [CrossRef] [PubMed]
23. Chen, X.; Walker, A.K.; Strahler, J.R.; Simon, E.S.; Tomanicek-Volk, S.L.; Nelson, B.B.; Hurley, M.C.; Ernst, S.; Williams, J.; Andrews, P.C. Organellar proteomics: Analysis of pancreatic zymogen granule membranes. *Mol. Cell. Proteom.* **2006**, *5*, 306–312. [CrossRef] [PubMed]
24. Yang, Q.S.; Wu, J.H.; Li, C.Y.; Wei, Y.R.; Sheng, O.; Hu, C.H.; Kuang, R.-B.; Huang, Y.-H.; Peng, X.-X.; McCardle, J.; et al. Quantitative proteomic analysis reveals that antioxidation mechanisms contribute to cold tolerance in plantain (*Musa paradisiaca* L.; ABB Group) seedlings. *Mol. Cell. Proteom.* **2012**, *11*, 1853–1869. [CrossRef] [PubMed]
25. Lan, P.; Li, W.F.; Wen, T.N.; Shiau, J.Y.; Wu, Y.C.; Lin, W.D.; Schmidt, W. iTRAQ protein profile analysis of arabidopsis roots reveals new aspects critical for iron homeostasis. *Plant Physiol.* **2011**, *155*, 821–834. [CrossRef] [PubMed]
26. Redding, A.M.; Mukhopadhyay, A.; Joyner, D.C.; Hazen, T.C.; Keasling, J.D. Study of nitrate stress in desulfovibrio vulgaris hildenborough using iTRAQ proteomics. *Brief. Funct. Genom. Proteom.* **2006**, *5*, 133–143. [CrossRef] [PubMed]
27. Nveawiah-Yoho, P.; Zhou, J.; Palmer, M.; Sauve, R.; Zhou, S.; Howe, K.J.; Fish, T.; Thannhauser, T.W. Identification of proteins for salt tolerance using a comparative proteomics analysis of tomato accessions with contrasting salt tolerance. *J. Am. Soc. Hort. Sci.* **2013**, *138*, 382–394.
28. Nveawiah-Yoho, P. Mechanisms for Salt Tolerance and Susceptibility in Tomato. Ph.D. Thesis, Tennessee State University, Nashville, TN, USA, 2012.
29. Ali, G.M.; Komatsu, S. Proteomic analysis of rice leaf sheath during drought stress. *J. Proteom. Res.* **2006**, *5*, 396–403. [CrossRef] [PubMed]
30. Ke, Y.; Han, G.; He, H.; Li, J. Differential regulation of proteins and phosphoproteins in rice under drought stress. *Biochem. Biophys. Res. Commun.* **2009**, *379*, 133–138. [CrossRef] [PubMed]
31. Salekdeh, G.; Siopongco, J.; Wade, L.; Ghareyazie, B.; Bennett, J. Proteomic analysis of rice leaves during drought stress and recovery. *Proteomics* **2002**, *2*, 1131–1145. [CrossRef]
32. Salekdeh, G.; Siopongco, J.; Wade, L.; Ghareyazie, B.; Bennett, J. A proteomic approach to analyzing drought- and salt-responsiveness in rice. *F. Crop. Res.* **2002**, *76*, 199–219. [CrossRef]
33. Shu, L.; Ding, W.; Wu, J.; Feng, F.; Luo, L. Proteomic analysis of rice leaves shows the different regulations to osmotic stress and stress signals. *J. Integr. Plant Biol.* **2010**, *52*, 981–995. [CrossRef] [PubMed]
34. Xiong, J.; Fu, B.; Xu, H.; Li, Y. Proteomic analysis of PEG-simulated drought stress-responsive proteins of rice leaves using a pyramiding rice line at the seedling stage. *Bot. Stud.* **2010**, *51*, 137–145.
35. De Vienne, D.; Leonardi, A.; Damerval, C.; Zivy, M. Genetics of proteome variation for QTL characterization: Application to drought-stress responses in maize. *J. Exp. Bot.* **1999**, *50*, 303–309. [CrossRef]
36. Mohammadkhani, N.; Heidari, R. Effects of drought stress on soluble proteins in two maize varieties. *Turkish J. Biol.* **2008**, *32*, 23–30.
37. Riccardi, F.; Gazeau, P.; de Vienne, D.; Zivy, M. Protein changes in response to progressive water deficit in maize. *Plant Physiol.* **1998**, *117*, 1253–1263. [CrossRef] [PubMed]
38. McCabe, M.S.; Garratt, L.C.; Schepers, F.; Jordi, W.J.; Stoopen, G.M.; Davelaar, E.; van Rhijn, J.H.; Power, J.B.; Davey, M.R. Effects of P(SAG12)-IPT gene expression on development and senescence in transgenic lettuce. *Plant Physiol.* **2001**, *127*, 505–516. [CrossRef] [PubMed]
39. Widjaja, I.; Naumann, K.; Roth, U.; Wolf, N.; Mackey, D.; Dangl, J.L.; Scheel, D.; Lee, J. Combining subproteome enrichment and Rubisco depletion enables identification of low abundance proteins differentially regulated during plant defense. *Proteomics* **2009**, *9*, 138–147. [CrossRef] [PubMed]
40. Fasoli, E.; D'Amato, A.; Kravchuk, A.V.; Boschetti, E.; Bachi, A.; Righetti, P.G. Popeye strikes again: The deep proteome of spinach leaves. *J. Proteom.* **2011**, *74*, 127–136. [CrossRef] [PubMed]

41. Von Toerne, C.; Kahle, M.; Schäfer, A.; Ispiryan, R.; Blindert, M.; Hrabe De Angelis, M.; Neschen, S.; Ueffing, M.; Hauck, S.M. Apoe, Mbl2, and Psp plasma protein levels correlate with diabetic phenotype in NZO mice-An optimized rapid workflow for SRM-based quantification. *J. Proteome Res.* **2013**, *12*, 1331–1343. [CrossRef] [PubMed]

42. Fonslow, B.R.; Carvalho, P.C.; Academia, K.; Freeby, S.; Xu, T.; Nakorchevsky, A.; Paulus, A.; Yates, J.R. Improvements in proteomic metrics of low abundance proteins through proteome equalization using ProteoMiner prior to MudPIT. *J. Proteome Res.* **2011**, *10*, 3690–3700. [CrossRef] [PubMed]

43. Polley, H.W. Implications of atmospheric and climatic change for crop yield and water use efficiency. *Crop Sci.* **2002**, *42*, 131–140. [CrossRef] [PubMed]

44. Finkelstein, R.R.; Lynch, T.J. The *Arabidopsis* abscisic acid response gene ABI5 encodes a basic leucine zipper transcription factor. *Plant Cell* **2000**, *12*, 599–609. [CrossRef] [PubMed]

45. Rösti, J.; Barton, C.J.; Albrecht, S.; Dupree, P.; Pauly, M.; Findlay, K.; Roberts, K.; Seifert, G.J. UDP-glucose 4-epimerase isoforms UGE2 and UGE4 cooperate in providing UDP-galactose for cell wall biosynthesis and growth of *Arabidopsis thaliana*. *Plant Cell* **2007**, *19*, 1565–1579. [CrossRef] [PubMed]

46. Uozu, S.; Tanaka-Ueguchi, M.; Kitano, H.; Hattori, K.; Matsuoka, M. Characterization of XET-related genes of rice. *Plant Physiol.* **2000**, *122*, 853–859. [CrossRef] [PubMed]

47. Nambara, E.; Marion-Poll, A. Abscisic acid biosynthesis and catabolism. *Annu. Rev. Plant Biol.* **2005**, *56*, 165–185. [CrossRef] [PubMed]

48. Leckie, C.P.; McAinsh, M.R.; Allen, G.J.; Sanders, D.; Hetherington, A.M. Abscisic acid-induced stomatal closure mediated by cyclic ADP-ribose. *Proc. Natl. Acad. Sci. USA* **1998**, *95*, 15837–15842. [CrossRef] [PubMed]

49. Wang, Y.; Ries, A.; Wu, K.; Yang, A.; Crawford, N.M. The *Arabidopsis* prohibitin gene PHB3 functions in nitric oxide-mediated responses and in hydrogen peroxide-induced nitric oxide accumulation. *Plant Cell* **2010**, *22*, 249–259. [CrossRef] [PubMed]

50. García-Mata, C.; Lamattina, L. Nitric oxide induces stomatal closure and enhances the adaptive plant responses against drought stress. *Plant Physiol.* **2001**, *126*, 1196–1204. [CrossRef] [PubMed]

51. Lefebvre, V.; North, H.; Frey, A.; Sotta, B.; Seo, M.; Okamoto, M.; Nambara, E.; Marion-Poll, A. Functional analysis of *Arabidopsis* NCED6 and NCED9 genes indicates that ABA synthesized in the endosperm is involved in the induction of seed dormancy. *Plant J.* **2006**, *45*, 309–319. [CrossRef] [PubMed]

52. Huo, H.; Dahal, P.; Kunusoth, K.; McCallum, C.M.; Bradford, K.J. Expression of 9-*cis*-EPOXYCAROTENOID DIOXYGENASE4 is essential for thermoinhibition of lettuce seed germination but not for seed development or stress tolerance. *Plant Cell* **2013**, *25*, 884–900. [CrossRef] [PubMed]

53. Tan, B.C.; Joseph, L.M.; Deng, W.T.; Liu, L.; Li, Q.B.; Cline, K.; McCarty, D.R. Molecular characterization of the Arabidopsis 9-*cis*-epoxycarotenoid dioxygenase gene family. *Plant J.* **2003**, *35*, 44–56. [CrossRef] [PubMed]

54. Seo, M.; Kanno, Y.; Frey, A.; North, H.M.; Marion-Poll, A. Dissection of Arabidopsis NCED9 promoter regulatory regions reveals a role for ABA synthesized in embryos in the regulation of GA-dependent seed germination. *Plant Sci.* **2016**, *246*, 91–97. [CrossRef] [PubMed]

55. Iuchi, S.; Kobayashi, M.; Taji, T.; Naramoto, M.; Seki, M.; Kato, T. Regulation of drought tolerance by gene manipulation of 9-*cis*-epoxycarotenoid dioxygenase, a key enzyme in abscisic acid biosynthesis in *Arabidopsis*. *Plant J.* **2001**, *27*, 325–333. [CrossRef] [PubMed]

56. Kim, S.; Kang, J.; Cho, D.-I.; Park, J.H.; Kim, S.Y. ABF2, an ABRE-binding bZIP factor, is an essential component of glucose signaling and its overexpression affects multiple stress tolerance. *Plant J.* **2004**, *40*, 75–87. [CrossRef] [PubMed]

57. Hillwig, M.S.; LeBrasseur, N.D.; Green, P.J.; MacIntosh, G.C. Impact of transcriptional, ABA-dependent, and ABA-independent pathways on wounding regulation of RNS1 expression. *Mol. Genet. Genom.* **2008**, *280*, 249–261. [CrossRef] [PubMed]

58. Jiang, T.; Zhang, X.F.; Wang, X.F.; Zhang, D.P. Arabidopsis 3-Ketoacyl-CoA Thiolase-2 (KAT2), an enzyme of fatty acid β-Oxidation, is involved in ABA signal transduction. *Plant Cell Physiol.* **2011**, *52*, 528–538. [CrossRef] [PubMed]

59. Knight, M.R.; Campbell, A.K.; Smith, S.M.; Trewavas, A.J. Transgenic plant aequorin reports the effects of touch and cold-shock and elicitors on cytoplasmic calcium. *Nature* **1991**, *352*, 524–526. [CrossRef] [PubMed]

60. Knight, M.; Read, N.; Campbell, A.; Trewavas, A. Imaging calcium dynamics in living plants using semi-synthetic recombinant aequorins. *J. Cell Biol.* **1993**, *121*, 83–90. [CrossRef] [PubMed]

61. Knight, M.R.; Smith, S.M.; Trewavas, A.J. Wind-induced plant motion immediately increases cytosolic calcium. *Proc. Natl. Acad. Sci. USA* **1992**, *89*, 4967–4971. [CrossRef] [PubMed]

62. Braam, J. Regulated expression of the calmodulin-related TCH genes in cultured *Arabidopsis* cells: Induction by calcium and heat shock. *Proc. Natl. Acad. Sci. USA* **1992**, *89*, 3213–3216. [CrossRef] [PubMed]

63. Haley, A.N.N.; Russell, A.J.; Wood, N.; Allan, A.C.; Knight, M.; Campbell, A.K.; Trewavas, A.J.; Ryan, C.A.; Lamb, C.J.; Jagendorf, A.T.; et al. Effects of mechanical signaling on plant cell cytosolic calcium. *Proc. Natl. Acad. Sci. USA* **1995**, *92*, 4124–4128. [CrossRef] [PubMed]

64. Klee, C.; Vanaman, T. Calmodulin. *Adv. Protein Chem.* **1982**, *35*, 213–322. [PubMed]

65. Roberts, D.; Lukas, T.; Watterson, D. Structure, function, and mechanisms of action of calmodulin. *CRC Crit. Rev. Plant Sci.* **1986**, *4*, 311–339. [CrossRef]

66. Cohen, P.; Klee, C. *Calmodulin*; Elsevier Biomedical Press: Amsterdam, The Netherlands, 1988.

67. Allan, E.; Hepler, P. Calmodulin and calcium-binding proteins. In *The Biochemistry of Plants*; Stumpf, P., Lonn, E., Eds.; Academic Press: New York, NY, USA, 1989; pp. 455–484.

68. Roberts, D.M.; Harmon, A.C. Calcium modulated proteins targets of intracellular calcium signals in higher plants. *Annu. Rev. Plant Physiol. Plant Mol. Biol.* **1992**, *43*, 375–414. [CrossRef]

69. Maruyama, K.; Urano, K.; Yoshiwara, K.; Morishita, Y.; Sakurai, N.; Suzuki, H.; Kojima, M.; Sakakibara, H.; Shibata, D.; Saito, K.; et al. Integrated analysis of the effects of cold and dehydration on rice metabolites, phytohormones, and gene transcripts. *Plant Physiol.* **2014**, *164*, 1759–1771. [CrossRef] [PubMed]

70. Madden, T.; Bally, M.; Hope, M.; Cullis, P.; Schieren, H.; Janoff, A. Protection of large unilamellar vesicles by trehalose during dehydration: Retention of vesicle contents. *Biochim. Biophys. Acta* **1985**, *817*, 67–74. [CrossRef]

71. Kaplan, F.; Guy, C.L. β-amylase induction and the protective role of maltose during temperature shock. *Plant Physiol.* **2004**, *135*, 1674–1684. [CrossRef] [PubMed]

72. Basu, P.S.; Ali, M.; Chaturvedi, S.K. Osmotic adjustment increases water uptake, remobilization of assimilates and maintains photosynthesis in chickpea under drought. *Indian J. Exp. Biol.* **2007**, *45*, 261–267. [PubMed]

73. Kempa, S.; Krasensky, J.; Dal Santo, S.; Kopka, J.; Jonak, C. A Central Role of Abscisic Acid in Stress-Regulated Carbohydrate Metabolism. *PLoS ONE* **2008**, *3*, e3935. [CrossRef] [PubMed]

74. Castonguay, Y.; Nadeau, P. Enzymatic control of soluble carbohydrate accumulation in cold-acclimated crowns of alfalfa. *Crop Sci.* **1998**, *38*, 1183–1189. [CrossRef]

75. Gilmour, S.; Sebolt, A.; Salazar, M.; Everard, J.; Thomashow, M. Overexpression of the Arabidopsis CBF3 transcriptional activator mimics multiple biochemical changes associated with cold acclimation. *Plant Physiol.* **2000**, *124*, 1854–1865. [CrossRef] [PubMed]

76. Taji, T.; Ohsumi, C.; Iuchi, S.; Seki, M.; Kasuga, M.; Kobayashi, M.; Yamaguchi-Shinozaki, K.; Shinozaki, K. Important roles of drought- and cold-inducible genes for galactinol synthase in stress tolerance in Arabidopsis thaliana. *Plant J.* **2002**, *29*, 417–426. [CrossRef] [PubMed]

77. Cook, D.; Fowler, S.; Fiehn, O.; Thomashow, M.F. A prominent role for the CBF cold response pathway in conFigure uring the low-temperature metabolome of Arabidopsis. *Proc. Natl. Acad. Sci. USA* **2004**, *101*, 15243–15248.

78. Peters, S.; Mundree, S.; Thomson, J.; Farrant, J.; Keller, F. Protection mechanisms in the resurrection plant Xerophyta viscosa (Baker): Both sucrose and raffinose family oligosaccharides (RFOs) accumulate in leaves in response to water deficit. *J. Exp. Bot.* **2007**, *58*, 1947–1956. [CrossRef] [PubMed]

79. Usadel, B.; Bläsing, O.E.; Gibon, Y.; Poree, F.; Höhne, M.; Günter, M.; Trethewey, R.; Kamlage, B.; Poorter, H.; Stitt, M. Multilevel genomic analysis of the response of transcripts, enzyme activities and metabolites in *Arabidopsis* rosettes to a progressive decrease of temperature in the non-freezing range. *Plant Cell Environ.* **2008**, *31*, 518–547. [CrossRef] [PubMed]

80. Hincha, D.K. Effects of calcium-induced aggregation on the physical stability of liposomes containing plant glycolipids. *Biochim. Biophys. Acta* **2003**, *1611*, 180–186. [CrossRef]

81. Nishizawa, A.; Yabuta, Y.; Shigeoka, S. Galactinol and raffinose constitute a novel function to protect plants from oxidative damage. *Plant Physiol.* **2008**, *147*, 1251–1263. [CrossRef] [PubMed]

82. Peterbauer, T.; Richter, A. Biochemistry and physiology of raffinose family oligosaccharides and galactosyl cyclitols in seeds. *Seed Sci. Res.* **2001**, *11*, 185–197.

83. Rhodes, D.; Handa, S.; Bressan, R. Metabolic changes associated with adaptation of plant cells to water stress. *Plant Physiol.* **1986**, *82*, 890–903. [CrossRef] [PubMed]

84. Girousse, C.; Bournoville, R.; Bonnemain, J.L. Water deficit-induced changes in concentrations in proline and some other amino acids in the phloem sap of alfalfa. *Plant Physiol.* **1996**, *111*, 109–113. [PubMed]

85. Hong, Z. Removal of feedback inhibition of delta 1-pyrroline-5-carboxylate synthetase results in increased proline accumulation and protection of plants from osmotic stress. *Plant Physiol.* **2000**, *122*, 1129–1136. [CrossRef] [PubMed]

86. Joshi, V.; Jander, G. *Arabidopsis* methionine garmma-lyase is regulated according to isoleucine biosynthesis needs but Plays a subordinate. *Plant Physiol.* **2009**, *151*, 367–378. [CrossRef] [PubMed]

87. Bischoff, F.R.; Ponstingl, H. Catalysis of guanine nucleotide exchange on Ran by the mitotic regulator RCC1. *Nature* **1991**, *354*, 80–82. [CrossRef] [PubMed]

88. Nemergut, M.E.; Lindsay, M.E.; Brownawell, A.M.; Macara, I.G. Ran-binding protein 3 links Crm1 to the Ran guanine nucleotide exchange factor. *J. Biol. Chem.* **2002**, *277*, 17385–17388. [CrossRef] [PubMed]

89. Cho, S.K.; Kim, J.E.; Park, J.A.; Eom, T.J.; Kim, W.T. Constitutive expression of abiotic stress-inducible hot pepper CaXTH3, which encodes a xyloglucan endotransglucosylase/hydrolase homolog, improves drought and salt tolerance in transgenic Arabidopsis plants. *FEBS Lett.* **2006**, *580*, 3136–3144. [CrossRef] [PubMed]

90. Thannhauser, T.; Shen, M.; Sherwood, R.; Howe, K.; Fish, T.; Yang, Y.; Chen, W.; Zhang, S. A workflow for large-scale empirical identification of cell wall N-linked glycoproteins of tomato (*Solanum lycopersicum*). *Electrophoresis* **2013**, *34*, 2417–2431. [CrossRef] [PubMed]

91. Yang, Y.; Qiang, X.; Owsiany, K.; Zhang, S.; Thannhauser, T.W.; Li, L. Evaluation of different multidimensional LC-MS/MS pipelines for iTRAQ-based proteomic analysis of potato tubers in response to cold storage. *J. Proteome Res.* **2011**, *10*, 4647–4660. [CrossRef] [PubMed]

92. Vizcaíno, J.A.; Csordas, A.; del-Toro, N.; Dianes, J.A.; Griss, J.; Lavidas, I.; Mayer, G.; Perez-Riverol, Y.; Reisinger, F.; Ternent, T.; et al. 2016 update of the PRIDE database and related tools. *Nucleic Acids Res.* **2016**, *44*, D447–D456. [CrossRef] [PubMed]

93. Zhou, S.; Okekeogbu, I.; Sangireddy, S.; Ye, Z.; Li, H.; Bhatti, S.; Hui, D.; Mcdonald, D.W.; Yang, Y.; Giri, S.; et al. Proteome modification in tomato plants upon long-term aluminum treatment. *J. Proteome Res.* **2016**, *15*, 1670–1684. [CrossRef] [PubMed]

94. MapMan, Version 3.5.1R2 2015. Available online: http://mapman.gabipd.org (accessed on 10 April 2016).

International Journal of
Molecular Sciences

Review

Drought-Responsive Mechanisms in Plant Leaves Revealed by Proteomics

Xiaoli Wang, Xiaofeng Cai, Chenxi Xu, Quanhua Wang and Shaojun Dai *

Development Centre of Plant Germplasm Resources, College of Life and Environmental Sciences,
Shanghai Normal University, Shanghai 200234, China; wangxl@shnu.edu.cn (X.W.); xfcai@shnu.edu.cn (X.C.);
chenxixu@shnu.edu.cn (C.X.); wangquanhuan@shnu.edu.cn (Q.W.)
* Correspondence: daishaojun@hotmail.com; Tel.: +86-21-6432-4576

Academic Editors: Setsuko Komatsu and Zahed Hossain
Received: 30 June 2016; Accepted: 22 September 2016; Published: 18 October 2016

Abstract: Plant drought tolerance is a complex trait that requires a global view to understand its underlying mechanism. The proteomic aspects of plant drought response have been extensively investigated in model plants, crops and wood plants. In this review, we summarize recent proteomic studies on drought response in leaves to reveal the common and specialized drought-responsive mechanisms in different plants. Although drought-responsive proteins exhibit various patterns depending on plant species, genotypes and stress intensity, proteomic analyses show that dominant changes occurred in sensing and signal transduction, reactive oxygen species scavenging, osmotic regulation, gene expression, protein synthesis/turnover, cell structure modulation, as well as carbohydrate and energy metabolism. In combination with physiological and molecular results, proteomic studies in leaves have helped to discover some potential proteins and/or metabolic pathways for drought tolerance. These findings provide new clues for understanding the molecular basis of plant drought tolerance.

Keywords: proteomics; leaves; drought stress; molecular mechanism

---

1. Introduction

Drought is an inevitable and recurring feature of world climate. Despite our efforts to forecast its onset and modify its impact, drought remains the single most important factor affecting worldwide crop growth and productivity [1]. More importantly, although plant drought response has been extensively studied [2–4], there are still no economically practical technological means to facilitate crop production under drought [5]. Therefore, there is an urgent need to further understand and enhance crop tolerance to drought stress.

Drought stress induces a number of changes at the morphological, physiological and biochemical level in all plant organs [6]. Plants have evolved several strategies to cope with drought stress, including drought escape via a short life cycle or developmental plasticity, drought avoidance via enhanced water uptake and reduced water loss, as well as drought tolerance via osmotic adjustment, antioxidant capacity, and desiccation tolerance [7]. With the development of high-throughput sequencing and various omic technologies, large amounts of sequence data sets and global changes in gene expressions have been reported. For example, using genomics and quantitative trait loci (QTL) mapping approaches, a number of major genes or QTL for drought stress resistance were identified in *Arabidopsis* [8], pearl millet (*Pennisetum glaucum*) [9], *Festuca pratensis* (Huds.) [10], wheat (*Triticum aestivum*) [11], maize (*Zea mays*) [12], and common bean (*Phaseolus vulgaris*) [13]. Besides, microarray analysis of drought response in shoots of two *T. aestivum* cultivars (i.e., *TAM 111* and *TAM 112*) have revealed that 1657 transcripts were commonly altered in both cultivars, but 474 and 1540 transcripts were unique to each cultivar, respectively [14]. In addition, genome-wide transcriptional analysis of *Populus euphratica* under four different drought treatments (watering with 70%, 50%, 20%, and 5% of daily evaporation for seven weeks) have identified 952, 1354, 2138 and 2360 transcripts, respectively. Among them, some candidate drought-responsive genes have a high potential to be used for crop breeding with drought tolerance. Those genes are involved in the biosynthesis of plant hormones, various signaling pathways (e.g., *ZEP*, *PYL*, *PP2C*, *SnRK2*, *ACO*, *ACS*, *ETR1*, and *ETO1*), osmoprotective pathways (e.g., *BADH*, *P5CS*, *PDH1*, *TPS1*, and *LEA*), as well as the detoxification of reactive oxygen species (e.g., *APX*, *SOD*, *GR*, and *ALDH7*) [15].

However, the mRNA levels usually do not correlate well with the protein abundances and functions, due to various post-translational modifications. High-throughput proteomics has proved to be a powerful tool for the comprehensive identification of drought-responsive proteins in plants [16,17]. In previous investigations, more than 2200 drought-responsive protein species have been identified in leaves from 25 plant species, including 18 herbs, 3 shrubs and 4 trees (Table 1, Supplementary Table S1). In the present review, these drought-responsive protein species are defined as 440 unique proteins on the basis of their protein sequence homology and functional domain similarity. These proteins are mainly involved in signaling, transcription, stress and defense, protein synthesis, folding and degradation, photosynthesis and photorespiration, carbohydrate and energy metabolism, membrane and transport, cell structure and cell cycle, nitrogen assimilation and amino acid metabolism, as well as fatty acid metabolism (Figure 1). In addition, we also found some drought-responsive phosphoproteins in four phosphoproteomics studies (Table 1, Supplementary Table S2). Phosphorylation is one of the most important post-translational modifications (PTMs) that modulates protein activity, protein–protein interaction and cellular localization. These drought-responsive phosphoproteins were mainly involved in signaling, transcription, photosynthesis and carbon metabolism, as well as in protein synthesis and turnover. The integrative analysis of physiological, molecular, and proteomic characteristics provides new clues for further understanding plant drought tolerance.

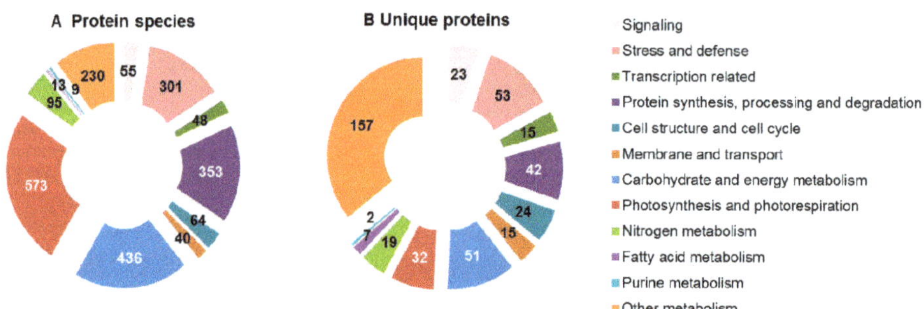

Figure 1. Functional categories of drought-responsive protein species and unique proteins in leaves revealed by proteomic studies. (**A**) Protein species include all the protein isoforms generated from gene variable splicing and post-translational modifications in the original publications; (**B**) Unique proteins indicate the protein family whose members have similar protein sequence homology and functional domains.

Table 1. Summary of current publications on drought-responsive proteomics in leaves.

Plant Types	No.	Plant Species	Drought Treatment Condition	Protein Species	Unique Proteins	Reference
Herbs	1	*Agrostis stolonifera*	35% soil water	81	57	[18]
	2	*Cynodon dactylon*	withholding irrigation, 15 days	54	42	[19]
	3	*Elymus elongatum*	75%, 50%, 25% field capacity, rewater 3, 14 days	11	9	[20]
	4	*Boea hygrometrica*	50% relative humidity (0.5, 8, 48 h); rehydrated 8, 48 h	8	8	[21]
	5	*Poa pratensis*	TE foliar spray, 14 days; withholding irrigation, 0, 10 and 15 days	58	32	[22]
			withholding irrigation, 10 and 15 days	88	36	[23]
	6	*Sporobolus stapfianus*	30% relative water content	108	55	[24]
	7	*Hordeum spontaneum*	withhold water to −2.5 MPa soil water potential, 3 weeks	32	22	[25]
		Hordeum vulgare	*P. indica*-inoculation 28 days, 25% field capacity	45	28	[26]
			10% field capacity, 5 days	37	23	[27]
			stopping water, 7 days	24	23	[28]
	8	*Oryza sativa*	30% relative humidity, 48, 72, 96, 120, 144, 168 and 192 h	94	64	[29]
			withholding water, 38, 43 days; rewater 5, 10 days	17	13	[30]
			50% reposition of the water lost daily, 20 days	15	15	[31]
			stop irrigation, 35 days	53	31	[32]
			no irrigation, 20 days	20	18	[33]
			withholding water, 6 days	10	10	[34] *
	9	*Triticum aestivum*	without water 5, 14, 24 days; rewater 25 days	159	113	[35]
			1/3 of field capacity, 14 h	13	13	[36]
			PEG 6000, 48 h	23	19	[37]
			20%–25% soil relative water content, 3 days	29	16	[38]
			withholding water, 9 days	30	21	[39]
			20% PEG for 48 h	31	29	[40] *
	10	*Zea mays*	12.5% volumetric soil water content, 6 days	220	142	[41]
			withholding water, 9 days	29	27	[42]
			withholding water for 1, 4–5 days	138	111	[43] *
			PEG solution (−0.7 MPa) for 8 h	149	41	[44] *
	11	*Saccharum officinarum*	10% PEG 6000,14 h	4	4	[45]
	12	*Musa paradisiaca*	0.21 M sorbitol, 48 days	24	20	[46]
	13	*Brassica napus*	no watering, 3, 7, 10 and 14 days	417	253	[47]
	14	*Citrullus lanatus*	no irrigation, 3 days	29	13	[48]
	15	*Glycine max*	10% PEG 6000 or exposed to drought conditions, 4 days	51	39	[17]
	16	*Phaseolus vulgaris*	no irrigation, 17 days	81	40	[49]
	17	*Medicago sativa*	water withholding, 7 days	29	20	[50]
	18	*Gossypium herbaceum*	35% relative water content	18	17	[51]
Shrubs	19	*Carissa spinarum*	25% relative humidity, 48, 120 h, rewater 24 h	23	12	[52]
	20	*Cistus albidus*	no irrigation, 107 days; rewater, 26 days	56	35	[53]
	21	*Hippophae rhamnoides*	25% field capacity, 20 days	13	11	[54]
Trees	22	*Eucalyptus* sp.	dry season, 6 months	46	19	[55]
	23	*Populus* spp.	25% field capacity, 18 days then rewater 28 days in glasshouse; no irrigation, 86 days in open field	33	29	[56]

Table 1. *Cont.*

Plant Types	No.	Plant Species	Drought Treatment Condition	Protein Species	Unique Proteins	Reference
Trees	23	*Populus* spp.	withholding water 4 and 7 days	13	5	[57]
			withholding water 8, 12 days; rewater 7 days	52	25	[58]
			120 ppb ozone, 13 h; 35% soil water, 7, 14, 21 and 28 days	25	19	[59]
			10% soil water content, 45 days	40	33	[60]
	24	*Quercus robur*	15% soil water content 3, 8, 56 and 65 days	41	28	[61]
		Quercus ilex	no irrigation 14 days; no irrigation 7 days, rewater 7 days	14	11	[62]
			not watering, 28 days	18	11	[63]
	25	*Malus domestica*	45%–55% field capacity	81	33	[64]

The references labelled with * are phosphoproteomic studies. The information of these phosphoproteins is listed in Supplementary Table S2. TE: trinexapac-ethyl; PEG: polyethylene glycol.

2. Drought Sensing and Signaling

2.1. Possible Drought Receptor

Although no specific receptors for drought sensing have been found in plants, a drought-responsive photoreceptor, phytochrome C1, was identified in *Z. mays* [41] (Figure 2A). Phytochrome is widely believed to regulate the transcription of light-responsive genes by modulating the activity of several transcription factors under both biotic and abiotic stresses [65]. In *Arabidopsis*, three phytochrome genes (i.e., *PHYA*, *PHYB*, and *PHYE*) are involved in suppressing drought tolerance [66]. These results imply a possible function of phytochrome C in mediating osmotic stress.

2.2. G Proteins

G proteins constitute one of the most important cell signaling cascades. Proteomic studies revealed that two G protein subunits (alpha subunit and beta subunit), several small G proteins (e.g., Ras-related protein Rab7 and Ras-related nuclear protein Ran), and a Ran-binding protein 1 were increased in drought-treated leaves (Supplementary Table S1, Figure 2A). Their functions in drought tolerance have been previously reported. The G protein alpha subunit may play a positive role in regulation of drought stress [67,68], whereas the G protein beta subunit in *Arabidopsis* may negatively regulate drought tolerance [69,70]. Besides, Rab proteins are the largest branch of the small G protein superfamily, which are involved in vesicle trafficking, intracellular signaling events and many important physiological processes, such as polar growth, plant hormone signal cross-talk, and stress response [71]. One of the Rab proteins, Rab 7, has been suggested to be involved in drought stress tolerance. The *PgRab7* gene was upregulated by dehydration in *P. glaucum* [72], while overexpression of the peanut *AhRabG3f* exhibited an enhanced tolerance to drought stress in transgenic peanut (*Arachis hypogaea* L.) [73]. Additionally, two isoforms of GDP dissociation inhibitor (GDI), involved in the regulation of Rab family activity, were increased in drought-stressed leaves of *Oryza sativa* [29]. The expression of the *MiRab-GDI* gene was induced in *Mangifera indica* under drought stress conditions [74], but the exact function of GDI in response to drought stress remains to be elucidated. In addition, Ran is another member of the small G protein group, which is involved in nucleo-cytoplasmic transportation of proteins and RNA, the formation of spindle asters, and the reassembly of the nuclear envelope in mitotic cells [71]. Ran-binding protein 1 is the major effector of Ran. The significant increase of Ran in *Poa pratensis* [23] and Ran-binding protein 1 in *Z. mays* under drought stress conditions [41] indicated the important role of cell cycle and DNA synthesis in drought tolerance. However, the drought-decreased Ran/TC4 in *Cistus albidus* under no irrigation for 107 days [53] implies their function is dependent on the drought intensity.

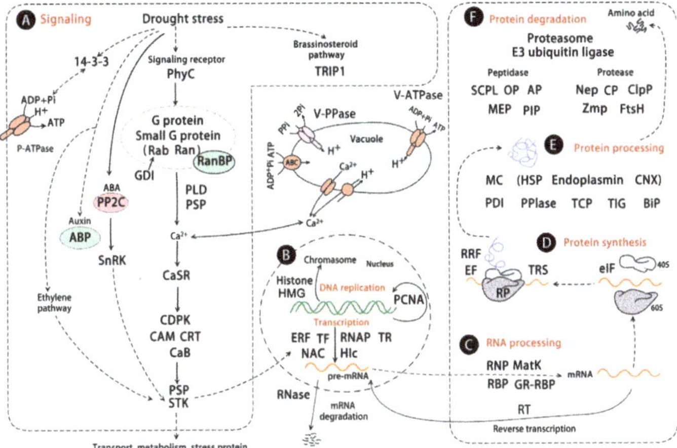

Figure 2. Schematic representation of drought responsive proteins involved in signaling, gene expression regulation, protein synthesis and degradation. The solid line indicates single-step reactions, and the dashed line indicates multi-step reactions. (**A**) Signaling; (**B**) DNA replication and transcription; (**C**) RNA processing; (**D**) Protein synthesis; (**E**) Protein processing; (**F**) Protein degradation. 14-3-3: 14-3-3 protein; Nep: aspartic proteinase; ClpP: ATP-dependent Clp protease; ABP: auxin-binding protein; BiP: endoplasmic reticulum-luminal binding protein; CaB: calcium ion binding protein; CaSR: calcium sensing receptor; CDPK: calcium-dependent protein kinase; CAM: calmodulin; CNX: calnexin; CRT: calreticulin; CP: cysteine proteinase; EF: elongation factor; ERF: ethylene-responsive transcription factor; GDI: GDP dissociation inhibitor; GR-RBP: glycine-rich RNA binding protein; HSP: heat shock protein; HMG: high mobility group protein; HOP: Hsp70-Hsp90 organizing protein; AP: leucine aminopeptidase; MatK: maturase K; MEP: metalloendopeptidase; MC: molecular chaperone; OP: oligopeptidase A-like; PPIase: peptidyl-prolyl *cis-trans* isomerase; PLD: phospholipase D; PhyC: phytochrome C; PCNA: proliferating cell nuclear antigen; PIP: proline iminopeptidase, putative; PDI: protein disulphide isomerase; PP2C: protein phosphatase 2C; RanBP: Ran-binding protein; RT: retrotransposon protein; RNase: ribonuclease; RNP: ribonucleoprotein; RP: ribosomal protein; RBP: RNA binding protein; Hlc: RNA helicase; RNAP: RNA polymerase; SCPL: serine carboxypeptidase-like protein; STK: serine/threonine kinase; PSP: serine/threonine-protein phosphatase; SnRK: sucrose non-fermenting 1-related protein kinase; TCP: T-complex protein; TRIP1: TGF-β receptor-interacting protein 1; TF: transcription factor; TR: transcription regulator; TIG: trigger factor-like protein; V-PPase: vacuolar H^+-pyrophosphatase; V-ATPase: vacuolar H^+-ATPase; Zmp: zinc metalloprotease.

2.3. Ca^{2+} Signaling and Protein Kinase

Under drought stress, calcium acts as a second messenger, which is employed to regulate specific protein kinase activity and downstream gene expression [75]. In proteomic studies, the abundances of several calcium binding proteins (CaBs), such as calmodulin (CaM), calcium sensing receptor (CaSR), calreticulin (CRT), and calcium-dependent protein kinase (CDPK), were changed in response to drought (Figure 2A). Among them, CRT is a unique endoplasmic reticulum (ER) luminal Ca^{2+} binding chaperone, which plays a role in many cellular processes [76]. The increase of CRT abundance enhanced the survival of *T. aestivum* plants under drought condition [77], and the *TaCRT*-overexpressing tobacco (*Nicotiana benthamiana*) plants exhibited enhanced drought resistance [78]. Proteomic studies revealed a drought-induced CRT1 in *Glycine max* [17], but a drought-reduced CRT1 in *Quercus robur* under prolonged (65 days) drought stress [61]. These results indicate that CRT functions in drought tolerance in a stress intensity-dependent manner. Besides, CDPKs regulate the downstream components in calcium-mediated signal transduction. In proteomic studies, the protein abundance and phosphorylation level of CDPKs were generally decreased in plants under drought. However, the abundance of CDPKs increased in the drought-tolerant genotype of *Z. mays* [41]. It has been reported that overexpression of the *CDPK* gene can enhance drought tolerance in transgenic *Arabidopsis* [79,80] and *O. sativa* [81]. These reports imply the important role of CDPKs in plant stress tolerance. The phosphorylation states of several protein kinases (e.g., serine/threonine-protein kinase, germinal center kinase (GCK)-like kinase MIK, receptor-like protein kinase HERK 1-like, phototropin family protein kinase, and salt-inducible protein kinase) were also changed in response to drought stress, implying their regulation in the drought response signaling pathway. In addition, the dephosphorylation mediated by protein phosphatases is an important event in the signal transduction process that regulates various cellular activities [82]. In proteomic studies, the phosphorylation level of protein phosphatase 2C (PP2C) and phospholipase D (PLD) in *Z. mays* [41] were drought-increased (Figure 2A). PP2C is known to be a negative regulator for plant drought tolerance in the abscisic acid (ABA) signaling pathway, which can inhibit the activity of SnRK, leading to a decrease of the phosphorylation of its substrates in the signaling cascade [83–85]. Both phosphorylation level [40,44] and abundance of PP2C [41,86] were obviously affected by drought stress. These data suggest that the PP2C-involved ABA signaling pathway is crucial for drought response. Additionally, a dual function of PLD in plant drought response was reported, for example, in drought-sensitive cultivars, the activity and expression of PLD increased more obviously than in drought-tolerant ones [87,88], whereas overexpression of the *PLDα* gene can significantly enhance drought tolerance in transgenic *Arabidopsis* [89] and *Populus tomentosa* [90]. In addition, a high level of PLD promotes stomatal closure at earlier stages, but disrupts membranes in prolonged drought stress [91]. These data imply that, in drought, PLD functions in a condition-dependent manner.

2.4. 14-3-3 Proteins

In proteomic studies, some members of the 14-3-3 protein family and 14-3-3-like proteins were increased in drought-treated leaves from many plant species (Figure 2A, Supplementary Table S1). The 14-3-3 proteins are intracellular dimeric phosphoserine/threonine binding molecules that participate in a wide range of vital regulatory processes, including signaling, transmembrane receptors, and transcription activation [92]. 14-3-3 proteins are also known as positive regulators of H^+-ATPase activity to control the electrochemical gradient across the plasma membrane, which contributes to the initiation of stress responses and other signal transduction pathways [93]. It has been reported that drought stress can directly alter the abundance of 14-3-3 proteins [94–96]. In addition, overexpression or silencing of the 14-3-3 protein genes can modulate drought tolerance of transgenic plants (e.g., *Gossypium hirsutum* and *Arabidopsis*) [97,98]. These results imply that 14-3-3 proteins may have diverse regulatory roles in leaves under drought stress, and the involvement of each 14-3-3 protein in drought-response needs to be further elucidated.

2.5. Ethylene and Auxin Signaling Pathways

It is well known that ABA is a key phytohormone that mediates the adaptive response to drought stress. Besides, other hormonal signals, such as ethylene, gibberellic acid, and jasmonic acid are also important cellular regulators in signal transduction pathway under drought conditions [99–101]. Proteomic studies revealed a drought-increased ethylene-responsive transcription factor (ERF) in *Gossypium herbaceum* [51] and some members of drought-responsive auxin-binding protein (ABP) family in *Q. robur* [61], *Z. mays* [41], and polar clones [59]. The *ERF* gene was found to be induced in *G. herbaceum* under drought stress [102,103] (Figure 2A). Overexpression of *ERFs* in various plants, such as sugarcane *SodERF3* overexpression in tobacco, tomato *TERF1* in rice, and *Brassica rapa BrERF4* in *Arabidopsis*, can improve plant drought tolerance. However, careful attention should be taken in the definition of ERF roles in plant drought tolerance. Different ERF members may play different roles in plants in response to drought stress. For example, *BpERF11* was found to negatively regulate osmotic tolerance in *Betula platyphylla* [104]. In addition, ABP functions as an auxin receptor, being involved in many development processes and drought response. However, as far as we know, little information is available for ABP members (i.e., ABP2, ABP20, and ABP19a) in response to drought stress. Additionally, a drought-increased TGF-β-receptor interacting protein 1 (TRIP1) was found in *Sporobolus stapfianus* [24]. TRIP-1 was phosphorylated by the brassinosteroid (BR)-insensitive I (BRI-1) protein, which is a serine/threonine kinase receptor essential for BR perception and signal transduction. The increase of TRIP-1 in *S. stapfianus* suggests that the BR signaling pathway would be triggered in response to water deficit.

3. Drought-Responsive Gene Expression Regulation

Drought-responsive gene expression is crucial in the transcriptional regulatory network [105]. Proteomic studies have found some gene expression-related proteins in the complex regulatory networks in leaves (Figure 2B). Chromatin structure modification is a prerequisite to sustain the transcriptional regulation which is involved in cell cycle progression. Significant abundance changes were observed in two chromatin structure modification-related proteins, which were histones and high mobility group proteins (HMG). Histones are the major proteins of chromatin, and the dynamic association of histones and their variants can regulate gene expression [106]. In proteomic studies, several histones (e.g., H1 and H2B) appeared to cause diverse abundance changes in different plant species in response to drought stress. For example, H2Bs were decreased in *C. albidus* [53] and *Brassica napus* [47], while histone H1 was decreased in a drought-sensitive *Z. mays* cultivar, but increased in a drought-tolerant one [41]. Similarly, the transcript and protein of histone H1 variant were all induced specifically in the tolerant genotype of *G. herbaceum* [107]. Besides, HMG is a highly conserved nuclear DNA-binding protein, which functions in DNA repair and chromatin modification after DNA damage [108]. A proteomic study revealed that two isoforms of HMG were drought-increased in drought-tolerant maize cultivar, but decreased in a sensitive cultivar [41]. Interestingly, the phosphorylation level of HMG was significantly decreased in a drought-tolerant wheat cultivar, but increased in a drought-sensitive one [40]. Previous studies have reported that phosphorylation of HMG reduced its binding to DNA, inhibiting replication and transcription [109]. These data imply that PTM (e.g., phosphorylation) of HMG is crucial for its function in plant drought tolerance.

In addition, RNA processing was also critical for plants to cope with drought stress. The abundances of several RNA processing-related proteins were changed in plants under drought stress (Figure 2C). Among them, five glycine-rich RNA binding proteins (GR-RBPs) were drought-increased, while three GR-RBPs were drought-decreased. RNA binding proteins (RBPs) can bind to RNA molecules, which are involved in almost all aspects of post-transcriptional gene regulation. The *GR-RBP* genes have been noted to be upregulated in response to water stress [110], which was suspected to function in the regulation of specific gene expression in response to stress. For example, the expression of *GR-RBP* gene in transgenic rice has been shown to have much higher recovery rates and grain yields when compared with that in wild-type plants under drought conditions [111]. However, the transgenic *Arabidopsis* and camellia plants that had an overexpressed *GR-RBPa* gene from *Camelina sativa* appeared to have a reduced drought tolerance [112]. Besides, proteomic studies revealed that the abundance and phosphorylation level of S-like ribonucleases (RNases) were significantly increased in rice under drought stress [30,34]. S-like RNases have acquired specialized functions such as stress regulation, defense against microorganisms, phosphate scavenging, and even nitrogen storage [113]. It has been reported that S-like RNases participated in the responses to salt, polyethylene glycol (PEG), and ABA [113]. All these have attracted great attention as researchers have attempted to further confirm their post-transcriptional regulation in response to drought stress. Additionally, an intron splicing related protein, and maturase K (MatK) appeared to show a drought decrease in poplar [60], and multiple organelle RNA editing factor 9, involved in RNA editing in mitochondria and plastids, was increased firstly, and then decreased in *B. napus* with the extension of drought stress [47]. Both of them are RNA processing-related proteins. Their changes indicated that the transcription regulation is diverse and complicates the study of plants' ability to cope with drought stress.

4. Drought-Responsive Protein Synthesis and Turnover

Protein synthesis and turnover is one of the fundamental metabolic processes for plants to cope with drought stress. Proteomic investigations revealed that 16% of the drought-responsive proteins in leaves are attributed to protein synthesis and turnover functions (Figure 1). Several proteins are involved in protein biosynthesis, such as ribosomal protein (RP), elongation factor (EF), translation initiation factor (TIF), tRNA synthase (TRS), and ribosome recycling factor (RRF) (Figure 2D). Most of them exhibited an increase under drought stress, which would be beneficial for protein synthesis in response to specific drought conditions. Besides, the proteins functioning in protein folding and processing showed diverse changes among different plant species and cultivars (Figure 2E). For example, peptidyl-prolyl *cis-trans* isomerases (PPIases) were significantly increased in *G. max* [17], *T. aestivum* [35], *O. sativa* [29], and *Q. robur* [61], but decreased in a drought-sensitive cultivar of *Phaseolus vulgaris* [49]. Protein disulfide isomerases (PDIs) were increased in barley and *B. napus*, but decreased in *Agrostis stolonifera*, *Q. robur*, and poplar. Additionally, ER-luminal binding protein (BiP), trigger factor-like protein (TIG), most heat shock proteins (HSPs), and other molecular chaperones (i.e., calnexin, endoplasmin) were increased, but T-complex protein and HSP70-HSP90 organizing protein were decreased in drought-treated leaves (Supplementary Table S1). These proteins function to maintain normal protein folding, repairing, and renaturation of the stress-damaged proteins. Among them, HSPs function in protein folding for drought tolerance, which has been widely discussed [114–116]. *HSP* genes have been transferred into *Arabidopsis* and yeast to improve their drought tolerance [117,118]. These data suggest that maintaining correct protein folding is important for leaves to cope with drought stress [35].

Besides, protein degradation is important to remove abnormal or damaged proteins and to control the levels of certain regulatory proteins during drought stress. Proteomic studies also revealed that some protein degradation-related proteins increased in response to drought stress, including proteasomes, proteases, and peptidases (Figure 2F). Previous studies have reported that some components in the protein degradation pathway, such as ubiquitin/26S proteasomes, small ubiquitin-like modifier (E3 SUMO) ligase, and proteases/peptidases were involved in plant drought tolerance [119–123]. For the ubiquitin/26S proteasome system, 7 out of 11 20S proteasomes (the core regulatory particle of 26S proteasome) were increased in leaves of *P. vulgaris* [49], *Hordeum vulgare* [26], *B. napus* [47], and *Medicago sativa* [50], respectively. Importantly, the phosphorylation level of E3 ubiquitin ligase, which is one of the key enzymes involved in ubiquitination, exhibited significantly increased values in leaves under drought stress [40,44]. Many studies have shown that E3 ubiquitin ligases were positively related to plant drought tolerance [124–126]. These findings indicate that the enhancement of the ubiquitin/26S proteasome system is important for plants to cope with drought. In addition, most proteases were drought-increased in plants, such as ATP-dependent Clp protease in *H. vulgare*, cysteine proteinase in *P. vulgaris*, zinc metalloprotease in *B. napus*, and aspartic proteinase in *Z. mays*. Consistently, some peptidases, such as serine carboxypeptidase in *G. herbaceum* and oligopeptidase in *O. sativa*, were increased in drought-stressed leaves. Among them, aminopeptidases (APs), catalyzing the hydrolysis of amino acids from the N-terminus of proteins, were generally increased in drought-tolerant plant species, and decreased in drought-sensitive plant species or cultivars. The essential role of APs in plant drought tolerance has been well addressed previously [123].

5. Reactive Oxygen Species (ROS) Scavenging Pathways

5.1. Superoxide Dismutase (SOD), Catalase (CAT), and Peroxidase (POD) Pathway

Water deficit interrupts normal cellular metabolism that results in the production of ROS. Plants have evolved diverse mechanisms to keep ROS homeostasis in cells, including antioxidative enzymes (e.g., SOD and CAT) and chemical antioxidants (e.g., glutathione and ascorbate) (Figure 3A). Among them, SOD acts as the first line of defense by converting $O_2^{\bullet-}$ into H_2O_2, and CAT converts H_2O_2 into H_2O and O_2. They are both involved in plants' drought tolerance. For example, the transgenic alfalfa expressing *MnSOD* gene from *Nicotiana plumbaginifolia* improved its survival and vigor after exposure to water deficit [127]. Overexpression of a cytosolic copper-zinc SOD from the mangrove plant *Avicennia* in rice can enhance its drought tolerance [128]. In addition, the increase of CAT activity is positively related to the drought degree [129], although the functional analysis of *CAT* gene on drought tolerance in transgenic plants is still scarce. The diverse abundances of SODs indicated that their functional state is dependent on subcellular location (e.g., plastid, peroxisome, or cytosol), drought conditions, and the plant's drought adaptation ability. For example, the cytosolic Cu-Zn SODs were increased in two *O. sativa* cultivars (i.e., drought avoidance CT9993 and drought tolerance IR62266), while the chloroplast Cu-Zn SODs were increased in CT9993, but decreased in IR62266 [30]. Additionally, in a drought-sensitive cultivar of *Malus domestica*, the abundance of Cu-Zn SOD was decreased, but the FeSOD was increased [64].

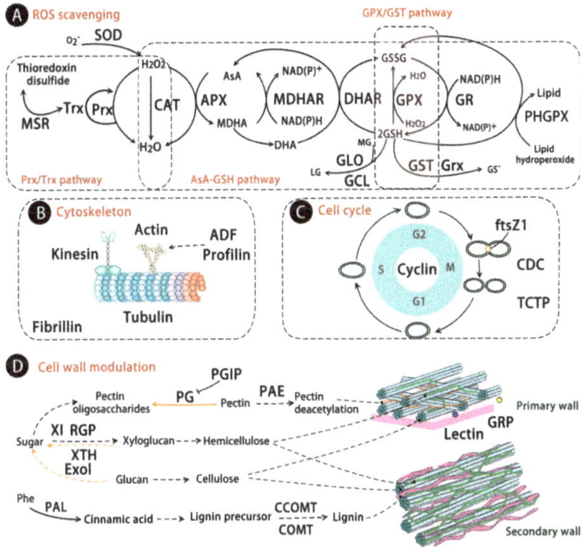

Figure 3. Schematic representation of drought responsive proteins involved in reactive oxygen species (ROS) scavenging, cell structure and cycle. The solid line indicates single-step reactions, and the dashed line indicates multi-step reactions. (**A**) ROS scavenging; (**B**) Cytoskeleton; (**C**) Cell cycle; (**D**) Cell wall modulation. ADF: actin depolymerizing factor; ALDH: aldehyde dehydrogenase; AsA: ascorbate; APX: ascorbate peroxidase; Exol: glucan exohydrolase; COMT: caffeic acid 3-*O*-methyltransferase; CCOMT: caffeoyl-CoA *O*-methyl-transferase; CAT: catalase; CDC: cell division cycle protein; DHAR: dehydroascorbate reductase; DHA: dehydroascorbate; GME: GDP-D-mannose-3′,5′-epimerase; GCL: glutamate-cysteine ligase; Grx: glutaredoxin; GPX: glutathione peroxidase; GR: glutathione reductase; GST: glutathione *S*-transferase; GSSG: glutathione disulfide; GSH: glutathione; GRP: glycine-rich protein; GLO: glyoxalase; LG: (*R*)-*S*-lactoylglutathione; MSR: methionine sulfoxide reductase; MG: methylglyoxal; MDHAR: monodehydroascorbate reductase; MDHA: monodehydroascorbate; ftsZ1: plastid division protein ftsZ1 precursor; PAE: pectin acetylesterase; PAL: phenylalanine ammonia-lyase; PG: pectin depolymerase; PGIP: polygalacturonase inhibitor; PHGPX: phospholipid hydroperoxide glutathione peroxidase; POD: peroxidase; Prx: peroxiredoxin; RGP: reversibly glycosylated polypeptide; SOD: superoxide dismutase; Trx: thioredoxin; TCTP: translationally-controlled tumor protein homolog; XI: xylanase inhibitor; XTH: xyloglucan endotransglycosylase.

POD catalyzes the reduction of H_2O_2 using various electron donors such as phenolic compounds, lignin precursors, auxin, and secondary metabolites. The drought-induced activities of POD were observed in leaves from *Arabidopsis* [130], *Ramonda serbica* [131], and *M. sativa* [132]. In proteomic studies, the abundances of PODs were increased in leaves of *Quercus ilex* [62] and a drought-tolerant *O. sativa* cultivar [32]. Moreover, the expression of POD genes was also increased in *Tamarix hispida* under drought stress [133]. These results indicate that PODs are critical for ROS scavenging in drought-stressed plants.

5.2. Ascorbate-Glutathione (AsA-GSH) Pathway

The AsA-GSH pathway is another key antioxidant pathway in response to drought [134,135]. During this process, the ascorbate peroxidase (APX) reduces H_2O_2 to H_2O using ascorbate (AsA) as an electron donor, then the oxidized AsA is restored by monodehydroascorbate reductase (MDHAR), dehydroascorbate reductase (DHAR), and glutathione reductase (GR) [136] (Figure 3A). Proteomic studies revealed that most proteins in this pathway were increased under drought stress conditions. For example, the APXs localized in thylakoids and mitochondria were drought-increased in many herbaceous plants (e.g., *Cynodon dactylon* [19], *G. max* [17], *T. aestivum* [35,37,38], *O. sativa* [43,44,46], *H. vulgare* [37], and *B. napus* [47]). The function of APX under drought stress has been well addressed [129]. APX1-deficient mutant (*apx1*) of *A. thaliana* was significantly sensitive to drought stress [137]. Transgenic tobacco plants overexpressing cytosolic APX alleviated the damage from water stress. In addition, rice *Osapx2* mutants had lower APX activity and were sensitive to drought, whereas overexpression of *Osapx2* in rice enhanced its stress tolerance [138].

Another AsA-GSH pathway-related enzyme, GR, catalyzes the reduction of glutathione disulfide (GSSG) to the sulfhydryl form GSH. Consistently with the drought-increased abundance of GR, the drought-induced *GR* genes were also found in cowpea and *P. vulgaris* under drought stress [139,140]. Overexpression of *B. rapa BrGR* in *E. coli* showed an increase of GR activity and tolerance to H_2O_2 [141]. Thus, the positive function of GR in plant drought tolerance can be presumed [142]. Besides, two enzymes that can maintain ascorbate in its reduced state, dehydroascorbate reductase (DHAR) and monodehydroascorbate reductase (MDHAR), were found to be increased in some plants (i.e., *C. dactylon* and *O. sativa*) under drought stress. The expression of DHAR is generally associated with plant drought tolerance [143,144]. Overexpression of DHAR has been shown to increase drought tolerance or biomass in transgenic potato [145], rice [146], and tobacco [147]. These imply that DHAR can be a very promising target to improve plant drought stress tolerance.

In addition, proteomic studies revealed that some proteins were involved in glutathione-mediated ROS scavenging: glyoxalase (GLO), phospholipid hydroperoxide glutathione peroxidase (PHGPX), glutamate-cysteine ligase (GCL), glutaredoxin (Grx), and monothiol Grx. GLO catalyzes the detoxification of methylglyoxyl, which is involved in the indirect scavenging of ROS. Among them, the abundances of GLOs were not always increased in plants under drought stress. Their response to drought was dependent on the plant genotype and duration of drought. Besides, GCL is the first enzyme in the GSH biosynthetic pathway which was increased in drought-stressed *B. napus* [47]. Similar profiles were also observed in monothiol Grx in *B. napus* [47], *Z. mays* [41], as well as PHGPX in *Citrullus lanatus* [48]. All these suggest that the GSH-mediated antioxidative defense pathway is increased in leaves as a drought adaptation mechanism.

5.3. Peroxiredoxin/Thioredoxin (Prx/Trx) and Glutathione Peroxidase/Glutathione S-Transferase (GPX/GST) Pathway

The Prx/Trx pathway plays a central role in detoxification of H_2O_2. In this process, Prx catalyzes the reduction of H_2O_2, and then uses Trx to restore its catalytic activity (Figure 3A). The abundances of Prx and Trx were increased in response to drought, indicating the enhanced detoxification in drought-stressed leaves. Besides, a Trx-linked enzyme, methionine sulfoxide reductase (MSR), was also increased in some plants [57]. MSR is involved in the enzymatic conversion of methionine sulfoxide to methionine, which is involved in the protection of cells and tissues from H_2O_2-induced stress [148]. It has been reported that various oxidative stresses, including drought stress, can enhance the expression of numerous *Msr* genes [149]. This implies that MSR plays an important role in response to drought stress.

Besides, the GPX/GST pathway is enhanced in some plants under water deficit conditions. GPX catalyzes the reduction of H_2O_2 using Trx [150], and GST catalyzes conjugation reactions between GSH and a number of xenobiotics, playing a crucial role in the degradation of toxic substances. Previous studies have reported that overexpression of the *GST* gene enhanced drought tolerance in tobacco [151,152] and *Arabidopsis* [153]. Similarly, overexpression of *P. glaucum GPX* in rice resulted in an increased tolerance to drought stress [154]. In proteomic studies, the GPXs were drought-increased in *Boea hygrometrica* [21], *E. elongatum* [20], and *B. napus* [47], while drought-increased GSTs were widely found in the plant species discussed in this review. The increases of GPX and GST reinforced the functional evidence for their potential detoxification role in drought tolerance.

6. Pathogenesis Related Proteins

In proteomics results, several pathogenesis-related proteins were increased in response to drought stress, such as chitinase, disease resistance protein (DRP), polyphenol oxidase (PPO), oryzacystain, pathogen defense-related protein 10 (PR10), and disease resistance gene analog PIC15. Chitinase functions in the determent of herbivory and pathogen attack by acting on insect exoskeletons and fungal cell walls [155]. Five basic chitinase species and an acidic chitinase were increased in two tree species, i.e., *Eucalyptus* sp. [55] and *Musa paradisiaca* [42], respectively. The induced transcript of basic chitinase was also observed in *Eucalyptus* sp. [55]. Besides, PPO catalyzes the oxygen-dependent oxidation of phenols to quinones during plant defense against pests and pathogens. It was reported that the abundances and activities of PPOs were also increased in leaves of *B. hygrometrica* under desiccation [21], indicating its substantial role in the adaptation to severe water deficit. Moreover, an oryzacystain was increased in drought-stressed *O. sativa* leaves. Oryzacystain is a cysteine proteinase inhibitor in the phytocystatin family of proteinase inhibitors, which plays an important role in plant defense against pathogen attacks and oxidative stress [156]. The tobacco overexpressing the oryzacystain gene displayed an increase of drought tolerance by improving total SOD and guaiacol POD activities [157]. In addition, the drought-induced PR10 and disease resistance gene analog PIC15 were also found in leaves (Supplementary Table S1). It was found that overexpression of *PR10* can improve drought tolerance in tobacco [158]. The response of these pathogenesis-related proteins to drought suggests that there is a cross-talk between biotic and abiotic stresses.

7. Osmotic Regulation

Osmotic regulation is crucial for plant drought resistance. Several important osmotic homeostasis-related proteins, such as late embryogenesis abundant (LEA) protein, dehydrin (DHN), and betaine aldehyde dehydrogenase (BADH), were accumulated in leaves under drought stress. Among them, LEA proteins are highly hydrophilic proteins which function in plant abiotic stress as cellular protectants to stabilize cellular components in response to water loss [159]. Similarly, in physiological studies [160], DHNs (also known as group 2 LEA proteins) were widely drought-accumulated in several plant species, including *Z. mays* [41], *C. dactylon* [19], *T. aestivum* [35], and *B. napus* [47]. Higher hydrophilicity and thermostability of DHNs suggest that they can stabilize the protein structure through detergent- and chaperone-like properties [161]. Besides, the DHN in *Z. mays* exhibited a significantly increased phosphorylation level under drought stress [43]. The phosphorylation of LEA2 may facilitate its binding to calcium. LEA2 acts as calcium buffer, and has calcium-dependent chaperone-like activity, which is similar to that of calreticulin and calnexin [162]. Besides, group 3 LEA proteins were also increased in *Z. mays* [41] and *B. napus* [47] under certain drought conditions. It has been reported that transformation of the *LEA* gene into a number of plant species can confer tolerance to drought stress. For example, the transgenic calli overexpressing sweet potato *LEA14* (*IbLEA14*) enhanced the plant tolerance to drought stress, whereas RNA interference (RNAi) calli exhibited increased stress sensitivity [163]. All these indicate that *LEA* can be taken as a candidate gene for improving plant drought tolerance. Another osmotic regulation-related enzyme, BADH, showed drought accumulation in leaves of *H. vulgare* [27] and *Hordeum spontaneum* [25]. BADH converts betaine aldehyde to glycine betaine, which mediates osmotic homeostasis, and has positive effects on plant adaptation to drought stress [164]. RNAi-directed downregulation of *OsBADH1* resulted in a decrease of stress tolerance and an increase of oxidative stress [165], implying its important role in plant drought tolerance.

8. Modulation of Cell Structure and Cell Cycle

8.1. Cytoskeleton and Cell Cycle

In leaves, water deficit rapidly decreases cell division rate [166,167]. Proteomic studies revealed that the abundances and phosphorylation levels of several proteins involved in cell division (i.e., cell division cycle protein, division protein ftsZ1, and cyclin A2) were decreased under water deficit [22,28,44,47] (Figure 3C). Because cytokinesis requires new cytoskeleton and cell wall components, it is predictable that the cell wall and cytoskeleton-related proteins would change in response to drought. The cytoskeleton (i.e., microtubules and actin filaments) is a highly dynamic component, which is crucial for cell division, movement, morphogenesis, and signal transduction [168]. In proteomic studies, several cytoskeleton proteins, such as actin, kinesin motor protein, tubulin, profilin, actin depolymerizing factor, and fibrillin were decreased in response to drought stress (Figure 3B). The *actin* genes were also reduced in drought-treated *H. vulgare* leaves [168]. The decrease of the cytoskeleton and cell cycle-related proteins implies that cell growth is suppressed during drought stress.

In response to drought, the translationally-controlled tumor protein homolog (TCTP) was significantly drought-increased in *H. vulgare* [26], *T. aestivum* [37], and *B. napus* [47], which would facilitate plant adaptation to drought stress. TCTP is a Ca^{2+}-binding protein with important functions in some cellular processes (e.g., protection against stress and apoptosis, cell growth, cell cycle progression, and microtubule organization) [169]. The protein abundance and gene expression of TCTP were increased in *P. indica*-colonized plants under drought stress. Additionally, PEG treatment for 48 h also increased the abundance of TCTP in leaves of *T. aestivum* [37]. Overexpression of *Arabidopsis* TCTP enhanced drought tolerance by rapid ABA-mediated stomata closure [170], while a TCTP knockdown mutant of *Arabidopsis* showed severe growth defects [171]. These results imply the importance of TCTP for plant drought tolerance.

8.2. Cell Wall Modulation

Water loss in plant tissues controls turgor pressure and directly affects the extensibility of the cell wall. Some drought-responsive enzymes involved in cell wall polysaccharide synthesis/hydrolysis, lignin biosynthesis, and cell wall loosening in leaves have been identified in proteomic studies (Figure 3D). Two enzymes involved in cell wall polysaccharide synthesis, reversibly glycosylated polypeptide and pectinacetylesterase, were drought-increased in *M. sativa* [50] and *B. napus* [47], respectively. Two enzyme inhibitors involved in polysaccharide hydrolysis inhibition, the xylanase inhibitor and polygalacturonase inhibitor, were decreased in drought-sensitive *Z. mays* cultivars, but increased in leaves of a tolerant cultivar [41]. Their changes in response to drought stress were in contrast to that of glucan exohydrolase, which is an enzyme involved in polysaccharide hydrolysis. These results indicate that the cell wall synthesis is enhanced in a drought-tolerant maize cultivar under drought stress, which may be associated with drought adaptation. Besides, three lignin biosynthesis related proteins, phenylalanine ammonia-lyase (PAL), caffeic acid 3-*O*-methyltransferase, and caffeoyl-CoA *O*-methyl-transferase, were generally increased under drought stress (Figure 3D). PAL catalyzes the transformation of phenylalanine to cinnamylate in the first step of lignin biosynthesis [172]. The activity of PAL was also obviously increased in the leaves of *Trifolium repens* under the early stages of drought stress (0–14 days), but decreased gradually as the period of stress was extended [173]. In addition, two drought-increased cell wall structural proteins (i.e., glycine-rich protein and fasciclin-like arabinogalactan protein) were found in the leaves of *B. napus* [47], which would enhance cell wall synthesis in response to drought stress. Interestingly, a significantly increased abundance and phosphorylation level of sucrose synthase was found in *Z. mays* [27,32,41,44], which would facilitate the synthesis of cell wall components by providing UDP-glucose directly to the cellulose synthases and/or callose synthases [174]. The drought-increased cell wall synthesis would enhance the mechanical strength for minimizing water loss and cell dehydration, which is crucial for plants to resist and recover from drought.

Besides, cell wall loosening is also important for growth adaptation in expanding leaves [166]. Two cell wall loosening/expansion-related enzymes, polygalacturonase/pectin depolymerase (PG) in *O. sativa* [29] and xyloglucan endotransglycosylase (XTH) in *Z. mays* [41], were increased under drought stress. PG can degrade pectin, while XTH can cleave and reform the bonds between xyloglucan chains to regulate cell wall rigidity. Previous studies have found drought-induced *XTH* genes upregulated in *Arabidopsis* and rice [175,176], which implies that XTH might serve as a stress marker gene in leaves. Taken together, the cell wall modulation, which results in either growth arrest in drought-sensitive cultivars or a continuation of growth with a reduced rate [177], would contribute to drought adaption by cell size adjustment for cell turgor maintenance [17].

9. Membrane Trafficking

Plant membrane transport system plays a significant role in response to water scarcity. In proteomic studies, several membrane trafficking proteins localized in mitochondrion, plasma, and vacuole were changed in response to drought. Two mitochondrial carrier proteins (i.e., dicarboxylate/tricarboxylate carrier (DTC) and 2-oxoglutarate/malate carrier protein (OMC)) were decreased in *T. aestivum* [35] and *G. max* [17], respectively. DTC and OMC can catalyze the transport of various metabolites (e.g., dicarboxylates, tricarboxylates, amino acids, and keto acids) across the inner mitochondrial membrane, and play an important role in several metabolic processes, such as the gluconeogenesis, nitrogen metabolism, as well as biotic stress [178]. However, three voltage-dependent anion channel proteins (VDAC) in *T. aestivum* [35] and two mitochondrial outer membrane porin 1-like proteins in *B. napus* [47] were significantly increased under drought stress. The VDAC localized in the outer membrane of mitochondria can regulate Ca^{2+} fluxes, ATP/ADP exchange, and metabolites. The abundance changes of these mitochondrion transport-related proteins under drought stress indicated that the ion/metabolite exchange between mitochondria and cytosol was modulated in leaves to cope with the stress. Among them, the function of VDAC has been proved to be involved in plant drought tolerance [179,180]. The transcriptional level of *VDAC* was induced in *Brassica rapa* by drought stress [179], and the overexpression of the *AtVDAC2* gene can confer drought resistance in *Arabidopsis* [180].

Remorin is a plant-specific plasma membrane protein that plays important role in plant-microbe interaction and signal transduction [181]. Some drought-increased remorins were found in *Z. mays* [41], *O. sativa* [44], and tolerant cultivars of *T. aestivum* [35]. Overexpressing heterologous remorin in *Arabidopsis* enhanced the tolerance to dehydration and salinity at the germination and seedling stages [182]. Besides, aquaporin (AQP), which transports water and other small molecules, is crucial for plants to combat drought stress. PEG treatment upregulated the expression of a PIP2 subgroup gene, *AQP* (*TaAQP7*), and the overexpression of *TaAQP7* increased drought tolerance in tobacco [183]. Similarly, a proteomic study also found an increased abundance of AQP in *Z. mays* under 12.5% soil water stress for 6 days [41]. Additionally, drought-induced abundance of lipid transfer protein (LTP) was observed in leaves of *C. albidus* [53]. LTP has been proved to be involved in abiotic stress [184]. In *Arabidopsis*, the loss-of-function mutant of LTP3 was sensitive to drought stress, whereas overexpressing plants appeared drought tolerant [185]. Moreover, the *OsDIL*-overexpressing transgenic *O. sativa* plants were more tolerant to drought stress during vegetative and reproductive development [186]. All these demonstrated that LTP was involved in plant tolerance to drought stress, and is probably a candidate gene for genetic improvement of crop yield in adaption to drought stress.

In addition, vacuolar H^+-pyrophosphatase (V-PPase), vacuolar-ATPase (V-ATPase), and ABC transporter ATPase showed dynamic changes in response to drought. V-ATPase and V-PPase are two tonoplast proton pumps for translocating H^+ into the vacuoles to generate a gradient of H^+, which provide a driving force for the accumulation of ions and other solutes in the vacuole. Their functions in plant abiotic stress tolerance have been widely discussed [187,188]. For example, an *M. domestica* vacuolar H^+-ATPase (VHA) encoding gene (*MdVHA-A*) was induced in shoots under PEG treatment, and overexpression of *MdVHA-A* in transgenic tobacco seedlings improved drought tolerance [189]. In *Z. mays*, the abundances of V-ATPase and V-PPase were increased in a drought tolerant cultivar, but decreased in an intolerant one [41]. The differential expression patterns of ATPase and V-PPase further indicate their important functions in plants' response to osmotic stress. Taken together, all the changes of membrane trafficking-related proteins in mitochondrion, plasma, and vacuole highlight that ion transport and membrane trafficking are crucial in leaves in order to cope with water deficit.

10. Photosynthesis and Photorespiration

It is well known that photosynthetic inhibition is one of the primary detrimental effects of water stress due to stomatal closure [190,191]. Thus, it is predictable that the universal decrease trends of photosynthesis-related proteins would be found in drought-stressed leaves (Figure 4A,B). Plants have developed many strategies to cope with drought stress, one important aspect is the recovery of photosynthesis. The drought-increased proteins involved in photoreaction and Calvin cycle were observed in leaves. For example, light-harvesting chlorophyll a/b-binding proteins (LHCB) were increased in tolerant genotypes of *Z. mays* [41], and *M. domestica* [64], but decreased or stable in sensitive genotypes. The LHCBs have been predicted to be involved in ABA signaling partially by modulating ROS homeostasis [192], and may be taken as interesting targets for crop breeding [4]. Besides, the increased abundance of sedoheptulose-1,7-bisphosphatase (SBPase) and carbonic anhydrase (CA) were found in drought-stressed *P. pratensis* [23] and *M. domestica* [64]. SBPase has a key role in regulating the photosynthetic Calvin cycle. CA catalyzes the reversible hydration of CO_2, and has a relevant role in CO_2 exchange by influencing the internal conductance [193].

In addition, the abundances of proteins involved in photorespiration were also decreased in some plant species, reflecting the drought-inhibited photorespiration process (Figure 4C). However, significant increases of four key enzymes (e.g., glycolate oxidase, glycine dehydrogenase, serine glyoxylate aminotransferase, and serine transhydroxymethyl transferase) were found in a high water-use efficiency *M. domestica* variety under moderate drought conditions [64]. Additionally, the protein abundances of aminomethyl transferase (AMT) and glycine dehydrogenase were decreased in a drought-sensitive cultivar of *C. dactylon*, but increased or stable in a tolerant cultivar [19]. Similarly, the downregulation of the gene encoding AMT was reported in dehydration-sensitive *H. vulgare*, but not in a dehydration-tolerant cultivar [194]. It is well known that photorespiration can protect photosynthesis from photoinhibition and prevent ROS accumulation in green tissues [195]. The proteomic findings provided further evidences that the regulation of photorespiration was important for plant drought tolerance [64].

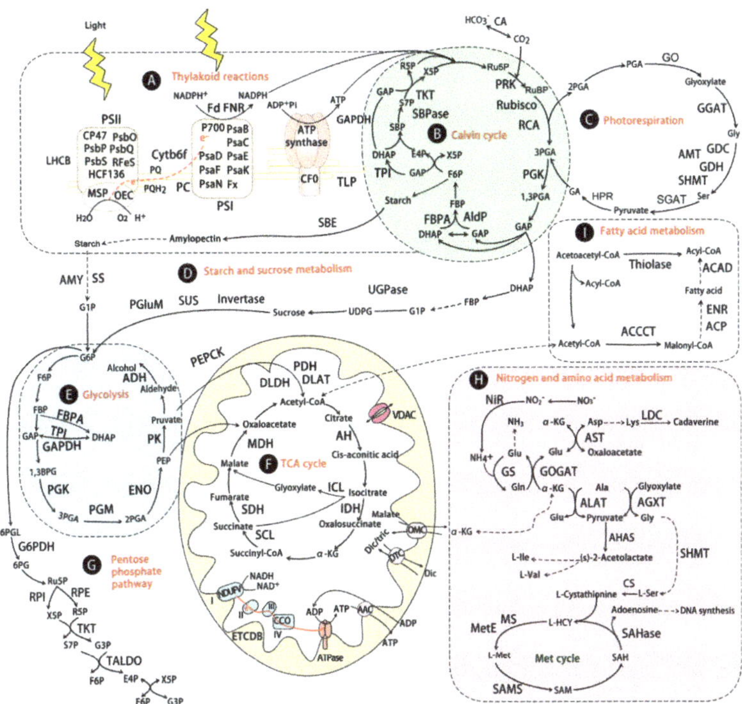

Figure 4. Schematic model of drought responsive proteins involved in photosynthesis, carbon metabolism, fatty acid metabolism, and nitrogen and amino acid metabolism. The solid line indicates single-step reactions, and the dashed line indicates multi-step reactions. (**A**) Thylakoid reaction; (**B**) Calvin cycle; (**C**) Photorespiration; (**D**) Starch and sucrose metabolism; (**E**) Glycolysis; (**F**) Tricarboxylic acid (TCA) cycle; (**G**) Pentose phosphate pathway; (**H**) Nitrogen and amino acid metabolism; (**I**) Fatty acid metabolism. 1,3BPG: 1,3-bisphosphoglycerate; OMC: 2-oxoglutarate/malate carrier protein; MetE: 5-methyltetrahydropteroyltriglutamate-homocysteine methyltransferase; 6PG: 6-phosphogluconate; G6PDH: 6-phosphogluconate dehydrogenase; 6PGL: 6-phosphoglucono-δ-lactone; AHAS: acetohydroxyacid synthase; AH: aconitate hydratase; ACAD: acyl-CoA dehydrogenase; SAHase: adenosylhomocysteinase; ALAT: alanine aminotransferase; AGXT: alanine-glyoxylate aminotransferase; ADH: alcohol dehydrogenase; AldP: aldolase; AMT: aminomethyl transferase; AMY: amylase; AST: aspartate aminotransferase; ATPase: ATP synthase; AAC: ATP/ADP translocase; CA: carbonic anhydrase; FNR: chloroplast ferredoxin-NADP⁺ oxidoreductase; MSP: chloroplast manganese stabilizing protein; CS: cysteine synthase; Cyt b6f: cytochrome b6f complex; CCO: cytochrome c oxidase; Dic: dicarboxylate; DTC: dicarboxylate/tricarboxylate carrier; DLAT: dihydrolipoamide acetyltransferase; DHAP: dihydroxyacetone phosphate; ETCDB: electron carrier/iron ion binding protein; ENO: enolase; E4P: erythrose-4-phosphate; Fd: ferredoxin; F6P: fructose 6-phosphate; FBP: fructose-1,6-bisphosphate; FBPA: fructose-bisphosphate aldolase; G6P: glucose 6-phosphate; G1P: glucose-1-phosphate; GOGAT: glutamate synthase; GGAT: glutamate-glyoxylate aminotransferase; GS: glutamine synthetase; GAP: glyceraldehyde 3-phosphate; GAPDH: glyceraldehyde-3-phosphate dehydrogenase; GA: glyceric acid; GCS: glycine cleavage system; GDC: glycine decarboxylase; GO: glycolate oxidase; HCY: homocysteine; HPR: hydroxypyruvate reductase; IDH: isocitrate dehydrogenase; ICL: isocitrate lyase; LHCB: light-harvesting chlorophyll a/b-binding protein; LDH: lipoamide dehydrogenase; LDC: lysine decarboxylase; MDH: malate dehydrogenase; MS: methionine synthase; NDUFV: NADH dehydrogenase (ubiquinone) flavoprotein; NIR: nitrite reductase; OEC: oxygen-evolving complex protein; PEPCK: phosphoenol pyruvate carboxykinase; PEP: phosphoenolpyruvate; PGluM: phosphoglucomutase; PGA: phosphoglycerate;

PGK: phosphoglycerate kinase; PGM: phosphoglycerate mutase; PRK: phosphoribulokinase; PSI: photosystem I reaction center protein; PSII: photosystem II reaction center protein; PC: plastocyanin; PDH: pyruvate dehydrogenase; DLDH: pyruvate dehydrogenase complex; PK: pyruvate kinase; R5P: ribose-5-phosphate; RPI: ribose-5-phosphate isomerase; Ru5P: ribulose 5-phosphate; Rubisco: ribulose-1,5-bisphosphate carboxylase/oxygenase; RCA: ribulose-1,5-bisphosphate carboxylase/oxygenase activase; RuBP: ribulose-1,5-disphosphate; RPE: ribulose-phosphate 3-epimerase; SAH: *S*-adenosyl-L-homocysteine; SAM: *S*-adenosylmethionine; SAMS: *S*-adenosylmethionine synthase; S7P: sedoheptulose 7-phosphate; SBPase: sedoheptulose-1,7-bisphosphatase; SBP: sedoheptulose-1,7-bisphosphate; SGAT: serine glyoxylate aminotransferase; SHMT: serine transhydroxymethyltransferase; SBE: starch branching enzyme; SS: starch synthase; SDH: succinate dehydrogenase; SCL: succinyl-CoA ligase; SUS: sucrose synthase; TLP: thylakoid lumen protein; TALDO: transaldolase; TKT: transketolase; Tric: tricarboxylate; TPI: triosephosphate isomerase; UDPG: uridine diphosphate glucose; UGPase: UTP-glucose pyrophosphorylase; VDAC: voltage dependent anion channel; X5P: xylulose 5-phosphate; α-KG: α-ketoglutarate.

11. Carbohydrate and Energy Metabolism

Proteomic studies have revealed that nearly 20% of total drought-responsive proteins were involved in carbohydrate and energy metabolism (e.g., glycolysis, tricarboxylic acid (TCA) cycle, electron transport chain, and ATP synthesis) in leaves to cope with drought stress (Figure 4E–G). For example, phosphoglucomutases (PGluMs) involved in glycolysis/gluconeogenesis were decreased in leaves of *S. stapfianus* [24] and *O. sativa* (stop irrigation for 35 days) [32], but increased in leaves of *C. lanatus* [48], *O. sativa* (under 30% relative humidity treatment for 144, 168 and 192 h) [29], and *M. paradisiaca* [46] in response to drought. Fructose-bisphosphate aldolase (FBPA), another glycolysis-related protein, was decreased in many aforementioned plant species, but also found to be drought-increased in other plant species such as *O. sativa* [29,32], *S. stapfianus* [24], and *M. paradisiaca* [46]. Besides, FBPA abundance was decreased in drought-sensitive cultivars of *M. domestica* [64] and *P. pratensis* [23], but was increased in tolerant cultivars. Similarly, aconitate hydratases involved in the TCA cycle were increased in leaves of tolerant cultivars of *Z. mays* and *P. pratensis* [23], whereas they were decreased in sensitive cultivars of *Z. mays* and *P. pratensis* [23]. The inhibition would be a mechanism for accumulating sugars as an osmolyte or energy source for recovery [62], while the increase of glycolysis and TCA may act as a strategy for providing energy during the activation of stress defenses, especially when the photosynthesis was inhibited. These apparently contradictory results could also be related to the different degrees and duration of water shortage, as well as to the tolerance ability of the plant species [62]. Among these proteins, cytosolic FBPA and glyceraldehyde-3-phosphate dehydrogenase (GAPDH) have a significantly positive correlation with drought tolerance [196], and overexpressing GAPDH in potato displayed an improvement of drought tolerance [197]. In addition, the reverse phosphorylation of some carbohydrate metabolism-related proteins (e.g., phosphoenolpyruvate carboxykinase [44], FBPA [44], and NAD-malate dehydrogenase (MDH) [34]) (Supplementary Table S2) further demonstrated that the carbon metabolism was modulated in plants in response to water stress. Although many drought-responsive carbohydrate metabolism-related proteins were revealed by proteomic studies, only a few of them have been functionally characterized to evaluate their role in drought stress tolerance.

Consistent with carbohydrate metabolism changes, mitochondrial electron transport chain and ATP synthesis-related proteins were changed in plants under drought stress. They were comprised of ATP synthase subunits (e.g., ATPase F0, F1, α, β, γ, δ, and ε subunits) and electron transport chain-related proteins (NADH dehydrogenase [47,86], quinone oxidoreductase [35,49,62], and cytochrome c oxidase [18,39,47]). Their changes imply that diverse changes of energy metabolism take place in leaves in response to water deficit. In some cases, plants may have the ability to enhance energy production to maintain major physiological activity and inhibit stress damage.

12. Nitrogen Assimilation and Amino Acid Metabolism

Nitrogen assimilation is affected by abiotic stress in plants. In this process, exogenous absorbed nitrate is transformed to ammonium by nitrate reductase (NR) and nitrite reductase (NIR), and then assimilated by glutamine synthetase (GS) and glutamate synthase (GOGAT) into amino acids. Proteomic studies revealed that several enzymes in nitrogen assimilation were drought-decreased in leaves (Figure 4H). Long-term (3 weeks) drought stress decreased the abundance of NIR in *H. vulgare* [25], but short-term (2–5 days) drought stress increased NIR in *O. sativa* [29]. Besides, drought significantly decreased the protein abundances of GS1, GS2, and GOGAT in many plant species (Supplementary Table S1), and the activities of NR and GS were reduced in *Leymus chinensis* [198], *T. aestivum* [199], and tomato [200] under water deficit. Similarly, the decrease at transcriptional level of *Brassica juncea GS1.1* and other key genes involved in nitrogen assimilation (e.g., *BjGDH1*, *BjGDH2*, and *BjASN2*) were also well addressed [201]. The reduction of NR, GS, and GOGAT indicates that the nitrogen assimilation is significantly inhibited by drought stress, which is supposed to be the main reason for the reduction of crop yield under drought stress.

Following the GS/GOGAT cycle, the amino group was transferred into other amino acids by aminotransferases (Figure 4H). Two aminotransferases (e.g., aspartate aminotransferase (AST) and alanine aminotransferase (ALAT)) were altered under drought stress. Among them, AST was drought-decreased in *P. pratensis* [23], and ALAT was decreased in *P. pratensis* [23] and *Z. mays* [41]. AST catalyzes the amino transfer from glutamate to oxaloacetate to form aspartate, which is part of ammonium assimilation. ALAT catalyzes the interconversion of L-glutamate and L-alanine. The abundance decline of these enzymes indicates that the amino acid metabolism and the synthesis of other metabolites derived from amino acids are inhibited under drought stress.

Interestingly, several key enzymes involved in *S*-adenosyl-L-methionine (SAM) cycle were generally increased in leaves in response to drought stress, including drought-increased 5-methyltetrahydropteroyltriglutamate-homocysteine methyltransferase (MetE), *S*-adenosyl-L-homocysteine hydrolase (SAHase), *S*-adenosylmethionine synthase (SAMS), and methionine synthase (MS) (Figure 4H). Similarly, MS in *Z. mays* [41], *Sorghum bicolor* [202], a drought-tolerant *C. dactylon* cultivar, as well as SAMS in *Populus* [59,86], *Q. ilex* [63], and *T. aestivum* [35] were increased under drought conditions. The metabolites in the SAM cycle were known to play major roles in methylation of DNA, proteins, and other metabolites, being involved in control of gene expression, cell wall metabolism, as well as the biosynthesis of polyamine and ethylene for stress tolerance [18,203,204]. In addition, the *MS* gene was also strongly induced in barley leaves under drought stress [205]. Transgenic *Arabidopsis* lines overexpressing *SAMS* from *Solanum brevidens* enhanced the ability of plant drought tolerance [206]. This implies that the increases of MS and SAMS could enhance the methionine and osmoregulant metabolism for plants to cope with drought stress.

13. Fatty Acid Metabolism and Other Metabolisms

Four proteins involved in fatty acid biosynthesis (i.e., acetyl-coenzyme A carboxylase carboxyl transferase, acyl carrier protein, enoyl-acyl carrier protein reductase, and lipoxygenase 6) and three involved in fatty acid degradation (i.e., thiolase I, thiolase II, and acyl-CoA dehydrogenase) were altered under dehydration conditions (Figure 4I). Physiological and molecular investigations have revealed that the fatty acid/lipid metabolism was changed in response to drought stress [207–209]. For example, a greater composition of unsaturated fatty acids in membrane lipids may contribute to superior leaf dehydration tolerance in *C. dactylon* [209]. In the leaves of *Parkinsonia aculeata*, the significantly increased contents of chloroplast lipids (i.e., monogalactosyldiacylglycerol and digalactosyldiacylglycerol) and extra-plastidial lipids (i.e., phosphatidylcholine and phosphatidylethanolamine) maintained *Parkinsonia* plants in a state of rapid recovery of lipid synthesis during rehydration [209]. The changes in fatty acid/lipid composition may help to maintain membrane integrity and preserve cell compartmentation under water stress [210].

In addition, two flavonoid biosynthesis related proteins (i.e., chalcone isomerase (CHI) and dihydroflavonol-4-reductase) involved in secondary metabolism were also changed in response to drought (Supplementary Table S1). It has been suggested that flavonoids have antioxidant capacity to protect plants against abiotic stress [211,212], and flavonoid accumulation was induced rapidly by abiotic stresses [213]. In proteomic studies, CHI, a key enzyme in flavonoid biosynthesis, was decreased in *O. sativa* under drought stress [29]. Similarly, the *CsCHI* gene in leaves of tea (*Camellia sinensis*) was also downregulated under drought treatment [214]. Besides, dihydroflavonol-4-reductase, another key enzyme in the flavonoid biosynthesis pathway, was decreased in a drought sensitive *Z. mays* cultivar [41], but increased in a tolerant one. The abundance changes of these enzymes currently attract great attention from researchers seeking to evaluate their potential protective roles in drought stress response.

14. Concluding Remarks

By integrative analysis of proteomics results, it has been found that, in total, 2216 protein species and more than 200 phosphoproteins were drought-responsive in plant leaves. The abundance changes of these proteins provide new clues towards further understanding the complex cellular and molecular processes in plant drought tolerance. The quantitative patterns of protein in various plants give rise to diverse drought-responsive mechanisms, including drought signal transduction, ROS scavenging, osmotic regulation, specific gene expression regulation, protein synthesis and turnover, cell structure modulation, as well as carbohydrate and energy metabolism (Figure 5). Exploring these protein functions will enable a holistic understanding of plant-environment interaction. Moreover, the potential proteins screened out by proteomics will provide candidate genes/proteins for genetic improvement in plant drought tolerance. However, some proteins showed dynamic changes depending on plant species and stress intensity, which makes it difficult to give a clear interpretation of the mechanism in plant drought response. Importantly, proteomic investigations also revealed that protein PTMs are crucial for signaling and metabolic mechanisms in plant drought response. The integration of those findings from physiological, gene expression, and other large-scale "omics" will help us to establish molecular networks of drought response and tolerance.

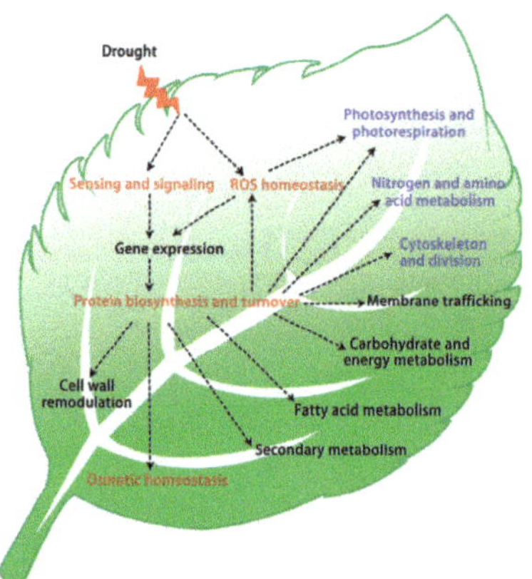

Figure 5. A summary of signaling and metabolic pathways in leaves in response to drought stress. Drought stress activates several signaling cascades, regulates gene expression, and promotes protein biosynthesis and turnover in drought-stressed leaves. Besides, drought stress inhibits photosynthesis, photorespiration, nitrogen and amino acid metabolism, as well as the cytoskeleton and cell division. Importantly, plants enhance ROS scavenging, osmotic regulation and cell wall remodulation. In addition, plants also modulate membrane trafficking, fatty acid metabolism, carbohydrate and energy metabolism, as well as secondary metabolism to cope with drought stress.

Supplementary Materials: Supplementary materials can be found at www.mdpi.com/1422-0067/17/10/1706/s1.

Acknowledgments: This work was supported by the Natural Science Foundation of Shanghai (No. 15ZR1431300), Capacity Construction Project of Local Universities, Shanghai, (No. 14390502700), and Shanghai Municipal Agricultural Commission (No. 201509).

Author Contributions: Xiaoli Wang and Shaojun Dai designed this work, collected the data, and co-wrote the manuscript. Xiaofeng Cai, Chengxi Xu, and Quanhua Wang collected the data and helped in editing the manuscript.

Conflicts of Interest: The authors declare no conflict of interest.

References

1. Chaves, M.M.; Pereira, J.S.; Maroco, J.; Rodrigues, M.L.; Ricardo, C.P.P.; Osorio, M.L.; Carvalho, I.; Faria, T.; Pinheiro, C. How plants cope with water stress in the field. Photosynthesis and growth. *Ann. Bot.* **2002**, *89*, 907–916. [PubMed]

2. Dinakar, C.; Bartels, D. Desiccation tolerance in resurrection plants: New insights from transcriptome, proteome, and metabolome analysis. *Front. Plant Sci.* **2013**, *4*, 482. [CrossRef] [PubMed]

3. Jogaiah, S.; Govind, S.R.; Tran, L.S.P. Systems biology-based approaches toward understanding drought tolerance in food crops. *Crit. Rev. Biotechnol.* **2013**, *33*, 23–39. [CrossRef] [PubMed]

4. Abreu, I.A.; Farinha, A.P.; Negrao, S.; Goncalves, N.; Fonseca, C.; Rodrigues, M.; Batista, R.; Saibo, N.J.M.; Oliveira, M.M. Coping with abiotic stress: Proteome changes for crop improvement. *J. Proteom.* **2013**, *93*, 145–168. [CrossRef] [PubMed]

5. Farooq, M.; Wahid, A.; Kobayashi, N.; Fujita, D.; Basra, S.M.A. Plant drought stress: Effects, mechanisms and management. *Agron. Sustain. Dev.* **2009**, *29*, 185–212. [CrossRef]

6. Osmond, C.B.; Grace, S.C. Perspectives on photoinhibition and photorespiration in the field: Quintessential inefficiencies of the light and dark reactions of photosynthesis? *J. Exp. Bot.* **1995**, *46*, 1351–1362. [CrossRef]

7. Zhang, Q. Strategies for developing Green Super Rice. *Proc. Natl. Acad. Sci. USA* **2007**, *104*, 16402–16409. [CrossRef] [PubMed]

8. El-Soda, M.; Kruijer, W.; Malosetti, M.; Koornneef, M.; Aarts, M.G.M. Quantitative trait loci and candidate genes underlying genotype by environment interaction in the response of *Arabidopsis thaliana* to drought. *Plant Cell Environ.* **2015**, *38*, 585–599. [CrossRef] [PubMed]

9. Yadav, R.S.; Sehgal, D.; Vadez, V. Using genetic mapping and genomics approaches in understanding and improving drought tolerance in pearl millet. *J. Exp. Bot.* **2011**, *62*, 397–408. [CrossRef] [PubMed]

10. Alm, V.; Busso, C.S.; Ergon, A.; Rudi, H.; Larsen, A.; Humphreys, M.W.; Rognli, O.A. QTL analyses and comparative genetic mapping of frost tolerance, winter survival and drought tolerance in meadow fescue (*Festuca pratensis* Huds.). *Theor. Appl. Genet.* **2011**, *123*, 369–382. [CrossRef] [PubMed]

11. Acuna-Galindo, M.A.; Mason, R.E.; Subramanian, N.K.; Hays, D.B. Meta-analysis of wheat QTL regions associated with adaptation to drought and heat stress. *Crop Sci.* **2015**, *55*, 477–492. [CrossRef]

12. Hao, Z.; Li, X.; Liu, X.; Xie, C.; Li, M.; Zhang, D.; Zhang, S. Meta-analysis of constitutive and adaptive QTL for drought tolerance in maize. *Euphytica* **2010**, *174*, 165–177. [CrossRef]

13. Asfaw, A.; Blair, M.W. Quantitative trait loci for rooting pattern traits of common beans grown under drought stress versus non-stress conditions. *Mol. Breed.* **2012**, *30*, 681–695. [CrossRef]

14. Reddy, S.K.; Liu, S.; Rudd, J.C.; Xue, Q.; Payton, P.; Finlayson, S.A.; Mahan, J.; Akhunova, A.; Holalu, S.V.; Lu, N. Physiology and transcriptomics of water-deficit stress responses in wheat cultivars TAM 111 and TAM 112. *J. Plant Physiol.* **2014**, *171*, 1289–1298. [CrossRef] [PubMed]

15. Krannich, C.T.; Maletzki, L.; Kurowsky, C.; Horn, R. Network candidate genes in breeding for drought tolerant crops. *Int. J. Mol. Sci.* **2015**, *16*, 16378–16400. [CrossRef] [PubMed]

16. Baerenfaller, K.; Massonnet, C.; Walsh, S.; Baginsky, S.; Buehlmann, P.; Hennig, L.; Hirsch-Hoffmann, M.; Howell, K.A.; Kahlau, S.; Radziejwoski, A.; et al. Systems-based analysis of *Arabidopsis* leaf growth reveals adaptation to water deficit. *Mol. Syst. Biol.* **2012**, *8*, 606. [CrossRef] [PubMed]

17. Mohammadi, P.P.; Moieni, A.; Hiraga, S.; Komatsu, S. Organ-specific proteomic analysis of drought-stressed soybean seedlings. *J. Proteom.* **2012**, *75*, 1906–1923. [CrossRef] [PubMed]

18. Merewitz, E.B.; Gianfagna, T.; Huang, B. Protein accumulation in leaves and roots associated with improved drought tolerance in creeping bentgrass expressing an *ipt* gene for cytokinin synthesis. *J. Exp. Bot.* **2011**, *62*, 5311–5333. [CrossRef] [PubMed]

19. Zhao, Y.; Du, H.; Wang, Z.; Huang, B. Identification of proteins associated with water-deficit tolerance in C-4 perennial grass species, *Cynodon dactylon* × *Cynodon transvaalensis* and *Cynodon dactylon*. *Physiol. Plant.* **2011**, *141*, 40–55. [CrossRef] [PubMed]

20. Gazanchian, A.; Hajheidari, M.; Sima, N.K.; Salekdeh, G.H. Proteome response of *Elymus elongatum* to severe water stress and recovery. *J. Exp. Bot.* **2007**, *58*, 291–300. [CrossRef] [PubMed]

21. Jiang, G.; Wang, Z.; Shang, H.; Yang, W.; Hu, Z.; Phillips, J.; Deng, X. Proteome analysis of leaves from the resurrection plant *Boea hygrometrica* in response to dehydration and rehydration. *Planta* **2007**, *225*, 1405–1420. [CrossRef] [PubMed]

22. Xu, C.; Huang, B. Proteins and metabolites regulated by trinexapac-ethyl in relation to drought tolerance in Kentucky bluegrass. *J. Plant Growth Regul.* **2012**, *31*, 25–37. [CrossRef]

23. Xu, C.; Huang, B. Comparative analysis of drought responsive proteins in Kentucky bluegrass cultivars contrasting in drought tolerance. *Crop Sci.* **2010**, *50*, 2543–2552. [CrossRef]

24. Oliver, M.J.; Jain, R.; Balbuena, T.S.; Agrawal, G.; Gasulla, F.; Thelen, J.J. Proteome analysis of leaves of the desiccation-tolerant grass, *Sporobolus stapfianus*, in response to dehydration. *Phytochemistry* **2011**, *72*, 1273–1284. [CrossRef] [PubMed]

25. Ashoub, A.; Baeumlisberger, M.; Neupaertl, M.; Karas, M.; Brueggemann, W. Characterization of common and distinctive adjustments of wild barley leaf proteome under drought acclimation, heat stress and their combination. *Plant Mol. Biol.* **2015**, *87*, 459–471. [CrossRef] [PubMed]

26. Ghabooli, M.; Khatabi, B.; Ahmadi, F.S.; Sepehri, M.; Mirzaei, M.; Amirkhani, A.; Jorrin-Novo, J.V.; Salekdeh, G.H. Proteomics study reveals the molecular mechanisms underlying water stress tolerance induced by *Piriformospora indica* in barley. *J. Proteom.* **2013**, *94*, 289–301. [CrossRef] [PubMed]

27. Ashoub, A.; Beckhaus, T.; Berberich, T.; Karas, M.; Bruggemann, W. Comparative analysis of barley leaf proteome as affected by drought stress. *Planta* **2013**, *237*, 771–781. [CrossRef] [PubMed]

28. Wendelboe-Nelson, C.; Morris, P.C. Proteins linked to drought tolerance revealed by DIGE analysis of drought resistant and susceptible barley varieties. *Proteomics* **2012**, *12*, 3374–3385. [CrossRef] [PubMed]

29. Pandey, A.; Rajamani, U.; Verma, J.; Subba, P.; Chakraborty, N.; Datta, A.; Chakraborty, S.; Chakraborty, N. Identification of extracellular matrix proteins of rice (*Oryza sativa* L.) Involved in dehydration-responsive network: A proteomic approach. *J. Proteome Res.* **2010**, *9*, 3443–3464. [PubMed]

30. Salekdeh, G.H.; Siopongco, J.; Wade, L.J.; Ghareyazie, B.; Bennett, J. Proteomic analysis of rice leaves during drought stress and recovery. *Proteomics* **2002**, *2*, 1131–1145. [CrossRef]

31. Rabello, F.R.; Villeth, G.R.C.; Rabello, A.R.; Rangel, P.H.N.; Guimaraes, C.M.; Huergo, L.F.; Souza, E.M.; Pedrosa, F.O.; Ferreira, M.E.; Mehta, A. Proteomic analysis of upland rice (*Oryza sativa* L.) exposed to intermittent water deficit. *Protein J.* **2014**, *33*, 221–230. [CrossRef] [PubMed]

32. Raorane, M.L.; Pabuayon, I.M.; Varadarajan, A.R.; Mutte, S.K.; Kumar, A.; Treumann, A.; Kohli, A. Proteomic insights into the role of the large-effect QTL qDTY12.1 for rice yield under drought. *Mol. Breed.* **2015**, *35*, 1–14. [CrossRef]

33. Ji, K.X.; Wang, Y.Y.; Sun, W.N.; Lou, Q.J.; Mei, H.W.; Shen, S.H.; Chen, H. Drought-responsive mechanisms in rice genotypes with contrasting drought tolerance during reproductive stage. *J. Plant Physiol.* **2012**, *169*, 336–344. [CrossRef] [PubMed]

34. Ke, Y.Q.; Han, G.Q.; He, H.Q.; Li, J.X. Differential regulation of proteins and phosphoproteins in rice under drought stress. *Biochem. Biophys. Res. Commun.* **2009**, *379*, 133–138. [CrossRef] [PubMed]

35. Ford, K.L.; Cassin, A.; Bacic, A. Quantitative proteomic analysis of wheat cultivars with differing drought stress tolerance. *Front. Plant Sci.* **2011**, *2*, 44. [CrossRef] [PubMed]

36. Peremarti, A.; Mare, C.; Aprile, A.; Roncaglia, E.; Cattivelli, L.; Villegas, D.; Royo, C. Transcriptomic and proteomic analyses of a pale-green durum wheat mutant shows variations in photosystem components and metabolic deficiencies under drought stress. *BMC Genom.* **2014**, *15*, 125. [CrossRef] [PubMed]

37. Zhang, H.M.; Zhang, L.S.; Lv, H.; Yu, Z.Y.; Zhang, D.P.; Zhu, W.N. Identification of changes in *Triticum aestivum* L. leaf proteome in response to drought stress by 2D-PAGE and MALDI-TOF/TOF mass spectrometry. *Acta Physiol. Plant.* **2014**, *36*, 1385–1398. [CrossRef]

38. Wang, X.; Vignjevic, M.; Jiang, D.; Jacobsen, S.; Wollenweber, B. Improved tolerance to drought stress after anthesis due to priming before anthesis in wheat (*Triticum aestivum* L.) var. Vinjett. *J. Exp. Bot.* **2014**, *65*, 6441–6456. [CrossRef] [PubMed]

39. Budak, H.; Akpinar, B.A.; Unver, T.; Turktas, M. Proteome changes in wild and modern wheat leaves upon drought stress by two-dimensional electrophoresis and nanoLC-ESI-MS/MS. *Plant Mol. Biol.* **2013**, *83*, 89–103. [CrossRef] [PubMed]

40. Zhang, M.; Lv, D.W.; Ge, P.; Bian, Y.W.; Chen, G.X.; Zhu, G.R.; Li, X.H.; Yan, Y.M. Phosphoproteome analysis reveals new drought response and defense mechanisms of seedling leaves in bread wheat (*Triticum aestivum* L.). *J. Proteom.* **2014**, *109*, 290–308. [CrossRef] [PubMed]

41. Benesova, M.; Hola, D.; Fischer, L.; Jedelsky, P.L.; Hnilicka, F.; Wilhelmova, N.; Rothova, O.; Kocova, M.; Prochazkova, D.; Honnerova, J.; et al. The physiology and proteomics of drought tolerance in maize: Early stomatal closure as a cause of lower tolerance to short-term dehydration? *PLoS ONE* **2012**, *7*, e38017. [CrossRef] [PubMed]

42. Sanggon, K.; Jinseok, L.; Jungtae, K.; Youngsang, K.; Dongwon, B.; Hwanhee, B.; Beomyoung, S.; Seongbum, B.; Youngup, K.; Miok, W. Physiological and proteomic analysis of the response to drought stress in an inbred Korean maize line. *Plant Omics* **2015**, *8*, 159–168.

43. Bonhomme, L.; Valot, B.; Tardieu, F.; Zivy, M. Phosphoproteome dynamics upon changes in plant water status reveal early events associated with rapid growth adjustment in maize leaves. *Mol. Cell. Proteom.* **2012**, *11*, 957–972. [CrossRef] [PubMed]

44. Hu, X.L.; Wu, L.J.; Zhao, F.Y.; Zhang, D.Y.; Li, N.N.; Zhu, G.H.; Li, C.H.; Wang, W. Phosphoproteomic analysis of the response of maize leaves to drought, heat and their combination stress. *Front. Plant Sci.* **2015**, *6*, 289. [CrossRef] [PubMed]

45. Zhou, G.; Yang, L.T.; Li, Y.R.; Zou, C.L.; Huang, L.P.; Qiu, L.H.; Huang, X.; Srivastava, M.K. Proteomic analysis of osmotic stress-responsive proteins in sugarcane leaves. *Plant Mol. Biol. Rep.* **2012**, *30*, 349–359. [CrossRef]

46. Vanhove, A.-C.; Vermaelen, W.; Panis, B.; Swennen, R.; Carpentier, S.C. Screening the banana biodiversity for drought tolerance: Can an in vitro growth model and proteomics be used as a tool to discover tolerant varieties and understand homeostasis. *Front. Plant Sci.* **2012**, *3*, 176. [CrossRef] [PubMed]

47. Koh, J.; Chen, G.; Yoo, M.-J.; Zhu, N.; Dufresne, D.; Erickson, J.E.; Shao, H.; Chen, S. Comparative proteomic analysis of *Brassica napus* in response to drought stress. *J. Proteome Res.* **2015**, *14*, 3068–3081. [CrossRef] [PubMed]

48. Akashi, K.; Yoshida, K.; Kuwano, M.; Kajikawa, M.; Yoshimura, K.; Hoshiyasu, S.; Inagaki, N.; Yokota, A. Dynamic changes in the leaf proteome of a C-3 xerophyte, *Citrullus lanatus* (wild watermelon), in response to water deficit. *Planta* **2011**, *233*, 947–960. [CrossRef] [PubMed]

49. Zadraznik, T.; Hollung, K.; Egge-Jacobsen, W.; Meglic, V.; Sustar-Vozlic, J. Differential proteomic analysis of drought stress response in leaves of common bean (*Phaseolus vulgaris* L.). *J. Proteom.* **2013**, *78*, 254–272. [CrossRef] [PubMed]

50. Aranjuelo, I.; Molero, G.; Erice, G.; Avice, J.C.; Nogues, S. Plant physiology and proteomics reveals the leaf response to drought in alfalfa (*Medicago sativa* L.). *J. Exp. Bot.* **2011**, *62*, 111–123. [CrossRef] [PubMed]

51. Deeba, F.; Pandey, A.K.; Ranjan, S.; Mishra, A.; Singh, R.; Sharma, Y.K.; Shirke, P.A.; Pandey, V. Physiological and proteomic responses of cotton (*Gossypium herbaceum* L.) to drought stress. *Plant Physiol. Biochem.* **2012**, *53*, 6–18. [CrossRef] [PubMed]

52. Zhang, M.H.; Li, G.W.; Huang, W.; Bi, T.; Chen, G.Y.; Tang, Z.C.; Su, W.A.; Sun, W.N. Proteomic study of *Carissa spinarum* in response to combined heat and drought stress. *Proteomics* **2010**, *10*, 3117–3129. [CrossRef] [PubMed]

53. Brossa, R.; Pinto-Marijuan, M.; Francisco, R.; Lopez-Carbonell, M.; Chaves, M.M.; Alegre, L. Redox proteomics and physiological responses in *Cistus albidus* shrubs subjected to long-term summer drought followed by recovery. *Planta* **2015**, *241*, 803–822. [CrossRef] [PubMed]

54. Xu, G.; Li, C.Y.; Yao, Y.N. Proteomics analysis of drought stress-responsive proteins in *Hippophae rhamnoides* L. *Plant Mol. Biol. Rep.* **2009**, *27*, 153–161. [CrossRef]

55. Bedon, F.; Villar, E.; Vincent, D.; Dupuy, J.W.; Lomenech, A.M.; Mabialangoma, A.; Chaumeil, P.; Barre, A.; Plomion, C.; Gion, J.M. Proteomic plasticity of two *Eucalyptus* genotypes under contrasted water regimes in the field. *Plant Cell Environ.* **2012**, *35*, 790–805. [CrossRef] [PubMed]

56. Bonhomme, L.; Monclus, R.; Vincent, D.; Carpin, S.; Claverol, S.; Lomenech, A.-M.; Labas, V.; Plomion, C.; Brignolas, F.; Morabito, D. Genetic variation and drought response in two *Populus* × *euramericana* genotypes through 2-DE proteomic analysis of leaves from field and glasshouse cultivated plants. *Phytochemistry* **2009**, *70*, 988–1002. [CrossRef] [PubMed]

57. He, C.Y.; Zhang, J.G.; Duan, A.G.; Zheng, S.X.; Sun, H.G.; Fu, L.H. Proteins responding to drought and high-temperature stress in *Populus* × *euramericana* cv. "74/76". *Trees Struct. Funct.* **2008**, *22*, 803–813. [CrossRef]

58. Durand, T.C.; Sergeant, K.; Renaut, J.; Planchon, S.; Hoffmann, L.; Carpin, S.; Label, P.; Morabito, D.; Hausman, J.F. Poplar under drought: Comparison of leaf and cambial proteomic responses. *J. Proteom.* **2011**, *74*, 1396–1410. [CrossRef] [PubMed]

59. Bohler, S.; Sergeant, K.; Jolivet, Y.; Hoffmann, L.; Hausman, J.-F.; Dizengremel, P.; Renaut, J. A physiological and proteomic study of poplar leaves during ozone exposure combined with mild drought. *Proteomics* **2013**, *13*, 1737–1754. [CrossRef] [PubMed]

60. Xiao, X.; Yang, F.; Zhang, S.; Korpelainen, H.; Li, C. Physiological and proteomic responses of two contrasting *Populus cathayana* populations to drought stress. *Physiol. Plant.* **2009**, *136*, 150–168. [CrossRef] [PubMed]

61. Sergeant, K.; Spiess, N.; Renaut, J.; Wilhelm, E.; Hausman, J.F. One dry summer: A leaf proteome study on the response of oak to drought exposure. *J. Proteom.* **2011**, *74*, 1385–1395. [CrossRef] [PubMed]

62. Echevarria-Zomeno, S.; Ariza, D.; Jorge, I.; Lenz, C.; del Campo, A.; Jorrin, J.V.; Navarro, R.M. Changes in the protein profile of *Quercus ilex* leaves in response to drought stress and recovery. *J. Plant Physiol.* **2009**, *166*, 233–245. [CrossRef] [PubMed]

63. Valero-Galvan, J.; Gonzalez-Fernandez, R.; Navarro-Cerrillo, R.M.; Gil-Pelegrin, E.; Jorrin-Novo, J.V. Physiological and proteomic analyses of drought stress response in Holm oak provenances. *J. Proteome Res.* **2013**, *12*, 5110–5123. [CrossRef] [PubMed]

64. Zhou, S.; Li, M.; Guan, Q.; Liu, F.; Zhang, S.; Chen, W.; Yin, L.; Qin, Y.; Ma, F. Physiological and proteome analysis suggest critical roles for the photosynthetic system for high water-use efficiency under drought stress in *Malus*. *Plant Sci.* **2015**, *236*, 44–60. [CrossRef] [PubMed]

65. Carvalho, R.F.; Campos, M.L.; Azevedo, R.A. The role of phytochrome in stress tolerance. *J. Integr. Plant Biol.* **2011**, *53*, 920–929. [CrossRef] [PubMed]

66. Boggs, J.Z.; Loewy, K.; Bibee, K.; Heschel, M.S. Phytochromes influence stomatal conductance plasticity in *Arabidopsis thaliana*. *Plant Growth Regul.* **2010**, *60*, 77–81. [CrossRef]

67. Yadav, D.K.; Shukla, D.; Tuteja, N. Rice heterotrimeric G-protein alpha subunit (RGA1): In silico analysis of the gene and promoter and its upregulation under abiotic stress. *Plant Physiol. Biochem.* **2013**, *63*, 262–271. [CrossRef] [PubMed]

68. Gao, Y.; Li, T.; Liu, Y.; Ren, C.; Zhao, Y.; Wang, M. Isolation and characterization of gene encoding G protein α subunit protein responsive to plant hormones and abiotic stresses in *Brassica napus*. *Mol. Biol. Rep.* **2010**, *37*, 3957–3965. [CrossRef] [PubMed]

69. Xu, D.-B.; Chen, M.; Ma, Y.N.; Xu, Z.S.; Li, L.C.; Chen, Y.F.; Ma, Y.Z. A G-protein β subunit, AGB1, negatively regulates the ABA response and drought tolerance by down-regulating AtMPK6-related pathway in *Arabidopsis*. *PLoS ONE* **2015**, *10*, e0116385. [CrossRef] [PubMed]

70. Zhang, L.; Hu, G.; Cheng, Y.; Huang, J. Heterotrimeric G protein α, and β subunits antagonistically modulate stomatal density in *Arabidopsis thaliana*. *Dev. Biol.* **2008**, *324*, 68–75. [CrossRef] [PubMed]

71. Ma, Q.-H. Small GTP-binding Proteins and their Functions in Plants. *J. Plant Growth Regul.* **2007**, *26*, 369–388. [CrossRef]

72. Choudhary, M.; Jayanand Padaria, J.C. Transcriptional profiling in pearl millet (*Pennisetum glaucum* L.R. Br.) for identification of differentially expressed drought responsive genes. *Physiol. Mol. Biol. Plants* **2015**, *21*, 187–196. [CrossRef] [PubMed]

73. Song, L.; Li, R.; Xiang, X.; Wang, J.; Qiao, L.; Song, X.; Pei, Y.; Sui, J. Overexpression of stress-inducible small GTP-binding protein AhRab7 (AhRabG3f) in peanut (*Arachis hypogaea* L.) enhances abiotic stress tolerance. *J. Food Agric. Environ.* **2012**, *10*, 888–894.

74. Liu, Z.; Luo, C.; Li, L.; Doug, L.; Can, V.; Wei, P.; He, X. Isolation, characterization and expression analysis of the GDP dissociation inhibitor protein gene *MiRab-GDI* from *Mangifera indica* L. *Sci. Hortic.* **2015**, *185*, 14–21. [CrossRef]

75. Huang, G.T.; Ma, S.L.; Bai, L.P.; Zhang, L.; Ma, H.; Jia, P.; Liu, J.; Zhong, M.; Guo, Z.F. Signal transduction during cold, salt, and drought stresses in plants. *Mol. Biol. Rep.* **2012**, *39*, 969–987. [CrossRef] [PubMed]
76. Garg, G.; Yadav, S.; Ruchi Yadav, G. Key roles of calreticulin and calnexin proteins in plant perception under stress conditions: A Review. *Adv. Life Sci.* **2015**, *5*, 18–26.
77. Islam, M.; Begum, M.C.; Kabir, A.H.; Alam, M.F. Molecular and biochemical mechanisms associated with differential responses to drought tolerance in wheat (*Triticum aestivum* L.). *J. Plant Interact.* **2015**, *10*, 195–201. [CrossRef]
78. Jia, X.Y.; Xu, C.Y.; Jing, R.L.; Li, R.Z.; Mao, X.G.; Wang, J.P.; Chang, X.P. Molecular cloning and characterization of wheat calreticulin (*CRT*) gene involved in drought-stressed responses. *J. Exp. Bot.* **2008**, *59*, 739–751. [CrossRef] [PubMed]
79. Jiang, S.; Zhang, D.; Wang, L.; Pan, J.; Liu, Y.; Kong, X.; Zhou, Y.; Li, D. A maize calcium-dependent protein kinase gene, *ZmCPK4*, positively regulated abscisic acid signaling and enhanced drought stress tolerance in transgenic *Arabidopsis*. *Plant Physiol. Biochem.* **2013**, *71*, 112–120. [CrossRef] [PubMed]
80. Zou, J.J.; Wei, F.J.; Wang, C.; Wu, J.J.; Ratnasekera, D.; Liu, W.X.; Wu, W.H. Arabidopsis calcium-dependent protein kinase CPK10 functions in abscisic acid- and Ca^{2+}-mediated stomatal regulation in response to drought stress. *Plant Physiol.* **2010**, *154*, 1232–1243. [CrossRef] [PubMed]
81. Wei, S.; Hu, W.; Deng, X.; Zhang, Y.; Liu, X.; Zhao, X.; Luo, Q.; Jin, Z.; Li, Y.; Zhou, S.; et al. A rice calcium-dependent protein kinase OsCPK9 positively regulates drought stress tolerance and spikelet fertility. *BMC Plant Biol.* **2014**, *14*, 133. [CrossRef] [PubMed]
82. You, J.; Chan, Z. ROS regulation during abiotic stress responses in crop plants. *Front. Plant Sci.* **2015**, *6*, 1092. [CrossRef] [PubMed]
83. Liu, L.X.; Hu, X.L.; Song, J.; Zong, X.j.; Li, D.P.; Li, D.Q. Over-expression of a *Zea mays* L. protein phosphatase 2C gene (*ZmPP2C*) in *Arabidopsis thaliana* decreases tolerance to salt and drought. *J. Plant Physiol.* **2009**, *166*, 531–542. [CrossRef] [PubMed]
84. Gosti, F.; Beaudoin, N.; Serizet, C.; Webb, A.A.; Vartanian, N.; Giraudat, J. ABI1 protein phosphatase 2C is a negative regulator of abscisic acid signaling. *Plant Cell* **1999**, *11*, 1897–1910. [CrossRef] [PubMed]
85. Li, F.H.; Fu, F.L.; Sha, L.N.; He, L.; Li, W.C. Differential expression of serine/threonine protein phosphatase type-2C under drought stress in maize. *Plant Mol. Biol. Rep.* **2009**, *27*, 29–37. [CrossRef]
86. Plomion, C.; Lalanne, C.; Claverol, S.; Meddour, H.; Kohler, A.; Bogeat-Triboulot, M.-B.; Barre, A.; Le Provost, G.; Dumazet, H.; Jacob, D.; et al. Mapping the proteome of poplar and application to the discovery of drought-stress responsive proteins. *Proteomics* **2006**, *6*, 6509–6527. [CrossRef] [PubMed]
87. Guo, B.Z.; Xu, G.; Cao, Y.G.; Holbrook, C.C.; Lynch, R.E. Identification and characterization of phospholipase D and its association with drought susceptibilities in peanut (*Arachis hypogaea*). *Planta* **2006**, *223*, 512–520. [CrossRef] [PubMed]
88. El Maarouf, H.; Zuily-Fodil, Y.; Gareil, M.; d'Arcy-Lameta, A.; Pham-Thi, A.T. Enzymatic activity and gene expression under water stress of phospholipase D in two cultivars of *Vigna unguiculata* L. Walp. differing in drought tolerance. *Plant Mol. Biol.* **1999**, *39*, 1257–1265. [CrossRef] [PubMed]
89. Wang, J.; Ding, B.; Guo, Y.; Li, M.; Chen, S.; Huang, G.; Xie, X. Overexpression of a wheat phospholipase D gene, *TaPLDα*, enhances tolerance to drought and osmotic stress in *Arabidopsis thaliana*. *Planta* **2014**, *240*, 103–115. [CrossRef] [PubMed]
90. Zhang, T.; Song, Y.; Liu, Y.; Guo, X.; Zhu, C.; Wen, F. Overexpression of phospholipase *Dα* gene enhances drought and salt tolerance of *Populus tomentosa*. *Chin. Sci. Bull.* **2008**, *53*, 3656–3665. [CrossRef]
91. Hong, Y.; Zheng, S.; Wang, X. Dual functions of phospholipase Dα1 in plant response to drought. *Mol. Plant* **2008**, *1*, 262–269. [CrossRef] [PubMed]
92. Fu, H.; Subramanian, R.R.; Masters, S.C. 14-3-3 proteins: Structure, function, and regulation. *Annu. Rev. Pharmacol.* **2000**, *40*, 617–647. [CrossRef] [PubMed]
93. Malakshah, S.N.; Rezaei, M.; Heidari, M.; Salekdeh, G.H. Proteomics reveals new salt responsive proteins associated with rice plasma membrane. *Biosci. Biotechnol. Biochem.* **2014**, *71*, 2144–2154. [CrossRef] [PubMed]
94. Chandna, R.; Augustine, R.; Kanchupati, P.; Kumar, R.; Kumar, P.; Arya, G.C.; Bisht, N.C. Class-specific evolution and transcriptional differentiation of 14-3-3 family members in mesohexaploid *Brassica rapa*. *Front. Plant Sci.* **2016**, *7*, 12. [CrossRef] [PubMed]
95. Chen, F.; Li, Q.; Sun, L.; He, Z. The rice 14-3-3 gene family and its involvement in responses to biotic and abiotic stress. *DNA Res.* **2006**, *13*, 53–63. [CrossRef] [PubMed]

96. Li, R.H.; Jiang, X.T.; Jin, D.H.; Dhaubhadel, S.; Bian, S.M.; Li, X.Y. Identification of 14-3-3 family in common bean and their response to abiotic stress. *PLoS ONE* **2015**, *10*, e0143280. [CrossRef] [PubMed]

97. Sun, G.; Xie, F.; Zhang, B. Transcriptome-wide identification and stress properties of the 14-3-3 gene family in cotton (*Gossypium hirsutum* L.). *Funct. Integr. Genom.* **2011**, *11*, 627–636. [CrossRef] [PubMed]

98. Sun, X.; Luo, X.; Sun, M.; Chen, C.; Ding, X.; Wang, X.; Yang, S.; Yu, Q.; Jia, B.; Ji, W. A *Glycine Soja* 14-3-3 Protein *GsGF14o* participates in stomatal and root hair development and drought tolerance in *Arabidopsis thaliana*. *Plant Cell Physiol.* **2014**, *55*, 99–118. [CrossRef] [PubMed]

99. Skirycz, A.; de Bodt, S.; Obata, T.; de Clercq, I.; Claeys, H.; de Rycke, R.; Andriankaja, M.; van Aken, O.; van Breusegem, F.; Fernie, A.R. Developmental stage specificity and the role of mitochondrial metabolism in the response of *Arabidopsis* leaves to prolonged mild osmotic stress. *Plant Physiol.* **2010**, *152*, 226–244. [CrossRef] [PubMed]

100. Verelst, W.; Skirycz, A.; Inze, D. Abscisic acid, ethylene and gibberellic acid act at different developmental stages to instruct the adaptation of young leaves to stress. *Plant Signal. Behav.* **2010**, *5*, 473–475. [CrossRef] [PubMed]

101. Wasternack, C.; Parthier, B. Jasmonate-signalled plant gene expression. *Trends Plant Sci.* **1997**, *2*, 302–307. [CrossRef]

102. Jin, L.G.; Liu, J.-Y. Molecular cloning, expression profile and promoter analysis of a novel ethylene responsive transcription factor gene *GhERF4* from cotton (*Gossypium hirstum*). *Plant Physiol. Biochem.* **2008**, *46*, 46–53. [CrossRef] [PubMed]

103. Jin, L.; Huang, B.; Li, H.; Liu, J. Expression profiles and transactivation analysis of a novel ethylene-responsive transcription factor gene *GhERF5* from cotton. *Prog. Nat. Sci.* **2009**, *19*, 563–572. [CrossRef]

104. Zhang, W.; Yang, G.; Mu, D.; Li, H.; Zang, D.; Xu, H.; Zou, X.; Wang, Y. An ethylene-responsive factor BpERF11 negatively modulates salt and osmotic tolerance in *Betula platyphylla*. *Sci. Rep.* **2016**, *6*, 23085. [CrossRef] [PubMed]

105. Shinozaki, K.; Yamaguchi-Shinozaki, K. Gene networks involved in drought stress response and tolerance. *J. Exp. Bot.* **2007**, *58*, 221–227. [CrossRef] [PubMed]

106. Trivedi, I.; Rai, K.M.; Singh, S.K.; Kumar, V.; Singh, M.; Ranjan, A.; Lodhi, N.; Sawant, S.V. Analysis of histones and histone variants in plants. In *Chromatin Remodeling: Metthods and Protocols*; Morse, R.H., Ed.; Springer: Berlin/Heidelberg, Germany, 2012; pp. 225–236.

107. Trivedi, I.; Ranjan, A.; Sharma, Y.K.; Sawant, S. The histone H1 variant accumulates in response to water stress in the drought tolerant genotype of *Gossypium herbaceum* L. *Protein J.* **2012**, *31*, 477–486. [CrossRef] [PubMed]

108. Lange, S.; Mitchell, D.L.; Vasquez, K.M. High mobility group protein B1 enhances DNA repair and chromatin modification after DNA damage. *Proc. Natl. Acad. Sci. USA* **2008**, *105*, 10320. [CrossRef] [PubMed]

109. Zhao, J.; Paul, L.K.; Grafi, G. The maize HMGA protein is localized to the nucleolus and can be acetylated in vitro at its globular domain, and phosphorylation by CDK reduces its binding activity to AT-rich DNA. *BBA Gene Regul. Mech.* **2009**, *1789*, 751–757. [CrossRef] [PubMed]

110. Singh, U.; Deb, D.; Singh, A.; Grover, A. Glycine-rich RNA binding protein of *Oryza sativa* inhibits growth of M15 *E. coli* cells. *BMC Res. Notes* **2011**, *4*, 18.

111. Yang, D.H.; Kwak, K.J.; Kim, M.K.; Park, S.J.; Yang, K.Y.; Kang, H. Expression of Arabidopsis glycine-rich RNA-binding protein AtGRP2 or AtGRP7 improves grain yield of rice (*Oryza sativa*) under drought stress conditions. *Plant Sci.* **2014**, *214*, 106–112. [CrossRef] [PubMed]

112. Kwak, K.J.; Kim, H.S.; Jang, H.Y.; Kang, H.; Ahn, S.J. Diverse roles of glycine-rich RNA-binding protein 7 in the response of camelina (*Camelina sativa*) to abiotic stress. *Acta Physiol. Plant.* **2016**, *38*, 129. [CrossRef]

113. Zheng, J.; Wang, Y.Y.; He, Y.A.; Zhou, J.J.; Li, Y.P.; Liu, Q.Q.; Xie, X.Z. Overexpression of an S-like ribonuclease gene, *OsRNS4*, confers enhanced tolerance to high salinity and hyposensitivity to phytochrome-mediated light signals in rice. *Plant Sci.* **2014**, *214*, 99–105. [CrossRef] [PubMed]

114. Timperio, A.M.; Egidi, M.G.; Zolla, L. Proteomics applied on plant abiotic stresses: Role of heat shock proteins (HSP). *J. Proteom.* **2008**, *71*, 391–411. [CrossRef] [PubMed]

115. Swindell, W.R.; Huebner, M.; Weber, A.P. Transcriptional profiling of *Arabidopsis* heat shock proteins and transcription factors reveals extensive overlap between heat and non-heat stress response pathways. *BMC Genom.* **2007**, *8*, 125. [CrossRef] [PubMed]

116. Sabehat, A.; Lurie, S.; Weiss, D. Expression of small heat-shock proteins at low temperatures. A possible role in protecting against chilling injuries. *Plant Physiol.* **1998**, *117*, 651–658. [CrossRef] [PubMed]

117. Gao, C.Q.; Jiang, B.; Wang, Y.C.; Liu, G.F.; Yang, C.P. Overexpression of a heat shock protein (*ThHSP18.3*) from *Tamarix hispida* confers stress tolerance to yeast. *Mol. Biol. Rep.* **2012**, *39*, 4889–4897. [CrossRef] [PubMed]

118. Masand, S.; Yadav, S.K. Overexpression of *MuHSP70* gene from *Macrotyloma uniflorum* confers multiple abiotic stress tolerance in transgenic *Arabidopsis thaliana*. *Mol. Biol. Rep.* **2016**, *43*, 53–64. [CrossRef] [PubMed]

119. Guo, Q.F.; Zhang, J.; Gao, Q.; Xing, S.C.; Li, F.; Wang, W. Drought tolerance through overexpression of monoubiquitin in transgenic tobacco. *J. Plant Physiol.* **2008**, *165*, 1745–1755. [CrossRef] [PubMed]

120. Yao, X.; Xiong, W.; Ye, T.T.; Wu, Y. Overexpression of the aspartic protease *ASPG1* gene confers drought avoidance in *Arabidopsis. J. Exp. Bot.* **2012**, *63*, 2579–2593. [CrossRef] [PubMed]

121. Simova-Stoilova, L.; Vaseva, I.; Grigorova, B.; Demirevska, K.; Feller, U. Proteolytic activity and cysteine protease expression in wheat leaves under severe soil drought and recovery. *Plant Physiol. Biochem.* **2010**, *48*, 200–206. [CrossRef] [PubMed]

122. Catala, R.; Ouyang, J.; Abreu, I.A.; Hu, Y.; Seo, H.; Zhang, X.; Chua, N.H. The *Arabidopsis* E3 SUMO ligase SIZ1 regulates plant growth and drought responses. *Plant Cell* **2007**, *19*, 2952–2966. [CrossRef] [PubMed]

123. Miazek, A.; Zagdańska, B. Involvement of exopeptidases in dehydration tolerance of spring wheat seedlings. *Biol. Plant.* **2008**, *52*, 687–694. [CrossRef]

124. Seo, Y.S.; Choi, J.Y.; Kim, S.J.; Kim, E.Y.; Shin, J.S.; Kim, W.T. Constitutive expression of *CaRma1H1*, a hot pepper ER-localized RING E3 ubiquitin ligase, increases tolerance to drought and salt stresses in transgenic tomato plants. *Plant Cell Rep.* **2012**, *31*, 1659–1665. [CrossRef] [PubMed]

125. Kim, S.J.; Kim, W.T. Suppression of *Arabidopsis* RING E3 ubiquitin ligase AtATL78 increases tolerance to cold stress and decreases tolerance to drought stress. *FEBS Lett.* **2013**, *587*, 2584–2590. [CrossRef] [PubMed]

126. Bae, H.; Kim, S.K.; Cho, S.K.; Kang, B.G.; Kim, W.T. Overexpression of *OsRDCP1*, a rice RING domain-containing E3 ubiquitin ligase, increased tolerance to drought stress in rice (*Oryza sativa* L.). *Plant Sci.* **2011**, *180*, 775–782. [CrossRef] [PubMed]

127. Mckersie, B.D.; Bowley, S.R.; Harjanto, E.; Leprince, O. Water-deficit tolerance and field performance of transgenic alfalfa overexpressing superoxide dismutase. *Plant Physiol.* **1996**, *111*, 1177–1181. [CrossRef] [PubMed]

128. Prashanth, S.; Sadhasivam, V.; Parida, A. Over expression of cytosolic copper/zinc superoxide dismutase from a mangrove plant *Avicennia marina* in indica rice var Pusa Basmati-1 confers abiotic stress tolerance. *Transgenic Res.* **2007**, *17*, 281–291. [CrossRef] [PubMed]

129. Sofo, A.; Scopa, A.; Nuzzaci, M.; Vitti, A. Ascorbate peroxidase and catalase activities and their genetic regulation in plants subjected to drought and salinity stresses. *Int. J. Mol. Sci.* **2015**, *16*, 13561–13578. [CrossRef] [PubMed]

130. Jung, S. Variation in antioxidant metabolism of young and mature leaves of *Arabidopsis thaliana* subjected to drought. *Plant Sci.* **2004**, *166*, 459–466. [CrossRef]

131. Veljovic-Jovanovic, S.; Kukavica, B.; Stevanovic, B.; Navari-Izzo, F. Senescence- and drought-related changes in peroxidase and superoxide dismutase isoforms in leaves of *Ramonda serbica. J. Exp. Bot.* **2006**, *57*, 1759–1768. [CrossRef] [PubMed]

132. Wang, W.B.; Kim, Y.H.; Lee, H.S.; Kim, K.Y.; Deng, X.P.; Kwak, S.S. Analysis of antioxidant enzyme activity during germination of alfalfa under salt and drought stresses. *Plant Physiol. Biochem.* **2009**, *47*, 570–577. [CrossRef] [PubMed]

133. Gao, C.Q.; Wang, Y.C.; Liu, G.F.; Wang, C.; Jiang, J.; Yang, C.P. Cloning of ten peroxidase (POD) genes from *Tamarix hispida* and characterization of their responses to abiotic stress. *Plant Mol. Biol. Rep.* **2010**, *28*, 77–89. [CrossRef]

134. Foyer, C.H.; Noctor, G. Ascorbate and glutathione: The heart of the redox hub. *Plant Physiol.* **2011**, *155*, 2–18. [CrossRef] [PubMed]

135. Noctor, G.; Foyer, C.H. Ascorbate and glutathione: Keeping active oxygen under control. *Annu. Rev. Plant Biol.* **2003**, *49*, 249–279. [CrossRef] [PubMed]

136. Miller, G.; Suzuki, N.; Ciftciyilmaz, S.; Mittler, R. Reactive oxygen species homeostasis and signalling during drought and salinity stresses. *Plant Cell Environ.* **2010**, *33*, 453–467. [CrossRef] [PubMed]

137. Selote, D.S.; Khanna-Chopra, R. Antioxidant response of wheat roots to drought acclimation. *Protoplasma* **2010**, *245*, 153–163. [PubMed]

138. Zhang, Z.Y.; Zhang, Q.; Wu, J.; Zheng, X.L.; Zheng, S.; Sun, X.; Qiu, Q.; Lu, T. Gene knockout study reveals that cytosolic ascorbate peroxidase 2 (*OsAPX2*) plays a critical role in growth and reproduction in rice under drought, salt and cold stresses. *PLoS ONE* **2013**, *8*, e57472. [CrossRef] [PubMed]

139. Contour-Ansel, D.; Torres-Franklin, M.L.; Cruz de Carvalho, M.H.; D'Arcy-Lameta, A. Glutathione reductase in leaves of cowpea: Cloning of two cDNAs, expression and enzymatic activity under progressive drought stress, desiccation and abscisic acid treatment. *Ann. Bot.* **2006**, *98*, 1279–1287. [CrossRef] [PubMed]

140. Torres-Franklin, M.L.; Contour-Ansel, D.; Zuily-Fodil, Y.; Pham-Thi, A.-T. Molecular cloning of glutathione reductase cDNAs and analysis of GR gene expression in cowpea and common bean leaves during recovery from moderate drought stress. *J. Plant Physiol.* **2008**, *165*, 514–521. [CrossRef] [PubMed]

141. Kim, I.S.; Shin, S.Y.; Kim, Y.S.; Kim, H.Y.; Yoon, H.S. Expression of a glutathione reductase from *Brassica rapa* subsp *pekinensis* enhanced cellular redox homeostasis by modulating antioxidant proteins in *Escherichia coli*. *Mol. Cells* **2009**, *28*, 479–487. [CrossRef] [PubMed]

142. Pang, C.-H.; Wang, B.-S. Role of ascorbate peroxidase and glutathione reductase in ascorbate–glutathione cycle and stress tolerance in plants. In *Ascorbate-Glutathione Pathway and Stress Tolerance in Plants*; Anjum, A.N., Chan, M.-T., Umar, S., Eds.; Springer: Dordrecht, The Netherlands, 2010; pp. 91–113.

143. Kang, G.Z.; Li, G.Z.; Liu, G.Q.; Xu, W.; Peng, X.Q.; Wang, C.Y.; Zhu, Y.J.; Guo, T.C. Exogenous salicylic acid enhances wheat drought tolerance by influence on the expression of genes related to ascorbate-glutathione cycle. *Biol. Plant.* **2013**, *57*, 718–724. [CrossRef]

144. Osipova, S.V.; Permyakov, A.V.; Permyakova, M.D.; Pshenichnikova, T.A.; Borner, A. Leaf dehydroascorbate reductase and catalase activity is associated with soil drought tolerance in bread wheat. *Acta Physiol. Plant.* **2011**, *33*, 2169–2177. [CrossRef]

145. Eltayeb, A.E.; Yamamoto, S.; Habora, M.E.E.; Yin, L.N.; Tsujimoto, H.; Tanaka, K. Transgenic potato overexpressing *Arabidopsis* cytosolic AtDHAR1 showed higher tolerance to herbicide, drought and salt stresses. *Breed. Sci.* **2011**, *61*, 3–10. [CrossRef]

146. Kim, Y.S.; Kim, I.S.; Bae, M.J.; Choe, Y.H.; Kim, Y.H.; Park, H.M.; Kang, H.G.; Yoon, H.S. Homologous expression of cytosolic dehydroascorbate reductase increases grain yield and biomass under paddy field conditions in transgenic rice (*Oryza sativa* L. *japonica*). *Planta* **2013**, *237*, 1613–1625. [CrossRef] [PubMed]

147. Eltayeb, A.E.; Kawano, N.; Badawi, G.H.; Kaminaka, H.; Sanekata, T.; Morishima, I.; Shibahara, T.; Inanaga, S.; Tanaka, K. Enhanced tolerance to ozone and drought stresses in transgenic tobacco overexpressing dehydroascorbate reductase in cytosol. *Physiol. Plant.* **2006**, *127*, 57–65. [CrossRef]

148. Le, D.T.; Tarrago, L.; Watanabe, Y.; Kaya, A.; Lee, B.C.; Tran, U.; Nishiyama, R.; Fomenko, D.E.; Gladyshev, V.N.; Tran, L.S.P. Diversity of plant methionine sulfoxide reductases B and evolution of a form specific for free methionine sulfoxide. *PLoS ONE* **2013**, *8*, e65637. [CrossRef] [PubMed]

149. Rouhier, N.; Dos Santos, C.V.; Tarrago, L.; Rey, P. Plant methionine sulfoxide reductase A and B multigenic families. *Photosynth. Res.* **2006**, *89*, 247–262. [CrossRef] [PubMed]

150. Foyer, C.H.; Shigeoka, S. Understanding oxidative stress and antioxidant functions to enhance photosynthesis. *Plant Physiol.* **2011**, *155*, 93–100. [CrossRef] [PubMed]

151. Liu, D.; Liu, Y.; Rao, J.; Wang, G.; Li, H.; Ge, F.; Chen, C. Overexpression of the glutathione *S*-transferase gene from *Pyrus pyrifolia* fruit improves tolerance to abiotic stress in transgenic tobacco plants. *Mol. Biol.* **2013**, *47*, 515–523. [CrossRef]

152. Ji, W.; Zhu, Y.; Li, Y.; Yang, L.; Zhao, X.; Cai, H.; Bai, X. Over-expression of a glutathione *S*-transferase gene, *GsGST*, from wild soybean (*Glycine soja*) enhances drought and salt tolerance in transgenic tobacco. *Biotechnol. Lett.* **2010**, *32*, 1173–1179. [CrossRef] [PubMed]

153. Xu, J.; Xing, X.J.; Tian, Y.S.; Peng, R.H.; Xue, Y.; Zhao, W.; Yao, Q.H. Transgenic *Arabidopsis* plants expressing tomato glutathione *S*-transferase showed enhanced resistance to salt and drought stress. *PLoS ONE* **2015**, *10*, e0136960. [CrossRef] [PubMed]

154. Islam, T.; Manna, M.; Reddy, M.K. Glutathione peroxidase of *Pennisetum glaucum* (PgGPx) is a functional Cd^{2+} dependent peroxiredoxin that enhances tolerance against salinity and drought stress. *PLoS ONE* **2015**, *10*, e0143344. [CrossRef] [PubMed]

155. Renner, T.; Specht, C.D. Molecular and functional evolution of class I chitinases for plant carnivory in the caryophyllales. *Mol. Biol. Evol.* **2012**, *29*, 2971–2985. [CrossRef] [PubMed]

156. Samac, D.A.; Smigocki, A.C. Expression of oryzacystatin I and II in alfalfa increases resistance to the root-lesion nematode. *Phytopathology* **2003**, *93*, 799–804. [CrossRef] [PubMed]

157. Demirevska, K.; Simova-Stoilova, L.; Fedina, I.; Georgieva, K.; Kunert, K. Response of oryzacystatin I transformed tobacco plants to drought, heat and light stress. *J. Agron. Crop Sci.* **2010**, *196*, 90–99. [CrossRef]

158. Jain, S.; Kumar, D.; Jain, M.; Chaudhary, P.; Deswal, R.; Sarin, N.B. Ectopic overexpression of a salt stress-induced pathogenesis-related class 10 protein (PR10) gene from peanut (*Arachis hypogaea* L.) affords broad spectrum abiotic stress tolerance in transgenic tobacco. *Plant Cell Tissue Organ.* **2012**, *109*, 19–31. [CrossRef]

159. Chakrabortee, S.; Boschetti, C.; Walton, L.J.; Sarkar, S.; Rubinsztein, D.C.; Tunnacliffe, A. Hydrophilic protein associated with desiccation tolerance exhibits broad protein stabilization function. *Proc. Natl. Acad. Sci. USA* **2007**, *104*, 18073–18078. [CrossRef] [PubMed]

160. Hu, L.; Wang, Z.; Do, H.; Huang, B. Differential accumulation of dehydrins in response to water stress for hybrid and common bermudagrass genotypes differing in drought tolerance. *J. Plant Physiol.* **2010**, *167*, 103–109. [CrossRef] [PubMed]

161. Jangpromma, N.; Kitthaisong, S.; Lomthaisong, K.; Daduang, S.; Jaisil, P.; Thammasirirak, S. A proteomics analysis of drought stress-responsive proteins as biomarker for drought-tolerant sugarcane cultivars. *Am. J. Biochem. Biotechnol.* **2010**, *6*, 89–102. [CrossRef]

162. Alsheikh, M.; Heyen, B.J.; Randall, S.K. Ion binding properties of the dehydrin ERD14 are dependent upon phosphorylation. *J. Biol. Chem.* **2003**, *278*, 40882–40889. [CrossRef] [PubMed]

163. Park, S.C.; Kim, Y.H.; Jeong, J.C.; Kim, C.Y.; Lee, H.S.; Bang, J.W.; Kwak, S.-S. Sweetpotato late embryogenesis abundant 14 (*IbLEA14*) gene influences lignification and increases osmotic- and salt stress-tolerance of transgenic calli. *Planta* **2011**, *233*, 621–634. [CrossRef] [PubMed]

164. Ashraf, M.; Foolad, M.R. Roles of glycine betaine and proline in improving plant abiotic stress resistance. *Environ. Exp. Bot.* **2007**, *59*, 206–216. [CrossRef]

165. Tang, W.; Sun, J.; Liu, J.; Liu, F.; Yan, J.; Gou, X.; Lu, B.-R.; Liu, Y. RNAi-directed downregulation of betaine aldehyde dehydrogenase 1 (*OsBADH1*) results in decreased stress tolerance and increased oxidative markers without affecting glycine betaine biosynthesis in rice (*Oryza sativa*). *Plant Mol. Biol.* **2014**, *86*, 443–454. [CrossRef] [PubMed]

166. Skirycz, A.; Inze, D. More from less: Plant growth under limited water. *Curr. Opin. Biotechnol.* **2010**, *21*, 197–203. [CrossRef] [PubMed]

167. Tardieu, F.; Granier, C.; Muller, B. Water deficit and growth. Co-ordinating processes without an orchestrator? *Curr. Opin. Plant Biol.* **2011**, *14*, 283–289. [CrossRef] [PubMed]

168. Sniegowska-Swierk, K.; Dubas, E.; Rapacz, M. Drought-induced changes in the actin cytoskeleton of barley (*Hordeum vulgare* L.) leaves. *Acta Physiol. Plant.* **2015**, *37*. [CrossRef]

169. Bommer, U.; Thiele, B. The translationally controlled tumour protein (TCTP). *Int. J. Biochem. Cell B* **2004**, *36*, 379–385. [CrossRef]

170. Kim, Y.M.; Han, Y.J.; Hwang, O.J.; Lee, S.-S.; Shin, A.Y.; Kim, S.Y.; Kim, J.I. Overexpression of *Arabidopsis* translationally controlled tumor protein gene *AtTCTP* enhances drought tolerance with rapid ABA-induced stomatal closure. *Mol. Cells* **2012**, *33*, 617–626. [CrossRef] [PubMed]

171. Berkowitz, O.; Jost, R.; Pollmann, S.; Masle, J. Characterization of TCTP, the translationally controlled tumor protein, from *Arabidopsis thaliana*. *Plant Cell* **2008**, *20*, 3430–3447. [CrossRef] [PubMed]

172. Baxter, H.L.; Stewart, C.N. Effects of altered lignin biosynthesis on phenylpropanoid metabolism and plant stress. *Biofuels* **2013**, *4*, 635–650. [CrossRef]

173. Lee, B.R.; Kim, K.Y.; Jung, W.J.; Avice, J.C.; Ourry, A.; Kim, T.H. Peroxidases and lignification in relation to the intensity of water-deficit stress in white clover (*Trifolium repens* L.). *J. Exp. Bot.* **2007**, *58*, 1271–1279. [CrossRef] [PubMed]

174. Amor, Y.; Haigler, C.H.; Johnson, S.; Wainscott, M.; Delmer, D.P. A membrane-associated form of sucrose synthase and its potential role in synthesis of cellulose and callose in plants. *Proc. Natl. Acad. Sci. USA* **1995**, *92*, 9353–9357. [CrossRef] [PubMed]

175. Dong, J.L.; Jiang, Y.Y.; Chen, R.J.; Xu, Z.J.; Gao, X.L. Isolation of a novel xyloglucan endotransglucosylase (*OsXET9*) gene from rice and analysis of the response of this gene to abiotic stresses. *Afr. J. Biotechnol.* **2011**, *10*, 17424–17434.

176. Nishiyama, R.; Le, D.T.; Watanabe, Y.; Matsui, A.; Tanaka, M.; Seki, M.; Yamaguchi-Shinozaki, K.; Shinozaki, K.; Tran, L.S.P. Transcriptome analyses of a salt-tolerant cytokinin-deficient mutant reveal differential regulation of salt stress response by cytokinin deficiency. *PLoS ONE* **2012**, *7*, e32124. [CrossRef] [PubMed]

177. Tenhaken, R. Cell wall remodeling under abiotic stress. *Front. Plant Sci.* **2015**, *5*, 771. [CrossRef] [PubMed]

178. Spagnoletta, A.; De Santis, A.; Tampieri, E.; Baraldi, E.; Bachi, A.; Genchi, G. Identification and kinetic characterization of *Ht*DTC, the mitochondrial dicarboxylate-tricarboxylate carrier of Jerusalem artichoke tubers. *J. Bioenerg. Biomembr.* **2006**, *38*, 57–65. [CrossRef] [PubMed]

179. Wang, J.; Zhang, L.D.; Zuo, K.J.; Oian, H.M.; Cao, Y.F.; Tang, K.X. Cloning and expressional studies of the voltage-dependent anion channel gene from *Brassica rapa* L. *J. Integr. Plant Biol.* **2006**, *48*, 197–203. [CrossRef]

180. Wen, G.; Yang, Y.; Chai, L.; Liu, Z.; Wang, J. Overexpression of the voltage-dependent anion channel 2 (*VDAC2*) gene induces drought resistance in *Arabidopsis thaliana*. *Plant Omics* **2014**, *7*, 171–177.

181. Lefebvre, B.; Timmers, T.; Mbengue, M.; Moreau, S.; Herve, C.; Toth, K.; Bittencourt-Silvestre, J.; Klaus, D.; Deslandes, L.; Godiard, L.; et al. A remorin protein interacts with symbiotic receptors and regulates bacterial infection. *Proc. Natl. Acad. Sci. USA* **2010**, *107*, 2343–2348. [CrossRef] [PubMed]

182. Checker, V.G.; Khurana, P. Molecular and functional characterization of mulberry EST encoding remorin (*MiREM*) involved in abiotic stress. *Plant Cell Rep.* **2013**, *32*, 1729–1741. [CrossRef] [PubMed]

183. Zhou, S.; Hu, W.; Deng, X.; Ma, Z.; Chen, L.; Huang, C.; Wang, C.; Wang, J.; He, Y.; Yang, G.; et al. Overexpression of the wheat aquaporin gene, *TaAQP7*, enhances drought tolerance in transgenic tobacco. *PLoS ONE* **2012**, *7*, e52439. [CrossRef] [PubMed]

184. Tapia, G.; Morales-Quintana, L.; Parra, C.; Berbel, A.; Alcorta, M. Study of nsLTPs in *Lotus japonicus* genome reveal a specific epidermal cell member (LjLTP10) regulated by drought stress in aerial organs with a putative role in cutin formation. *Plant Mol. Biol.* **2013**, *82*, 485–501. [CrossRef] [PubMed]

185. Guo, L.; Yang, H.; Zhang, X.; Yang, S. Lipid transfer protein 3 as a target of MYB96 mediates freezing and drought stress in *Arabidopsis*. *J. Exp. Bot.* **2013**, *64*, 1755–1767. [CrossRef] [PubMed]

186. Guo, C.; Ge, X.; Ma, H. The rice *OsDIL* gene plays a role in drought tolerance at vegetative and reproductive stages. *Plant Mol. Biol.* **2013**, *82*, 239–253. [CrossRef] [PubMed]

187. Kabala, K.; Janicka-Russak, M.; Reda, M.; Migocka, M. Transcriptional regulation of the V-ATPase subunit c and V-PPase isoforms in *Cucumis sativus* under heavy metal stress. *Physiol. Plant.* **2014**, *150*, 32–45. [CrossRef] [PubMed]

188. Fukuda, A.; Tanaka, Y. Effects of ABA, auxin, and gibberellin on the expression of genes for vacuolar H^+-inorganic pyrophosphatase, H^+-ATPase subunit A, and Na^+/H^+ antiporter in barley. *Plant Physiol. Biochem.* **2006**, *44*, 351–358. [CrossRef] [PubMed]

189. Dong, Q.L.; Wang, C.R.; Liu, D.D.; Hu, D.G.; Fang, M.J.; You, C.X.; Yao, Y.X.; Hao, Y.J. MdVHA-A encodes an apple subunit A of vacuolar H^+-ATPase and enhances drought tolerance in transgenic tobacco seedlings. *J. Plant Physiol.* **2013**, *170*, 601–609. [CrossRef] [PubMed]

190. Patro, L.; Mohapatra, P.K.; Biswal, U.C.; Biswal, B. Dehydration induced loss of photosynthesis in *Arabidopsis* leaves during senescence is accompanied by the reversible enhancement in the activity of cell wall beta-glucosidase. *J. Photochem. Photobiol. B* **2014**, *137*, 49–54. [CrossRef] [PubMed]

191. Ghotbi-Ravandi, A.A.; Shahbazi, M.; Shariati, M.; Mulo, P. Effects of mild and severe drought stress on photosynthetic efficiency in tolerant and susceptible barley (*Hordeum vulgare* L.) genotypes. *J. Agron. Crop Sci.* **2014**, *200*, 403–415. [CrossRef]

192. Xu, Y.; Liu, R.; Yan, L.; Liu, Z.; Jiang, S.; Shen, Y.; Wang, X.; Zhang, D. Light-harvesting chlorophyll a/b-binding proteins are required for stomatal response to abscisic acid in *Arabidopsis*. *J. Exp. Bot.* **2012**, *63*, 1095–1106. [CrossRef] [PubMed]

193. Majeau, N.; Arnoldo, M.; Coleman, J.R. Modification of carbonic anhydrase activity by antisense and over-expression constructs in transgenic tobacco. *Plant Mol. Biol.* **1994**, *25*, 377–385. [CrossRef] [PubMed]

194. Abebe, T.; Melmaiee, K.; Berg, V.; Wise, R.P. Drought response in the spikes of barley: Gene expression in the lemma, palea, awn, and seed. *Funct. Integr. Genom.* **2010**, *10*, 191–205. [CrossRef] [PubMed]

195. Voss, I.; Sunil, B.; Scheibe, R.; Raghavendra, A.S. Emerging concept for the role of photorespiration as an important part of abiotic stress response. *Plant Biol.* **2013**, *15*, 713–722. [CrossRef] [PubMed]

196. Degenkolbe, T.; Do, P.T.; Kopka, J.; Zuther, E.; Hincha, D.K.; Koehl, K.I. Identification of drought tolerance markers in a diverse population of rice cultivars by expression and metabolite profiling. *PLoS ONE* **2013**, *8*, e63637. [CrossRef] [PubMed]

197. Kappachery, S.; Baniekal-Hiremath, G.; Yu, J.W.; Park, S.W. Effect of over-and under-expression of glyceraldehyde 3-phosphate dehydrogenase on tolerance of plants to water-deficit stress. *Plant Cell Tissue Organ.* **2015**, *121*, 97–107. [CrossRef]

198. Xu, Z.Z.; Zhou, G.S. Nitrogen metabolism and photosynthesis in *Leymus chinensis* in response to long-term soil drought. *J. Plant Growth Regul.* **2006**, *25*, 252–266. [CrossRef]

199. Fresneau, C.; Ghashghaie, J.; Cornic, G. Drought effect on nitrate reductase and sucrose-phosphate synthase activities in wheat (*Triticum durum* L.): Role of leaf internal CO_2. *J. Exp. Bot.* **2007**, *58*, 2983–2992. [PubMed]

200. Sánchez-Rodríguez, E.; Rubio-Wilhelmi, M.d.M.; Ríos, J.J.; Blasco, B.; Rosales, M.Á.; Melgarejo, R.; Romero, L.; Ruiz, J.M. Ammonia production and assimilation: Its importance as a tolerance mechanism during moderate water deficit in tomato plants. *J. Plant Physiol.* **2011**, *168*, 816–823. [CrossRef] [PubMed]

201. Goel, P.; Singh, A.K. Abiotic stresses downregulate key genes involved in nitrogen uptake and assimilation in *Brassica juncea* L. *PLoS ONE* **2015**, *10*, e0143645. [CrossRef] [PubMed]

202. Jedmowski, C.; Ashoub, A.; Beckhaus, T.; Berberich, T.; Karas, M.; Bruggemann, W. Comparative analysis of *Sorghum bicolor* proteome in response to drought stress and following recovery. *Int. J. Proteom.* **2014**, *2014*, 395905. [CrossRef] [PubMed]

203. Tabor, C.W.; Tabor, H. Methionine adenosyltransferase (*S*-Adenosylmethionine synthetase) and *S*-adenosylmethionine decarboxylase. In *Advances in Enzymology and Related Areas of Molecular Biology*; Meister, A., Ed.; John Wiley & Sons, Inc.: Hoboken, NJ, USA, 2006; pp. 251–282.

204. Boerjan, W.; Bauw, G.; Van Montagu, M.; Inze, D. Distinct phenotypes generated by overexpression and suppression of *S*-adenosyl-L-methionine synthetase reveal developmental patterns of gene silencing in tobacco. *Plant Cell* **1994**, *6*, 1401–1414. [CrossRef] [PubMed]

205. Narita, Y.; Taguchi, H.; Nakamura, T.; Ueda, A.; Shi, W.; Takabe, T. Characterization of the salt-inducible methionine synthase from barley leaves. *Plant Sci.* **2004**, *167*, 1009–1016. [CrossRef]

206. Kim, S.H.; Kim, S.H.; Palaniyandi, S.A.; Yang, S.H.; Suh, J.-W. Expression of potato *S*-adenosyl-L-methionine synthase (*SbSAMS*) gene altered developmental characteristics and stress responses in transgenic *Arabidopsis* plants. *Plant Physiol. Biochem.* **2015**, *87*, 84–91. [CrossRef] [PubMed]

207. Gasulla, F.; vom Dorp, K.; Dombrink, I.; Zaehringer, U.; Gisch, N.; Doermann, P.; Bartels, D. The role of lipid metabolism in the acquisition of desiccation tolerance in *Craterostigma plantagineum*: A comparative approach. *Plant J.* **2013**, *75*, 726–741. [CrossRef] [PubMed]

208. Benhassaine-Kesri, G.; Aid, F.; Demandre, C.; Kader, J.C.; Mazliak, P. Drought stress affects chloroplast lipid metabolism in rape (*Brassica napus*) leaves. *Physiol. Plant.* **2002**, *115*, 221–227. [CrossRef] [PubMed]

209. Zhong, D.; Du, H.; Wang, Z.; Huang, B. Genotypic variation in fatty acid composition and unsaturation levels in bermudagrass associated with leaf dehydration tolerance. *J. Am. Soc. Hortic. Sci.* **2011**, *136*, 35–40.

210. Gigon, A.; Matos, A.R.; Laffray, D.; Zuily-Fodil, Y.; Pham-Thi, A.T. Effect of drought stress on lipid metabolism in the leaves of *Arabidopsis thaliana* (ecotype Columbia). *Ann. Bot.* **2004**, *94*, 345–351. [CrossRef] [PubMed]

211. Castellarin, S.D.; Matthews, M.A.; di Gaspero, G.; Gambetta, G.A. Water deficits accelerate ripening and induce changes in gene expression regulating flavonoid biosynthesis in grape berries. *Planta* **2007**, *227*, 101–112. [CrossRef] [PubMed]

212. Liu, M.L.; Li, X.R.; Liu, Y.B.; Cao, B. Regulation of flavanone 3-hydroxylase gene involved in the flavonoid biosynthesis pathway in response to UV-B radiation and drought stress in the desert plant, *Reaumuria soongorica*. *Plant Physiol. Biochem.* **2013**, *73*, 161–167. [CrossRef] [PubMed]

213. Meng, C.; Zhang, S.; Deng, Y.S.; Wang, G.D.; Kong, F.Y. Overexpression of a tomato flavanone 3-hydroxylase-like protein gene improves chilling tolerance in tobacco. *Plant Physiol. Biochem.* **2015**, *96*, 388–400. [CrossRef] [PubMed]

214. Rani, A.; Singh, K.; Ahuja, P.S.; Kumar, S. Molecular regulation of catechins biosynthesis in tea (*Camellia sinensis* (L.) O. Kuntze). *Gene* **2012**, *495*, 205–210. [CrossRef] [PubMed]

International Journal of
Molecular Sciences

Article

Proteomic Analysis of *Hylocereus polyrhizus* Reveals Metabolic Pathway Changes

Qingzhu Hua [1], Qianjun Zhou [2], Susheng Gan [3], Jingyu Wu [1], Canbin Chen [1], Jiaqiang Li [4], Yaoxiong Ye [4], Jietang Zhao [1], Guibing Hu [1,*] and Yonghua Qin [1,*]

[1] State Key Laboratory for Conservation and Utilization of Subtropical Agro-Bioresources/Key Laboratory of Biology and Genetic Improvement of Horticultural Crops-South China, Ministry of Agriculture, College of Horticulture, South China Agricultural University, Guangzhou 510642, China; huaqingzhu1010@gmail.com (Q.H.); wujingyuscau@163.com (J.W.); nnchencanbin@163.com (C.C.); zhaojietang@gmail.com (J.Z.)

[2] General Station of the Administration of Seeds Guangdong Province, Guangzhou 510500, China; zzzzglk@163.com

[3] Plant Biology Section, School of Integrative Plant Science, Cornell University, Ithaca, NY 14853, USA; sg288@cornell.edu

[4] Dongguan Institute of Forest Science, Dongguan 523106, China; jqli618@163.com (J.L.); yxyeforest@163.com (Y.Y.)

* Correspondence: guibing@scau.edu.cn (G.H.); qinyh@scau.edu.cn (Y.Q.); Tel./Fax: +86-20-8528-6905 (G.H.); Tel.: +86-20-8528-0505 (Y.Q.); Fax: +86-20-8528-2107 (Y.Q.)

Academic Editor: Setsuko Komatsu
Received: 3 July 2016; Accepted: 13 September 2016; Published: 28 September 2016

Abstract: Red dragon fruit or red pitaya (*Hylocereus polyrhizus*) is the only edible fruit that contains betalains. The color of betalains ranges from red and violet to yellow in plants. Betalains may also serve as an important component of health-promoting and disease-preventing functional food. Currently, the biosynthetic and regulatory pathways for betalain production remain to be fully deciphered. In this study, isobaric tags for relative and absolute quantitation (iTRAQ)-based proteomic analyses were used to reveal the molecular mechanism of betalain biosynthesis in *H. polyrhizus* fruits at white and red pulp stages, respectively. A total of 1946 proteins were identified as the differentially expressed between the two samples, and 936 of them were significantly highly expressed at the red pulp stage of *H. polyrhizus*. RNA-seq and iTRAQ analyses showed that some transcripts and proteins were positively correlated; they belonged to "phenylpropanoid biosynthesis", "tyrosine metabolism", "flavonoid biosynthesis", "ascorbate and aldarate metabolism", "betalains biosynthesis" and "anthocyanin biosynthesis". In betalains biosynthesis pathway, several proteins/enzymes such as polyphenol oxidase, CYP76AD3 and 4,5-dihydroxy-phenylalanine (DOPA) dioxygenase extradiol-like protein were identified. The present study provides a new insight into the molecular mechanism of the betalain biosynthesis at the posttranscriptional level.

Keywords: *Hylocereus polyrhizus*; isobaric tags for relative and absolute quantitation (iTRAQ); transcriptome; betalain biosynthesis

1. Introduction

Betalains are red and yellow pigments in plants of Caryophyllales only. It is interesting that betalains and anthocyanins are naturally mutually exclusive in an individual plant. Betalain pigments have potential benefits in promoting health and preventing diseases of human being by serving as potent antioxidant and possessing anti-inflammatory and chemo-preventive activities in vitro and in vivo [1–6]. Betalains also contribute to the early-phase insulin response [7] and play a role in dye-sensitized solar cells (DSSCs) [8–10]. The biosynthesis of betalains has become one of the hot research topics due to its high nutritional properties and bioactivities.

Betalain biosynthesis is influenced by various factors such as light, temperature, nutrition, plant growth regulators and enzymatic activities. Light is important for betalain biosynthesis in *Amaranthus*, *Caudatus*, *Portulaca* and *Suaeda* [11–16]. In the presence of light, cytokinin, ethylene and methyl jasmonate (MeJA) can enhance betacyanin accumulation in species of *Amaranthus*, *Beta vulgaris*, *Portulaca*, *Suaeda* and *Mesembryanthemum* [14,17–22]. Betalains can also be accumulated in adverse circumstances such as salinity and drought stress [23–25], which are attributed to defense against biotic and abiotic stresses [26].

The key enzymes involved in the betalain biosynthetic pathways have been known. There are five pathways derived from tyrosine and four pathways derived from tyramine [27,28]. Except for enzymatic reactions, the conjugation reactions of pigment formation are assumed to occur spontaneously [28–30]. Enzymes involved in betalain biosynthesis in higher plants are usually divided into three classes, namely tyrosinase (TYR), 4,5-dihydroxy-phenylalanine (DOPA)-dioxygenase (DOD), and glucosyltransferases (GTs). TYR plays a key role in hydroxylation and oxidation to form L-DOPA, *cyclo*-DOPA, tyr-betaxanthin and dopaxanthin that are the pivotal precursors of betalain biosynthesis [12,29,31–33]. TYR sequences have been obtained from *Phytolacca americana* [34], *Spinacia oleracea* [35] and *Suaeda salsa* [36]. The first PPO promoter was identified in *B. vulgaris* subsp. *cicla* [37]. DOD is a key enzyme catalyzing the cleavage of L-DOPA needed for the formation of betalamic acid. Compared to other enzymes, DOD has been fully studied in betalain-producing plant species such as *Amaranthus hypochondriacus* [38], *Portulaca grandiflora* [39,40], *Suaeda salsa* [41,42], *B. vulgaris* [43], *Opuntia ficus-indica* [44], *Parakeelya mirabilis* [45], *Mirabilis jalapa* and *Bougainvillea glabra* [46]. CYP76AD, a novel cytochrome P450 enzyme, catalyzes oxidation of L-DOPA for the formation of cyclo-DOPA in *B. vulgaris* [47], *A. hypochondriacus* [38] and *M. jalapa* [48]. In addition, *CYP76AD* and *DOD* have also been considered as key genes involved in betalain biosynthesis based on phylogenetic analysis [49,50]. GTs are the key enzyme in betalain biosynthetic pathway to make the pigments stable and diversified. GT sequences have been obtained from *B. vulgaris* [51], *O. ficus-indica* [44], *P. americana* [52] and *A. hypochondriacus* [38], and there have been no further major advances in GTs since they were identified as B5GT (betanidin-5-O-GT occurred on betanidin 5-O location), B6GT (betanidin-6-O-GT occurred on betanidin 6-O location) and CDOPA5GT (cyclo-DOPA 5-O-GT occurred on cyclo-DOPA) [53–55]. Recently, the first regulatory gene (a R2R3 *MYB*) associated with regulation of the betalain biosynthetic pathway was identified in this step [56,57].

Dragon fruit or pitaya is one of the tropical fruits belonging to *Hylocereus* in the Cactaceae family. Pitaya is cultivated in a wide ecological range due to its tolerance to such environmental cues as drought, heat and poor soil [58,59]. Pitaya is characterized by its excellent nutritional, commercial and medical value [60,61]. It has become one of the newly cultivated fruits in Thailand, Philippines, Vietnam, Malaysia and China. Red pulp pitaya (*H. polyrhizus*) is the only fruit that contains betalains [62–64]. Great progress has been made in betalain purification, identification and biosynthesis-related compound analyses [65–67]. However, the key enzymes or genes involved in the betalain biosynthesis pathways in pitaya remain elusive. Recently, we obtained nine putative betalain biosynthesis-related genes by analyzing the transcriptomic data of *H. polyrhizus* [68]. Here we used isobaric tags for relative and absolute quantitation (iTRAQ) to identify key proteins/enzymes that are differentially expressed in pitaya fruit at white vs. red pulp stages in *H. polyrhizus*, with a focus on those proteins potentially involved in the betalain biosynthesis in pulp coloration of this fruit.

2. Results and Discussion

2.1. Overview of the Proteomics Analysis of the Pitaya Fruit Development

iTRAQ is a recently developed technique in quantitative proteomics that very accurately measures large-fold changes in protein expression within broad, dynamic ranges of protein abundance [69].

Here we used this technique to perform comparative proteomic analyses of the key enzymes involved in betalain biosynthetic pathways in pitaya. For the proteomic analysis, two proteomic libraries were constructed using the pitaya (*H. polyrhizus*) fruits at two developmental stages (white and red pulps). Total proteins and changes in the protein profile upon coloring process were explored using the iTRAQ technique. A total of 89,583 peptides were identified that belong to 33,548 peptide species. In total, 6725 proteins were identified and quantified under the condition of 1% false-discovery rate (FDR) and only those unique peptides were used for quantitative comparison. A threshold of $\geq$1.5 for fold changes (FC) was set to filter the comparative data sets, resulting in the identification of 1946 proteins as the differentially expressed between the two samples. Among the differentially expressed proteins, 936 were significantly up-expressed in the fruits at the red pulp stage compared with those at the white stage. iTRAQ has the advantages of high coverage, accuracy and sensitivity, and our proteomic analyses of *H. polyrhizus* fruits using the iTRAQ technique could provide useful information for understanding the pitaya fruit development in general and the betalain biosynthesis pathway in particular as discussed below (Tables 1 and 2).

Table 1. Kyoto encyclopedia of genes and genomes (KEGG) pathway enrichment of the genes/proteins with a positive correlation (*p*-value < 0.05, FC (pro) > 1.5, FC (RNA) > 2) between the levels of transcripts and proteins.

KEGG Pathway	No. All [a]	No. Up [b]	No. Down [c]
Starch and sucrose metabolism	6	3	3
Phenylpropanoid biosynthesis	5	5	0
Lipid biosynthesis proteins	5	3	2
Fatty acid metabolism	5	5	0
Cysteine methionine metabolism	5	4	1
Amino sugar and nucleotide sugar metabolism	5	1	4
Ascorbate and aldarate metabolism	4	4	0
Tyrosine metabolism	4	2	2
Cyanoamino metabolism	3	0	3
Chloroalkane and chloroalkene degradation	3	3	0
Glycerolipid metabolism	3	3	0
Valine leucine isoleucine biosynthesis	3	0	3
Isoquinoline alkaloid biosynthesis	2	0	2
Flavonoid biosynthesis	2	1	1
Naphthalene degradation	2	2	0

[a], the total number of differentially expressed on genes and proteins; [b], the number of significantly up-regulated expression in the pitaya fruit at the red pulp stage; [c], the number of significantly down-regulated expression in the pitaya fruit at the red pulp stage. No., number; FC, fold changes.

Table 2. Proteins enriched in several KEGG pathways with a positive correlation (*p*-value < 0.05, FC (pro) > 1.5, FC (RNA) > 2) between the levels of transcripts and proteins.

Accession No.	Annotation	Ratio (32rd day vs. 21st day)
	Starch and sucrose metabolism	
comp28763_c0_seq2_8	K00688 starch phosphorylase (EC:2.4.1.1)	2.218
comp31213_c0_seq1_12	K05350 β-glucosidase (EC:3.2.1.21)	0.444
comp34816_c0_seq1_6	K00695 sucrose synthase (EC:2.4.1.13)	2.163
comp34279_c0_seq1_6	K05350 β-glucosidase (EC:3.2.1.21)	0.435
comp35284_c0_seq1_11	K01051 pectinesterase (EC:3.1.1.11)	2.531
comp37752_c0_seq1_2	K00695 sucrose synthase (EC:2.4.1.13)	0.261

Table 2. *Cont.*

Accession No.	Annotation	Ratio (32rd day vs. 21st day)
Phenylpropanoid biosynthesis		
comp21956_c1_seq1_1	K12355 coniferyl-aldehyde dehydrogenase (EC:1.2.1.68)	0.437
comp21956_c2_seq1_2	K12355 coniferyl-aldehyde dehydrogenase (EC:1.2.1.68)	0.465
comp31213_c0_seq1_12	K05350 β-glucosidase (EC:3.2.1.21)	0.444
comp32862_c0_seq1_3	K09754 p-coumarate 3-hydroxylase (EC:1.14.13.-)	0.651
comp34279_c0_seq1_6	K05350 β-glucosidase (EC:3.2.1.21)	0.435
Ascorbate and aldarate metabolism		
comp21409_c0_seq1_2	K00434 L-ascorbate peroxidase (EC:1.11.1.11)	1.961
comp27881_c0_seq1_8	K00128 aldehyde dehydrogenase (NAD+) (EC:1.2.1.3)	2.211
comp28683_c0_seq1_3	K00434 L-ascorbate peroxidase (EC:1.11.1.11)	1.618
comp31481_c0_seq1_3	K08232 monodehydroascorbate reductase (NADH) (EC:1.6.5.4)	1.696
Tyrosine metabolism		
comp23153_c0_seq1_3	K00001 alcohol dehydrogenase (EC:1.1.1.1)	3.612
comp25940_c0_seq3_7	K00813 aspartate aminotransferase (EC:2.6.1.1)	0.589
comp26435_c0_seq1_2	K01592 tyrosine decarboxylase (EC:4.1.1.25)	0.637
comp27557_c0_seq1_3	K00001 alcohol dehydrogenase (EC:1.1.1.1)	7.142
Flavonoid biosynthesis		
comp29018_c0_seq1_2	K01859 chalcone isomerase (EC:5.5.1.6)	2.205
comp32862_c0_seq1_3	K09754 p-coumarate 3-hydroxylase (EC:1.14.13.-)	0.651
Anthocyanin biosynthesis		
comp29018_c0_seq1_2	K01859 chalcone isomerase (EC:5.5.1.6)	2.205
comp24963_c0_seq1_6	K00660 chalcone synthase (EC:2.3.1.74)	1.556
Betalains biosynthesis		
comp16058_c0_seq1_4	CYP76AD3 (*Mirabilis jalapa*)	2.373
comp37692_c0_seq1_7	4,5-DOPA dioxygenase extradiol	3.035
comp37375_c0_seq2_4	Polyphenol oxidase	0.279
comp37375_c0_seq1_4	polyphenol oxidase	0.262
comp25579_c0_seq2_3	multicopper oxidase, putative	0.419
comp26435_c0_seq1_2	Aromatic-L-amino-acid decarboxylase	0.637

2.2. Functional Classification of Differentially Expressed Proteins during Pitaya Fruit Coloring

The proteomes representing the protein profiles of the pitaya fruits at the white or red pulp stages were comparatively analyzed based on the assigned functions of the proteins in the public protein databases. Five hundred and seven (507) differentially accumulated proteins (p-value < 0.05, fold change (FC) > 1.5) in the red pulp fruits were annotated and classified into "biological process (BP)", "molecular function (MF)", and "cellular component (CC)" categories as well as their sub-categories. Based on the numbers of unique proteins identified in each of the functional categories, the two largest categories for each functional group were as follows: "translation" and "carbohydrate metabolic processes" for BP; "structural constituent of ribosome" and "DNA binding" for MF; and "ribosome" and "intracellular" for CC (Figure 1). Some of these proteins could be mapped to the following pathways: "phenylpropanoid biosynthesis", "phenylalanine metabolism", "tyrosine metabolism" and "flavonoid biosynthesis" (Figure 1). These pathways might have impacts on the formation of betalain. Two hundred eighteen (218) proteins were assigned to 23 kyoto encyclopedia of genes and genomes (KEGG) pathways. The most frequently detected or abundant proteins, representing 37.6% of all the protein identified, belonged to the pathway of "ribosome" (Figure 2). Further work is needed to establish the metabolic profiles in red pitaya at a series of developmental stages.

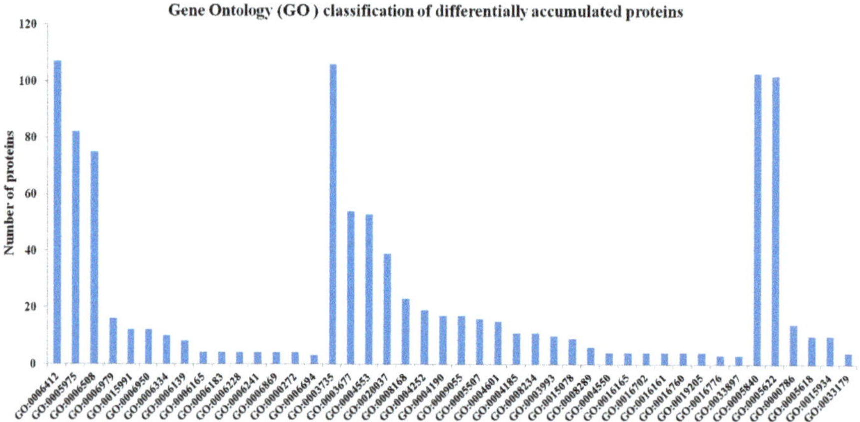

Figure 1. Gene Ontology (GO) classification of differentially accumulated proteins during pitaya fruit ripening. GO:0006412//translation, GO:0005975//carbohydrate metabolic process, GO:0006508// proteolysis, GO:0006979//response to oxidative stress, GO:0015991//ATP hydrolysis coupled proton transport, GO:0006950//response to stress, GO:0006334//nucleosome assembly, GO:0006139// nucleobase-containing compound metabolic process, GO:0006165//nucleoside diphosphate phosphorylation, GO:0006183//GTP, biosynthetic process, GO:0006228//UTP biosynthetic process, GO:0006241//CTP biosynthetic process, GO:0006869//lipid transport, GO:0000272//polysaccharide catabolic process, GO:0006694//steroid biosynthetic process, GO:0003735//structural constituent of ribosome, GO:0003677//DNA binding, GO:0004553//hydrolase activity, hydrolyzing *O*-glycosyl compounds, GO:0020037//heme binding, GO:0008168//methyltransferase activity, GO:0004252// serine-type endopeptidase activity, GO:0004190//aspartic-type endopeptidase activity, GO:0009055// electron carrier activity, GO:0005507//copper ion binding, GO:0004601//peroxidase activity, GO:0004185//serine-type carboxypeptidase activity, GO:0008234//cysteine-type peptidase activity, GO:0003993//acid phosphatase activity, GO:0015078//hydrogen ion transmembrane transporter activity, GO:0008289//lipid binding, GO:0004550//nucleoside diphosphate kinase activity, GO:0016165//linoleate 13S-lipoxygenase activity, GO:0016702//oxidoreductase activity, acting on single donors with incorporation of molecular oxygen, incorporation of two atoms of oxygen, GO:0016161//beta-amylase activity, GO:0016760//cellulose synthase (UDP-forming) activity, GO:0019205//nucleobase-containing compound kinase activity, GO:0016776//phosphotransferase activity, phosphate group as acceptor, GO:0033897//ribonuclease T2 activity, GO:0005840//ribosome, GO:0005622//intracellular, GO:0000786//nucleosome, GO:0005618//cell wall, GO:0015934//large ribosomal subunit, GO:0033179//proton-transporting V-type ATPase, V0 domain.

2.3. Integrated Analyses of Transcriptomic and Proteomic Datasets on the Pitaya Ripening

In our previous study, the high efficient RNA sequencing (RNA-Seq) technology was used to identify key genes related to betalain biosynthesis during pulp coloration of *H. polyrhizus*. A total of about 12 Gb raw RNA-Seq data were generated and de novo assembled into 122,677 transcripts, of which 122,668 were annotated [68]. Differentially expressed transcripts and proteins between the two stages were identified by comparing the RNA-seq and iTRAQ datasets. Although most of the differentially expressed were consistent at both the transcript and protein levels (see below), discrepancy was detected between the transcript and protein data, and such genes/proteins fell into the following two groups. The first group is those differentially expressed at the transcript level but not at the protein level (*p*-value < 0.05, FC (pro) < 1.3, FC (RNA) > 2). They functionally belonged to "oxidation-reduction process" (BP); "oxidoreductase activity, acting on paired donors, with incorporation or reduction of molecular oxygen" (MF); and "protein complex" (CC) (Table S1). These differentially expressed were further analyzed using the KEGG database, and fell into

12 pathways, with five differentially-accumulated proteins associated with the "arginine metabolism" (Table S2). The second group is those differentially expressed at the protein level but not at the transcript level (p-value < 0.05, FC (pro) > 1.5, FC (RNA) < 2). However, no protein matching with corresponding transcript were detected. Analyzing the Gene Ontology (GO) categories and KEGG pathways of the differentially expressed with a negative correlation between the transcript and protein levels allowed us to map 22 differentially expressed genes/proteins (Tables S3 and S4). The fact that the abundance of transcripts differed from that of respective proteins strongly suggested that there were posttranscriptional regulation involved in the metabolisms and other cellular processes associated with the pitaya ripening. In fact, similar findings have previously been found in many other biological processes such as in human [70], yeast [71] and plants [72]. Such findings became possible only when the transcriptomic and proteomic datasets were integrated.

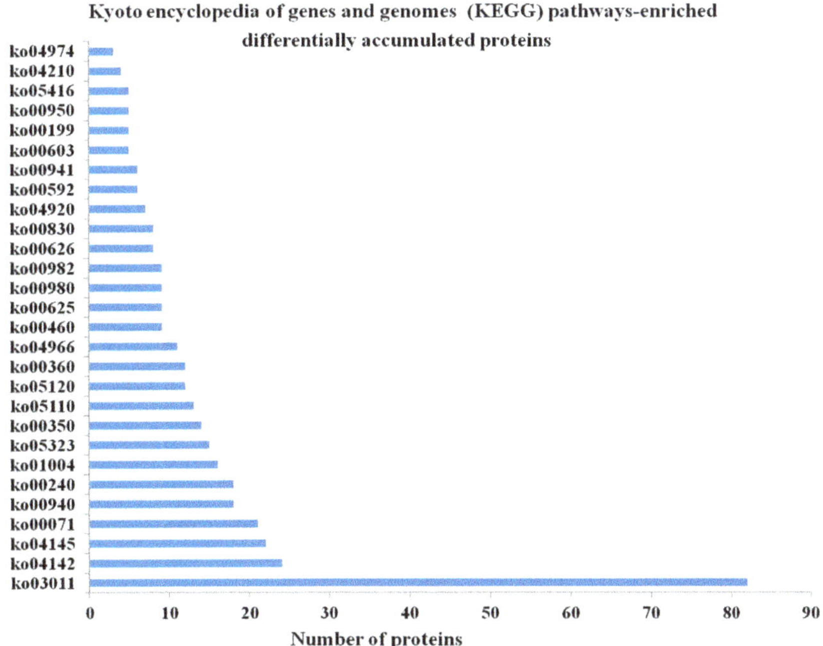

Figure 2. Kyoto encyclopedia of genes and genomes(KEGG) pathways-enriched differentially accumulated proteins during pitaya fruit ripening. ko03011 Ribosome, ko04142 Lysosome, ko04145 Phagosome, ko00071 Fatty acid metabolism, ko00940 Phenylpropanoid biosynthesis, ko00240 Pyrimidine metabolism, ko01004 Lipid biosynthesis proteins, ko05323 Rheumatoid arthritis, ko00350 Tyrosine metabolism, ko05110 Vibrio cholerae infection, ko05120 Epithelial cell signaling in Helicobacter pylori infection, ko00360 Phenylalanine metabolism, ko04966 Collecting duct acid secretion, ko00460 Cyanoamino acid metabolism, ko00625 Chloroalkane and chloroalkene degradation, ko00980 Metabolism of xenobiotics by cytochrome P450, ko00982 Drug metabolism—cytochrome P450, ko00626 Naphthalene degradation, ko00830 Retinol metabolism, ko04920 Adipocytokine signaling pathway, ko00592 alpha-Linolenic acid metabolism, ko00941 Flavonoid biosynthesis, ko00603 Glycosphingolipid biosynthesis - globo series, ko00199 Cytochrome P450, ko00950 Isoquinoline alkaloid biosynthesis, ko05416 Viral myocarditis, ko04210 Apoptosis, ko04974 Protein digestion and absorption.

Despite the discrepancy discussed above, most of the genes/proteins had a positive correlation (p-value < 0.05, FC (pro) > 1.5, FC (RNA) > 2), i.e., differentially expressed at both transcript and protein levels. These genes/proteins were mapped to such categories as "oxidation-reduction process"

(BP), "oxidoreductase activity" (MF) and "membrane" (CC) (Figure 3). There were totally 15 pathways were obtained, including "starch and sucrose metabolism", "phenylpropanoid biosynthesis", "lipid biosynthesis proteins", "ascorbate and aldarate metabolism", "tyrosine metabolism" and "flavonoid biosynthesis" (Table 1).

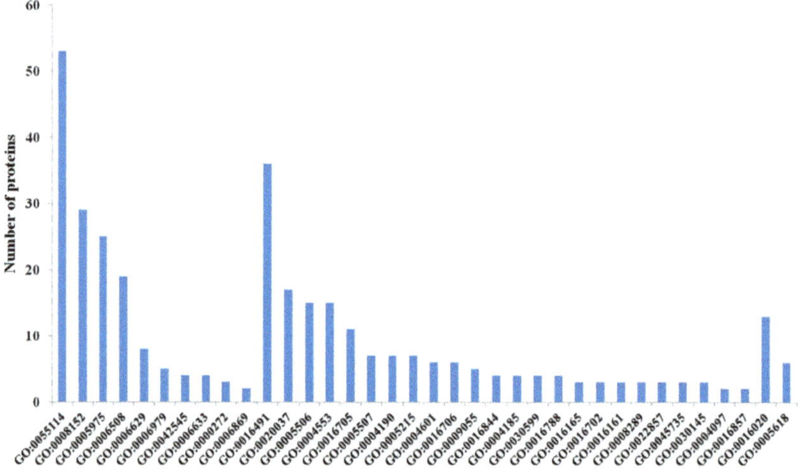

Figure 3. Gene Ontology (GO) classification of the genes/proteins whose transcript and protein levels were positively correlated (p-value < 0.05, FC (pro) > 1.5, FC (RNA) > 2) during the pitaya fruit ripening. GO:0055114//oxidation-reduction process, GO:0008152//metabolic process, GO:0005975// carbohydrate metabolic process, GO:0006508//proteolysis, GO:0006629//lipid metabolic process, GO:0006979//response to oxidative stress, GO:0042545//cell wall modification, GO:0006633//fatty acid biosynthetic process, GO:0000272//polysaccharide catabolic process, GO:0006869//lipid transport, GO:0016491//oxidoreductase activity, GO:0020037//heme binding, GO:0005506//iron ion binding, GO:0004553//hydrolase activity, hydrolyzing *O*-glycosyl compounds, GO:0016705// oxidoreductase activity, acting on paired donors, with incorporation or reduction of molecular oxygen, GO:0005507//copper ion binding, GO:0004190//aspartic-type endopeptidase activity, GO:0005215//transporter activity, GO:0004601//peroxidase activity, GO:0016706//oxidoreductase activity, acting on paired donors, with incorporation or reduction of molecular oxygen, 2-oxoglutarate as one donor, and incorporation of one atom each of oxygen into both donors, GO:0009055//electron carrier activity, GO:0016844//strictosidine synthase activity, GO:0004185// serine-type carboxypeptidase activity, GO:0030599//pectinesterase activity, GO:0016788//hydrolase activity, acting on ester bonds, GO:0016165//linoleate 13S-lipoxygenase activity, GO:0016702// oxidoreductase activity, acting on single donors with incorporation of molecular oxygen, incorporation of two atoms of oxygen, GO:0016161//beta-amylase activity, GO:0008289//lipid binding, GO:0022857//transmembrane transporter activity, GO:0045735//nutrient reservoir activity, GO:0030145//manganese ion binding, GO:0004097//catechol oxidase activity, GO:0016857//racemase and epimerase activity, acting on carbohydrates and derivatives, GO:0016020//membrane, GO:0005618//cell wall.

Among the positively correlated genes/proteins, three were up-regulated in "starch and sucrose metabolism". It was reported that soluble solids concentration (SSC) was closely related to the synthesis of betacyanin [73]. Five up-regulated genes/proteins at both the transcriptional and posttranscriptional levels were in "phenylpropanoid biosynthesis". Two up-regulated and two down-regulated genes/proteins belonged to the "tyrosine metabolism" pathway. Betalains were derived from tyrosine [28], and the "phenylpropanoid biosynthesis" had been considered to be the upstream pathway of the tyrosine pathway. Two transcripts and proteins involved in "flavonoid biosynthesis" were significantly up-regulated and down-regulated, respectively. Four up-regulated transcripts and proteins were enriched in "ascorbate and aldarate metabolism" (Table 2), which played a role in the biosynthesis of betalains [28] (Table 2). These proteins may be associated with the betalain biosynthesis in *H. polyrhizus*.

Naturally, betalains and anthocyanin could not co-exist simultaneously in one plant. However, they can exist together by genetic engineering strategy [30,31]. Chalcone isomerase (CHI) and chalcone synthase (CHS) were the upstream enzymes of the anthocyanin biosynthesis pathway. In this study, one protein annotated as CHI and another as CHS were identified. CHI and CHS had higher expressed levels in the pitaya fruit at the red stage than at the white stage. This result was consistent with the conjecture that betalain-producing plants could not produce anthocyanins due to lower levels of dihydroflavonol reductase (DFR), anthocyanidin synthase (ANS) and leuco anthocyanidin reductase (LAR) [74]. More interestingly, five proteins, namely comp37375_c0_seq2_4 and comp37375_c0_seq1_4 (polyphenol oxidase), comp37692_c0_ seq1_7 (4, 5-DOPA dioxygenase extradiol-like), comp16058_c0_seq1_4 (CYP76AD3), and comp26435_c0_seq1_2 (Aromatic-L-amino-acid decarboxylase), were identified as some of the key enzymes in the betalain biosynthesis pathway (Table 2). Not surprisingly, the transcriptional levels of these proteins were increasing with the progression of the betalain formation in the pitaya fruit as revealed in our previous study [68].

These results suggested that these pathways could be involved in the biosynthesis of pigments, and it was possible that polyphenol oxidase, CYP76AD3 and 4,5-DOPA dioxygenase extradiol-like protein were responsible for betalain biosynthesis in the pitaya fruit. These results were consistent with our previous findings from transcriptomic analysis that these enzymes might be involved in betalain biosynthesis in *H. polyrhizus* [68].

3. Materials and Methods

3.1. Plant Material

"7-1" (*H. polyrhizus*), a superior selection with red flesh color and excellent quality, were used as materials. Plants were cultivated in Dalingshan Forest Park (East longitude: 113°42′22″–113°48′12″, Northern latitudes: 22°50′00″–22°53′32″), Dongguan City, Guangdong Province, China. Two libraries from white and red pulp stages were constructed for iTRAQ analyses on the 21st and 32nd days after artificial pollination in July and August 2014 (Figure S1). Every library consisted of equal amounts of protein from three fruit at each fruit developmental stage.

3.2. Samples Preparation, Protein Extraction and Detection

Upon harvested, the samples described above were immediately frozen in liquid nitrogen and stored at −80 °C prior to protein extraction. Trichloroacetic acid (TCA)-acetone method was used for protein extraction. Briefly, pulps without seeds were ground to fine powder in liquid nitrogen. Cold TCA-acetone (ratio of material to liquid is 1:4) was added to the cold 15 mL tube and vortexed for 30 s. Total proteins were precipitated overnight at −20 °C. Precipitation was collected by spinning 12,000 rpm for 15 min at 4 °C and subsequently washed with cold acetone and 90% acetone, respectively. The precipitation was freeze-dried by vacuum and dissolved in lysis buffer [8 M ureophil, 0.5% SDS (sodium dodecyl sulfate), PMSF (phenylmethanesulfonyl fluoride), PI (protease inhibitor), 40 mM TEAB (Triethylamine borane)]. Proteins in the supernatant was pelleted by centrifuging 12,000 rpm for 15 min and extracted with 8 M ureophil extraction buffer (including 0.1% SDS, PI, PMSF). The concentration and quality of the protein were assayed by bicinchoninic acid (BCA) method and SDS-PAGE gel electrophoresis. At least 400 μg proteins (>2.0 μg/μL) were used for further experiment.

3.3. iTRAQ Labeling

Above extracted protein samples were prepared using the iTRAQ® Reagents 8plexMulti-plex kit (AB Sciex, Boston, MA, USA). Enriched phosphopeptides were labeled with isobaric tags for relative and absolute quantification reagents (AB Sciex). iTRAQ 113 and 115 were used to label white and red samples, respectively. The labeled samples were combined equably and graded with RP C18 chromatographic column.

3.4. Strong Cation Exchange Chromatography (SCX)

The mixed and labeled samples were fractionated by SCX fractionation using HPLC (high performance liquid chromatography). The mobile phases consisted of buffer A (2% acetonitrile and 98% H_2O (pH 10)) and B (90% acetonitrile and 10% H_2O (pH 10)). The labeled samples were concentrated by vacuum and dissolved in 100 μL buffer A (pH 10). After the mixture was centrifuged, the supernatant was loaded onto a reverse phase (RP) C18 precolumn (LC Packings) (Agilent Technologies, Palo Alto, CA, USA). Separation was performed using a linear gradient at a flow rate of 1 mL/min. The gradient of elution is shown in Table S5. The liquid effluent was collected at a speed of 1.5 mL/min. Multiple components were obtained by merging samples according to the chromatogram map.

3.5. LC-MS/MS Analysis

The liquid chromatography–mass spectrometry (LC-MS)/MS method was developed for the separation and analysis of samples. The mass spectrometer used for detection was Q Exactive mass spectrometer (Thermo Scientific, Waltham, MA, USA). The HPLC was NCS3500 system. The mobile phases consisted of buffer A (99.9% H_2O and 0.1% formic acid) and B (99.9% acetonitrile and 0.1% formic acid). The flow rate was 300 nL/min. The gradient of elution is as shown in Table S6.

Full MS scans range was 350–1600 *m*/*z*. The runtime was 75 min and the resolution was 70,000. The precursor ions were selected for the MS/MS scans using higher energy collision-induced dissociation (HCD) for each precursor ion. Then MS2 (secondary mass spectrum) sequences were determined. The dynamic exclusion option was implemented with a repeat count of 1 and exclusion duration of 15 s. The values of automated gain control (AGC) were set to 1×10^6 and 2×10^5 for full MS and MS2, respectively.

3.6. Proteomic Data Analysis

For peptide data analysis, raw mass data were processed and searched against the protein databases downloaded from the public databases using Proteome Discoverer software (Thermo Scientific, Waltham, MA, USA). Searches were performed using the following criteria: the precursor mass tolerance was set to 20 ppm, and fragment ion mass tolerance was set to 0.02 Da for HCD in addition to the general settings. The search parameters allowed two missed cleavage for tyrpsin. The maximum delta Cn was considered as 0.05, while 10 was thre maximum number of peptides reported. The score threshold for peptide identification was set at 0.01 false-discovery rate (FDR) in the iTRAQ experiment. The list of proteins obtained from the iTRAQ data was exported to Supplementary File S1, which contains such protein-specific information as accession numbers, percent coverage, protein scores, number of peptides matching individual proteins, etc. Proteins with a fold-change cutoff ≥1.5 between the two stages were identified as differentially expressed.

4. Conclusions

Betalains play a role in appearance quality and nutritional value of red pitaya, and they are also involved in stress-resistance. Study of the pitaya pulp pigments proteome is important to understand the correlative pathways contributed to the betalain biosynthesis. In this study, proteomic changes in *H. polyrhizus* were first investigated using iTRAQ. In total, 1946 differentially expressed proteins were identified to characterize the proteome of red pitaya. Based on integrated analyses of transcriptomic and proteomic datasets, several transcripts and proteins owning positive correlation were gathered in "phenylpropanoid biosynthesis", "tyrosine metabolism", "flavonoid biosynthesis", "ascorbate and aldarate metabolism", "betalains biosynthesis" and "anthocyanin biosynthesis". These pathways are related to the metabolites of betalain biosynthesis. Five proteins, which were annotated to polyphenol oxidase, CYP76AD3, and 4,5-DOPA dioxygenase extradiol-like, were found to be differentially expressed in the betalain biosynthesis pathway. The present study provides the first proteomic analysis of the red pitaya pulps by iTRAQ and could offer new insights into the molecular mechanism of the betalain biosynthesis at the posttranscriptional level.

Supplementary Materials: Supplementary materials can be found at www.mdpi.com/1422-0067/17/10/1606/s1.

Acknowledgments: This work was supported by the Science and Technology Planning Project of Guangdong Province (2014B020202010), Science and Technology Program of Guangzhou (2014Y2-00164), YangFan Innovative and Entepreneurial Research Team Project (2014YT02H013) and the earmarked fund for Guangdong Modern Agro-industry Technology Research System (2016LM1128 and 2016LM1130).

Author Contributions: Performed the experiments: Qingzhu Hua. Contributed field management/sample collection/technical assistance: Qianjun Zhou, Susheng Gan, Jingyu Wu, Canbin Chen, Jiaqiang Li, Yaoxiong Ye and Jietang Zhao. Wrote and revised the paper: Qingzhu Hua, Susheng Gan and Yonghua Qin. Conceived and designed the experiments: Yonghua Qin and Guibing Hu.

Conflicts of Interest: The authors declare no conflict of interest.

References

1. Clifford, T.; Howatson, G.; West, D.J.; Stevenson, E.J. The potential benefits of red beetroot supplementation in health and disease. *Nutrients* **2015**, *7*, 2801–2822. [CrossRef] [PubMed]

2. Nowacki, L.T.; Vigneron, P.; Rotellini, L.; Cazzola, H.L.N.; Merlier, F.; Prost, E.; Ralanairina, R.; Gadonna, J.; Rossi, C.; Vayssade, M. Betanin-enriched red beetroot (*Beta vulgaris* L.) extract induces apoptosis and autophagic cell death in MCF-7 cells. *Phytother. Res.* **2015**, *29*, 1964–1973. [CrossRef] [PubMed]

3. Martinez, R.M.; Longhi-Balbinot, D.T.; Zarpelon, A.C.; Staurengo-Ferrari, L.; Baracat, M.M.; Georgetti, S.R.; Sassonia, R.C.; Verri, W.A., Jr.; Casagrande, R. Anti-inflammatory activity of betalain-rich dye of *Beta vulgaris*: Effect on edema, leukocyte recruitment, superoxide anion and cytokine production. *Arch. Pharm. Res.* **2015**, *38*, 494–504. [CrossRef] [PubMed]

4. Naselli, F.; Tesoriere, L.; Caradonna, F.; Bellavia, D.; Attanzio, A.; Gentile, C.; Livrea, M.A. Anti-proliferative and pro-apoptotic activity of whole extract and isolated indicaxanthin from *Opuntia ficus-indica* associated with re-activation of the onco-suppressor p16(INK4a) gene in human colorectal carcinoma (Caco-2) cells. *Biochem. Biophys. Res. Commun.* **2014**, *450*, 652–658. [CrossRef] [PubMed]

5. Allegra, M.; Ianaro, A.; Tersigni, M.; Panza, E.; Tesoriere, L.; Livrea, M.A. Indicaxanthin from cactus pear fruit exerts anti-inflammatory effects in carrageenin-induced rat pleurisy. *J. Nutr.* **2014**, *144*, 185–192. [CrossRef] [PubMed]

6. Allegra, M.; Tesoriere, L.; Livrea, M.A. Betanin inhibits the myeloperoxidase/nitrite-induced oxidation of human low-density lipoproteins. *Free Radic. Res.* **2007**, *41*, 335–341. [CrossRef] [PubMed]

7. Wootton-Beard, P.C.; Brandt, K.; Fell, D.; Warner, S.; Ryan, L. Effects of a beetroot juice with high neobetanin content on the early-phase insulin response in healthy volunteers. *J. Nutr. Sci.* **2014**, *3*, e9. [CrossRef] [PubMed]

8. Sengupta, D.; Mondal, B.; Mukherjee, K. Visible light absorption and photo-sensitizing properties of spinach leaves and beetroot extracted natural dyes. *Spectrochim. Acta A* **2015**, *148*, 85–92. [CrossRef] [PubMed]

9. Hernandez-Martinez, A.R.; Estevez, M.; Vargas, S.; Quintanilla, F.; Rodriguez, R. New dye-sensitized solar cells obtained from extracted bracts of *Bougainvillea glabra* and *spectabilis* betalain pigments by different purification processes. *Int. J. Mol. Sci.* **2011**, *12*, 5565–5576. [CrossRef] [PubMed]

10. Calogero, G.; Bartolotta, A.; di Marco, G.; di Carlo, A.; Bonaccorso, F. Vegetable-based dye-sensitized solar cells. *Chem. Soc. Rev.* **2015**, *44*, 3244–3294. [CrossRef] [PubMed]

11. Boo, H.; Shin, K.; Heo, J.; Jeong, J.; Paek, K. Betalain synthesis by hairy root of red beet cultured in vitro under different light quality. In Proceedings of the 4th International ISHS Symposium on Artificial Lighting, Quebec, QC, Canada, 7–9 November 2000; pp. 209–214.

12. Wang, C.Q.; Song, H.; Gong, X.Z.; Hu, Q.G.; Liu, F.; Wang, B.S. Correlation of tyrosinase activity and betacyanin biosynthesis induced by dark in C3 halophyte *Suaeda salsa* seedlings. *Plant Sci.* **2007**, *173*, 487–494.

13. Kishima, Y.; Nozaki, K.; Akashi, R.; Adachi, T. Light-inducible pigmentation in *Portulaca* callus; selection of a high betalain producing cell line. *Plant Cell Rep.* **1991**, *10*, 304–307. [CrossRef] [PubMed]

14. Kishima, Y.; Shimaya, A.; Adachi, T. Evidence that blue light induces betalain pigmentation in *Portulaca* callus. *Plant Cell Tissue Org.* **1995**, *43*, 67–70. [CrossRef]

15. Kochhar, V.K.; Kochhar, S.; Mohr, H. An analysis of the action of light on betalain synthesis in the seedling of *Amaranthus caudatus*, var. viridis. *Planta* **1981**, *151*, 81–87. [CrossRef] [PubMed]

16. Khandaker, L.; Akond, A.S.M.G.; Ali, M.B.; Oba, S. Biomass yield and accumulations of bioactive compounds in red amaranth (*Amaranthus tricolor* L.) grown under different colored shade polyethylene in spring season. *Sci. Hortic.* **2010**, *123*, 289–294. [CrossRef]

17. Biddington, N.L.; Thomas, T.H. A modified *Amaranthus* betacyanin bioassay for the rapid determination of cytokinins in plant extracts. *Planta* **1973**, *111*, 183–186. [CrossRef] [PubMed]

18. Bianco-Colomas, J. Qualitative and quantitative aspects of betalains biosynthesis in *Amaranthus caudatus* L. var. pendula seedlings. *Planta* **1980**, *149*, 176–180. [CrossRef] [PubMed]

19. Nazmul, M.; Bhuiyan, H.; Adachi, T. Stimulation of betacyanin synthesis through exogenous methyl jasmonate and other elicitors in suspension-cultured cells of *Portulaca*. *J. Plant Physiol.* **2003**, *160*, 1117–1124.

20. Shin, K.S.; Murthy, H.N.; Heo, J.W.; Paek, K.Y. Induction of betalain pigmentation in hairy roots of red beet under different radiation sources. *Biol. Plant.* **2003**, *47*, 149–152. [CrossRef]

21. Thomas, V.; Mwafaq, I.; Juergen, S.; Victor, W.; Manfred, N.; Dieter, S. Light-induced betacyanin and flavonol accumulation in bladder cells of *Mesembryanthemum crystallinum*. *Phytochemistry* **1999**, *52*, 583–592.

22. Zhao, S.Z.; Sun, H.Z.; Chen, M.; Wang, B.S. Light-regulated betacyanin accumulation in euhalophyte *Suaeda salsa* calli. *Plant Cell Tissue Org.* **2010**, *102*, 99–107.

23. Jain, G.; Gould, K.S. Functional significance of betalain biosynthesis in leaves of *Disphyma australe* under salinity stress. *Environ. Exp. Bot.* **2015**, *109*, 131–140. [CrossRef]

24. Jain, G.; Schwinn, K.E.; Gould, K.S. Betalain induction by L-DOPA application confers photoprotection to saline-exposed leaves of *Disphyma australe*. *New Phytol.* **2015**, *207*, 1075–1083. [CrossRef] [PubMed]

25. Solovchenko, A.E.; Merzlyak, M.N. Screening of visible and UV radiation as a photoprotective mechanism in plants. *Russ. J. Plant Physiol.* **2008**, *55*, 719–737. [CrossRef]

26. Lakhotia, P.; Singh, K.P.; Singh, S.K.; Singh, M.C.; Prasad, K.V.; Swaroop, K. Influence of biotic and abiotic elicitors on production of betalain pigments in *bougainvillea* callus cultures. *Indian J. Hortic.* **2014**, *71*, 373–378.

27. Han, X.H.; Gao, Z.J.; Xiao, X.G. Enzymes and genes involved in the betalain biosynthesis in higher plants. *Afr. J. Biotechnol.* **2009**, *8*, 6735–6744.

28. Gandia-Herrero, F.; Garcia-Carmona, F. Biosynthesis of betalains: Yellow and violet plant pigments. *Trends Plant Sci.* **2013**, *18*, 334–343. [CrossRef] [PubMed]

29. Dieter, S.; Thomas, V.; Schliemann, W. Recent advances in betalain research. *Phytochemistry* **2003**, *62*, 247–269.

30. Harris, N.N.; Javellana, J.; Davies, K.M.; Lewis, D.H.; Jameson, P.E.; Deroles, S.C.; Calcott, K.E.; Gould, K.S.; Schwinn, K.E. Betalain production is possible in anthocyanin-producing plant species given the presence of DOPA-dioxygenase and L-DOPA. *BMC Plant Biol.* **2012**, *12*, 34. [CrossRef] [PubMed]

31. Nakatsuka, T.; Yamada, E.; Takahashi, H.; Imamura, T.; Suzuki, M.; Ozeki, Y.; Tsujimura, I.; Saito, M.; Sakamoto, Y.; Sasaki, N.; et al. Genetic engineering of yellow betalain pigments beyond the species barrier. *Sci. Rep.* **2013**, *3*, 1970. [CrossRef] [PubMed]

32. Gandia-Herrero, F.; Escribano, J.; Garcia-Carmona, F. Betaxanthins as substrates for tyrosinase. An approach to the role of tyrosinase in the biosynthetic pathway of betalains. *Plant Physiol.* **2005**, *138*, 421–432. [CrossRef] [PubMed]

33. Gandia-Herrero, F.; Escribano, J.; Garcia-Carmona, F. Characterization of the monophenolase activity of tyrosinase on betaxanthins: the tyramine-betaxanthin/dopamine-betaxanthin pair. *Planta* **2005**, *222*, 307–318. [CrossRef] [PubMed]

34. Joy, R.W.I.; Sugiyama, M.; Fukuda, H.; Komamine, A. Cloning and characterization of polyphenol oxidase cDNAs of *Phytolacca americana*. *Plant Physiol.* **1995**, *107*, 1083–1089. [CrossRef] [PubMed]

35. Hind, G.; Marshak, D.R.; Coughlan, S.J. Spinach thylakoid polyphenol oxidase: Cloning, characterization, and relation to a putative protein kinase. *Biochemistry* **1995**, *34*, 8157–8164. [CrossRef] [PubMed]

36. Ma, H.; Zhu, H.Y.; Li, L.L.; Chen, L.J.; Guo, Z.F.; Zhong, M. Cloning and sequence analysis of polyphenol oxidase gene of *Suaeda salsa*. *Guangdong Agric. Sci.* **2013**, *11*, 149–152.

37. Yu, Z.H.; Han, Y.N.; Xiao, X.G. A PPO promoter from betalain-producing red swiss chard, directs petiole- and root-preferential expression of foreign gene in anthocyanins-producing plants. *Int. J. Mol. Sci.* **2015**, *16*, 27032–27043. [CrossRef] [PubMed]

38. Casique-Arroyo, G.; Martinez-Gallardo, N.; de la Vara, L.G.; Delano-Frier, J.P. Betacyanin biosynthetic genes and enzymes are differentially induced by (a)biotic stress in *Amaranthus hypochondriacus*. *PLoS ONE* **2014**. [CrossRef] [PubMed]

39. Christinet, L.; Burdet, F.; Zaiko, M.; Hinz, U.; Zryd, J.P. Characterization and functional identification of a novel plant 4,5-extradiol dioxygenase involved in betalain pigment biosynthesis in *Portulaca grandiflora*. *Plant Physiol.* **2004**, *134*, 265–274. [CrossRef] [PubMed]

40. Takahashi, K.; Takamura, E.; Sakuta, M. Isolation and expression analysis of two DOPA dioxygenases in *Phytolacca americana*. *Z. Naturforsch. C* **2009**, *64*, 564–573. [CrossRef] [PubMed]

41. Zhao, S.Z.; Sun, H.Z.; Gao, Y.; Sui, N.; Wang, B.S. Growth regulator-induced betacyanin accumulation and dopa-4,5-dioxygenase (*DODA*) gene expression in euhalophyte *Suaeda salsa* calli. *In Vitro Cell. Dev. Biol.-Plant* **2011**, *47*, 391–398. [CrossRef]

42. Yuan, R. The Cloning and Functional Analysis of 4,5-DOPA-Dioxygenase in *Suaeda salsa*. Master's Thesis, Shandong Normal University, Ji'nan, China, 2008.

43. Gandia-Herrero, F.; Garcia-Carmona, F. Characterization of recombinant β *vulgaris* 4,5-DOPA-extradiol -dioxygenase active in the biosynthesis of betalains. *Planta* **2012**, *236*, 91–100. [CrossRef] [PubMed]

44. Stintzing, F.C.; Herbach, K.M.; Mosshammer, M.R.; Carle, R.; Yi, W.G.; Sellappan, S.; Akoh, C.C.; Bunch, R.; Felker, P. Color, betalain pattern, and antioxidant properties of cactus pear (*Opuntia spp.*) clones. *J. Agric. Food Chem.* **2005**, *53*, 442–451. [CrossRef] [PubMed]

45. Chung, H.; Schwinn, K.E.; Ngo, H.M.; Lewis, D.H.; Massey, B.; Calcottt, K.E.; Crowhurst, R.; Joyce, D.C.; Gould, K.S.; Davies, K.M.; et al. Characterisation of betalain biosynthesis in *Parakeelya* flowers identifies the key biosynthetic gene DOD as belonging to an expanded *LigB* gene family that is conserved in betalain-producing species. *Front. Plant Sci.* **2015**, *6*, 499. [CrossRef] [PubMed]

46. Sasaki, N.; Abe, Y.; Goda, Y.; Adachi, T.; Kasahara, K.; Ozeki, Y. Detection of DOPA 4,5-dioxygenase (DOD) activity using recombinant protein prepared from *Escherichia coli* cells harboring cDNA encoding *DOD* from *Mirabilis jalapa*. *Plant Cell Physiol.* **2009**, *50*, 1012–1016. [CrossRef] [PubMed]

47. Hatlestad, G.J.; Sunnadeniya, R.M.; Akhavan, N.A.; Gonzalez, A.; Goldman, I.L.; McGrath, J.M.; Lloyd, A.M. The beet R locus encodes a new cytochrome P450 required for red betalain production. *Nat. Genet.* **2012**, *44*, 130–816. [CrossRef] [PubMed]

48. Suzuki, M.; Miyahara, T.; Tokumoto, H.; Hakamatsuka, T.; Goda, Y.; Ozeki, Y.; Sasaki, N. Transposon-mediated mutation of *CYP76AD3* affects betalain synthesis and produces variegated flowers in four o'clock (*Mirabilis jalapa*). *J. Plant Physiol.* **2014**, *171*, 1586–1590. [CrossRef] [PubMed]

49. Yang, Y.; Moore, M.J.; Brockington, S.F.; Soltis, D.E.; Wong, G.K.S.; Carpenter, E.J.; Zhang, Y.; Chen, L.; Yan, Z.; Xie, Y.; et al. Dissecting molecular evolution in the highly diverse plant clade Caryophyllales using transcriptome sequencing. *Mol. Biol. Evol.* **2015**, *32*, 2001–2014. [CrossRef] [PubMed]

50. Brockington, S.F.; Yang, Y.; Gandia-Herrero, F.; Covshoff, S.; Hibberd, J.M.; Sage, R.F.; Wong, G.K.S.; Moore, M.J.; Smith, S.A. Lineage-specific gene radiations underlie the evolution of novel betalain pigmentation in Caryophyllales. *New Phytol.* **2015**, *207*, 1170–1180. [CrossRef] [PubMed]

51. Isayenkova, J.; Wray, V.; Manfred, N.; Strack, D.; Thomas, V. Cloning and functional characterisation of two regioselective flavonoid glucosyltransferases from *Beta vulgaris*. *Phytochemistry* **2006**, *67*, 1598–1612. [CrossRef] [PubMed]

52. Noguchi, A.; Kunikane, S.; Homma, H.; Liu, W.; Sekiya, T.; Hosoya, M.; Kwon, S.; Ohiwa, S.; Katsuragi, H.; Nishino, T.; et al. Identification of an inducible glucosyltransferase from *Phytolacca americana* L. cells that are capable of glucosylating capsaicin. *Plant Biotechnol. Nar.* **2009**, *26*, 285–292. [CrossRef]

53. Thomas, V.; Rudi, G.; Dieter, S. Cloning and expression of a cDNA encoding betanidin 5-*O*-glucosyltransferase, a betanidin- and flavonoid-specific enzyme with high homology to inducible glucosyltransferases from the Solanaceae. *Plant J.* **1999**, *19*, 509–519.

54. Thomas, V. Substrate specificity and sequence analysis define a polyphyletic origin of betanidin 5- and 6-*O*-glucosyltransferase from *Dorotheanthus bellidiformis*. *Planta* **2002**, *214*, 492–495.

55. Sasaki, N.; Wada, K.; Koda, T.; Kasahara, K.; Adachi, T.; Ozeki, Y. Isolation and characterization of cDNAs encoding an enzyme with glucosyltransferase activity for cyclo-DOPA from four o'clocks and feather cockscombs. *Plant Cell Physiol.* **2005**, *46*, 666–670. [CrossRef] [PubMed]
56. Hatlestad, G.J.; Akhavan, N.A.; Sunnadeniya, R.M.; Elam, L.; Cargile, S.; Hembd, A.; Gonzalez, A.; McGrath, J.M.; Lloyd, A.M. The beet Y locus encodes an anthocyanin MYB-like protein that activates the betalain red pigment pathway. *Nat. Genet.* **2015**, *47*, 92. [CrossRef] [PubMed]
57. Des Marais, D.L. To betalains and back again: A tale of two pigments. *New Phytol.* **2015**, *207*, 939–941. [CrossRef] [PubMed]
58. Lim, T.K. *Hylocereus polyrhizus.* In *Edible Medicinal and Non-Medicinal Plants*; Springer-Verlag Berlin: Berlin, Germany, 2012; Volume 1, pp. 643–649.
59. Lim, T.K. *Hylocereus undatus.* In *Edible Medicinal and Non-Medicinal Plants*; Springer-Verlag Berlin: Berlin, Germany, 2012; Volume 1, pp. 650–655.
60. Adnan, L.; Osman, A.; Hamid, A.A. Antioxidant activity of different extracts of red pitaya (*Hylocereus polyrhizus*) seed. *Int. J. Food Prop.* **2011**, *14*, 1171–1181. [CrossRef]
61. Zhuang, Y.L.; Zhang, Y.F.; Sun, L.P. Characteristics of fibre-rich powder and antioxidant activity of pitaya (*Hylocereus undatus*) peels. *Int. J. Food Sci. Technol.* **2012**, *47*, 1279–1285.
62. Garcia-Cruz, L.; Valle-Guadarrama, S.; Salinas-Moreno, Y.; Joaquin-Cruz, E. Physical, chemical, and antioxidant activity characterization of pitaya (*Stenocereus pruinosus*) fruits. *Plant Food Hum. Nutr.* **2013**, *68*, 403–410. [CrossRef] [PubMed]
63. Wu, L.C.; Hsu, H.W.; Chen, Y.C.; Chiu, C.C.; Lin, Y.I.; Ho, J.A. Antioxidant and antiproliferative activities of red pitaya. *Food Chem.* **2006**, *95*, 319–327. [CrossRef]
64. Tenore, G.C.; Novellino, E.; Basile, A. Nutraceutical potential and antioxidant benefits of red pitaya (*Hylocereus polyrhizus*) extracts. *J. Funct. Foods* **2012**, *4*, 129–136. [CrossRef]
65. Wybraniec, S.; Stalica, P.; Jerz, G.; Klose, B.; Gebers, N.; Winterhalter, P.; Sporna, A.; Szaleniec, M.; Mizrahi, Y. Separation of polar betalain pigments from cacti fruits of *Hylocereus polyrhizus* by ion-pair high-speed countercurrent chromatography. *J. Chromatogr. A* **2009**, *1216*, 6890–6899. [CrossRef] [PubMed]
66. Lim, S.D.; Yusof, Y.A.; Chin, N.L.; Talib, R.A.; Endan, J.; Aziz, M.G. Effect of extraction parameters on the yield of betacyanins from pitaya fruit (*Hylocereus polyrhizus*) pulps. *J. Food Agric. Environ.* **2011**, *9*, 158–162.
67. Suh, D.H.; Lee, S.; Heo, D.Y.; Kim, Y.; Cho, S.K.; Lee, S.; Lee, C.H. Metabolite profiling of red and white pitayas (*Hylocereus polyrhizus* and *Hylocereus undatus*) for comparing betalain biosynthesis and antioxidant activity. *J. Agric. Food Chem.* **2014**, *62*, 8764–8771. [CrossRef] [PubMed]
68. Hua, Q.Z.; Chen, C.J.; Chen, Z.; Chen, P.K.; Ma, Y.W; Wu, J.Y.; Zheng, J.; Hu, G.B.; Zhao, J.T.; Qin, Y.H. Transcriptomic analysis reveals key genes related to betalain biosynthesis in pulp coloration of *Hylocereus polyrhizus*. *Front. Plant Sci.* **2016**, *6*, 1179.
69. Casado-Vela, J.; Jose Martinez-Esteso, M.; Rodriguez, E.; Borras, E.; Elortza, F.; Bru-Martinez, R. iTRAQ-based quantitative analysis of protein mixtures with large fold change and dynamic range. *Proteomics* **2010**, *10*, 343–347. [CrossRef] [PubMed]
70. Tian, Q.; Stepaniants, S.B.; Mao, M.; Weng, L.; Feetham, M.C.; Doyle, M.J.; Yi, E.C.; Dai, H.Y.; Thorsson, V.; Eng, J.; et al. Integrated genomic and proteomic analyses of gene expression in mammalian cells. *Mol. Cell Proteom.* **2004**, *3*, 960–969. [CrossRef] [PubMed]
71. Rossouw, D.; van den Dool, A.H.; Jacobson, D.; Bauer, F.F. Comparative transcriptomic and proteomic profiling of industrial wine yeast strains. *Appl. Environ. Microb.* **2010**, *76*, 3911–3923. [CrossRef] [PubMed]
72. Carp, M.-J.; Gepstein, S. Genomics and proteomics of leaf senescence. In *Senescence Processes in Plants*; Gan, S., Ed.; Blackwell Publishing: Oxford, UK, 2007; pp. 202–230.
73. Jamaludin, N.A.; Ding, P.; Hamid, A.A. Physico-chemical and structural changes of red-fleshed dragon fruit (*Hylocereus polyrhizus*) during fruit development. *J. Sci. Food Agric.* **2011**, *91*, 278–285. [CrossRef] [PubMed]
74. Shimada, S.; Inoue, Y.T.; Sakuta, M. Anthocyanidin synthase in non-anthocyanin-producing caryophyllales species. *Plant J.* **2005**, *44*, 950–959. [CrossRef] [PubMed]

International Journal of
Molecular Sciences

Article

Comparative Proteomic Analysis of Mature Pollen in Triploid and Diploid *Populus deltoides*

Xiao-Ling Zhang [1,†], Jin Zhang [1,2,*,†], Ying-Hua Guo [1], Pei Sun [1], Hui-Xia Jia [1,2], Wei Fan [1], Meng-Zhu Lu [1,2] and Jian-Jun Hu [1,2,*]

1   State Key Laboratory of Tree Genetics and Breeding, Key Laboratory of Tree Breeding and Cultivation of State Forestry Administration, Research Institute of Forestry, Chinese Academy of Forestry, Beijing 100091, China; ls_wu@caf.ac.cn (X.-L.Z.); yh_guo@caf.ac.cn (Y.-H.G.); sun_caf@sina.com (P.S.); huixia__jia@126.com (H.-X.J.); atom201001@163.com (W.F.); lumz@caf.ac.cn (M.-Z.L.)
2   Collaborative Innovation Center of Sustainable Forestry in Southern China, Nanjing Forestry University, Nanjing 210037, China
*   Correspondence: zhang007jin@163.com (J.Z.); hujj@caf.ac.cn (J.-J.H.); Tel.: +86-10-6288-8862 (J.-J.H.)
†   These authors contributed equally to this work.

Academic Editor: Setsuko Komatsu
Received: 3 July 2016; Accepted: 29 August 2016; Published: 3 September 2016

Abstract: Ploidy affects plant growth vigor and cell size, but the relative effects of pollen fertility and allergenicity between triploid and diploid have not been systematically examined. Here we performed comparative analyses of fertility, proteome, and abundances of putative allergenic proteins of pollen in triploid poplar 'ZhongHuai1' ('ZH1', triploid) and 'ZhongHuai2' ('ZH2', diploid) generated from the same parents. The mature pollen was sterile in triploid poplar 'ZH1'. By applying two-dimensional gel electrophoresis (2-DE), a total of 72 differentially expressed protein spots (DEPs) were detected in triploid poplar pollen. Among them, 24 upregulated and 43 downregulated proteins were identified in triploid poplar pollen using matrix-assisted laser desorption/ionisation coupled with time of-flight tandem mass spectrometer analysis (MALDI-TOF/TOF MS/MS). The main functions of these DEPs were related with "S-adenosylmethionine metabolism", "actin cytoskeleton organization", or "translational elongation". The infertility of triploid poplar pollen might be related to its abnormal cytoskeletal system. In addition, the abundances of previously identified 28 putative allergenic proteins were compared among three poplar varieties ('ZH1', 'ZH2', and '2KEN8'). Most putative allergenic proteins were downregulated in triploid poplar pollen. This work provides an insight into understanding the protein regulation mechanism of pollen infertility and low allergenicity in triploid poplar, and gives a clue to improving poplar polyploidy breeding and decreasing the pollen allergenicity.

Keywords: poplar; pollen; diploid; triploid; allergenic protein; proteomics

---

1. Introduction

Polyploidy is having more than two sets of chromosomes present within a nucleus, which is a widespread phenomenon in the plant kingdom. Polyploidization frequently occurs in plants, and it has been regarded as one of the major speciation mechanisms [1]. Polyploidy has considerable effects on the expression of genes and abundance of proteins [2]. Compared with diploid, polyploidy shows advantages in many aspects, such as increasing in biomass and fruit yield [3]. Polyploidy also provides genome buffering, as it is helpful to increase the allelic diversity and then generate novel phenotypic variation [4]. However, most of the polyploidy appears to decrease pollen fertility and reduced seed amount. Meirmans et al. found that triploid dandelion did not produce more seeds or heavier seeds, and it was due to male sterility because of disorder of its nuclear genes [5].

As an important biomass and feedstock plant species, poplar provides a wide range of industrial construction woods [6]. Since Nilsson-Ehle found the giant natural European triploid aspen (*Populus tremula*) [7], more and more natural or artificial polyploidy poplar species were reported and applied into breeding. Cai et al. produced tetraploid *P. pseudo-simonii* from leaf explants of diploid *P. pseudo-simonii* by colchicine treatment. The size and density of leaf stomata in tetraploid poplar were significantly greater than that of diploid [8]. At present, the study of the low fertility of polyploidy has mainly focused on the morphological observation. The polyploid medicinal plant *Pinellia ternata* (Araceae) has low fertility and reduced amount of seeds because of abnormal meiotic division [9], which is similar to that of *Arabidopsis*, rice, and wheat [10–12]. However, the molecular characteristics of pollen in polyploid poplar have not been revealed.

Between April and May, male blossoming of Salicaceae results in amounts of pollen spread into the air, which seriously affects human health [13]. As the main allergens in pollen, the allergenic proteins will hydration quickly and to elicit the allergic reaction in a short time. Immune electron microscopy of dry and rehydrated birch pollen showed that Bet v I, a major allergen in birch pollen, can migrate into the exine and to the surface of pollen grains after brief hydration [14]. At present, the study on the pollen allergy is mainly diagnosed through testing the allergic reaction of the patients' serums. Several studies have indicated that the main allergies varied in different cities. In New York City, NY, U.S., 371 allergic patients' serums were tested and the highest rates of allergies were derived from *Quercus* spp. (34.3%), *Betula* spp. (32.9%), and *Acer* spp. (32.9%), followed by *Populus* spp. (20.6%) and *Ulmus* spp. (24.6%) [15]. Erkara surveyed the pollen grains in the atmosphere of Sivrihisar, Turkey for two continuous years, and the result showed that the majority of the allergenic pollen grains were from *Pinaceae*, *Cupressaceae*, *Fraxinus* spp., *Cedrus* spp., *Populus* spp., and so on [16]. In Talca, Chile, the highest number of airborne pollen grains in the atmosphere was from *Platanus acerifolia* (203 grains/m^3, day), and *Populus* spp. had a maximum weekly daily average 103 grains/m^3 [17].

Proteomics is a powerful tool to understand the dynamic changes of proteins [18]. In this study, we compared the differences of pollen germination and proteome between two *P. deltoides* varieties, 'ZhongHuai1' ('ZH1', triploid) and 'ZhongHuai2' ('ZH2', diploid), which were generated from the same parents (*P. deltoides* '55/56' × *P. deltoides* 'Imperial') [19–21]. In our previous study, a total of 28 putative allergenic proteins were identified from mature pollen of diploid *P. deltoides* CL. '2KEN8' using proteome research approach [22]. Here, the protein abundances and gene expression levels of the 28 putative allergenic proteins were compared among the three varieties ('ZH1', 'ZH2', and '2KEN8'). This study is helpful for understanding the molecular mechanism of differences in pollen fertility and allergenicity between triploid and diploid poplar.

2. Results

In our previous study, we identified a triploid *P. deltoides* variety 'ZH1' and a diploid variety 'ZH2' from the same parents (*P. deltoides* '55/56' × *P. deltoides* 'Imperial'), and the ploidy levels of the hybrids were determined using flow cytometric analysis [21]. To confirm the ploidy level of the two varieties, we further analyzed the karyotypes of the two materials. As shown in Figure 1A, the chromosome numbers of 'ZH1' and 'ZH2' were $2n = 3x = 57$ and $2n = 2x = 38$, respectively. The result showed that 'ZH1' was triploid and 'ZH2' was diploid.

2.1. Pollen Germination in Triploid and Diploid P. deltoides

To compare the pollen fertility of triploid and diploid poplar, the pollen of 'ZH1' and 'ZH2' was germinated in vitro (Figure 1B). After 2 h cultivation, the pollen of 'ZH2' began to germinate, and the germination rate was increased as a function of culture time. Over 88% pollen of diploid 'ZH2' was germinated when the pollen was cultivated for 48 h, while the pollen of triploid 'ZH1' was not germinated at all, even after 48 h cultivation (Figure 1B,C). The length of pollen tubes in different culture times of diploid 'ZH2' was measured, which increased with the elongation of culture time.

At the beginning 4 h, most pollen tubes of 'ZH2' were distributed, ranging from 0 to 200 μm. After 48 h cultivation, most pollen tubes of 'ZH2' had a length of 500–1000 μm (Figure 1D).

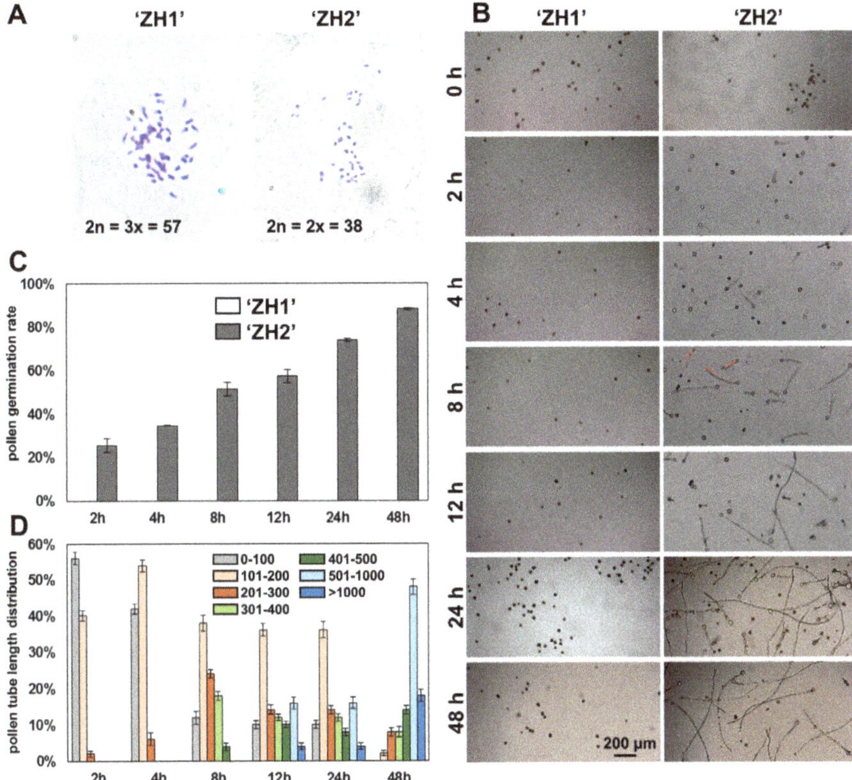

Figure 1. Pollen germination of triploid 'ZH1' and diploid 'ZH2' *P. deltoides*. (**A**) Somatic chromosome number in triploid poplar 'ZH1' and diploid poplar 'ZH2'; (**B**) pollen tube growth of 'ZH1' and 'ZH2' during the germination (2, 4, 8, 12, and 24 h); (**C**) pollen germination rate of 'ZH1' and 'ZH2'; (**D**) pollen tube length distribution of 'ZH2' pollen during the germination.

2.2. Two-Dimensional Gel Electrophoresis (2-DE) Analysis and Identification of Differentially Expressed Proteins

As shown in Figure 2, the pollen proteins of 'ZH1' and 'ZH2' were separated using 2-DE. A total of 557 and 598 repeatable protein spots were identified from 'ZH1' and 'ZH2' mature pollen, respectively. Among these protein spots, 488 were matched in the two varieties' mature pollen. The matched proteins covered the isoelectric point (pI) ranging from 5.0 to 6.5, and their molecular weight (MW) ranged from 17 to 63 kDa. To compare the differences of mature pollen proteins between triploid 'ZH1' and diploid 'ZH2', the intensity of each matched spot from three biological replicates was analyzed using ImageMaster 2D Platinum software version 6.0 (GE Healthcare, Little Chalfont, UK), and significantly ($p \leq 0.05$) altered spots were identified with the Student's *t*-test. A total of 72 repeatable differentially expressed protein spots (DEPs) were identified; 22 were upregulated and 50 were downregulated in the mature pollen of triploid poplar 'ZH1' (Table 1).

Table 1. Differentially expressed protein spots (DEPs) identified in the pollen of triploid 'ZH1' and diploid 'ZH2' *P. deltoides*.

Spot Number	Gene ID	Annotation	Score	Coverage (%)	Experimental M_W (kDa)/Isoelectric Point (pI)	Theoretial M_W (kDa)/pI	Average Fold Change ('ZH1'/'ZH2')	p-Value
		Carbohydrate Metabolism and Energy Metabolism Related Proteins						
29	Potri.010G027800	Pyruvate orthophosphate dikinase	663	57	28/6.07	39.78/5.64	−5.83	0.017
2	Potri.002G082100	PDI-like 1-2	396	41	50/4.58	43.88/4.64	−5.57	0.050
28	Potri.019G063600	pfkB-like carbohydrate kinase family protein	115	56	25/5.78	35.27/5.81	−5.04	0.049
15	Potri.015G131100	Enolase	184	71	42/5.88	47.74/5.66	−4.03	0.045
18	Potri.015G131100	Enolase	290	54	44/5.56	47.74/5.66	−4.03	0.046
41	Potri.002G181300	Pyrophosphorylase 1	400	46	23/5.98	24.88/5.92	−3.83	0.008
10	Potri.015G131100	Enolase	290	68	45/5.87	47.74/5.66	−3.64	0.041
49	Potri.002G134600	P-loop containing nucleoside triphosphate hydrolases superfamily protein	191	66	15/5.48	23.04/5.56	−3.40	0.012
43	Potri.006G082500	Pyrophosphorylase 4	322	48	20/6.66	24.92/5.92	−3.20	0.028
12	Potri.015G131100	Enolase	290	68	45/5.91	47.74/5.66	−2.91	0.044
31	Potri.010G117900	Aldolase superfamily protein	61	27	25/5.94	42.8/6.44	−2.83	0.030
27	Potri.001G061400	Transketolase family protein	396	47	24/5.53	38.85/5.87	−2.82	0.028
30	Potri.008G166800	Lactate/malate dehydrogenase family protein	206	49	29/6.55	36.3/6.61	−2.56	0.013
55	Potri.002G134600	P-loop containing nucleoside triphosphate hydrolases superfamily protein	63	41	12/5.57	12.18/5.45	−2.44	0.018
13	Potri.006G116800	Enolase	90	55	39/5.57	47.84/5.56	−2.17	0.041
14	Potri.006G116800	Enolase	175	60	39/5.85	47.84/5.56	−2.17	0.003
26	Potri.001G061400	Transketolase family protein	95	51	24/5.45	28.85/5.87	−2.05	0.016
52	Potri.003G060100	Rubredoxin-like superfamily protein	276	50	10/5.34	13.21/5.66	−2.03	0.015
11	Potri.017G144700	UDP-glucose pyrophosphorylase 2	118	47	34/6.44	51.74/6.78	−2.01	0.013
51	Potri.013G102100	Thioredoxin superfamily protein	173	46	11/4.97	11.92/5.32	−2.01	0.051
25	Potri.017G029000	pfkB-like carbohydrate kinase family protein	156	66	25/4.89	25.47/4.98	−2.00	0.030
50	Potri.001G173800	Rubredoxin-like superfamily protein	205	75	11/4.88	11.63/5.60	2.12	0.013
53	Potri.018G083500	Thioredoxin-dependent peroxidase 1	467	74	10/5.43	17.52/5.55	2.98	0.042
		Amino Acid Metabolism Related Proteins						
22	Potri.002G189000	S-adenosylmethionine synthetase 2	72	47	35/5.49	43.61/5.62	−5.24	0.022
9	Potri.006G123200	Methionine adenosyltransferase 3	617	65	35/5.65	43.00/5.76	−4.69	0.018
19	Potri.006G123200	Methionine adenosyltransferase 3	617	65	35/5.65	43.00/5.76	−3.40	0.018
17	Potri.008G099300	S-adenosylmethionine synthetase family protein	483	79	33/5.48	43.60/5.50	−2.31	0.036
48	Potri.010G003500	Cytidine/deoxycytidylate deaminase family protein	90	59	16/5.18	20.74/5.26	−2.01	0.007

Table 1. *Cont.*

Spot Number	Gene ID	Annotation	Score	Coverage (%)	Experimental M_W (kDa)/Isoelectric Point (pI)	Theoretial M_W (kDa)/pI	Average Fold Change ('ZH1'/'ZH2')	*p*-Value
		Protein Metabolism Related Proteins						
20	Potri.001G153000	Hyaluronan / mRNA binding family	65	28	35/6.21	39.79/6.25	−4.06	0.031
21	Potri.002G215900	GTP binding Elongation factor Tu family protein	70	43	35/5.78	49.29/7.66	−3.17	0.018
54	Potri.002G056200	Ribosomal protein L7Ae/L30e/S12e/Gadd45 family protein	311	65	10/5.44	15.66/5.48	−2.98	0.050
58	Potri.009G146200	60S acidic ribosomal protein family	127	30	11/4.33	11.39/4.36	−2.65	0.027
47	Potri.003G109200	Mitochondrion-localized small heat shock protein 23.6	118	52	16/5.13	23.96/5.36	−2.64	0.017
42	Potri.007G079500	Copper ion binding;cobalt ion binding;zinc ion binding	467	58	17/6.54	27.83/6.50	−2.56	0.015
7	Potri.003G006300	Chloroplast heat shock protein 70-2	593	39	75/5.14	75.41/5.24	−2.36	0.008
8	Potri.001G087500	Heat shock protein 70 (Hsp 70) family protein	142	39	75/5.18	84.74/5.04	−2.16	0.017
59	Potri.018G057600	Profilin 5	91	83	9/4.57	9.90/4.76	−2.14	0.011
44	Potri.018G063200	Chaperonin 20	227	68	18/6.34	26.87/6.64	−2.00	0.037
64	Potri.009G022300	Cystatin B	127	77	8/6.43	11.23/5.59	−2.00	0.026
67	Potri.009G039600	Tim10/DDP family zinc finger protein	109	66	6/5.45	9.86/5.56	2.00	0.026
36	Potri.011G089000	Co-chaperone GrpE family protein	79	48	18/5.12	34.58/6.04	2.01	0.010
37	Potri.015G122400	Proteasome subunit PAB1	147	54	17/5.23	21.73/5.95	2.01	0.012
38	Potri.015G122400	Proteasome subunit PAB1	147	63	17/6.09	21.73/5.95	2.02	0.012
57	Potri.004G187200	HSP20-like chaperones superfamily protein	121	36	11/5.34	13.05/5.54	2.06	0.035
40	Potri.006G008800	20S proteasome α subunit C1	404	74	20/5.87	27.55/5.96	2.08	0.038
62	Potri.012G039100	Tim10/DDP family zinc finger protein	161	40	8/5.79	11.23/5.81	2.10	0.001
23	Potri.009G018600	Glutathione S-transferase, C-terminal-like;Translation elongation factor EF1B/ribosomal protein S6	130	56	24/4.58	24.59/4.62	2.34	0.044
24	Potri.003G081000	Ubiquitin C-terminal hydrolase 3	313	43	24/4.65	21.64/4.78	2.71	0.036
60	Potri.003G047700	Profilin 3	411	58	9/4.72	14.21/4.71	4.10	0.021
61	Potri.001G190800	Profilin 1	132	51	8/5.26	14.22/4.75	5.59	0.013

Table 1. *Cont.*

Spot Number	Gene ID	Annotation	Score	Coverage (%)	Experimental M_W (kDa)/Isoelectric Point (pI)	Theoretial M_W (kDa)/pI	Average Fold Change ('ZH1'/'ZH2')	p-Value
		Defense or Stress Related Protein						
16	Potri.003G113400	Stress-inducible protein, putative	106	39	61/6.22	65.81/6.17	-2.90	0.034
56	Potri.003G107100	Lipase/lipooxygenase, PLAT/LH2 family protein	139	48	8/5.83	10.37/6.05	2.00	0.015
5	Potri.007G024000	Late embryogenesis abundant (LEA) protein	162	44	34/4.61	44.97/4.64	2.12	0.016
6	Potri.007G024000	Late embryogenesis abundant (LEA) protein	721	57	34/4.75	44.41/4.81	2.30	0.021
3	Potri.007G024000	Late embryogenesis abundant (LEA) protein	124	47	48/4.98	44.41/4.81	2.34	0.034
63	Potri.013G031100	Copper/zinc superoxide dismutase 1	154	35	8/6.33	21.07/7.34	2.77	0.028
4	Potri.007G024000	Late embryogenesis abundant (LEA) protein	101	40	50/4.89	44.97/4.64	3.00	0.030
33	Potri.002G034400	NmrA-like negative transcriptional regulator family protein	280	60	25/5.62	23.96/5.51	3.11	0.338
66	Potri.004G107100	Late embryogenesis abundant protein (LEA) family protein	73	64	8/5.28	7.11/6.18	3.94	
65	Potri.017G108400	Late embryogenesis abundant protein (LEA) family protein	124	80	7/5.23	6.93/6.13	4.79	
		Signal Transduction Proteins						
1	Potri.013G009500	Calreticulin 1b	86	39	47/4.02	51.46/4.92	−2.34	0.024
32	Potri.002G095600	Annexin 1	80	34	22/6.87	23.69/6.45	−2.10	0.019
		Allergy Related Proteins						
45	Potri.001G392400	Pollen Ole e 1 allergen and extensin family protein	79	36	12/4.89	18.15/4.78	−5.47	0.018
46	Potri.011G111300	Pollen Ole e 1 allergen and extensin family protein	483	51	15/4.55	18.32/4.85	2.59	0.015
		Unknown Proteins						
34	Potri.001G034400	Nascent polypeptide-associated complex (NAC), α subunit family protein	100	31	22/4.34	22.31/4.34	−3.84	0.038
35	Potri.004G135300	Catalytic LigB subunit of aromatic ring-opening dioxygenase family	171	69	23/4.87	29.67/5.88	−2.09	0.041
39	Potri.005G085500	Copper ion binding;cobalt ion binding;zinc ion binding	262	61	22/7.23	28.02/7.71	2.45	0.047

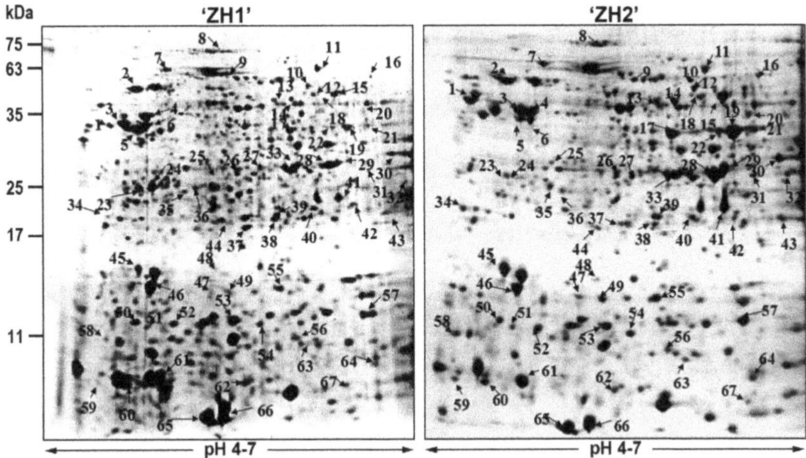

Figure 2. Two-dimensional gel electrophoresis (2-DE) image analysis of triploid 'ZH1' and diploid 'ZH2' *P. deltoides*. The numbers with arrows indicate the differentially expressed protein spots (DEPs) with $p < 0.05$.

2.3. Functional Classification of DEPs

Subsequently, 67 of the 72 DEPs were successfully identified using MALDI-TOF/TOF MS/MS (Table 1). For most of the identified proteins, the experimental MW and pI were basically consistent with the theoretical values of the identified proteins. These proteins are involved in various metabolic pathways or processes, such as carbohydrate-metabolism- and energy-metabolism-related proteins (34.3%), amino-acid-metabolism-related proteins (7.5%), protein-metabolism-related proteins (32.8%), defense- or stress-related proteins (14.9%), signal transduction proteins (3%), and allergy-related proteins (3%) (Table 1).

In addition, the 67 proteins were grouped by Gene Ontology (GO) system into biological process (BP), molecular function (MF), and cellular compartment (CC) classes. In terms of BP, 30% of the DEPs were involved in the biological process of energy provision and 19% DEPs were involved in the lipid metabolic process. For MF category, more than half of these proteins belonged to catalytic activity and transporter activity, accounting for 38.80% and 22.39%, respectively. On the basis of predicted subcellular location, the proteins were classified into 10 categories in cellular compartments; about 21% were located in endoplasmic reticulum (Figure 3).

In order to further understand functional categories of DEPs, REVIGO reduction analysis tool was used to summarize GO terms together with their *P*-values [23]. According to categories BP, gene clusters represent "S-adenosylmethionine metabolism", "actin cytoskeleton organization", "translational elongation", "cellular processes", and "catabolism". Among them, "S-adenosylmethionine metabolism" and "actin cytoskeleton organization" were the main significantly different biological processes between triploid and diploid poplar mature pollen. The MF indicated seven GO terms related with "cytochrome oxidase activity", "methionine adenosyltransferase activity", "inorganic diphosphatase activity", "cytoskeletal protein binding", "phosphopyruvate hydratase activity", "magnesium ion binding", and "translation elongation factor activity". For categories based on CC, the identified proteins were mainly related with actin cytoskeleton (Figure 4).

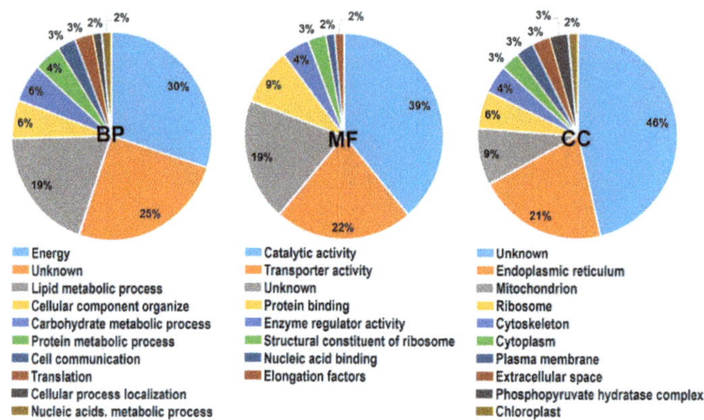

Figure 3. Functional annotation of identified DEPs based on Gene Ontology (GO) categorization. Results are summarized for three main GO categories: biological process (BP), molecular function (MF), and cellular component (CC).

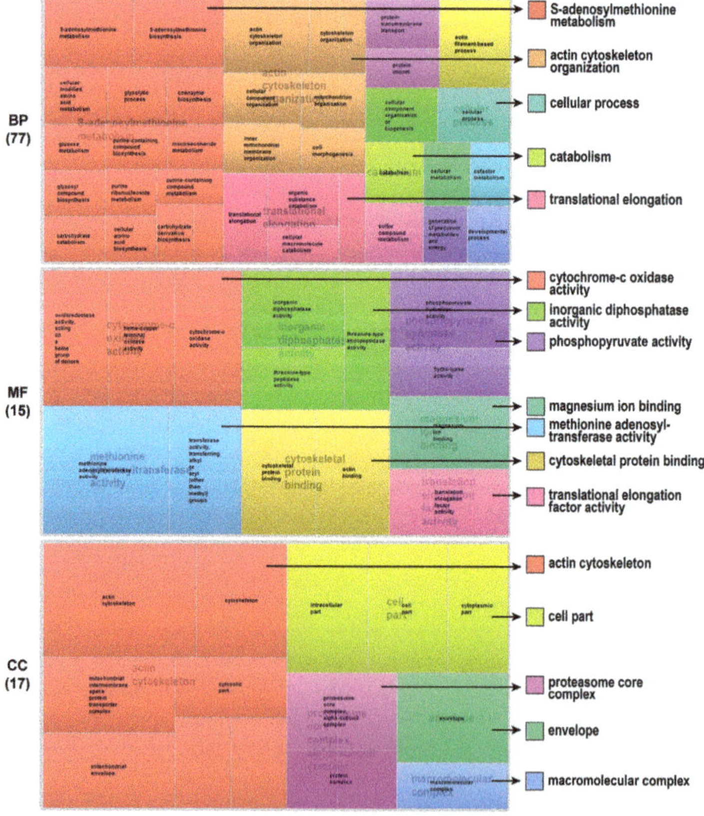

Figure 4. GO treemaps for the DEPs in 'ZH1' and 'ZH2' pollen. GO terms for proteins identified in 'ZH1' and 'ZH2' pollen are shown. The box size correlates to the −log10 *P*-value of the GO term. Boxes with the same color can be grouped together and correspond to the same upper-hierarchy GO-term.

2.4. Expression Patterns of Identified DEPs

To explore the potential roles of the DEPs in poplar developmental processes, we investigated the expression profiles of identified DEPs across various tissues. The majority of these genes showed tissue-specific expression patterns (Figure 5). Based on the expression patterns, the genes were classified into four clusters (Cluster 1–4). As shown in Figure 5, only genes in Cluster 3 were highly expressed in the female and male catkins. Genes in Cluster 1 had relatively low abundance in female and male catkins, genes in Cluster 2 were highly expressed in seedlings (including S.CL—seedling continuous light, ES.L—etiolated seedling 6 days transferred to light 3 h, and ES—etiolated seedling), while genes in Cluster 4 were highly expressed in differentiating xylem.

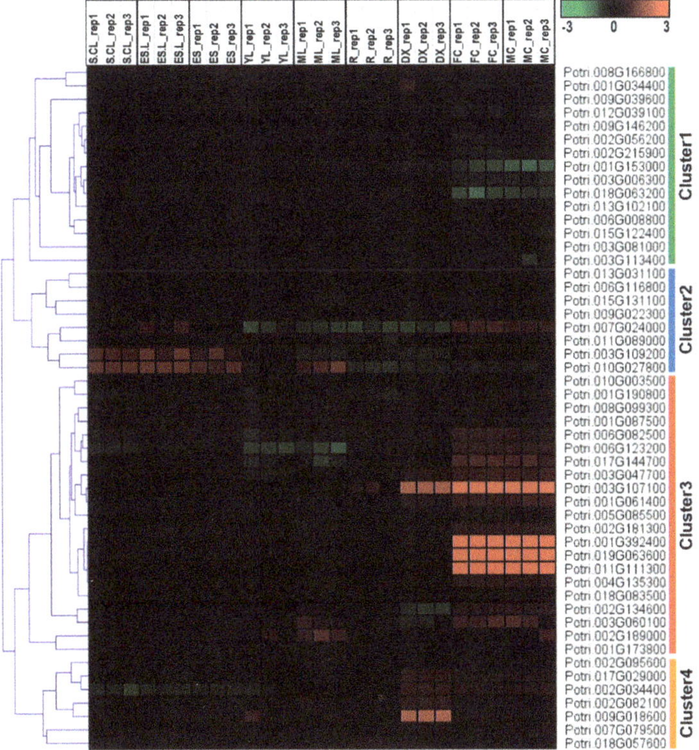

Figure 5. Expression profiles of genes coding DEPs between 'ZH1' and 'ZH2' mature pollen across different tissues. The Affymetric microarray data were obtained from the National Center for Biotechnology Information (NCBI) Gene Expression Omnibus (GEO) database under the series accession number GSE13990. S.CL—Seedling continuous light; ES.L—Etiolated seedling 6 days transferred to light 3 h; ES—Etiolated seedling; YL—Young leaf; ML—Mature leaf; R—Root; DX—Differentiating xylem; FC—Female catkin; MC—Male catkin. Background-corrected expression intensities were log-transformed and visualized as heatmaps. Color scale represents log2 expression values, green represent low level and red indicates high level of transcript abundance.

2.5. Comparative Analysis of Predicted Allergenic Proteins

In our previous study, we identified 28 candidate allergenic proteins in diploid *P. deltoides* '2KEN8' mature pollen [22]. Here, we compared the abundance of these 28 candidate allergenic proteins among the three varieties ('ZH1', 'ZH2', and '2KEN8') (Figure 6A and Table 2). The expression profiles of the predicted allergenic proteins in the two varieties ('ZH1' and 'ZH2') were similar compared with '2KEN8'. For example, 6 and 12 predicted allergenic proteins were up- and downregulated in both 'ZH1'/'2KEN8' and 'ZH2'/'2KEN8', respectively. Overall, most allergenic proteins in 'ZH1' and 'ZH2' mature pollen have relatively lower abundance than that in '2KEN8' (Figure 6B,C and Table 2). In the triploid 'ZH1', 14 predicted allergenic proteins were downregulated and 7 were upregulated when compared with the diploid 'ZH2' (Figure 6D). To confirm the expression profiles of the genes coding the putative allergenic proteins, qRT-PCR analysis was performed on 'ZH1' and 'ZH2' mature pollen for six genes (Figure 6E). Protein expression and gene expression of the six protein spots were analyzed; protein expression and gene expression were almost proportional in relationship.

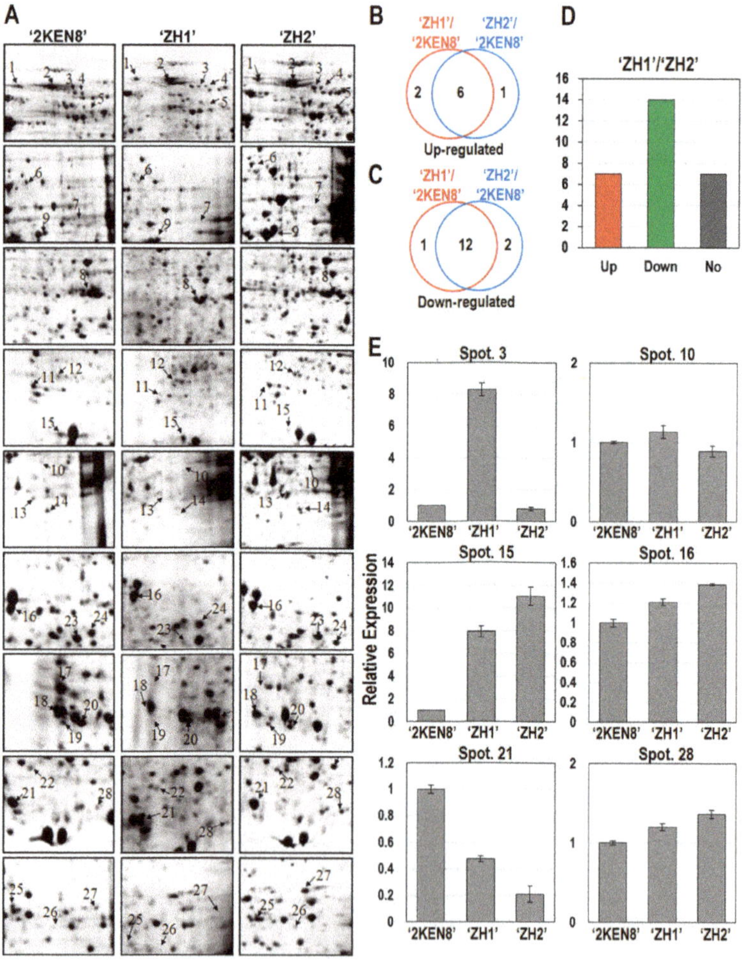

Figure 6. Predicted allergenic proteins in *P. deltoides* 'ZH1', 'ZH2', and '2KEN8' mature pollen. (**A**) The protein abundance of previously predicted 28 allergenic proteins [22] in three *P. deltoides* cultivars 'ZH1', 'ZH2', and '2KEN8' mature pollen; Venn diagram of upregulated (**B**) or downregulated (**C**) proteins in 'ZH1' and 'ZH2' compared with '2KEN8' from (**A**); (**D**) differentially expressed proteins among 28 predicted allergenic proteins between 'ZH1' and 'ZH2'; (**E**) expression analysis of six selected allergen related genes in the three cultivars using qRT-PCR.

Table 2. Differential expression of 28 candidate allergenic proteins.

Spot Number	Gene ID	Annotation	Average Fold Changes		
			'ZH1'/'2KEN8'	'ZH2'/'2KEN8'	'ZH1'/'ZH2'
1	Potri.003G006300	Chloroplast heat shock protein 70-2	−1.48	−1.30	−2.59
2	Potri.001G087500	Heat shock protein 70 (Hsp 70) family protein	−1.88	2.87	−5.57
3	Potri.001G285500	Mitochondrial HSO70 2	1.49	−1.72	3.33
4	Potri.009G079700	Mitochondrial HSO70 2	−1.42	−1.16	−1.31
5	Potri.006G116800	Enolase	−4.08	−1.17	−2.17
6	Potri.015G131100	Enolase	−2.06	−3.89	−2.91
7	Potri.019G067200	Pectin lyase-like superfamily protein	–	–	–
8	Potri.002G034400	NmrA-like negative transcriptional regulator family protein	2.19	3.48	−1.29
9	Potri.012G114900	Pectin lyase-like superfamily protein	4.81	6.55	−3.10
10	Potri.010G117900	Aldolase superfamily protein	–	–	–
11	Potri.001G034400	Nascent polypeptide-associated complex (NAC), α subunit family protein	−3.56	−1.98	2.88
12	Potri.009G018600	Glutathione S-transferase, C-terminal-like;Translation elongation factor EF1B/ribosomal protein S6	2.17	4.57	−2.37
13	Potri.008G056300	Triosephosphate isomerase	−3.61	−1.74	3.12
14	Potri.013G092600	Manganese superoxide dismutase 1	1.03	−1.52	0.77
15	Potri.001G392400	Pollen Ole e 1 allergen and extensin family protein	3.86	4.07	−1.37
16	Potri.011G111300	Pollen Ole e 1 allergen and extensin family protein	1.37	1.54	−1.62
17	Potri.009G146200	60S acidic ribosomal protein family	–	–	–
18	Potri.006G235200	Profilin 4	-4.10	−3.45	−1.99
19	Potri.018G057600	Profilin 4	–	–	–
20	Potri.003G047700	Profilin 3	−6.93	−3.48	4.10
21	Potri.001G190800	Profilin 5	−3.14	−1.36	5.59
22	Potri.007G018000	Thioredoxin H-type 1	−2.29	−1.33	−3.19
23	Potri.018G083500	Thioredoxin-dependent peroxidase 1	−1.11	−3.55	3.97
24	Potri.009G147900	HSP20-like chaperones superfamily protein	−4.63	−6.33	−2.79
25	Potri.001G254700	HSP20-like chaperones superfamily protein	–	–	–
26	Potri.006G093500	HSP20-like chaperones superfamily protein	–	–	–
27	Potri.001G254700	HSP20-like chaperones superfamily protein	–	–	–
28	Potri.005G232700	Thioredoxin H-type 1	0.36	0.89	−0.63

3. Discussion

3.1. The Mature Pollen of Triploid Poplar 'ZH1' Failed to Germinate

As a widespread biological process, polyploidization has provided much genetic variation for plant adaptive evolution. It not only provides extra gene copies, strengthening the robustness against malignant mutations, but also provides abundant genetic materials for neofunctionalization. Therefore, polyploidy has been considered as an important force in the evolution of plants [24–26]. However, polyploidy has always been accompanied by low fertility. Soltis et al. reported that the rates of pollen germination in the 187 triploid grapes ranged from 0% to 5.88%, and about 46% of the 187 triploids showed no germination [21]. Triploid clones of *Hypericum androsaemum* had 0%–6% pollen germination rate, and the seeds of the triploids were fewer than the diploids. In addition, the seeds from the triploids failed to germinate [27]. In this study, we observed that the mature pollen of triploid poplar 'ZH1' failed to germinate, which is similar with the findings in many other triploid species.

The ploidy of plant species will affect the protein abundance in proteomic level. An et al. analyzed the leaf proteomes of cassava diploid and autotetraploid genotypes; 47 upregulated proteins and 5 downregulated proteins were identified in autotetraploid genotype [25]. In our study, the proteomic analysis showed that significant differences in protein expression patterns between triploid and diploid poplar pollen. A total of 72 DEPs were identified, 22 were upregulated and 50 were downregulated in mature pollen of triploid poplar 'ZH1'. Most of the identified DEPs showed similar molecular weights and pI values between the experimental and the theoretical results (Table 1). The individuals with large differences between experimental and theoretical results (e.g., spot 21) might be caused by the different variants generated by alternative splicing events. Among the DEPs, most proteins were related with "carbohydrate metabolism and energy metabolism", "amino acid metabolism", "protein metabolism", and "defense or stress". However, in regard to proteins related to the "defense or stress", there were 9 upregulated and 1 downregulated protein in 'ZH1' compared with 'ZH2'. In addition, most of the DEPs related to defense or stress belong to Cluster 3 and Cluster 4, which were highly expressed in the female and male catkins.

In *Arabidopsis*, ploidy affects various morphological and fitness traits, such as stomata size, flower size, and seed weight [28]. To further understand the function of DEPs that were associated with the sterility of triploid, the annotation and enrichment of the GO was conducted. GO-enrichment analysis revealed that the GO terms related with actin cytoskeleton were significantly enriched in DEPs between triploid and diploid poplar mature pollen (Figure 4). As an important component of pollen grain, cytoskeleton controls not only how the pollen tube grows but also how the cytoplasm dynamically reorganizes during tube elongation [29]. During the fertilization, the pollen tube must be guided to enter the ovule via the micropyle with the involvement of actin filaments and actin-binding proteins. *Arabidopsis* microtubule-associated protein 18 (MAP18) modulates actin filaments to directional cell growth and cortical microtubule organization [30]. Generally, actin filaments in pollen tubes will arrange into higher-order longitudinal actin cables to generate the reverse fountain cytoplasmic streaming pattern. During the process, the actin-depolymerizing factor (ADF7) evolved to promote turnover of longitudinal actin cables by severing actin filaments [31]. The abnormal cytoskeletal systems might be related with low pollen viability and abnormal pollen tube elongation in pollen of triploid poplar 'ZH1'.

In this study, the ubiquitin proteasome carboxyl terminal hydrolase (spot 24) was upregulated in pollen of triploid poplar 'ZH1'. In eukaryotic organisms, the ubiquitin carboxyl terminal hydrolytic enzymes are involved in short-life proteins turn over and some abnormal protein degradation pathways, which play important control functions in cell growth, signal transduction, and aspects such as plant senescence [32–34]. Lin et al. found that ubiquitin carboxyl terminal hydrolase in sterile lines was highly expressed, and speculated that the ubiquitin–proteasome pathway is closely related to wheat male sterility [35]. Sequence alignment showed that the protein in spot 39 shares 71.1% similarity with a 24 kDa protein of FoF1-ATP complex in *Pyrus bretschneideri* (not shown). FoF1-ATP synthase is the key enzyme in the oxidative phosphorylation and phosphorylation in vivo [36]. Previous studies showed that pollen of mitochondrial ATP synthase subunit has an important role in the development of the male gametophyte. FoF1-ATP synthase function disorders affect mitochondrial energy output, leading to anther dysplasia [37,38]. In this study, FoF1-ATP synthase showed upregulated in pollen of triploid poplar 'ZH1'. It is suggested that the low fertility of 'ZH1' may be related to the high expression level of FoF1-ATP synthase.

3.2. Allergenic Proteins Were Differentially Expressed in Mature Pollen of 'ZH1' and 'ZH2'

Based on our previously identified 28 allergenic proteins in *P. deltoides* CL. '2KEN8' [22], the expression patterns of allergenic proteins in 'ZH1', 'ZH2', and '2KEN8' were compared. The expression of candidate allergenic proteins in 'ZH1' and 'ZH2' were lower than that in '2KEN8'. To a certain degree, it could be considered that the potential allergenicity of pollen in triploid poplar 'ZH1' and diploid poplar 'ZH2' may be lower than '2KEN8'. Compared with the diploid poplar 'ZH2', most allergenic proteins showed low expression in pollen of triploid poplar 'ZH1'. It implies that not only germination, but also potential allergenicity of the pollen, in triploid poplar 'ZH1' were lower than diploid poplar 'ZH2'. Cry j1 and Cry j2 are the major allergens in pollen of Japanese cedar (*Cryptomeria japonica*). Kondo et al. observed that the amounts of Cry j1 and Cry j2 proteins in the pollen of triploid Japanese cedar were less than diploid [39]. It could be deduced that the pollen in triploid might be less allergenic.

Based on the genome-wide prediction, Chen et al. predicted 145 and 107 pollen allergens from rice and *Arabidopsis*, respectively. These allergens are putatively involved in stress responses and metabolic processes such as cell wall metabolism during pollen development [40]. In the 28 candidate allergenic proteins from poplar mature pollen, several members belong to heat shock protein or profilin families. The class I small heat shock proteins (Hsps) were reported as one class of allergens in soybean [41]. Profilins are pan-allergen proteins present in various edible plant parts and pollen, such as birch pollen Bet v2, olive pollen Ole e2, grass pollen Phl p12, and soybean allergen Gly m3 [42,43]. Enolase is a glycolytic enzyme which was identified as a class of highly conserved fungal allergens, such as *Cladosporium herbarum*, *Alternaria alternata*, *Curvularia lunata*, *Penicillium citrinum*, and *Aspergillus fumigatus* [44–46]. Triosephosphate isomerase was described as allergen in wheat, latex, and lychee [47,48]. The remaining allergenic proteins are mainly involved in protein synthesis and degradation, but whether they can cause allergic reactions needs to be further studied. In this study, the expression of two Ole e1 were downregulated in pollen of triploid poplar 'ZH1' compared with the diploid poplar 'ZH2' (Table 2). Ole e1 is a major allergen which was first identified in olive tree pollen [49]. The proteins encoded by tomato *LAT52* gene and maize *Zmc13* gene also showed high similarity to Ole e1 [50,51]. Although Ole e1 plays important role in pollen physiology (e.g., hydration, pollen tube germination or growth, and other reproductive processes), its biological function is not yet known [52].

4. Materials and Methods

4.1. Plant Materials and Pollen Collection

Experimental materials in this study were collected from Chinese Academy of Forestry (Beijing, China). The triploid poplar 'ZH1' and diploid poplar 'ZH2' were obtained from the same parents by artificial cross-breeding. The cut flowering branches were cultured in water in greenhouse. At anthesis, fresh pollen was collected in the morning by shaking the tassel in a plastic bag, while old pollen and anthers were removed from tassels by vigorously shaking the evening of the day before.

4.2. Determination of the Ploidy Level

Root tips (0.5–1.0 cm) were collected from young plants and pretreated with saturated aqueous solution of *p*-dichlorobenzene (Sigma-Aldrich, Steinheim, Germany) for 3 h (room temperature), then fixed in Carnoy (glacial acetic acid:absolute ethanol 1:3) (Sigma-Aldrich, Steinheim, Germany) for 1 h. After treatments, the root tips were washed in distilled water, hydrolyzed in 1 mmol/l HCl (Sigma-Aldrich, Steinheim, Germany) for 45 min at 45 °C, stained using phenol-fuchsin solution (Sigma-Aldrich, Steinheim, Germany) and squashed [53,54]. Karyotype symmetry was classified according to Stebbins [55].

4.3. Pollen Germination

Pollen was germinated on liquid germination medium (15% sucrose, 100 mg/L H_3BO_3, 300 mg/L $CaCl_2$, 200 mg/L $MgSO_4$, 100 mg/L KNO_3, pH 6.0) (Sigma-Aldrich, Steinheim, Germany) at 22 °C in the dark. Pollen of 'ZH1' and 'ZH2' was germinated in the same conditions. Pollen germination rates and the lengths of pollen tubes were measured microscopically after 2, 4, 8, 12, 24, and 48 h of incubation. For germination rates, each sample was observed in 15 fields of view. At least 30 pollen grains were analyzed in each field. Pollen grains were considered to be germinated when the pollen tube grows longer than the diameter of the pollen grain. For pollen tube growth, 50 pollen tubes were measured using ImageJ software (National Institutes of Health, Bethesda, MD, USA) [56] at each time point. All the experiments were performed in three biological replicates.

4.4. Protein Extraction and 2-DE Gel Electrophoresis

Protein extraction and 2-DE gel electrophoresis were performed as previously described [15]. Briefly, the total protein of pollen was extracted according to trichloroacetic acid (TCA)-acetone precipitation method, and the protein concentration was determined using the Bradford method [57]. The 2-DE was performed using the IPGphor system and IPG dry strips (18 cm, pH 4–7, nonlinear gradient) (GE Healthcare, Buckinghamshire, UK). For each sample, 450 µg of protein was uploaded on a strip which was saturated in rehydration solution. After isoelectric focusing (IEF), strips were immediately equilibrated for 15 min in a buffer containing 0.1 M Tris-HCl (pH 8.8), 2% (w/v) SDS, 6 M urea, 30% (v/v) glycerol and 0.1 M DTT, and another 15 min in the same buffer containing 0.25 M iodoacetamide without DTT. The second dimension was performed using 12.5% SDS-PAGE at 20 mA/bloc until the dye front reached the end of the gel in a PROTEAN II xi multi-cell (Bio-Rad, Richmond, CA, USA). After electrophoresis, the gel was fixed overnight in the stationary solution (50% (v/v) ethanol with 10% (v/v) orthophosphoric acid). And then transferred the gel to the mixed dyeing solution in the shock stained for 13 h and washed with water. Each 2-DE was repeated at least 3 times to ensure the reliability of results. The 2-DE images were assembled in a matchset using the Imagemaster 2D platinum 7.0 software (GE Healthcare, Little Chalfont, UK). After automated spot detection, the matched spots were verified and adjusted manually.

4.5. MALDI-TOF/TOF MS/MS and Database Search

Images of the stained gels were captured with a scanner (UMAX Powerlook 2100 XL; UMAX, Taiwan, China). Spot detection, matching, and background subtraction were performed using the ImageMaster 2D Platinum software (version 6.0; Amersham Biosciences, Uppsala, Sweden), followed by manual editing. All the spots detected in each gel were matched with the corresponding spots from the reference gels. To exclude the likely differences introduced by sample loading or gel staining/destaining, the normalized relative percent volume values (% volume) of the protein spots from three replicates were used for further statistical analysis. Selected spots were digested with gold grade trypsin, and then analyzed by a MALDI-TOF/TOF tandem mass spectrometer ABI4800 proteomics analyzer (Applied Biosystems, Framingham, MN, USA). For protein identification, the acquired MS/MS data were uploaded on the Protein Pilot software (Applied Biosystems, Framingham, MN, USA) and compared against *P. trichocarpa* genome (V3.0) database (https://phytozome.jgi.doe.gov/pz/portal.html). Proteins identified with a Mowse score $\geq$60 ($p < 0.05$) were reported. To annotate the identified proteins, the Gene Ontology (GO) was used to classify the proteins into three main classes, biological process (BP), molecular function (MF), and cellular component (CC). In addition, the enriched GO terms were slimmed in REVIGO web server [23].

4.6. RNA Isolation and qRT-PCR

Total RNA was extracted using the RNeasy Plant Mini Kit (Qiagen, Hilden, Germany) with on-column treatment using RNase-free DNase I (Qiagen, Hilden, Germany) to remove genomic DNA contamination. First-strand cDNA was synthesized with approximately 1 µg RNA using the SuperScript III reverse transcription kit (Invitrogen, Carlsbad, CA, USA) and random primers according to the manufacturer's instructions. Primers for 28 predicted allergenic genes were according to Zhang et al. [22]. All the primer sequences used in this study are listed in Table A1. qRT-PCR was conducted on LightCycler 480 Detection System (Roche, Penzberg, Germany) using SYBR Premix Taq Kit (TaKaRa, Dalian, China) according to the manufacturer's procedure. The *PtActin* gene was used as reference gene.

5. Conclusions

In this study, we analyzed the fertility and proteome of pollen in triploid poplar 'ZH1' and diploid poplar 'ZH2'. Compared with the 'ZH2', the mature pollen in triploid poplar 'ZH1' failed to germinate. Through comparative proteomics, 67 of 72 DEPs were identified using MALDI-TOF/TOF MS/MS. The main functions of DEPs between triploid and diploid poplar pollen were "S-adenosylmethionine metabolism", "actin cytoskeleton organization", and "translational elongation". Furthermore, the abundances of 28 putative allergenic proteins in three varieties ('ZH1', 'ZH2', and '2KEN8') were compared. In short, not only fertility but also potential allergenicity of pollen were decreased in triploid poplar 'ZH1'. This study is helpful for understanding the molecular mechanism of differences in pollen fertility and allergenicity between triploid and diploid poplar.

Acknowledgments: This work was supported by Special Fund for Forestry Scientific Research in the Public Interest (201304103-4) and National High-tech R & D Program of China (2011AA100201) to Jian-Jun Hu. The funders had no role in study design, data collection and analysis, decision to publish, or preparation of the manuscript.

Author Contributions: Jin Zhang and Jian-Jun Hu conceived and designed the experiments; Xiao-Ling Zhang, Jin Zhang, Ying-Hua Guo and Wei Fan performed the experiments; Pei Sun and Hui-Xia Jia analyzed the data; Meng-Zhu Lu and Jian-Jun Hu contributed reagents/materials/analysis tools; Xiao-Ling Zhang and Jin Zhang wrote the paper.

Conflicts of Interest: The authors declare no conflict of interest.

Appendix A

Table A1. The sequences of qRT-PCR primers.

Spot Number	Gene ID	Forward Primer	Reverse Primer
3	Potri.001G285500	TTTGGAAAGAGCCCTAGCAA	CCTGAGTCTGGTTGTCAGCA
10	Potri.010G117900	TTTTTGTCTGGAGGGCAATC	TTTCACCCTCGGCAGAATAC
15	Potri.001G392400	CCAAAATCAGCGAGGGAATA	AGAATCAATGCTCCGGAATG
16	Potri.011G111300	TGCTTCTCCTCCGTTCTCAT	GTCAATGTGCCATTCTCACG
21	Potri.001G190800	ACATGGTGATCCAGGGAGAG	GAGCACCAGCATTACCCTTC
28	Potri.005G232700	TCGAGAAGGGAAAAGGGTCT	ATTGCCTCCACATTCCACTC
Reference	Actin	GTGCTTCTAAGTTCCGAACAGTGC	GACTACCAAAGTGTCTGACCACCA

References

1. Parmacek, M.S.; Epstein, J.A. Cardiomyocyte renewal. *N. Engl. J. Med.* **2009**, *361*, 86. [CrossRef] [PubMed]

2. Adams, K.L.; Wendel, J.F. Novel patterns of gene expression in polyploid plants. *Trends Genet.* **2005**, *21*, 539–543. [CrossRef] [PubMed]

3. Bingham, E.; Groose, R.; Woodfield, D.; Kidwell, K. Complementary gene interactions in alfalfa are greater in autotetraploids than diploids. *Crop. Sci.* **1994**, *34*, 823–829. [CrossRef]

4. Lewis, W.H. *Polyploidy: Biological Relevance*; Lewis, W.H., Ed.; Springer: Dordrecht, The Netherland, 2012. [CrossRef]

5. Meirmans, P.; Den Nijs, H.J.C.; van Tienderen, P. Male sterility in triploid dandelions: Asexual females vs. asexual hermaphrodites. *Heredity* **2006**, *96*, 45–52. [CrossRef] [PubMed]

6. Lambardi, M. Cryopreservation of germplasm of *Populus* (poplar) species. In *Cryopreservation of Plant Germplasm ii*; Springer: Berlin, Germany, 2002; pp. 269–286.

7. Nilsson-Ehle, H. Note regarding the gigas form of *Populus tremula* found in nature. *Hereditas* **1936**, *21*, 272–382.

8. Cai, X.; Kang, X.-Y. In vitro tetraploid induction from leaf explants of populus pseudo-simonii kitag. *Plant Cell Rep.* **2011**, *30*, 1771–1778. [CrossRef] [PubMed]

9. Liu, Y.; Hui, R.; Deng, R.; Wang, J.; Wang, M.; Li, Z. Abnormal male meiosis explains pollen sterility in the polyploid medicinal plant *Pinellia ternata* (araceae). *Genet. Mol. Res.* **2012**, *11*, 112–120. [CrossRef] [PubMed]

10. Madlung, A.; Tyagi, A.P.; Watson, B.; Jiang, H.; Kagochi, T.; Doerge, R.W.; Martienssen, R.; Comai, L. Genomic changes in synthetic arabidopsis polyploids. *Plant J.* **2005**, *41*, 221–230. [CrossRef] [PubMed]

11. Kumar, P.; Singhal, V.; Kaur, D.; Kaur, S. Cytomixis and associated meiotic abnormalities affecting pollen fertility in clematis orientalis. *Biol. Plant.* **2010**, *54*, 181–184. [CrossRef]

12. Kaul, M.L. *Male Sterility in Higher Plants*; Springer: Dordrecht, The Netherland, 2012; Volume 10.

13. Kim, K.-H.; Jahan, S.A.; Kabir, E. A review on human health perspective of air pollution with respect to allergies and asthma. *Environ. Int.* **2013**, *59*, 41–52. [CrossRef] [PubMed]

14. Vrtala, S.; Grote, M.; Duchene, M.; Vanree, R.; Kraft, D.; Scheiner, O.; Valenta, R. Properties of tree and grass pollen allergens: Reinvestigation of the linkage between solubility and allergenicity. *Int. Arch. Allergy Immunol.* **1993**, *102*, 160–169. [CrossRef] [PubMed]

15. Lin, R.Y.; Clauss, A.E.; Bennett, E.S. Hypersensitivity to common tree pollens in new york city patients. *Allergy Asthma Proc.* **2002**, *23*, 253–258. [PubMed]

16. Erkara, I.P. Concentrations of airborne pollen grains in Sivrihisar (Eskisehir), Turkey. *Environ. Monit. Assess.* **2008**, *138*, 81–91. [CrossRef] [PubMed]

17. Mardones, P.; Grau, M.; Araya, J.; Córdova, A.; Pereira, I.; Peñailillo, P.; Silva, R.; Moraga, A.; Aguilera-Insunza, R.; Yepes-Nuñez, J. First annual register of allergenic pollen in Talca, Chile. *Allergol. Immunopathol.* **2013**, *41*, 233–238. [CrossRef] [PubMed]

18. Wu, W.W.; Wang, G.; Baek, S.J.; Shen, R.-F. Comparative study of three proteomic quantitative methods, DIGE, cICAT, and iTRAQ, using 2D gel-or LC-MALDI TOF/TOF. *J. Proteome Res.* **2006**, *5*, 651–658. [CrossRef]

19. Hu, J.; Li, L.; Li, S.; Zhao, Z.; Han, Y. A new insect-resistant poplar variety 'Zhonghuai1'. *Sci. Silvae Sin.* **2008**, *44*, 169.

20. Li, L.; Hu, J.; Li, S.; Zhao, Z.; Han, Y. A new insect resistant poplar variety 'Zhonghuai2'. *Sci. Silvae Sin.* **2008**, *44*, 174.

21. Jia, H.; Ji, H.; Hu, J.; Lu, M. Fingerprints of SSR markers and ploidy detection for new populus varieties. *Sci. Silvae Sin.* **2015**, *2*, 009. [CrossRef]

22. Zhang, J.; Wu, L.-S.; Fan, W.; Zhang, X.-L.; Jia, H.-X.; Li, Y.; Yin, Y.-F.; Hu, J.-J.; Lu, M.-Z. Proteomic analysis and candidate allergenic proteins in populus deltoides CL.'2KEN8' mature pollen. *Front. Plant Sci.* **2015**, *6*. [CrossRef] [PubMed]

23. Supek, F.; Bošnjak, M.; Škunca, N.; Šmuc, T. Revigo summarizes and visualizes long lists of gene ontology terms. *PLoS ONE* **2011**, *6*, e21800. [CrossRef] [PubMed]

24. Soltis, D.E.; Albert, V.A.; Leebens-Mack, J.; Bell, C.D.; Paterson, A.H.; Zheng, C.; Sankoff, D.; Wall, P.K.; Soltis, P.S. Polyploidy and angiosperm diversification. *Am. J. Bot.* **2009**, *96*, 336–348. [CrossRef] [PubMed]

25. An, F.; Fan, J.; Li, J.; Li, Q.X.; Li, K.; Zhu, W.; Wen, F.; Carvalho, L.; Chen, S. Comparison of leaf proteomes of cassava (Manihot esculenta Crantz) cultivar NZ199 diploid and autotetraploid genotypes. *PLoS ONE* **2014**, *9*, e85991. [CrossRef] [PubMed]

26. Saminathan, T.; Nimmakayala, P.; Manohar, S.; Malkaram, S.; Almeida, A.; Cantrell, R.; Tomason, Y.; Abburi, L.; Rahman, M.A.; Vajja, V.G. Differential gene expression and alternative splicing between diploid and tetraploid watermelon. *J. Exp. Bot.* **2015**, *66*, 1369–1385. [CrossRef] [PubMed]

27. Trueblood, C.E.; Ranney, T.G.; Lynch, N.P.; Neal, J.C.; Olsen, R.T. Evaluating fertility of triploid clones of Hypericum androsaemum L. for use as non-invasive landscape plants. *HortScience* **2010**, *45*, 1026–1028.

28. Miller, M.; Zhang, C.; Chen, Z.J. Ploidy and hybridity effects on growth vigor and gene expression in arabidopsis thaliana hybrids and their parents. *G3: Genes Genomes Genet.* **2012**, *2*, 505–513. [CrossRef] [PubMed]

29. Cai, G.; del Casino, C.; Romagnoli, S.; Cresti, M. Pollen cytoskeleton during germination and tube growth. *Curr. Sci. Bangalore* **2005**, *89*, 1853.

30. Zhu, L.; Zhang, Y.; Kang, E.; Xu, Q.; Wang, M.; Rui, Y.; Liu, B.; Yuan, M.; Fu, Y. MAP 18 regulates the direction of pollen tube growth in arabidopsis by modulating F-actin organization. *Plant Cell* **2013**, *25*, 851–867. [CrossRef] [PubMed]

31. Zheng, Y.; Xie, Y.; Jiang, Y.; Qu, X.; Huang, S. Arabidopsis actin-depolymerizing factor7 severs actin filaments and regulates actin cable turnover to promote normal pollen tube growth. *Plant Cell* **2013**, *25*, 3405–3423. [CrossRef] [PubMed]

32. Hochstrasser, M. Ubiquitin-dependent protein degradation. *Annu. Rev. Genet.* **1996**, *30*, 405–439. [CrossRef] [PubMed]

33. Ciechanover, A. Intracellular protein degradation: From a vague idea thru the lysosome and the ubiquitin-proteasome system and onto human diseases and drug targeting. *Biochim. Biophys. Acta (BBA) Proteins Proteom.* **2012**, *1824*, 3–13. [CrossRef] [PubMed]

34. Ciechanover, A.; Stanhill, A. The complexity of recognition of ubiquitinated substrates by the 26S proteasome. *Biochim. Biophys. Acta (BBA) Mol. Cell Res.* **2014**, *1843*, 86–96. [CrossRef]

35. Lin, W.; Peng, S.; Ai, L.; Chen, S.; Lin, L. Advances in proteomics research of plant male sterility. *Chin. Agric. Sci. Bull.* **2013**, *29*, 277–280.

36. Okuno, D.; Iino, R.; Noji, H. Rotation and structure of FoF1-ATP synthase. *J. Biochem.* **2011**, *149*, 655–664. [CrossRef] [PubMed]

37. Linke, B.; Börner, T. Mitochondrial effects on flower and pollen development. *Mitochondrion* **2005**, *5*, 389–402. [CrossRef] [PubMed]

38. Hanson, M.R.; Bentolila, S. Interactions of mitochondrial and nuclear genes that affect male gametophyte development. *Plant Cell* **2004**, *16*, S154–S169. [CrossRef]

39. Kondo, Y.; Ipsen, H.; Løwenstein, H.; Karpas, A.; Hsieh, L.S. Comparison of concentrations of Cry j 1 and Cry j 2 in diploid and triploid japanese cedar (Cryptomeria japonica) pollen extracts. *Allergy* **1997**, *52*, 455–459. [CrossRef]

40. Chen, M.; Xu, J.; Devis, D.L.; Shi, J.; Ren, K.; Searle, I.; Zhang, D. Origin and functional prediction of pollen allergens in plants. *Plant Physiol.* **2016**, *00625*. [CrossRef] [PubMed]

41. Gagnon, C.; Poysa, V.; Cober, E.R.; Gleddie, S. Soybean allergens affecting north american patients identified by 2D gels and mass spectrometry. *Food Anal. Methods* **2010**, *3*, 363–374. [CrossRef]

42. Valenta, R.; Duchene, M.; Ebner, C.; Valent, P.; Sillaber, C.; Deviller, P.; Ferreira, F.; Tejkl, M.; Edelmann, H.; Kraft, D. Profilins constitute a novel family of functional plant pan-allergens. *J. Exp. Med.* **1992**, *175*, 377–385. [CrossRef] [PubMed]

43. Alvarado, M.; Jimeno, L.; de La Torre, F.; Boissy, P.; Rivas, B.; Lázaro, M.; Barber, D. Profilin as a severe food allergen in allergic patients overexposed to grass pollen. *Allergy* **2014**, *69*, 1610–1616. [CrossRef]

44. Sharma, V.; Gupta, R.; Jhingran, A.; Singh, B.P.; Sridhara, S.; Gaur, S.N.; Arora, N. Cloning, recombinant expression and activity studies of a major allergen "enolase" from the fungus Curvularia lunata. *J. Clin. Immunol.* **2006**, *26*, 360–369. [CrossRef] [PubMed]

45. Lai, H.-Y.; Tam, M.F.; Tang, R.-B.; Chou, H.; Chang, C.-Y.; Tsai, J.-J.; Shen, H.-D. cDNA cloning and immunological characterization of a newly identified enolase allergen from Penicillium citrinum and Aspergillus fumigatus. *Int. Arch. Allergy Immunol.* **2002**, *127*, 181–190. [CrossRef] [PubMed]

46. Simon-Nobbe, B.; Probst, G.; Kajava, A.V.; Oberkofler, H.; Susani, M.; Crameri, R.; Ferreira, F.; Ebner, C.; Breitenbach, M. IgE-binding epitopes of enolases, a class of highly conserved fungal allergens. *J. Allergy Clin. Immunol.* **2000**, *106*, 887–895. [CrossRef] [PubMed]

47. Sander, I.; Flagge, A.; Merget, R.; Halder, T.M.; Meyer, H.E.; Baur, X. Identification of wheat flour allergens by means of 2-dimensional immunoblotting. *J. Allergy Clin. Immunol.* **2001**, *107*, 907–913. [CrossRef] [PubMed]

48. Posch, A.; Chen, Z.; Dunn, M.J.; Wheeler, C.H.; Petersen, A.; Leubner-Metzger, G.; Baur, X. Latex allergen database. *Electrophoresis* **1997**, *18*, 2803–2810. [CrossRef] [PubMed]

49. Villalba, M.; Batanero, E.; Monsalve, R.I.; de La Pena, M.G.; Lahoz, C.; Rodríguez, R. Cloning and expression of ole e I, the major allergen from olive tree pollen. Polymorphism analysis and tissue specificity. *J. Biol. Chem.* **1994**, *269*, 15217–15222. [PubMed]

50. Twell, D.; Wing, R.; Yamaguchi, J.; McCormick, S. Isolation and expression of an anther-specific gene from tomato. *Mol. Gen. Genet.* **1989**, *217*, 240–245. [CrossRef] [PubMed]

51. Hanson, D.D.; Hamilton, D.A.; Travis, J.L.; Bashe, D.M.; Mascarenhas, J.P. Characterization of a pollen-specific cdna clone from zea mays and its expression. *Plant Cell* **1989**, *1*, 173–179. [CrossRef] [PubMed]

52. Jiménez-López, J.C.; de Dios Alche, J.; Rodríguez-García, M.I. *Systematic and Phylogenetic Analysis of the Ole e 1 Pollen Protein Family Members in Plants*; InTECH: Rijeka, Croatia, 2011.

53. Li, M.-X.; Chen, R.-Y. A suggestion on the standardization of karyotype analysis in plants. *J. Wuhan Bot. Res.* **1985**, *3*, 297–302.

54. Fu, W.; Ma, X.; Tang, Q.; Mo, C. Karyotype analysis and genetic variation of a mutant in siraitia grosvenorii. *Mol. Biol. Rep.* **2012**, *39*, 1247–1252. [CrossRef] [PubMed]

55. Stebbins, G. *Chromosomal Evolution in Higher Plants*; Edward Arnold: London, UK, 1971; pp. 87–89.

56. Abràmoff, M.D.; Magalhães, P.J.; Ram, S.J. Image processing with imageJ. *Biophotonics Int.* **2004**, *11*, 36–42.
57. Bradford, M.M. A rapid and sensitive method for the quantitation of microgram quantities of protein utilizing the principle of protein-dye binding. *Anal. Biochem.* **1976**, *72*, 248–254. [CrossRef]

International Journal of
Molecular Sciences

Article

Comparative Proteomic and Physiological Analysis Reveals the Variation Mechanisms of Leaf Coloration and Carbon Fixation in a Xantha Mutant of *Ginkgo biloba* L.

Xinliang Liu [1,†], Wanwen Yu [1,†], Guibin Wang [1], Fuliang Cao [1,*], Jinfeng Cai [1,2] and Huanli Wang [3]

1   Co-Innovation Center for Sustainable Forestry in Southern China, Nanjing Forestry University, Nanjing 210037, China; xliu25@lakeheadu.ca (X.L.); youeryuww@163.com (W.Y.); gbwang@njfu.edu.cn (G.W.); caijinfeng1984@126.com (J.C.)
2   The Jiangsu Provincial Platform for Conservation and Utilization of Agricultural Germplasm, Nanjing 210037, China
3   Institute of Botany, Jiangsu Province and Chinese Academy of Sciences, Nanjing 210037, China; wanghuanlis@163.com
*   Correspondence: fuliangcaonjfu@163.com; Tel.: +86-25-8542-7994
†   These authors contributed equally to this work.

Academic Editor: Setsuko Komatsu
Received: 10 August 2016; Accepted: 18 October 2016; Published: 27 October 2016

Abstract: Yellow-green leaf mutants are common in higher plants, and these non-lethal chlorophyll-deficient mutants are ideal materials for research on photosynthesis and plant development. A novel xantha mutant of *Ginkgo biloba* displaying yellow-colour leaves (YL) and green-colour leaves (GL) was identified in this study. The chlorophyll content of YL was remarkably lower than that in GL. The chloroplast ultrastructure revealed that YL had less dense thylakoid lamellae, a looser structure and fewer starch grains than GL. Analysis of the photosynthetic characteristics revealed that YL had decreased photosynthetic activity with significantly high nonphotochemical quenching. To explain these phenomena, we analysed the proteomic differences in leaves and chloroplasts between YL and GL of ginkgo using two-dimensional gel electrophoresis (2-DE) coupled with MALDI-TOF/TOF MS. In total, 89 differential proteins were successfully identified, 82 of which were assigned functions in nine metabolic pathways and cellular processes. Among them, proteins involved in photosynthesis, carbon fixation in photosynthetic organisms, carbohydrate/energy metabolism, amino acid metabolism, and protein metabolism were greatly enriched, indicating a good correlation between differentially accumulated proteins and physiological changes in leaves. The identifications of these differentially accumulated proteins indicates the presence of a specific different metabolic network in YL and suggests that YL possess slower chloroplast development, weaker photosynthesis, and a less abundant energy supply than GL. These studies provide insights into the mechanism of molecular regulation of leaf colour variation in YL mutants.

Keywords: *Ginkgo biloba* L.; xantha mutant; comparative proteomics; chloroplast; photosynthesis

---

1. Introduction

Chlorophyll (Chl) is the most important pigment in plants and is usually embedded in the thylakoid membranes of chloroplasts [1,2]. Chl is a green pigment, essential for photosynthesis, that absorbs energy from sunlight in antenna systems and transfers the energy to the reaction centre [3]. The absorbed light energy is then used to synthesize carbohydrates from carbon dioxide and water, a fundamental life process in plants. In higher plants, Chl is mainly biosynthesized in plastids, and its metabolic pathway has been extensively studied using genetic and biochemical methods in various organisms, particularly *Arabidopsis thaliana* [4–6]. Mutations in Chl biosynthesis, degradation or other related pathways lead to Chl-deficient mutants or leaf colour mutants. These mutants are widespread in nature and yield various mutant leaf colours, such as albino, virescent, chlorina, xanthas, maculate, stripe and dark green [7–9]. A number of yellow-green leaf colour mutants have been identified in model plants, including *A. thaliana*, *Oryza sativa*, *Solanum lycopersicum*, *Zea mays* and *Triticum aestivum* [10–14].

Yellow-green leaf colour mutants are induced by multiple genetic and environmental factors, among which genetic change plays a decisive role. In *O. sativa*, mutations in the genes that encode glutamyl-tRNA, Mg-chelatase, heme oxygenase and geranylgeranyl reductase lead to yellow-green leaves, whereas T-DNA insertional mutagenesis in rice can generate a chlorina phenotype [7,15–18]. Gene mutations involved in chloroplast development can also affect leaf colour. In *Arabidopsis*, two leaf-variegated mutants, *yellow variegated1* (*var1*) and *var2*, are caused by the loss of FtsH5 and FtsH2 (ATP-dependent zinc metalloprotease), respectively; these proteins are involved in the repair of photo-damaged proteins in thylakoid membranes [19,20]. The genetic changes in mutants can lead to changes in leaf phenotype, microstructures, fluorescence and physiological properties. These leaf colour mutants are thus ideal materials to study photosynthesis, photomorphogenesis, Chl metabolism, and chloroplast development [13,17,21].

Proteomics analysis deals directly with large-scale changes in proteins, the main components of physiological metabolic pathways in living cells. Proteins are responsible for most biological processes. Proteomics is a powerful approach for investigating the complete proteome at specific development stages or identifying changes in the proteome under different environmental conditions [22]. Proteome analysis, using two-dimensional polyacrylamide gel electrophoresis (2-DE), is one of the most sensitive and potent methods in proteomics studies [23]. Subcellular proteomic analysis has been widely studied in plants, including for the characterization of the network of cellular processes in a particular organelle, such as mitochondria, chloroplasts, nuclei, and plasma membranes [24–26]. Chloroplasts are the site of photosynthesis in plants and the largest metabolically active compartments in mature leaves. In *A. thaliana*, proteomics approaches have revealed 73% of unique organelle stress-responsive proteins belong to chloroplasts, indicating the specific sensitivity of chloroplasts to environmental stress [27]. Many proteomic studies have focused on chloroplasts or chloroplast fractions, providing significant insight into chlorophyll metabolism, photosynthesis, or the chloroplast response to environmental stress [25–28]. A proteomic analysis of the thylakoid membrane of a Chl b-less rice mutant and wild type by Chen et al. suggested that the reduction of Chl b affects light-harvesting complex I (LHC-I) assembly more severely than LHC-II [28]. Wang et al. performed transcriptomic and proteomic analyses of a Chl-deficient tea plant cultivar and suggested a complementary approach to better understand the mechanisms responsible for the chlorina phenotype [29]. Zhou et al. identified two colour patterns of flower buds in an ornamental peach and studied the differential expression of proteins, which provided important insights into the molecular mechanism of flower petal coloration [30]. Therefore, the analysis of the differential expression of proteins between mutant and normal tissue, particularly at the subcellular level, may provide new insights into the regulatory mechanism responsible for the mutant phenotype.

Ginkgo biloba, often called a "living fossil", is the only remaining species of *Ginkgoaceae* and is well-known for its medicinal value and ornamental beauty [31]. Ginkgo is planted throughout China as a multi-value deciduous tree species of ornamental due to its unique leaf pattern and tree form. Recently, we discovered a pigment-deficient mutant of *G. biloba* that exhibited a yellow-green leaf phenotype on a main branch and was initially identified as a xantha mutant in Jiujiang City, Jiangxi Province, China (29°49′ N, 116°40′ E). The mutant is an ancient tree with an estimated age of 150 years, a height of 18.8 m, and a diameter of 1.6 m at 2 m above ground. The branch is supposed to be a bud mutation and constitutes one-fourth of the crown of the tree, with the rest of the tree having green leaves. During the early growth stages, leaves of the xantha mutant are yellow and are remarkably different from green leaves until early July. As the mutant leaves mature, the colour gradually turns yellow-green until October, and finally the leaves turn yellow again. This type of bright and stable leaf colour phenotype is rare in ginkgo, and this mutation is considered a better ornamental germplasm resource for cultivation than wild type. At present, little is known regarding the molecular basis of this leaf mutant.

In this study, we used a proteomic approach to compare the total leaf protein and chloroplast protein profiles of the yellow-colour leaf (YL) and the green-colour leaf (GL) of *G. biloba*. Using a 2-DE proteomic approach and MALDI-TOF/TOF MS analysis, we successfully identified 89 differential proteins, which were predicted to play potential roles in particular cellular processes in the YL mutant. To elucidate the relationship between the variation of photosynthesis and specific proteins with altered expression, we performed photosynthesis characterization and chloroplast ultrastructure observation, leading to a further understanding of the molecular mechanisms of pigment biosynthesis and photosynthesis alteration in the YL mutant.

2. Results

2.1. Chlorophyll Concentration, Chloroplast Ultrastructure and Photosynthesis Performance

The leaf morphologies of GL and YL are shown in Figure 1A,B. No difference in phenotype was observed between YL and GL, except leaf chlorosis in YL. The chlorophyll a and b contents were markedly higher in GL than in YL, but the chlorophyll a/b ratio was significantly lower in GL than in YL ($p < 0.01$). The gas exchange parameters of the leaves of two colours are shown in Figure 1G–J, and the net photosynthetic rate (P_n), transportation rate (E) and stomatal conductance (g_s) were significantly higher in GL than YL ($p < 0.01$). There were no significant differences in the internal CO_2 concentration (C_i) between the two types of leaves. As shown in Figure 1K–O, the effective quantum yield of photosystem II electron transport (ΦPSII) and photochemical quenching (q_p) were significantly higher in GL than in YL ($p < 0.05$). The patterns of the efficiency of excitation energy capture by open photosystem II reaction centres (Fv′/Fm′) was similar in GL and YL, whereas the maximum quantum yield of photosystem II (Fv/Fm) and the nonphotochemical quenching (NPQ) were considerably lower in GL than in YL ($p < 0.01$).

To identify morphological changes in photosynthetic organelles and leaves of GL and YL, the chloroplast ultrastructure was examined. As shown in Figure 1C, the chloroplasts of GL exhibited an elliptical shape, with a typical lamellar grana structure consisting of thylakoid and numerous starch grains and few osmiophilic granules. By contrast, the chloroplasts in YL leaves were abnormal in appearance, with a slim spindle shape with no obvious boundary and simple distributed grana structure, and usually contained no well-defined starch grains and a few small plastoglobules. YL had a reduced number of thylakoids in grana and contained diffuse internal membranes and a dilated lamellae system compared to GL (Figure 1D). In addition, the chloroplasts were smaller in size in YL than in GL.

Figure 1. *Cont.*

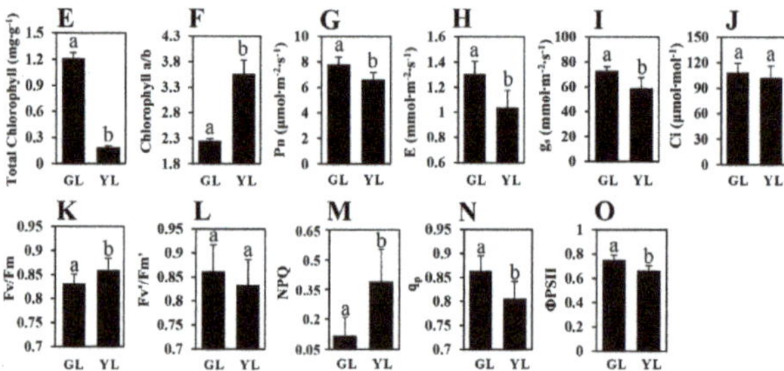

Figure 1. Photosynthetic parameters and chloroplast ultrastructure of ginkgo yellow-colour leaves (YL) and green-colour leaves (GL). (**A,B**) Phenotypes of the GL (left) and YL mutant (right); (**C,D**) Chloroplast ultrastructure in GL (**C**) and YL (**D**). S, starch grain; T, thylakoid; O, osmiophilic granule; (**E,F**) Total chlorophyll content and chlorophyll a/b; (**G–J**) Gas exchange parameters in P_n (**G**), E (**H**), g_s (**I**), C_i (**J**); (**K–O**) Changes in chlorophyll fluorescence parameters, including maximum quantum efficiency of photosystem II (PSII) (Fv/Fm) (**K**), efficiency of excitation energy capture by open PSII centres (Fv′/Fm′) (**L**), nonphotochemical quenching (NPQ) (**M**), photochemical quenching (q_p) (**N**), effective quantum yield of PSII electron transport (ΦPSII) (**O**). The values are presented as means ± SD (*n* = 15). Different letters indicate significant differences at $p < 0.05$.

2.2. 2-DE Analysis and Identification of Differentially Accumulated Proteins

To analyse and compare the changes in protein profiles, comparative proteomics of ginkgo leaves and chloroplasts of the two leaves of different colours were performed using 2-DE and MALDI-TOF/TOF MS. The representative 2-DE maps with immobilized pH gradient (IPG) strips of pH 4–7 and 3–10 are shown in Figure 2 and Figure S1 respectively. As more proteins are detectable clearly in the IPG 4–7 gel than in the IPG 3–10 gel, IPG 4–7 gel were used in this study. The 2-DE profile of the total protein in the leaves demonstrated that more than 1300 protein spots were detected with good reproducibility in the range of molecular masses of 15–120 kDa and pH 4–7. A total of 28 protein spots in YL exhibited significant changes in abundance of less than 0.66-fold or more than 1.5-fold ($p < 0.05$) compared to GL (Figure 2A,B). Similar protein patterns were revealed in the 2-DE profile of total leaf proteins and chloroplast proteins. The 2-DE profile of chloroplast proteins revealed that more than 1100 protein spots were detected, and a total of 61 protein spots exhibited significant changes in abundance (Figure 2C,D). After analysis by MALDI-TOF/TOF MS, a total of 89 differentially accumulated protein spots were successfully identified (Table 1). Compared to GL, 42 (47%) of the differential protein spots identified were increased in YL, whereas 47 (53%) were decreased. Sub-cellular localization, predicted by WolfPsort, revealed that the majority of identified proteins were located in the cytoplasm and chloroplast. More than 46% of the identified total leaf proteins were predicted to be located in the chloroplasts, implying that chloroplast proteome is virtually changed in YL. Whereas in the chloroplast, nearly 82% of the identified proteins from chloroplast were predicted to be typical chloroplast proteins.

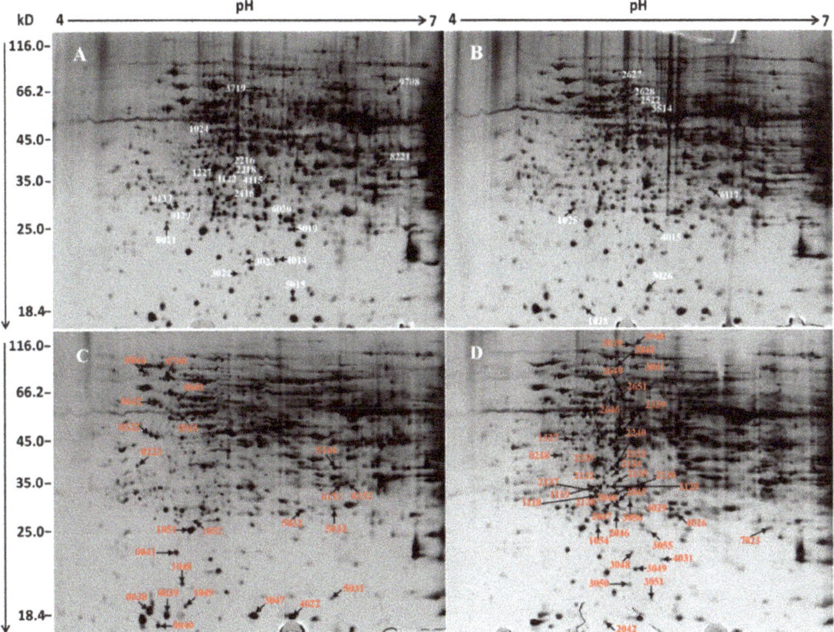

Figure 2. Representative 2-DE profiles of proteins in *G. biloba*. The molecular mass (M_r) in kDa and pI of proteins are indicated on the left and top, respectively. The identified proteins are annotated by a number and arrow. (**A**) Total proteins from YL; (**B**) Total proteins from GL; (**C**) Chloroplast proteins from YL; (**D**) Chloroplast proteins from GL. The arrows indicate increased protein spots in ginkgo.

Table 1. Identification and database search of differentially accumulated proteins in G. biloba.

Spot No.[a]	Accession No. (GI)[b]	Protein Name[c]	Species[c]	Exp kD/pI[d]	Theor kD/pI[e]	Score[f]	SC[g]	MP[h]	FC[i]	E[j]	Sub-CL[k]
Amino acid metabolism											
9708	gi|356508448	5-methyltetrahydropteroyltriglutamate-homocysteine methyltransferase-like	Glycine max	66.2/6.88	89.1/6.41	59	4%	2	1.80	T	Chlo
4026	gi|563247	Acetolactate synthase precursor	Xanthium sp.	33.9/5.64	70.8/7.03	91	5%	2	0.33	C	Chlo
0129	gi|2493895	Cysteine synthase	Citrullus lanatus	35.1/4.78	34.5/6.25	79	7%	2	2.06	T	Cyto
4115	gi|99698	Glutamate-ammonia ligase	Arabidopsis thaliana	36.8/5.62	40.9/5.40	225	8%	2	1.68	T	Cyto
2048	gi|225432496	Glutamine synthetase leaf isozyme, chloroplastic	Vitis vinifera	34.1/5.26	48.3/7.57	114	4%	1	0.18	C	Chlo
3048	gi|121334	Glutamine synthetase PR-1	Phaseolus vulgaris	25.1/5.44	39.3/5.78	117	4%	2	0.28	C	Cyto
1049	gi|357144704	Glycerate dehydrogenase-like	Brachypodium distachyon	19.9/4.61	42.2/6.68	64	5%	1	5.37	C	Chlo
2046	gi|288063	Ketol-acid reductoisomerase	Arabidopsis thaliana	33.6/5.29	64.3/6.55	112	5%	2	0.18	C	Chlo
1054	gi|1709006	S-adenosylmethionine synthase 3	Actinidia chinensis	31.3/5.13	39.8/6.20	185	13%	3	0.19	C	Cyto
2116	gi|356505256	S-adenosylmethionine synthase 3-like isoform 1	Glycine max	34.8/5.21	43.2/6.13	312	14%	4	1.53	T	Cyto
5032	gi|2821961	Spermidine synthase	Arabidopsis thaliana	33.3/5.81	32.7/4.97	82	11%	3	2.03	C	Chlo
Biosynthesis of secondary metabolites											
2132	gi|53830379	Anthocyanidin reductase	Ginkgo biloba	37.8/5.02	37.6/5.63	89	4%	2	0.17	C	Chlo
2218	gi|356566889	LOW QUALITY PROTEIN: (+)-neomenthol dehydrogenase-like	Glycine max	39.1/5.15	57.9/7.49	55	2%	2	1.51	T	Cyto
1048	gi|225380888	GDP-D-mannose-3′,5′-epimerase	Malus domestica	21.2/4.71	42.9/6.25	79	8%	2	2.31	C	Cyto
2239	gi|1170029	Glutamate-1-semialdehyde 2,1-aminomutase, chloroplastic	Hordeum vulgare	31.1/5.07	49.7/6.39	148	11%	4	0.10	C	Chlo
Carbohydrate/Energy metabolism											
5031	gi|344190186	Enolase	Corylus heterophylla	20.4/5.88	49.4/5.62	119	7%	2	12.43	C	Chlo
2137	gi|15221107	Enolase 1	Arabidopsis thaliana	36.2/5.15	51.8/5.79	119	6%	2	0.61	C	Chlo
7023	gi|2982328	Pyruvate dehydrogenase E1 β subunit	Picea mariana	32.4/6.49	32.0/5.73	66	5%	2	0.32	C	Chlo
4022	gi|111660950	ADP-glucose pyrophosphorylase small subunit	Citrus sinensis	19.2/5.62	57.3/6.73	171	7%	3	2.31	C	Chlo
3841	gi|228210	Granule-bound starch synthase	Solanum tuberosum	80.9/5.32	67.1/6.92	138	2%	1	0.32	C	Chlo
3049	gi|15223331	Granule-bound starch synthase 1	Arabidopsis thaliana	24.4/5.34	67.5/8.76	175	5%	2	0.31	C	Chlo
0740	gi|17939849	Mitochondrial F1 ATP synthase β subunit	Arabidopsis thaliana	66.3/4.82	63.6/6.52	308	13%	5	1.80	C	Cyto
2133	gi|351767989	Myo-inositol-1-phosphate synthase	Triticum aestivum	39.1/5.08	56.3/5.64	67	4%	2	0.30	C	Cyto
1025	gi|115440691	Os01g0817700	Oryza sativa Japonica Group	35.2/4.66	61.0/5.42	134	4%	3	0.62	T	Cyto
4029	gi|356567630	2-methylene-furan-3-one reductase	Glycine max	34.1/5.64	42.0/8.96	337	10%	4	0.11	C	Chlo
3056	gi|255544584	phosphoglycerate kinase, chloroplastic	Glycine max	35.1/5.23	50.1/8.74	293	13%	4	0.12	C	Chlo
3051	gi|1172159	Starch synthase	Ipomoea batatas	21.8/5.48	67.9/7.47	171	13%	5	0.26	C	Chlo

Table 1. *Cont.*

Spot No.[a]	Accession No. (GI)[b]	Protein Name[c]	Species[c]	Exp kD/pI[d]	Theor kD/pI[e]	Score[f]	SC[g]	MP[h]	FC[i]	E[j]	Sub-CL[k]
Carbon fixation in photosynthetic organisms											
0218	gi｜118175929	Chloroplast sedoheptulose-1,7-bisphosphatase	*Morus alba var. multicaulis*	43.2/4.73	42.8/6.06	92	7%	2	0.63	C	Chlo
2129	gi｜120664	Glyceraldehyde-3-phosphate dehydrogenase B, chloroplastic	*Spinacia oleracea*	35.4/5.11	48.6/6.72	140	6%	2	0.58	C	Chlo
3050	gi｜15223484	Phosphoglycerate kinase	*Arabidopsis thaliana*	23.7/5.32	50.0/8.27	398	12%	5	0.62	C	Chlo
2522	gi｜132000	Ribulose bisphosphate carboxylase large chain	*Nicotiana acuminata*	53.2/5.09	53.4/6.41	322	6%	3	0.64	T	Cyto
3047	gi｜1006698	Rubisco subunit binding protein, β subunit	*Pseudotsuga menziesii*	19.2/5.31	9.7/4.37	151	18%	1	9.31	C	Chlo
1024	gi｜161777955	Ribulose-1,5-bisphosphate carboxylase/oxygenase large subunit	*Luculia pinceana*	42.8/4.89	52.2/6.17	171	7%	2	3.16	T	Chlo
2216	gi｜161777955	Ribulose-1,5-bisphosphate carboxylase/oxygenase large subunit	*Luculia pinceana*	40.8/5.18	52.2/6.17	238	7%	3	2.10	T	Chlo
2047	gi｜12098	Ribulose-1,5-bisphosphate carboxylase/oxygenase large subunit	*Afrocarpus falcatus*	34.9/5.29	53.1/5.91	121	3%	2	0.21	C	Chlo
0223	gi｜337263422	Chloroplast rubisco activase	*Ophiopogon japonicus*	40.5/4.46	48.0/6.04	87	3%	1	5.05	C	Chlo
0021	gi｜10720249	Rubisco activase	*Vigna radiata var. radiata*	33.4/4.51	48.0/7.57	177	13%	4	1.79	T	Chlo
5144	gi｜4261547	Rubisco activase	*Spinacia oleracea*	40.6/6.06	47.8/7.67	141	3%	1	2.08	C	Chlo
2627	gi｜170129	Rubisco activase precursor	*Spinacia oleracea*	59.1/5.03	51.7/6.28	171	5%	2	0.61	T	Chlo
2646	gi｜170129	Rubisco activase precursor	*Spinacia oleracea*	55.2/5.22	51.7/6.28	93	3%	1	0.20	C	Chlo
2651	gi｜170129	Rubisco activase precursor	*Spinacia oleracea*	55.2/5.32	51.7/6.28	167	5%	2	0.25	C	Chlo
0561	gi｜170129	Rubisco activase precursor	*Spinacia oleracea*	52.8/4.86	51.7/6.28	160	5%	2	7.04	C	Chlo
1122	gi｜255541252	Transketolase, putative	*Ricinus communis*	38.8/5.01	81.6/6.52	199	8%	5	1.55	T	Chlo
Cellular process											
2339	gi｜20329	Actin	*Oryza sativa* Indica Group	44.2/5.09	42.2/5.72	81	4%	1	0.48	C	Cyto
8221	gi｜227069391	Actin 4	*Picea abies*	41.2/6.59	41.9/5.31	298	16%	4	2.29	T	Cyto
3940	gi｜399213	ATP-dependent Clp protease ATP-binding subunit clpA homolog CD4B, chloroplastic	*Solanum lycopersicum*	110.2/5.35	102.4/5.86	207	4%	3	0.29	C	Chlo
3939	gi｜399213	ATP-dependent Clp protease ATP-binding subunit clpA homolog CD4B, chloroplastic	*Solanum lycopersicum*	110.2/5.34	102.4/5.86	230	8%	5	0.41	C	Chlo
4031	gi｜42561751	ATP-dependent zinc metalloprotease FtsH 8	*Arabidopsis thaliana*	24.8/5.51	73.3/5.72	202	11%	6	0.33	C	Chlo
2240	gi｜2492515	ATP-dependent zinc metalloprotease FtsH, chloroplastic	*Capsicum annuum*	30.2/5.29	71.2/6.55	131	11%	4	0.41	C	Chlo
2130	gi｜728744	Auxin-induced protein PCNT115	*Nicotiana tabacum*	35.2/5.03	34.3/7.10	120	9%	3	0.35	C	Nucl
3055	gi｜6692685	F12K11.22	*Arabidopsis thaliana*	30.2/5.42	71.0/5.81	248	12%	6	0.16	C	Chlo
Lipid metabolism											
3026	gi｜210110834	β-hydroxyacyl-ACP dehydrase 1	*Arachis hypogaea*	24.3/5.31	24.1/9.10	112	12%	3	0.27	T	Chlo
0137	gi｜29367475	Fibrillin-like protein	*Oryza sativa* Japonica Group	35.3/4.65	33.9/5.04	126	6%	3	2.02	T	Chlo
0322	gi｜339697596	Stearoyl-ACP desaturase	*Ginkgo biloba*	44.6/4.55	47.2/6.14	157	14%	4	1.83	C	Chlo

Table 1. *Cont.*

Spot No.[a]	Accession No. (GI)[b]	Protein Name[c]	Species[c]	Exp kD/pI[d]	Theor kD/pI[e]	Score[f]	SC[g]	MP[h]	FC[i]	E[j]	Sub-CL[k]
Photosynthesis											
2134	gi丨380356155	ATP synthase CF1 α chain (chloroplast)	*Ginkgo biloba*	38.2/5.13	55.8/5.02	440	18%	8	0.25	C	Chlo
3514	gi丨3913118	ATP synthase subunit β, chloroplastic	*Picea abies*	53.3/5.12	52.6/5.18	224	10%	3	0.65	T	Chlo
2649	gi丨3913118	ATP synthase subunit β, chloroplastic	*Picea abies*	55.3/5.29	52.6/5.18	523	20%	6	0.39	C	Chlo
6151	gi丨119904	Nicotinamide adenine dinucleotide phosphate (NADP) reductase, leaf isozyme, chloroplastic	*Pisum sativum*	37.4/6.09	40.5/8.56	119	6%	1	4.41	C	Chlo
6152	gi丨119904	NADP reductase, leaf isozyme, chloroplastic	*Pisum sativum*	37.5/6.11	40.5/8.56	146	6%	1	8.67	C	Chlo
1052	gi丨131390	Oxygen-evolving enhancer protein 2, chloroplastic	*Pisum sativum*	29.8/4.81	28.2/8.29	158	18%	3	1.79	C	Chlo
1118	gi丨225423755	Photosystem II stability/assembly factor HCF136, chloroplastic	*Vitis vinifera*	36.1/5.09	44.5/6.92	370	19%	6	0.50	C	Chlo
3024	gi丨351726724	Rieske iron-sulphur protein precursor	*Glycine max*	24.8/5.23	24.5/9.01	95	18%	3	2.37	T	Chlo
Protein metabolism											
0641	gi丨15222729	Chaperonin 60 subunit β 1	*Arabidopsis thaliana*	57.9/4.79	64.2/6.21	57	2%	1	1.94	C	Chlo
6020	gi丨123601	Heat shock 70 kDa protein	*Glycine max*	31.3/5.71	71.3/5.37	415	10%	6	2.49	T	Cyto
5033	gi丨585273	Heat shock 70 kDa protein	*Solanum tuberosum*	33.4/6.02	73.3/6.37	190	3%	2	7.63	C	Mito
3719	gi丨357112870	Heat shock cognate 70 kDa protein 2-like	*Brachypodium distachyon*	66.7/5.45	71.6/5.09	259	10%	5	1.51	T	Cyto
0846	gi丨357112870	Heat shock cognate 70 kDa protein 2-like	*Brachypodium distachyon*	70.2/4.49	71.6/5.09	355	11%	6	2.26	C	Cyto
2628	gi丨24637539	Heat shock protein 60	*Prunus dulcis*	58.8/5.04	58.1/5.26	102	5%	2	0.66	T	Cyto
2848	gi丨166919370	Chloroplast heat shock protein 70-1	*Ipomoea nil*	81.3/5.18	74.5/5.14	497	9%	5	0.48	C	Chlo
2045	gi丨124245039	Chloroplast heat shock protein 70 (HSP70)	*Cucumis sativus*	35.1/5.26	75.5/5.18	721	16%	10	0.34	C	Chlo
0039	gi丨242076604	Hypothetical protein SORBIDRAFT_06g023840	*Sorghum bicolour*	19.3/4.55	85.3/5.42	151	8%	4	7.75	C	Chlo
3023	gi丨224084924	Proteasome subunit β type 3-2 family protein	*Populus trichocarpa*	27.2/5.23	23.1/5.18	215	17%	3	2.41	T	Cyto
0038	gi丨255576477	Plastid-specific 30S ribosomal protein 3, chloroplast precursor	*Ricinus communis*	19.8/4.41	20.4/7.79	116	11%	3	4.20	C	Chlo
2042	gi丨13094963	Initiation factor eIF5-A	*Manihot esculenta*	18.5/5.09	17.8/5.60	73	10%	2	0.54	C	Nucl
1119	gi丨402753	Translation elongation factor EF-G	*Glycine max*	37.1/5.12	77.9/5.04	69	4%	2	0.35	C	Chlo
2135	gi丨218312	chloroplast elongation factor TuB (EF-TuB)	*Nicotiana sylvestris*	37.4/5.16	46.8/5.70	240	10%	3	0.05	C	Chlo
Redox homeostasis											
4015	gi丨224091909	2-cys peroxiredoxin	*Populus trichocarpa*	32.5/5.35	28.9/6.84	191	19%	6	0.09	T	Chlo
1051	gi丨220898265	Ascorbate peroxidase	*Ginkgo biloba*	29.1/4.77	27.7/5.81	118	11%	2	2.38	C	Chlo
0040	gi丨294861514	Cytosolic ascorbate peroxidase 2	*Rubia cordifolia*	18.4/4.49	16.8/5.34	135	15%	1	5.36	C	Cyto
0041	gi丨220898261	FeSOD	*Ginkgo biloba*	25.2/4.68	27.2/6.76	312	38%	5	1.94	C	Chlo
1028	gi丨373842096	Peroxiredoxin	*Tamarix hispida*	19.8/4.74	17.6/6.08	83	10%	1	0.43	T	Cyto
5015	gi丨45643751	Copper-zinc superoxide dismutase	*Citrullus lanatus*	20.2/5.75	15.1/5.05	120	17%	2	1.53	T	Cyto

Table 1. *Cont.*

Spot No.[a]	Accession No. (GI)[b]	Protein Name[c]	Species[c]	Exp kD/pI[d]	Theor kD/pI[e]	Score[f]	SC[g]	MP[h]	FC[i]	E[j]	Sub-CL[k]
Unclassified											
2128	gi | 242079005	Hypothetical protein SORBIDRAFT_07g019320	*Sorghum bicolour*	35.5/5.09	46.7/4.83	195	8%	2	0.33	C	Chlo
4014	gi | 297720697	Os01g0915900	*Oryza sativa* Japonica Group	26.6/5.74	28.3/8.75	57	8%	1	1.71	T	Nucl
5019	gi | 224143607	Predicted protein	*Populus trichocarpa*	32.1/5.71	31.8/5.26	262	18%	4	1.53	T	Chlo
6117	gi | 224080984	Predicted protein	*Populus trichocarpa*	34.9/5.62	42.5/5.69	113	15%	5	0.66	T	Chlo
1227	gi | 116791600	Unknown	*Picea sitchensis*	40.9/5.05	21.0/8.48	143	7%	2	1.51	T	Cyto
1327	gi | 116787373	Unknown	*Picea sitchensis*	44.9/4.86	65.8/5.69	159	7%	3	0.63	C	Chlo
0642	gi | 148909901	Unknown	*Picea sitchensis*	57.6/4.59	63.4/5.12	118	5%	2	2.29	C	Chlo

[a] Spot number corresponds to the differentially accumulated proteins indicated in Figure 2; [b] Accession number according to the National Center for Biotechnology Information (NCBI) database; [c] The names and species of the proteins obtained by MASCOT software; [d] The experimental mass (kDa) and pI of the identified proteins were calculated by PDQuest; [e] The theoretical mass (kDa) and pI values of the identified proteins were retrieved from the protein database; [f] MASCOT score after searching against the database; [g] The sequence coverage percentage of identified proteins; [h] Number of identified peptides (Peptide sequences were shown in Table S1); [i] The protein abundance ratio (YL/GL); [j] Type of extracts. T: total leaf proteins; C: chloroplast; [k] Sub-cellular localization. Chlo: Chloroplast, Cyto: Cytoplasmic, Nucl: Nuclear, Mito: Mitochondrial.

2.3. Functional Categorization of Identified Proteins

The functions of the identified proteins were annotated by searching against the NCBI database and grouped according to functional categories based on analysis of KEGG pathways and the literature. Among the 89 identified proteins, 82 proteins (92%) had assigned functions or sequences similar to those of known proteins, and the other seven (8%) were identified as novel with no assigned function (Figure 3). The 82 proteins with assigned functions in both types of leaves were classified into nine metabolic pathways and cellular processes, including amino acid metabolism (11 proteins), biosynthesis of secondary metabolites (4 proteins), carbohydrate/energy metabolism (12 proteins), carbon fixation in photosynthetic organisms (16 proteins), cellular processes (8 proteins), lipid metabolism (3 proteins), photosynthesis (8 proteins), protein metabolism (14 proteins), and redox homeostasis (6 proteins) (Figure 3A). The distribution of increased and decreased proteins in the different functional groups is shown in Figure 3B.

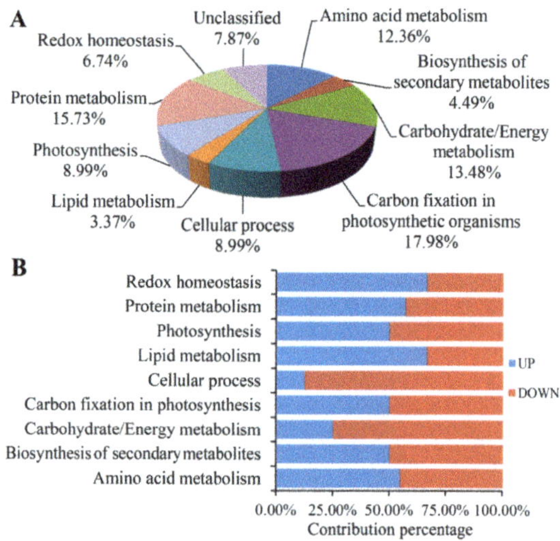

Figure 3. Function classification of all identified proteins in *G. biloba*. (**A**) Functional categorization of identified proteins; (**B**) Contributions to molecular functions from increased (blue) and decreased (red) proteins.

2.4. Protein-Protein Interaction Analysis

To further understand the functional associations of the identified proteins from both types of leaves, a protein-protein interaction network was generated using the STRING database (Figure 4). The interacting proteins have important functions in the clusters of amino acid metabolism, carbohydrate/energy metabolism, photosynthesis, protein metabolism and cell processes. As shown in Figure 4, the differentially expressed proteins from different clusters connected tightly, revealing direct or indirect functional links in the network. The chloroplast heat shock protein 70-1 (cpHsc70-1) plays an important role in protein precursor import into chloroplasts and had the highest number of interactions (12) in the network. Two proteins involved in carbon fixation in photosynthetic organisms, glyceraldehyde-3-phosphate dehydrogenase B (GAPB) and phosphoglycerate kinase (AT1G56190), also had a high number of interactions (10 and 8, respectively). These proteins were connected tightly to the other enzymes involved in the glycolysis pathway, such as enolase 1 (ENO1) and pyruvate dehydrogenase E1 β subunit (T2H7.8). Moreover, these proteins with a high number of connections were present in lower amounts in the YL mutant. The interactions among these proteins may play important roles in carbohydrate metabolism and photosynthetic performance.

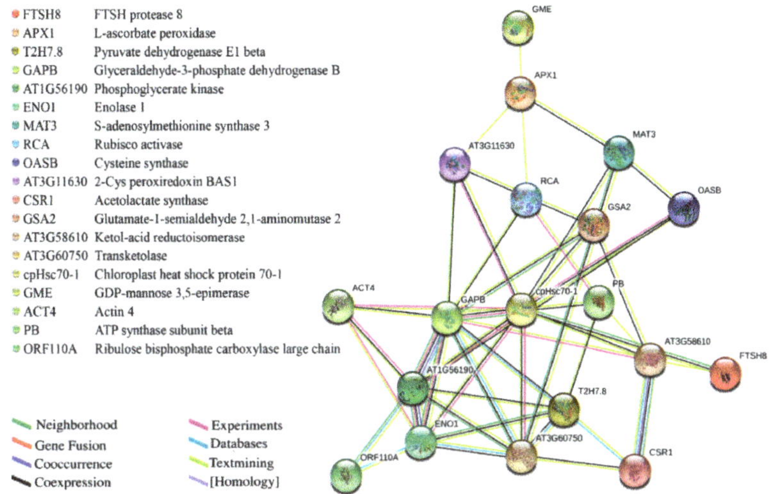

Figure 4. Protein-protein interaction network of differentially accumulated proteins analysed by STRING 10.

2.5. Gene Expression Analysis by qRT-PCR

To further verify the changes of proteins between the two colour leaves, we examined the transcript levels of selected protein-coding genes by qRT-PCR. Seven proteins involved in amino acid metabolism, cellular processes, photosynthesis, protein metabolism and chlorophyll metabolism were selected, and the expression levels of their encoded genes were analysed (Figure 5). The results showed that the expression levels of spermidine synthase (SPDS), heat shock 70 kDa protein (HSP70) and plastid-specific 30S ribosomal protein 3 (PSRP3) increased significantly in YL, and the expression levels of ATP synthase CF1 α chain (atpA), ATP-dependent Clp protease ATP-binding subunit clpA homolog CD4B (CD4B) and elongation factor TuB (EF-TuB) decreased, which is consistent with the expression patterns of their corresponding proteins. In contrast, the protein and mRNA expression patterns of glutamate-1-semialdehyde aminotransferase (GSA) differed, which indicated that some of the identified proteins are regulated at the transcriptional level and others are regulated post-transcriptionally.

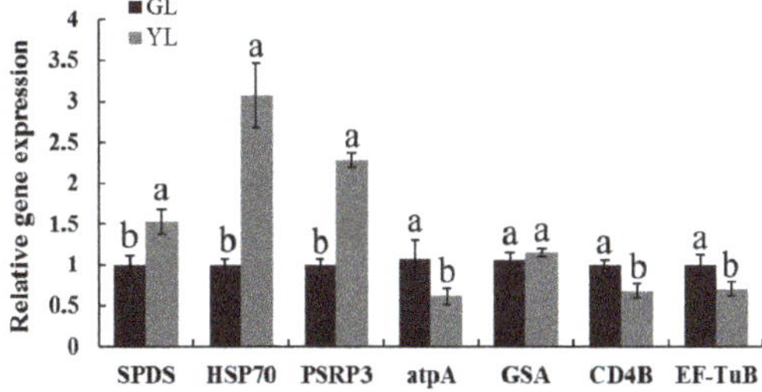

Figure 5. Expression levels of selected protein-encoding genes in *G. biloba*. SPDS, HSP70, PSRP3, atpA, GSA, CD4B and EF-Tu are spermidine synthase (spot 5032), heat shock 70 kDa protein (spot 5033), plastid-specific 30S ribosomal protein 3(spot 0038), ATP synthase CF1 α chain (spot 2134), glutamate-1-semialdehyde aminotransferase (spot 2239), ATP-dependent Clp protease ATP-binding subunit clpA homolog CD4B (spot 3940), and elongation factor TuB (spot 2135), respectively. Different letters indicate significant differences at $p < 0.05$.

3. Discussion

Yellow-green leaf colour mutants are common in plants and always exhibit low photosynthetic capacity and weak growth. Consequently, these mutants are considered valuable materials for photosynthesis research [13,32,33]. The photosynthesis and a wide range of other metabolisms occur in the chloroplasts, making them the largest metabolically active compartments in the mature leaf cell. Leaf colour mutants lead to impaired chloroplast development, and this damage is often considered a cause of the alteration in photosynthesis [33]. In the present study, we examined a xantha mutant of *G. biloba* and its photosynthetic and anatomic characteristics. We also performed a comparative analysis of the protein expression patterns in both leaves and chloroplasts of YL and GL as a first step towards understanding the molecular mechanisms.

3.1. Photosynthesis and Anatomical Characteristics

In higher plants, photosynthesis is a complex biological process in which light energy absorption and transformation occur on the thylakoid membrane of chloroplasts. The chlorophyll-protein complexes are embedded in a lipid matrix of thylakoid membranes, and its contents in chloroplasts play a regulatory role in the amount of light energy absorption, which directly determines the photosynthetic potential and primary production of plant leaves [2,34]. Chl biosynthesis is associated with the formation of thylakoid membranes and is also coordinated with chloroplast development [35,36]. The anatomical characteristics of chloroplasts have been described in several Chl-deficient mutant species, such as *O. sativa*, *A. thaliana*, *Brassica napus*, and *Triticum turgidum* [19,28,33,36,37]. In general, the chloroplasts of chlorotic mutants are not well developed, with abnormal chloroplast ultrastructures. The chloroplasts of chlorotic mutants often contain fewer, thinner and irregularly arranged grana lamellae compared with wild type, and the thylakoids usually swell at different levels. In the YL mutant, anatomical characterization revealed that this ginkgo mutant had less dense thylakoid lamellae, irregularly shaped chloroplasts, and few or no starch grains. These features are consistent with those of three previously studied rice chlorina mutants and indicate an underdeveloped chloroplast ultrastructure in the YL mutant [17,33]. These structures might lead to decreased Chl contents and abnormal Chl a/b, resulting in delayed leaf greening.

The structure of chloroplasts determines the photosynthetic capacity of plant leaves. Abnormal Chl can reduce the filling of light-harvesting complexes, resulting in low photosynthetic efficiency in Chl-deficient mutants. Most Chl-deficient mutants have lower photosynthetic efficiency than wild type, with the exception of a very few specific mutants. For example, a xantha rice mutant named *Huangyu B* has higher photosynthetic efficiency than its wild-type parent, and the photosynthetic rate of a pale-green durum wheat mutant is similar to that of wild type [37,38]. In our study, YL had lower P_n, E and g_s but not C_i than GL, consistent with most Chl-deficient mutant plants. The Chl content in YL was approximately one-sixth of that in GL, a much more pronounced difference than those observed in other species. Leafing out and growth remained normal in the mutant because it possesses the fundamental chloroplast structure to perform photosynthesis. The fluorescence kinetics parameter ΦPSII was significantly lower in YL than in GL, indicating its low photochemical efficiency. The apparent inhibition of the photosynthetic property of the YL mutant was probably due to the low chlorophyll content and aberrant chloroplast development. Remarkably, the mutant had significantly higher NPQ than GL. NPQ reflects the nonphotochemical dissipation of the excess excitation energy, which protects the tissue from the damaging effects of reactive oxygen species (ROS) [39]. Based on these results, we suggest that PSII is inhibited in YL, resulting in increased effectiveness of the consumption and dissipation of absorbed light energy compared to GL.

3.2. Proteins Involved in Photosynthesis and Carbon Fixation

Light-dependent reactions are one set of reactions in photosynthesis and are catalysed by four major protein complexes in the thylakoid membrane: photosystem I (PSI), photosystem II (PSII), cytochrome b6f complex, and ATP synthase. These complexes work together to transform light energy into chemical forms, ATP and NADPH. In the present study, 16 differentially accumulated proteins involved in carbon fixation in photosynthetic organisms and 8 protein species involved in photosynthesis were identified in the leaves of two different colours of ginkgo. These proteins function directly in the electron transport and carbon fixation pathways (Figure 6, Table 1). Several studies have reported that some photosystem proteins in xantha mutants are deficient or present at low levels [40,41]. However, we observed that three different protein species, Rieske iron–sulphur protein precursor (PetC, spot 3024), oxygen-evolving enhancer protein 2 (OEC, spot 1052), and ferredoxin NADP reductase (FNR, spot 6151 and spot 6152), which originate from the cytochrome b6f complex PSII and PSI, respectively, were increased in YL (Figure 6). These complexes create and transfer electrons from H_2O to $NADP^+$ and ADP, executing the initial energy conversion reactions. We also detected one photosystem II stability/assembly factor HCF136 (HCF136, spot 1118) in our study, which was decreased in YL. HCF136 is an assembly factor of the photosystem II reaction centre that is localized in the lumen of stroma thylakoids [42]. Three different subunits of ATP synthase (spot 2134, atpA; spot 3514 and spot 2649, ATP synthase subunit β), however, obviously decreased. These subunits are essential to ATP synthesis and play a key role in light-reactions of photosynthesis. These fundamental metabolic alterations might be due to the low level of Chl, a component of the light-harvesting complexes. The complex changes also indicated that the thylakoids in YL were obviously affected, consistent with the chloroplast ultrastructure.

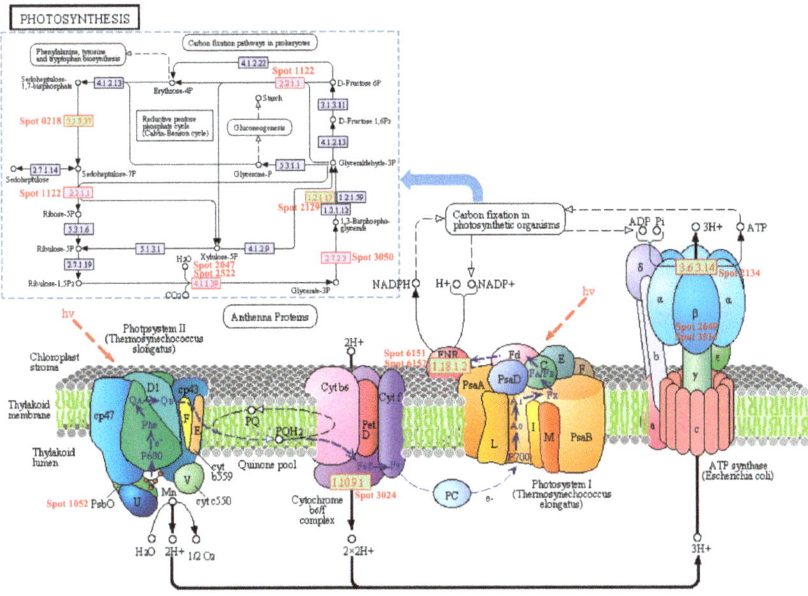

Figure 6. Identified proteins involved in photosynthesis in *G. biloba*. The blue dashed outline on the wireframe indicates the proteins that are involved in carbon fixation in photosynthetic organisms.

Ribulose-1,5-bisphosphate carboxylase oxygenase (Rubisco) is the initial carbon fixation enzyme of photosynthesis and catalyses the first step of the Calvin cycle for carbon assimilation [43]. Rubisco is the most abundant protein in leaves and constitutes nearly 50% of the soluble leaf protein in C3 plants. There are multiple forms of this protein in plant leaves, and there is a dynamic balance between Rubisco and its degraded forms, complicating the determination of the abundance of Rubisco by 2-DE [44]. In this study, four spots were identified as Rubisco large subunits (RLS), two of which (spot 1024 and spot 2216) were increased and the others (spot 2522 and spot 2047) were decreased.

Rubisco activase is a catalytic chaperone for Rubisco; it changes the conformation of inactive Rubisco to the active form using the energy from ATP hydrolysis [45]. In this study, three types of Rubisco activase (RCA, spot 0223, spot 0021, spot 5144) were markedly increased in YL, indicating that the activation of Rubisco was increased. We also observed that the other four protein species participating in the Calvin cycle declined in abundance, including sedoheptulose-1,7-bisphosphatase precursor (SBP, spot 0218), glyceraldehyde-3-phosphate dehydrogenase B (GAPB, spot 2129), and phosphoglycerate kinase (PGK, spot 3050). These changes indicate that YL has a low CO_2 fixation capacity, which slows the photosynthetic process.

3.3. Proteins Involved in Cellular Processes

As a principal component of microfilaments of cytoskeletons in eukaryotic cells, actin plays an important role in cell development, movement, differentiation and cellular components. Actin filaments associated with the surface of chloroplasts are supposed to function as an anchor for the chloroplast and to maintain cell structure; therefore, actin can be used as marker of cytoplasmic protein when examining the purity of isolated chloroplasts [25,46]. The ATP-dependent zinc metalloprotease FtsH family is a large, well-characterized family of chloroplast proteases that is involved in the degradation of unassembled proteins and the turnover of photosynthetic protein complexes in thylakoid membranes [20,47]. FtsH proteins are also suggested to play a direct or indirect role in chloroplast biogenesis, photosystem II repair, and complex formation [48,49]. The disruption of either FtsH2 or FtsH5 results in some degree of leaf variegation, and the double mutant of FtsH2 and FtsH8 exhibits an albino-like phenotype in *Arabidopsis*, indicating that the subunits are necessary for the pool of FtsH isomers that are important for chloroplast biogenesis [20,49]. A previous study also demonstrated that the levels of FtsHs, including FtsH2, FtsH8, and FtsH5, were significantly reduced in a leaf variegation thf1 mutant compared with wild type but were increased in a virescent ClpR4-3 thf1 mutation [50]. ClpA is an ATP-dependent chaperone of caseinolytic proteases (Clp) that mainly exists in chloroplasts and is essential for the biogenesis and function of chloroplasts [51]. In this study, spot 4031, spot 2240 and spot 3055 were identified as FtsH proteases, spot 3940 and spot 3939 were identified as CD4B, and all were reduced in YL, which might be attributable to the underdeveloped chloroplasts. Additionally, the decrease in CD4B in the mutant might decrease the formation and maintenance of functional thylakoid electron transport [52].

ROS are produced by aerobic metabolic processes in chloroplasts, mitochondria, and peroxisomes, such as respiration and photosynthesis. ROS can cause oxidative damage to the membrane system but simultaneously also plays many signalling roles in plants in regulating development and mediating stress responses [53]. The scavenging of ROS in cells is regulated by antioxidants such as ascorbate peroxidase (APX), superoxide dismutase (SOD), and catalase (CAT). In the present study, three enzymatic antioxidants, ascorbate peroxidase (APx, spot 1051 and spot 0040), FeSOD (spot 0041) and copper–zinc superoxide dismutase (Cu/ZnSOD, spot 5015), involved in redox homeostasis were identified as increased. As members of the defence system against ROS in chloroplasts, APx and SOD are the key enzymes catalysing the conversion of H_2O_2 to H_2O. FeSOD and Cu/ZnSOD are a different class of SODs that is identified by their metal co-factor, and both are considered ROS-scavenging enzymes [54,55]. The YL mutant may be more sensitive to photo-induced damage and ROS accumulation and therefore had a higher NPQ and an increased abundance of APx and SOD in chloroplasts to detoxify the ROS damage and maintain cellular homeostasis. However, two peroxiredoxins (Prx, spot 1028 and spot 4015) decreased in the YL mutant. In YL, most Prx might be chloroplastic, whereas in GL, most Prx might be cytosolic.

3.4. Proteins Involved in Carbohydrate Metabolism

In addition to the 24 specific proteins discussed in Section 3.2 related to the photosynthetic pathway in chloroplasts, 12 proteins associated with carbohydrate metabolism were identified in this study. Four enzymes (spot 5031, spot 2137, spot 7023, spot 1025) were involved in the glycolysis pathway, three of which (spot 2137, spot 7023, spot 1025) exhibited a decreased expression pattern in YL. The pyruvate dehydrogenase multienzyme complex (PDH) plays a key role in carbohydrate metabolism and catalyses the oxidative decarboxylation of pyruvate to generate acetyl-CoA. As a subunit of PDH, pyruvate dehydrogenase E1 (PDH E1) catalyses the oxidative decarboxylation of pyruvate and transfers the hydroxyethyl group to thiamine diphosphate (ThDP) [56,57]. PDH E1 (spot 7023) was less abundant in YL, which might suggest that the mutant maintained a low level of acetyl-CoA. Myo-inositol-1-phosphate synthase (MIPS, spot 2133) also declined in abundance. MIPS catalyses the conversion of D-glucose 6-phosphate to 1-L-myo-inositol-1-phosphate (MIP), providing the immediate precursor to myo-inositol, which is an intermediate carbon skeleton of cell wall uronic acids and pentoses derived from D-glucose [58]. MIPS has been found to play a key role in the biosynthesis of inositol and inositol phosphate and to be important in plant development and environmental stress tolerance [59].

In chloroplasts, starch is the major energy storage compound in the form of granules and is used as a primary store of excess carbohydrate produced during photosynthesis. We identified four protein species associated with starch metabolism that differed in abundance, including one protein that increased, ADP-glucose pyrophosphorylase small subunit (AGPase, spot 4022), and three proteins that decreased: starch synthase (SS, spot 3051) and granule-bound starch synthase (GBSS, spot 3049 and spot 3841). They play a major role in the regulation of starch synthesis. AGPase catalyses the conversion of glucose-1-phosphate to ADP-glucose, the substrate for starch synthesis [60]. GBSS is a type of SS and is a key enzyme in the formation of amylose found in starches [61]. The decreased abundance of SS and GBSS might suggest that energy is stored at low levels in YL. This conclusion is supported by the chloroplast ultrastructure: starch grain accumulation was rare in YL compared with GL, as shown in Figure 1C,D. The decrease in these differentially accumulated proteins might suggest that carbohydrate metabolism is reduced in YL.

3.5. Proteins Involved in Protein Metabolism

The biosynthesis, metabolism and transport of proteins in chloroplasts are crucial for chloroplast biogenesis and development [62]. A total of 14 identified proteins in our study are involved in protein metabolism. Three protein species related to protein biosynthesis, initiation factor eIF5-A (eIF5-A, spot 2042), translation elongation factor EF-G (EF-G, spot 1119) and EF-TuB (spot 2135), were decreased in the YL mutant. The protein eIF5-A is thought to play an important role in the translation machinery; it stimulates methionyl-puromycin synthesis for the formation of the first peptide bond [63]. EF-G and EF-Tu catalyse the elongation cycle of translation, which functions in the sequential addition of amino acids to the growing polypeptide chain [64]. A *snowy cotyledon 1* mutant of *Arabidopsis* containing a point mutation of the *EF-G* gene exhibits white cotyledons during the early development stage [65]. The protein species eIF5-A, EF-G and EF-Tu work together to perform initiation and elongation of the newly growing peptide chains. The decreased abundance of these proteins in YL might inhibit protein biosynthesis in the chloroplasts. Two heat shock protein 60 proteins (Hsp60, spot 2628 and spot 0641) and six heat shock 70 kDa proteins (Hsp70, spot 6020, spot 5033, spot 3719, spot 0846, spot 2848 and spot 2045), which belong to the chaperone family and are involved in protein folding and assembly, were differentially expressed. Hsp60 was originally identified as the Rubisco ligase protein and may combine the large and small subunits of newly synthesized Rubisco to prevent incorrect aggregation. Hsp60 is also supposed to assist with the folding and assembly of proteins within chloroplasts [66]. Hsp70 is a type of molecular chaperone that has the essential function of assisting with protein folding processes, preventing protein aggregation, and targeting unstable proteins for proteolytic degradation. These chaperones appear to play specific roles according their subcellular location. For instance, cytosolic Hsp70 assists protein folding and prevents aggregation, and chloroplastic Hsp70 is involved in the import and translocation of a precursor protein [67]. We observed four Hsp70s that increased significantly and two chloroplast HSP70s that decreased in the YL, which might imply that protein folding and assembly in were enhanced in YL, whereas translation decreased. Additionally, proteasome subunit β (spot 3023) increased in YL, which may indicate enhanced protein proteolysis in the mutant. Based on these findings, the identified proteins involved in protein metabolism were largely implicated in the inhibition of protein biosynthesis and the enhancement of protein folding and assembly in the YL mutant.

3.6. Proteins Involved in Other Metabolic Pathways

Glutamine synthetase (GS) is a key enzyme in the plant primary nitrogen assimilatory process because it catalyses the conversion of ammonium to form glutamine and other related nitrogenous compounds, which are important substrates for protein synthesis. GS has different subcellular localizations, and the major role of chloroplastic GS is thought to be the reassimilation of ammonia generated in photorespiration [68]. In this study, two chloroplastic GS (spot 2048 and spot 3048) decreased significantly, implying that N assimilation was inhibited in the chloroplasts of YL. Cysteine synthase (CSase) catalyses the synthesis of cysteine from *O*-acetylserine and disulphides. Cysteine is the only amino acid that contains disulphide bonds (S–S) and protects cellular environments from oxidative stress. The formation of S–S is a key step in the activation of molecular chaperones activation and is an essential substance for the correct folding and stability of some proteins [69]. Spermidine synthase (SPDS) is a key enzyme that participates in polyamine biosynthesis in plants and plays a pivotal role in plant defence against environmental stresses as a signalling regulator [70]. Here, we identified two protein species, CSase (spot 0129) and SPDS (spot 5032), that were more abundant in YL than in GL, which might be related to the increase in molecular chaperone activity and the decrease in the effect of oxidative stress.

GDP-D-mannose-3′,5′-epimerase (GME) catalyses the conversion of GDP-L-galactose from GDP-D-mannose, which precedes the critical step of the ascorbate biosynthesis pathway. GME also plays an important role at the intersection between ascorbate and non-cellulosic cell-wall polysaccharide biosynthesis in higher plants [71]. Ascorbate is another major antioxidant in plants that prevents cells from ROS damage generated by cellular processes as well as in vitro stress. Spot 1048 was identified as GME and exhibited higher abundance in the YL mutant. Glutamate-1-semialdehyde 2,1-aminomutase (GSA) is a key enzyme in plant tetrapyrrole biosynthesis that catalyses the synthesis of δ-aminolevulinic acid, a precursor in chlorophyll and heme biosynthesis [72]. In the present experiment, GSA (spot 2239) decreased significantly in YL, which might suggest the likely reason for the xanthas phenotype was the reduction of a precursor in the chlorophyll biosynthesis pathway. By contrast, there were no significant differences in the mRNA abundance of GSA between these two types of leaves, which might suggest that this identified protein is regulated at the post-transcriptional level.

In addition, our results indicated that the 2-DE profiles of total leaf proteins and chloroplast proteins presented similar protein patterns, which was in accordance with that found in *Glycine max* and *Kandelia candel* [73,74]. It is most likely because chloroplasts in leaf cell are numerous and are rich in soluble proteins, such as Rubisco, Rubisco activase, and ATP synthase, which account for more than half of the total soluble proteins in mature leaf cells [43,75,76]. The bioinformatic analysis with WolfPsort revealed that nearly 82% of the identified proteins from chloroplast were predicted to be typical chloroplast proteins, indicating the high purity of chloroplast isolation. Since proteins were searched against the viridiplantae database of NCBI, many differential proteins such as Rubisco large subunit were identified as orthologous complexes in other organisms [77,78]. Our results also revealed that 82 of the 89 identified proteins appeared to be 60 unique kinds of proteins, suggesting these sets of proteins might be isoforms due to the post-translational modifications or belong to the same protein families. These findings provided evidence on how the isoforms or family proteins in chloroplast modulate the complex metabolic network in YL.

4. Materials and Methods

4.1. Plant Materials

Similar sized leaves of two ginkgo leaf types (YL and GL) with uniform genetic backgrounds were collected from the 150-year old mutant tree in April. The sample leaves were stored in the dark at 4 °C for 24 h until chloroplast isolation. Leaves for protein and RNA extraction were collected from one-year-old lateral branches on the sun-exposed side, immediately frozen in liquid nitrogen and stored in a −70 °C freezer until use. Each sample consisted of 20 leaves.

4.2. Pigment Determination and Photosynthetic Characteristics

To investigate the changes in the photosynthetic physiology, we analysed the chlorophyll content, gas-exchange parameters and chlorophyll fluorescence in the two leaves of different colours. Chlorophyll was extracted and quantified using a spectrophotometer as previously described by Tang et al. [79]. Green and yellow leaves (0.2 g fresh weight) were prepared and extracted with a mixed extraction solution (acetone:alcohol:distilled water = 4.5:4.5:1, volume ratio) until complete bleaching. The concentrations of chlorophyll were calculated based on the absorbance of the solution at 645 and 663 nm. The chlorophyll content (chlorophyll a and b) was measured with six replicates.

The net photosynthesis rate (P_n), transportation rate (E), stomatal conductance (g_s) and internal CO_2 concentration (C_i) were measured with a CIRAS-2 portable open system gas analyser (PP Systems, Haverhill, MA, USA) at an ambient temperature of approximately 25 °C, a CO_2 concentration of 400 $\mu mol \cdot mol^{-1}$, relative humidity of 60% and photosynthetic photon flux density (PPFD) of 1500 $\mu mol \cdot m^{-2} \cdot s^{-1}$. The chlorophyll fluorescence was measured with a FMS-2 pulse modulated fluorometer (Hansatech Instruments, King's Lynn, UK) as described by Xu et al. [80]. To measure the maximum quantum yield of photosystem II (Fv/Fm), leaves were dark-adapted for at least 30 min. The effective quantum yield of photosystem II (PSII) electron transport ($\Phi PSII = (Fm' - Fs)/Fm'$), the efficiency of excitation energy capture by open PSII centres (Fv'/Fm'), nonphotochemical quenching ($NPQ = (Fm - Fm')/Fm'$), photochemical quenching ($q_p = (Fm' - Fs)/(Fm' - Fo')$) were calculated. The gas exchange parameters and chlorophyll fluorescence were recorded between 09:00 and 11:00 on sun-exposed leaves of similar sizes on one-year-old lateral branches.

4.3. Transmission Electron Microscopy Analysis

To observe the ultrastructure of chloroplasts, samples of GL and YL were fixed in 2.5% glutaraldehyde solution in 100 mM phosphate buffer (pH 7.2) for 24 h at 4 °C and postfixed with 1% osmium tetroxide (OsO_4) in deionized water. The postfixed tissues were rinsed three times with phosphate buffer, followed by dehydration in a graded ethanol series (30%, 50%, 70%, 80%, 90%, 100%, 100%) of 20 min each. After embedding in an Epoxy resin composed of a mixed solution of Epon812, the samples were sectioned with a glass knife using an LKB-V ultramicrotome (LKB, Bromma, Sweden). Ultra-thin sections were deposited on copper grids and stained with uranyl acetate and lead citrate. TEM experiments were conducted on a JEM-2100 transmission electron microscope (JEOL Ltd., Tokyo, Japan).

4.4. Isolation of Chloroplasts

Chloroplasts were isolated according to Diekmann et al. [81], with minor modifications. Leaves (100 g fresh weight) were homogenized in 600 mL of an ice-cold buffer (1.25 M NaCl, 50 mM Tris-HCl pH 8.0, 7 mM EDTA, 5% PVP-40, 0.1% BSA, 1 mM DTT) using a commercial blender, and all subsequent operations were performed at 4 °C. Chloroplasts were purified using 30% sucrose solution (30% sucrose in 50 mM Tris-HCl pH 8.0, 25 mM EDTA) and dissolved in wash buffer (0.35 M sorbitol, 50 mM Tris-HCl pH 8.0, 25 mM EDTA). The chloroplast solution was layered carefully on top of pre-chilled sucrose gradients (prepared using 30% and 52% sucrose solution) and centrifuged at 14,600× *g* for 70 min (4 °C). Intact chloroplasts were collected from the interphase of the 52% to

30% layers using a wide-bore pipette, diluted with wash buffer, and then directly used in protein extraction or stored at −80 °C. The purified chloroplasts were visualized under a DM5000B fluorescence microscope (Leica, Wetzlar, Germany).

4.5. Protein Extraction and Quantification

The procedure for extracting proteins from chloroplasts was modified from the improved phenol (BPP) method [82]. Chloroplast pellets were re-suspended uniformly in ice-cold extraction buffer containing 100 mM Tris (pH 8.0), 100 mM EDTA, 50 mM borax, 50 mM vitamin C, 1% Triton X-100, 2% β-mercaptoethanol, and 30% sucrose. Two volumes of Tris-saturated phenol (pH 8.0) were added to the solution, followed by vortexing for 10 min at room temperature. After the sample was centrifuged at $15,000\times g$ for 15 min (4 °C), the upper phase was transferred to a new centrifuge tube. The chloroplast proteins were centrifuged at $15,000\times g$ for 15 min (4 °C) and precipitated by adding ammonium sulphate-saturated methanol. The protein pellets were washed twice with ice-cold acetone and then lyophilized on an ALPHA 2-4 LD plus freezer dryer (Christ, Osterode, Germany).

Total proteins were extracted from leaves according to the acetone/trichloroacetic acid (TCA) precipitation method [83]. The leaf sample powder was suspended in acetone buffer containing 10% TCA and 0.07% β-mercaptoethanol and incubated at −20 °C for 2 h. After centrifugation at $15,000\times g$ for 20 min (4 °C), the resultant precipitate was washed three times with ice-cold acetone and then lyophilized. The obtained protein powder was dissolved in lysis buffer containing 7 M urea, 2 M thiourea, 4% CHAPS, 1% dithiothreitol (DTT), 0.5% IPG buffer pH 3–10. The protein concentration was quantified according to the Bradford method [84].

4.6. 2-D Electrophoresis and Gel Imaging Analysis

Two-dimensional electrophoresis was performed using a GE Healthcare 2-DE system (GE Healthcare, Piscataway, NJ, USA) according to the manufacturer's instructions, with some modifications. Approximately 350 µg of protein were loaded onto a 24 cm nonlinear gradient IPG strip (pH 4–7 and 3–10), and the strips were loaded into an Ettan IPGpho 3 IEF System (GE). The separation of proteins in the second dimension was performed on an SDS-PAGE gel (12.5% polyacrylamide) using the Ettan Daltsix apparatus (GE). The signal was visualized by silver staining. Three biological replicates were prepared for each type of leaf.

The gel images were digitized with an Image scanner III (GE), and analysed with the PDQuest software package (Version 8.0.1, Bio-Rad, New York, NY, USA). Spots were detected, matched, and normalized on the basis of the total density of gels with the parameter of percentage volume according to the manufacturer's instructions. For each protein spot, the average spot quantity value and its variance coefficient at each sample were determined. One-way analysis of variance (ANOVA) was performed at $p < 0.05$ to assess for absolute protein changes between the two leaves of different colours. Spots with a mean abundance that changed more than 1.5-fold or less than 0.66-fold in different leaves were considered differentially accumulated proteins and selected for protein identification.

4.7. Protein Digestion and MALDI-TOF-TOF Analysis

Protein spots were rehydrated in digestion solution with sequencing grade modified trypsin as described previously [85]. After digestion, the tryptic peptides were extracted, lyophilized and stored at −80 °C. All peptide mass fingerprint data were obtained using an UltrafleXtreme MALDI TOF/TOF mass spectrometer (Bruker Daltonics, Bremen, Germany) with flexAnalysis software. All spectra of proteins were submitted to database searching using the MASCOT search program (available online: http://www.matrixscience.com), against NCBI database (Viridiplantae). The searching parameters were as follows: 100 ppm mass accuracy, trypsin cleavage one missed cleavage allowed, carbamidomethylation of Cys as fixed modification, oxidation of Met was allowed as variable modification, and MS/MS fragment tolerance was set to 0.6 Da. The subcellular locations of the differentially accumulated proteins in this study were predicted using the publicly available program,

WolfPsort (available online: http://www.genscript.com/tools/wolf-psort/) [86]. Differentially accumulated proteins were functionally classified based on Kyoto Encyclopedia of Genes and Genomes (KEGG) pathway analysis (available online: http://www.kegg.jp/kegg/pathway.html) and the literature, and the protein–protein interaction network was evaluated using STRING 10 (STRING is a database of known and predicted protein-protein interactions, which is available online: http://string-db.org) against the *A. thaliana* database.

4.8. RNA Extraction and qRT-PCR Analysis

Total RNA was extracted from leaves using the MiniBEST Plant RNA Extraction Kit (Takara, Otsu, Japan) according to the manufacturer's instructions. The quality of the RNA was determined by agarose gel electrophoresis, and the concentration was detected by UV spectrophotometric analysis. The reverse transcription-polymerase chain reaction was performed using a Transcriptor First Stand cDNA Synthesis Kit (Roche, Mannheim, Germany). Real-time quantitative PCR (qRT-PCR) was performed on ABI PRISM 7500 Sequence Detection System (Applied Biosystems, Foster City, CA, USA) using SYBR Green PCR Master Mix (Roche). The glyceraldehyde-3-phosphate dehydrogenase gene (*GbGAPDH*, GenBank Accession no. L26924) was used as the reference gene [87]. The gene-specific primers for *GbGAPDH* and selected protein-coding genes were designed based on the transcriptome sequencing performed previously (Table 2). The qRT-PCR data were calibrated using the $2^{-\Delta\Delta Ct}$ method for relative quantification between different colours of leaves [88]. Three biological replicates were prepared for each sample.

Table 2. Primers for qRT-PCR.

Gene	Primer Forward Sequence (5′–3′)	Primer Reverse Sequence (5′–3′)
GAPDH	GGTGCCAAAAAGGTGGTCAT	CAACAACGAACATGGGAGCAT
SPDS	ACATCTTCCACTTTGCTCTATTCCA	CGAGGGTCTTCATAGCCTACTGC
HSP70	ACTCAGAAGGGGCACGAACA	AAATCGCCTTCCTATCAACCG
PSRP3	TCATTGCCCACTTCATCCGC	CGCTCACTTCCTCTTCTGCTGC
atpA	TTATTGGGGACAGGCAGACCG	GGAGCGAGATATTGTAATGTAGCG
GSA	TGGCATCACTCCAGACCTTACA	GCAACCATCTCCATTATCTCCC
CD4B	AAGGCAGCCACAAATAGAACGG	CAAGACCCTCAGCAATAGCCG
EF-Tu	ATTTCCTGGAGACGATGTGCC	TCAGTCTGTCTCCGAGGAATGG

4.9. Statistical Analysis

Parametric data are presented as means $\pm$ SD. Statistical analyses were performed by one-way analysis of variance (ANOVA), and mean differences were compared by the lowest standard deviations (L.S.D.) test using SPSS 17.0.

5. Conclusions

In the present study, the morphological, photosynthetic and proteomic characteristics of YL and GL variations of ginkgo were analysed. The YL mutant had a relatively low chlorophyll content, photosynthetic properties and undeveloped chloroplasts, with few or no starch grains. Proteomic analysis of both leaves and chloroplasts of the two different colours of leaves was performed using 2-DE technology coupled with MALDI-TOF/TOF MS. A total of 89 proteins were successfully identified, and 82 of these proteins were assigned clear functions belonging to nine metabolic pathways and processes. Notably, these identified proteins were largely enriched in protein and carbohydrate metabolism, including photosynthesis, carbon fixation in photosynthetic organisms, carbohydrate/energy metabolism, amino acid metabolism, and protein metabolism, suggesting the ginkgo mutant has specific differences in regulating cellular processes. The proteomic comparison of the two leaves of different colours revealed that the YL mutant had a severe effect on the photosynthetic machinery, with a reduction of CO_2 fixation efficiency and reduced ATP synthesis, and thus caused a decrease in primary carbon metabolism. The identification of proteins involved in protein metabolism indicated that the mutant has an inhibitory effect on protein biosynthesis and enhanced protein folding and assembly. Furthermore, proteomic analysis identified several antioxidant proteins that are involved in redox homeostasis and might improve the defensive ability of cells against ROS damage. The identification of these differentially accumulated proteins reveals the presence of a complex metabolic network in YL and the possible reason for the variation in leaf colour. Future studies should examine more leaf developmental stages to elucidate the regulation of pigment biosynthesis or the changes in the colour pattern and to evaluate the alteration of comprehensive metabolic networks of this leaf colour mutant.

Supplementary Materials: Supplementary materials can be found at www.mdpi.com/1422-0067/17/11/1794/s1.

Acknowledgments: This work was supported by the Forestry Industry Research Special Funds for Public Welfare Projects (201504105), China Postdoctoral Science Foundation (2013M541680), Colleges and Universities in Jiangsu Province Plans to Graduate Research and Innovation (CXZZ13_0552), Jiangsu Postdoctoral Science Foundation (1401062B), and Priority Academic Program Development of Jiangsu Higher Education Institutions (PAPD).

Author Contributions: Xinliang Liu, Wanwen Yu, and Fuliang Cao conceived and designed the experiments; Xinliang Liu and Wanwen Yu performed the experiments; Xinliang Liu and Guibin Wang analysed the data; Jinfeng Cai and Huanli Wang contributed reagents/materials/analysis tools; Xinliang Liu and Wanwen Yu wrote the paper.

Conflicts of Interest: The authors declare no conflict of interest.

References

1. Eckhardt, U.; Grimm, B.; Hörtensteiner, S. Recent advances in chlorophyll biosynthesis and breakdown in higher plants. *Plant Mol. Biol.* **2004**, *56*, 1–14. [CrossRef] [PubMed]

2. Kobayashi, K.; Fujii, S.; Sasaki, D.; Baba, S.; Ohta, H.; Masuda, T.; Wada, H. Transcriptional regulation of thylakoid galactolipid biosynthesis coordinated with chlorophyll biosynthesis during the development of chloroplasts in *Arabidopsis*. *Front. Plant Sci.* **2014**, *5*. [CrossRef] [PubMed]

3. Fromme, P.; Melkozernov, A.; Jordan, P.; Krauss, N. Structure and function of photosystem I: Interaction with its soluble electron carriers and external antenna systems. *FEBS Lett.* **2003**, *555*, 40–44. [CrossRef]

4. Grimm, B. Novel insights in the control of tetrapyrrole metabolism of higher plants. *Curr. Opin. Plant Biol.* **1998**, *1*, 245–250. [CrossRef]

5. Hörtensteiner, S.; Kräutler, B. Chlorophyll breakdown in higher plants. *Biochim. Biophys. Acta* **2011**, *1807*, 977–988. [CrossRef] [PubMed]

6. Nagata, N.; Tanaka, R.; Satoh, S.; Tanaka, A. Identification of a vinyl reductase gene for chlorophyll synthesis in *Arabidopsis thaliana* and implications for the evolution of prochlorococcus species. *Plant Cell* **2005**, *17*, 233–240. [CrossRef] [PubMed]

7. Jung, K.H.; Hur, J.; Ryu, C.H.; Choi, Y.; Chung, Y.Y.; Miyao, A.; Hirochika, H.; An, G. Characterization of a rice chlorophyll-deficient mutant using the T-DNA gene-trap system. *Plant Cell Physiol.* **2003**, *44*, 463–472. [CrossRef] [PubMed]

8. Manjaya, J. Genetic improvement of soybean variety VLS-2 through induced mutations. *Small* **2009**, *38*, 106–109.

9. Singh, R.; Ikehashi, H. Monogenic male-sterility in rice: Induction, identification and inheritance. *Crop Sci.* **1981**, *21*, 286–289. [CrossRef]

10. Falbel, T.G.; Meehl, J.B.; Staehelin, L.A. Severity of mutant phenotype in a series of chlorophyll-deficient wheat mutants depends on light intensity and the severity of the block in chlorophyll synthesis. *Plant Physiol.* **1996**, *112*, 821–832. [CrossRef] [PubMed]

11. Greene, B.A.; Staehelin, L.A.; Melis, A. Compensatory alterations in the photochemical apparatus of a photoregulatory, chlorophyll b-deficient mutant of maize. *Plant Physiol.* **1988**, *87*, 365–370. [CrossRef] [PubMed]

12. Runge, S.; van Cleve, B.; Lebedev, N.; Armstrong, G.; Apel, K. Isolation and classification of chlorophyll-deficient xantha mutants of *Arabidopsis thaliana*. *Planta* **1995**, *197*, 490–500. [CrossRef] [PubMed]

13. Terry, M.J.; Kendrick, R.E. Feedback inhibition of chlorophyll synthesis in the phytochrome chromophore-deficient aurea and yellow-green-2 mutants of tomato. *Plant Physiol.* **1999**, *119*, 143–152. [CrossRef] [PubMed]

14. Wu, Z.; Zhang, X.; He, B.; Diao, L.; Sheng, S.; Wang, J.; Guo, X.; Su, N.; Wang, L.; Jiang, L. A chlorophyll-deficient rice mutant with impaired chlorophyllide esterification in chlorophyll biosynthesis. *Plant Physiol.* **2007**, *145*, 29–40. [CrossRef] [PubMed]

15. Li, Q.; Zhu, F.-Y.; Gao, X.; Sun, Y.; Li, S.; Tao, Y.; Lo, C.; Liu, H. Young leaf chlorosis 2 encodes the stroma-localized heme oxygenase 2 which is required for normal tetrapyrrole biosynthesis in rice. *Planta* **2014**, *240*, 701–712. [CrossRef] [PubMed]

16. Liu, W.; Fu, Y.; Hu, G.; Si, H.; Zhu, L.; Wu, C.; Sun, Z. Identification and fine mapping of a thermo-sensitive chlorophyll deficient mutant in rice (*Oryza sativa* L.). *Planta* **2007**, *226*, 785–795. [CrossRef] [PubMed]

17. Zhang, H.T.; Li, J.; Yoo, J.H.; Yoo, S.C.; Cho, S.H.; Koh, H.J.; Seo, H.S.; Paek, N.C. Rice *Chlorina-1* and *Chlorina-9* encode ChlD and ChlI subunits of Mg-chelatase, a key enzyme for chlorophyll synthesis and chloroplast development. *Plant Mol. Biol.* **2006**, *62*, 325–337. [CrossRef] [PubMed]

18. Zhou, Y.; Gong, Z.Y.; Yang, Z.F.; Yuan, Y.; Zhu, J.Y.; Wang, M.; Yuan, F.H.; Wu, S.J.; Wang, Z.Q.; Yi, C.D. Mutation of the light-induced yellow leaf 1 gene, which encodes a geranylgeranyl reductase, affects chlorophyll biosynthesis and light sensitivity in rice. *PLoS ONE* **2013**, *8*, e75299. [CrossRef] [PubMed]

19. Miura, E.; Kato, Y.; Matsushima, R.; Albrecht, V.; Laalami, S.; Sakamoto, W. The balance between protein synthesis and degradation in chloroplasts determines leaf variegation in *Arabidopsis* yellow variegated mutants. *Plant Cell* **2007**, *19*, 1313–1328. [CrossRef] [PubMed]

20. Sakamoto, W.; Zaltsman, A.; Adam, Z.; Takahashi, Y. Coordinated regulation and complex formation of YELLOW VARIEGATED$_1$ and YELLOW VARIEGATED$_2$, chloroplastic FtsH metalloproteases involved in the repair cycle of photosystem II in arabidopsis thylakoid membranes. *Plant Cell* **2003**, *15*, 2843–2855. [CrossRef] [PubMed]

21. Yoshioka-Nishimura, M.; Nanba, D.; Takaki, T.; Ohba, C.; Tsumura, N.; Morita, N.; Sakamoto, H.; Murata, K.; Yamamoto, Y. Quality control of photosystem II: Direct imaging of the changes in the thylakoid structure and distribution of FtsH proteases in spinach chloroplasts under light stress. *Plant Cell Physiol.* **2014**, *55*, 1255–1265. [CrossRef] [PubMed]

22. Tomanek, L. Environmental proteomics: Changes in the proteome of marine organisms in response to environmental stress, pollutants, infection, symbiosis, and development. *Annu. Rev. Mar. Sci.* **2011**, *3*, 373–399. [CrossRef] [PubMed]

23. Wittmann-Liebold, B.; Graack, H.R.; Pohl, T. Two-dimensional gel electrophoresis as tool for proteomics studies in combination with protein identification by mass spectrometry. *Proteomics* **2006**, *6*, 4688–4703. [CrossRef] [PubMed]

24. Canovas, F.M.; Dumas-Gaudot, E.; Recorbet, G.; Jorrin, J.; Mock, H.P.; Rossignol, M. Plant proteome analysis. *Proteomics* **2004**, *4*, 285–298. [CrossRef] [PubMed]

25. Fan, P.; Feng, J.; Jiang, P.; Chen, X.; Bao, H.; Nie, L.; Jiang, D.; Lv, S.; Kuang, T.; Li, Y. Coordination of carbon fixation and nitrogen metabolism in *Salicornia europaea* under salinity: Comparative proteomic analysis on chloroplast proteins. *Proteomics* **2011**, *11*, 4346–4367. [CrossRef] [PubMed]

26. Joyard, J.; Ferro, M.; Masselon, C.; Seigneurin-Berny, D.; Salvi, D.; Garin, J.; Rolland, N. Chloroplast proteomics and the compartmentation of plastidial isoprenoid biosynthetic pathways. *Mol. Plant* **2009**, *2*, 1154–1180. [CrossRef] [PubMed]

27. Taylor, N.L.; Tan, Y.F.; Jacoby, R.P.; Millar, A.H. Abiotic environmental stress induced changes in the *Arabidopsis thaliana* chloroplast, mitochondria and peroxisome proteomes. *J. Proteom.* **2009**, *72*, 367–378. [CrossRef] [PubMed]

28. Chen, X.; Zhang, W.; Xie, Y.; Lu, W.; Zhang, R. Comparative proteomics of thylakoid membrane from a chlorophyll b-less rice mutant and its wild type. *Plant Sci.* **2007**, *173*, 397–407. [CrossRef]

29. Wang, L.; Cao, H.; Chen, C.; Yue, C.; Hao, X.; Yang, Y.; Wang, X. Complementary transcriptomic and proteomic analyses of a chlorophyll-deficient tea plant cultivar reveal multiple metabolic pathway changes. *J. Proteom.* **2016**, *130*, 160–169. [CrossRef] [PubMed]

30. Zhou, Y.; Wu, X.X.; Zhang, Z.; Gao, Z.H. Comparative proteomic analysis of floral color variegation in peach. *Biochem. Biophys. Res. Commun.* **2015**, *464*, 1101–1106. [CrossRef] [PubMed]

31. Shen, G.A.; Pang, Y.Z.; Wu, W.S.; Liao, Z.H.; Zhao, L.X.; Sun, X.F.; Tang, K.X. Cloning and characterization of a root-specific expressing gene encoding 3-hydroxy-3-methylglutaryl coenzyme a reductase from *Ginkgo biloba*. *Mol. Biol. Rep.* **2006**, *33*, 117–127. [CrossRef] [PubMed]

32. Raveh, E.; Wang, N.; Nobel, P.S. Gas exchange and metabolite fluctuations in green and yellow bands of variegated leaves of the monocotyledonous cam species *Agave americana*. *Physiol. Plant.* **1998**, *103*, 99–106. [CrossRef]

33. Wu, Z.; Zhang, X.; Wang, J.; Wan, J. Leaf chloroplast ultrastructure and photosynthetic properties of a chlorophyll-deficient mutant of rice. *Photosynthetica* **2014**, *52*, 217–222. [CrossRef]

34. Gitelson, A.A.; Gritz, Y.; Merzlyak, M.N. Relationships between leaf chlorophyll content and spectral reflectance and algorithms for non-destructive chlorophyll assessment in higher plant leaves. *J. Plant Physiol.* **2003**, *160*, 271–282. [CrossRef] [PubMed]

35. Von Wettstein, D.; Gough, S.; Kannangara, C.G. Chlorophyll biosynthesis. *Plant Cell* **1995**, *7*, 1039–1057. [CrossRef] [PubMed]

36. Zhu, L.X.; Zeng, X.H.; Chen, Y.L.; Yang, Z.H.; Qi, L.P.; Pu, Y.Y.; Yi, B.; Wen, J.; Ma, C.Z.; Shen, J.X. Genetic characterisation and fine mapping of a chlorophyll-deficient mutant (*Bnac. ygl*) in *Brassica napus*. *Mol. Breed.* **2014**, *34*, 603–614. [CrossRef]

37. Peremarti, A.; Marè, C.; Aprile, A.; Roncaglia, E.; Cattivelli, L.; Villegas, D.; Royo, C. Transcriptomic and proteomic analyses of a pale-green durum wheat mutant shows variations in photosystem components and metabolic deficiencies under drought stress. *BMC Genom.* **2014**, *15*. [CrossRef] [PubMed]

38. Zhou, X.S.; Wu, D.X.; Shen, S.Q.; Sun, J.W.; Shu, Q.Y. High photosynthetic efficiency of a rice (*Oryza sativa* L.) xantha mutant. *Photosynthetica* **2006**, *44*, 316–319. [CrossRef]

39. Hideg, É.; Kós, P.B.; Schreiber, U. Imaging of NPQ and ROS formation in tobacco leaves: Heat inactivation of the water-water cycle prevents down-regulation of PSII. *Plant Cell Physiol.* **2008**, *49*, 1879–1886. [CrossRef] [PubMed]

40. Yu, J.J.; Zhang, J.Z.; Zhao, Q.; Liu, Y.L.; Chen, S.X.; Guo, H.L.; Shi, L.; Dai, S.J. Proteomic analysis reveals the leaf color regulation mechanism in chimera *Hosta* "gold standard" leaves. *Int. J. Mol. Sci.* **2016**, *17*. [CrossRef] [PubMed]

41. Yuan, M.; Xu, M.Y.; Yuan, S.; Chen, Y.E.; Du, J.B.; Xu, F.; Zhang, Z.W.; Guo, Z.C.; Zhao, Z.Y.; Lin, H.H. Light regulation to chlorophyll synthesis and plastid development of the chlorophyll-less golden-leaf privet. *J. Integr. Plant Biol.* **2010**, *52*, 809–816. [CrossRef] [PubMed]

42. Plücken, H.; Müller, B.; Grohmann, D.; Westhoff, P.; Eichacker, L.A. The HCF136 protein is essential for assembly of the photosystem II reaction center in *Arabidopsis thaliana*. *FEBS Lett.* **2002**, *532*, 85–90. [CrossRef]

43. Spreitzer, R.J.; Salvucci, M.E. Rubisco: Structure, regulatory interactions, and possibilities for a better enzyme. *Plant Biol.* **2002**, *53*, 449–475. [CrossRef] [PubMed]

44. Zhao, C.F.; Wang, J.Q.; Cao, M.L.; Zhao, K.; Shao, J.M.; Lei, T.T.; Yin, J.N.; Hill, G.G.; Xu, N.Z.; Liu, S.Q. Proteomic changes in rice leaves during development of field-grown rice plants. *Proteomics* **2005**, *5*, 961–972. [CrossRef] [PubMed]

45. Carmo-Silva, A.E.; Salvucci, M.E. The activity of rubisco's molecular chaperone, rubisco activase, in leaf extracts. *Photosynth. Res.* **2011**, *108*, 143–155. [CrossRef] [PubMed]

46. Kumatani, T.; Sakurai-Ozato, N.; Miyawaki, N.; Yokota, E.; Shimmen, T.; Terashima, I.; Takagi, S. Possible association of actin filaments with chloroplasts of spinach mesophyll cells in vivo and in vitro. *Protoplasma* **2006**, *229*, 45–52. [CrossRef] [PubMed]

47. Ostersetzer, O.; Adam, Z. Light-stimulated degradation of an unassembled Rieske FeS protein by a thylakoid-bound protease: The possible role of the FtsH protease. *Plant Cell* **1997**, *9*, 957–965. [CrossRef] [PubMed]

48. Zaltsman, A.; Feder, A.; Adam, Z. Developmental and light effects on the accumulation of FtsH protease in *Arabidopsis* chloroplasts-implications for thylakoid formation and photosystem II maintenance. *Plant J.* **2005**, *42*, 609–617. [CrossRef] [PubMed]

49. Zaltsman, A.; Ori, N.; Adam, Z. Two types of FtsH protease subunits are required for chloroplast biogenesis and photosystem ii repair in *Arabidopsis*. *Plant Cell* **2005**, *17*, 2782–2790. [CrossRef] [PubMed]

50. Wu, W.J.; Zhu, Y.; Ma, Z.X.; Sun, Y.; Quan, Q.; Li, P.; Hu, P.Z.; Shi, T.L.; Lo, C.; Chu, I.K. Proteomic evidence for genetic epistasis: ClpR4 mutations switch leaf variegation to virescence in *Arabidopsis*. *Plant J.* **2013**, *76*, 943–956. [CrossRef] [PubMed]

51. Adam, Z.; Rudella, A.; van Wijk, K.J. Recent advances in the study of Clp, FtsH and other proteases located in chloroplasts. *Curr. Opin. Plant Biol.* **2006**, *9*, 234–240. [CrossRef] [PubMed]

52. Rasool, K.; Khan, M.; Aldawood, A.; Tufail, M.; Mukhtar, M.; Takeda, M. Optimization of protein isolation from date palm plants and its utilization in differential proteomics associated with red palm weevil infestation. *Pak. J. Agric. Sci.* **2014**, *51*, 907–917.

53. Karuppanapandian, T.; Moon, J.C.; Kim, C.; Manoharan, K.; Kim, W. Reactive oxygen species in plants: Their generation, signal transduction, and scavenging mechanisms. *Aust. J. Crop Sci.* **2011**, *5*, 709–725.

54. Asada, K. Ascorbate peroxidase—A hydrogen peroxide-scavenging enzyme in plants. *Physiol. Plant.* **1992**, *85*, 235–241. [CrossRef]

55. Van Breusegem, F.; Slooten, L.; Stassart, J.-M.; Moens, T.; Botterman, J.; Van Montagu, M.; Inzé, D. Overproduction of *Arabidopsis thaliana* FeSOD confers oxidative stress tolerance to transgenic maize. *Plant Cell Physiol.* **1999**, *40*, 515–523. [CrossRef] [PubMed]

56. Fries, M.; Jung, H.-I.; Perham, R.N. Reaction mechanism of the heterotetrameric ($\alpha_2\beta_2$) E1 component of 2-oxo acid dehydrogenase multienzyme complexes. *Biochemistry* **2003**, *42*, 6996–7002. [CrossRef] [PubMed]

57. Soo, P.C.; Horng, Y.T.; Lai, M.J.; Wei, J.R.; Hsieh, S.C.; Chang, Y.L.; Tsai, Y.H.; Lai, H.C. Pirin regulates pyruvate catabolism by interacting with the pyruvate dehydrogenase E1 subunit and modulating pyruvate dehydrogenase activity. *J. Bacteriol.* **2007**, *189*, 109–118. [CrossRef] [PubMed]

58. Funkhouser, E.A.; Loewus, F.A. Purification of *myo*-inositol 1-phosphate synthase from rice cell culture by affinity chromatography. *Plant Physiol.* **1975**, *56*, 786–790. [CrossRef] [PubMed]

59. Abreu, E.F.; Aragão, F.J. Isolation and characterization of a *myo*-inositol-1-phosphate synthase gene from yellow passion fruit (*Passiflora edulis* f. *flavicarpa*) expressed during seed development and environmental stress. *Ann. Bot.* **2007**, *99*, 285–292. [PubMed]

60. Tiessen, A.; Hendriks, J.H.; Stitt, M.; Branscheid, A.; Gibon, Y.; Farré, E.M.; Geigenberger, P. Starch synthesis in potato tubers is regulated by post-translational redox modification of ADP-glucose pyrophosphorylase a novel regulatory mechanism linking starch synthesis to the sucrose supply. *Plant Cell* **2002**, *14*, 2191–2213. [CrossRef] [PubMed]

61. Visser, R.; Somhorst, I.; Kuipers, G.; Ruys, N.; Feenstra, W.; Jacobsen, E. Inhibition of the expression of the gene for granule-bound starch synthase in potato by antisense constructs. *Mol. Gen. Genet.* **1991**, *225*, 289–296. [CrossRef] [PubMed]

62. Pogson, B.J.; Albrecht, V. Genetic dissection of chloroplast biogenesis and development: An overview. *Plant Physiol.* **2011**, *155*, 1545–1551. [CrossRef] [PubMed]

63. Park, M.H. The post-translational synthesis of a polyamine-derived amino acid, hypusine, in the eukaryotic translation initiation factor 5A (eIF5A). *J. Biochem.* **2006**, *139*, 161–169. [CrossRef] [PubMed]

64. Gao, Y.G.; Selmer, M.; Dunham, C.M.; Weixlbaumer, A.; Kelley, A.C.; Ramakrishnan, V. The structure of the ribosome with elongation factor G trapped in the posttranslocational state. *Science* **2009**, *326*, 694–699. [CrossRef] [PubMed]

65. Albrecht, V.; Ingenfeld, A.; Apel, K. Characterization of the *snowy cotyledon 1* mutant of *Arabidopsis thaliana*: The impact of chloroplast elongation factor G on chloroplast development and plant vitality. *Plant Mol. Biol.* **2006**, *60*, 507–518. [CrossRef] [PubMed]

66. Vierling, E. The roles of heat shock proteins in plants. *Annu. Rev. Plant Biol.* **1991**, *42*, 579–620. [CrossRef]

67. Wang, W.X.; Vinocur, B.; Shoseyov, O.; Altman, A. Role of plant heat-shock proteins and molecular chaperones in the abiotic stress response. *Trends Plant Sci.* **2004**, *9*, 244–252. [CrossRef] [PubMed]

68. Miflin, B.J.; Habash, D.Z. The role of glutamine synthetase and glutamate dehydrogenase in nitrogen assimilation and possibilities for improvement in the nitrogen utilization of crops. *J. Exp. Bot.* **2002**, *53*, 979–987. [CrossRef] [PubMed]

69. Hidaka, Y.; Shimamoto, S. Folding of peptides and proteins: Role of disulfide bonds, recent developments. *Biomol. Concepts* **2013**, *4*, 597–604. [CrossRef] [PubMed]

70. Kasukabe, Y.; He, L.; Nada, K.; Misawa, S.; Ihara, I.; Tachibana, S. Overexpression of spermidine synthase enhances tolerance to multiple environmental stresses and up-regulates the expression of various stress-regulated genes in transgenic *Arabidopsis thaliana*. *Plant Cell Physiol.* **2004**, *45*, 712–722. [CrossRef] [PubMed]

71. Gilbert, L.; Alhagdow, M.; Nunes-Nesi, A.; Quemener, B.; Guillon, F.; Bouchet, B.; Faurobert, M.; Gouble, B.; Page, D.; Garcia, V. GDP-D-mannose 3, 5-epimerase (GME) plays a key role at the intersection of ascorbate and non-cellulosic cell-wall biosynthesis in tomato. *Plant J.* **2009**, *60*, 499–508. [CrossRef] [PubMed]

72. Grimm, B. Primary structure of a key enzyme in plant tetrapyrrole synthesis: Glutamate 1-semialdehyde aminotransferase. *Proc. Natl. Acad. Sci. USA* **1990**, *87*, 4169–4173. [CrossRef] [PubMed]

73. Ahsan, N.; Nanjo, Y.; Sawada, H.; Kohno, Y.; Komatsu, S. Ozone stress-induced proteomic changes in leaf total soluble and chloroplast proteins of soybean reveal that carbon allocation is involved in adaptation in the early developmental stage. *Proteomics* **2010**, *10*, 2605–2619. [CrossRef] [PubMed]

74. Wang, L.; Pan, D.; Li, J.; Tan, F.; Hoffmann-Benning, S.; Liang, W.; Chen, W. Proteomic analysis of changes in the *Kandelia candel* chloroplast proteins reveals pathways associated with salt tolerance. *Plant Sci.* **2015**, *231*, 159–172. [CrossRef] [PubMed]

75. Behrens, C.; Blume, C.; Senkler, M.; Eubel, H.; Peterhänsel, C.; Braun, H.-P. The "protein complex proteome" of chloroplasts in *Arabidopsis thaliana*. *J. Proteom.* **2013**, *91*, 73–83. [CrossRef] [PubMed]

76. Van Wijk, KJ.; Baginsky, S. Plastid proteomics in higher plants: Current state and future goals. *Plant Physiol.* **2011**, *155*, 1578–1588. [CrossRef] [PubMed]

77. Liska, AJ.; Shevchenko, A. Expanding the organismal scope of proteomics: Cross-species protein identification by mass spectrometry and its implications. *Proteomics* **2003**, *3*, 19–28. [CrossRef] [PubMed]

78. Carpentier, S.C.; Panis, B.; Vertommen, A.; Swennen, R.; Sergeant, K.; Renaut, J.; Laukens, K.; Witters, E.; Samyn, B.; Devreese, B. Proteome analysis of non-model plants: A challenging but powerful approach. *Mass Spectrom. Rev.* **2008**, *27*, 354–377. [CrossRef] [PubMed]

79. Tang, Y.; Huang, J.; Wang, R. Change law of hyperspectral data in related with chlorophyll and carotenoid in rice at different developmental stages. *Rice Sci.* **2004**, *11*, 274–282.

80. Xu, S.; Chen, W.; Huang, Y.; He, X. Responses of growth, photosynthesis and VOC emissions of *Pinus tabulaeformis* Carr. Exposure to elevated CO_2 and/or elevated O_3 in an urban area. *Bull. Environ. Contam. Toxicol.* **2012**, *88*, 443–448. [CrossRef] [PubMed]

81. Diekmann, K.; Hodkinson, T.R.; Fricke, E.; Barth, S. An optimized chloroplast DNA extraction protocol for grasses (Poaceae) proves suitable for whole plastid genome sequencing and SNP detection. *PLoS ONE* **2008**, *3*, e2813. [CrossRef] [PubMed]

82. Fan, P.; Wang, X.; Kuang, T.; Li, Y. An efficient method for the extraction of chloroplast proteins compatible for 2-DE and MS analysis. *Electrophoresis* **2009**, *30*, 3024–3033. [CrossRef] [PubMed]

83. Xiong, H.; Shen, H.; Zhang, L.; Zhang, Y.; Guo, X.; Wang, P.; Duan, P.; Ji, C.; Zhong, L.; Zhang, F.; et al. Comparative proteomic analysis for assessment of the ecological significance of maize and peanut intercropping. *J. Proteom.* **2013**, *78*, 447–460. [CrossRef] [PubMed]

84. Bradford, M.M. A rapid and sensitive method for the quantitation of microgram quantities of protein utilizing the principle of protein-dye binding. *Anal. Biochem.* **1976**, *72*, 248–254. [CrossRef]

85. Ma, H.; Song, L.; Huang, Z.; Yang, Y.; Wang, S.; Wang, Z.; Tong, J.; Gu, W.; Ma, H.; Xiao, L. Comparative proteomic analysis reveals molecular mechanism of seedling roots of different salt tolerant soybean genotypes in responses to salinity stress. *EuPA Open Proteom.* **2014**, *4*, 40–57. [CrossRef]

86. Horton, P.; Park, KJ.; Obayashi, T.; Fujita, N.; Harada, H.; Adams-Collier, C.; Nakai, K. Wolf PSORT: Protein localization predictor. *Nucleic Acids Res.* **2007**, *35*, W585–W587. [CrossRef] [PubMed]

87. Xu, F.; Cheng, H.; Cai, R.; Li, L.L.; Chang, J.; Zhu, J.; Zhang, F.X.; Chen, L.J.; Wang, Y.; Cheng, S.H. Molecular cloning and function analysis of an anthocyanidin synthase gene from *Ginkgo biloba*, and its expression in abiotic stress responses. *Mol. Cells* **2008**, *26*, 536–547. [PubMed]

88. Livak, K.J.; Schmittgen, T.D. Analysis of relative gene expression data using real-time quantitative PCR and the $2^{-\Delta\Delta Ct}$ method. *Methods* **2001**, *25*, 402–408. [CrossRef] [PubMed]

MDPI AG

St. Alban-Anlage 66

4052 Basel, Switzerland

Tel. +41 61 683 77 34

Fax +41 61 302 89 18

http://www.mdpi.com

IJMS Editorial Office

E-mail: ijms@mdpi.com

http://www.mdpi.com/journal/ijms